T0184067

Lecture Notes in Computer Science   11651

Commenced Publication in 1973
Founding and Former Series Editors:
Gerhard Goos, Juris Hartmanis, and Jan van Leeuwen

Advanced Research in Computing and Software Science
Subline of Lecture Notes in Computer Science

More information about this series at http://www.springer.com/series/7407

Leszek Antoni Gąsieniec ·
Jesper Jansson ·
Christos Levcopoulos (Eds.)

Fundamentals
of Computation Theory

22nd International Symposium, FCT 2019
Copenhagen, Denmark, August 12–14, 2019
Proceedings

 Springer

Editors
Leszek Antoni Gąsieniec (iD)
University of Liverpool
Liverpool, UK

Jesper Jansson (iD)
Hong Kong Polytechnic University
Hong Kong, China

Christos Levcopoulos (iD)
Lund University
Lund, Sweden

ISSN 0302-9743   ISSN 1611-3349 (electronic)
Lecture Notes in Computer Science
ISBN 978-3-030-25026-3   ISBN 978-3-030-25027-0 (eBook)
https://doi.org/10.1007/978-3-030-25027-0

LNCS Sublibrary: SL1 – Theoretical Computer Science and General Issues

This Springer imprint is published by the registered company Springer Nature Switzerland AG
The registered company address is: Gewerbestrasse 11, 6330 Cham, Switzerland

Preface

This volume contains the papers presented at the 22nd International Symposium on Fundamentals of Computation Theory (FCT 2019), held in Copenhagen, Denmark, during August 12–14, 2019.

FCT is a biennial conference that circulates between various European countries on a regular basis. The first FCT conference took place in Poznan-Kórnik, Poland in 1977, and the most recent ones in Liverpool, UK (FCT 2013), Gdańsk, Poland (FCT 2015), and Bordeaux, France (FCT 2017). FCT focuses on theoretical computer science; in particular, algorithms, complexity, and formal methods.

FCT 2019 received 45 submissions. Each submission was reviewed by at least three Program Committee members, in many cases with the assistance of external expert reviewers. After a thorough review process followed by a discussion phase, 21 of the submitted papers were accepted for presentation at the conference.

Invited talks were given at FCT 2019 by Libor Barto (Charles University), Bernard Chazelle (Princeton University), Kousha Etessami (University of Edinburgh), and Torben Hagerup (University of Augsburg). The papers describing the invited talks are included in these proceedings.

The Program Committee selected the following paper for the best paper award, as well as the best student paper award:

- Bireswar Das, Shivdutt Sharma, and P. R. Vaidyanathan: "Succinct Representation of Finite Groups"

To be eligible for the best student paper award, at least one of the authors had to be a full-time student at the time of the submission, and the student(s) were required to have made a significant contribution to the paper.

We would like to thank the FCT Steering Committee, the FCT 2019 Organizing Committee, the invited speakers, all authors who submitted their work, the FCT 2019 Program Committee members, and the external reviewers. We also thank Jyrki Katajainen for all his help with practical matters when, due to unforeseen circumstances, he became unable to act as the Program Committee chair and Organizing Committee chair, and we were asked by the Steering Committee to step in. Furthermore, we would like to thank Springer for publishing the proceedings, and the Easy-Chair conference management system for simplifying the entire process of reviewing the submitted papers and generating the proceedings. Finally, we acknowledge the support of the Department of Computer Science at the University of Copenhagen (DIKU) and the European Association for Theoretical Computer Science (EATCS).

August 2019

Leszek Gąsieniec
Jesper Jansson
Christos Levcopoulos

Organization

Steering Committee

Bogdan Chlebus	University of Colorado, USA
Marek Karpinski (Chair)	University of Bonn, Germany
Andrzej Lingas	Lund University, Sweden
Miklos Santha	CNRS and Paris Diderot University, France
Eli Upfal	Brown University, USA

Program Committee

Marthe Bonamy	CNRS, LaBRI, University of Bordeaux, France
Irene Finocchi	Sapienza University of Rome, Italy
Leszek Gąsieniec (Co-chair)	University of Liverpool, UK
William Harris	Galois Inc., USA
Mika Hirvensalo	University of Turku, Finland
Štěpán Holub	Charles University, Czech Republic
Jesper Jansson (Co-chair)	The Hong Kong Polytechnic University, SAR China
Jyrki Katajainen (Original chair)	University of Copenhagen, Denmark
Ralf Klasing	CNRS, LaBRI, University of Bordeaux, France
Rastislav Kralovic	Comenius University, Bratislava, Slovakia
Stefan Kratsch	University of Bonn, Germany
Erik Jan van Leeuwen	Utrecht University, The Netherlands
Christos Levcopoulos (Co-chair)	Lund University, Sweden
Florin Manea	Kiel University, Germany
Toby Murray	University of Melbourne, Australia
Aris Pagourtzis	National Technical University of Athens, Greece
Nitin Saxena	Indian Institute of Technology, Kanpur, India
Jeffrey Shallit	University of Waterloo, Canada
Jesper Larsson Träff	TU Wien (Vienna University of Technology), Austria
Peter Widmayer	ETH Zurich, Switzerland

Organizing Committee

Thomas Hildebrandt (Chair)	University of Copenhagen, Denmark
Nanna Højholt	University of Copenhagen, Denmark
Jyrki Katajainen (Original chair)	University of Copenhagen, Denmark

Additional Reviewers

Andronikos, Theodore
Arseneva, Elena
Aspnes, James
Balaji, Nikhil
Casel, Katrin
Chatterjee, Prerona
Chau, Vincent
Chlebus, Bogdan
Choi, Jee
Cunial, Fabio
Czumaj, Artur
Damaschke, Peter
Das, Bireswar
De, Minati
Delbot, François
Escobar, Santiago
Foucaud, Florent
Ganian, Robert
Gawrychowski, Pawel
Goto, Keisuke
Heinrich, Marc
Himmel, Anne-Sophie
Kim, Hyunjin
Kociumaka, Tomasz
Lauria, Massimo
Lingas, Andrzej
Lucarelli, Giorgio
Mahajan, Meena

Mampentzidis, Konstantinos
Mann, Stephen
Mayordomo, Elvira
Meer, Klaus
Miyauchi, Atsushi
Muskalla, Sebastian
Nagao, Atsuki
Nakanishi, Masaki
Niskanen, Reino
Ochem, Pascal
Papaspyrou, Nikolaos
Pilipczuk, Marcin
Pinna, G. Michele
Pitassi, Toniann
Robinson, Peter
Ruzicka, Pavel
Saarela, Aleksi
Sakavalas, Dimitris
Smith, Shaden
Snoeyink, Jack
Sommer, Frank
Srivastav, Abhinav
Suderland, Martin
T. P., Sandhya
Thapen, Neil
Thierauf, Thomas
Villagra, Marcos
Yamakami, Tomoyuki

Algorithms for Some Classes of Infinite-State MDPs and Stochastic Games (Abstract of Invited Talk)

Kousha Etessami

School of Informatics, University of Edinburgh, Edinburgh, UK
kousha@inf.ed.ac.uk

Abstract. I will survey a body of work developed over the past 15 years or so, on algorithms for, and the computational complexity of, analyzing and model checking some important families of countably infinite state Markov chains, Markov decision processes (MDPs), and stochastic games. I will also highlight some of the open questions remaining in this area, including some algorithmic questions regarding arithmetic circuits.

Contents

Algorithms

Invited Papers

Algebraic Theory of Promise Constraint Satisfaction Problems, First Steps

Libor Barto[(✉)] [iD]

Department of Algebra, Faculty of Mathematics and Physics, Charles University,
Sokolovská 83, 18675 Praha 8, Czechia
libor.barto@gmail.com
http://www.karlin.mff.cuni.cz/~barto

Abstract. What makes a computational problem easy (e.g., in P, that is, solvable in polynomial time) or hard (e.g., NP-hard)? This fundamental question now has a satisfactory answer for a quite broad class of computational problems, so called fixed-template constraint satisfaction problems (CSPs) – it has turned out that their complexity is captured by a certain specific form of symmetry. This paper explains an extension of this theory to a much broader class of computational problems, the promise CSPs, which includes relaxed versions of CSPs such as the problem of finding a 137-coloring of a 3-colorable graph.

Keywords: Computational complexity ·
Promise constraint satisfaction · Polymorphism

1 Introduction

In Computational Complexity we often try to place a given computational problem into some familiar complexity class, such as P, NP-complete, etc. In other words, we try to determine the image of a computational problem under the following mapping Φ.

$$\Phi : \text{computational problems} \to \text{complexity classes}$$
$$\text{problem} \mapsto \text{its complexity class}$$

When we try to achieve this goal for a whole class of computational problems, say $\mathscr{S}$, it is a natural idea to look for some intermediate collection $\mathfrak{I}$ of "invariants" and a decomposition of Φ through $\mathfrak{I}$:

$$\mathscr{S} \xrightarrow{\Psi} \mathfrak{I} \to \text{complexity classes}$$

Members of $\mathfrak{I}$ are then objects that exactly capture the computational complexity of problems in $\mathscr{S}$. The larger $\mathscr{S}$ is and the more objects Ψ glues together, the better such a decomposition is.

Libor Barto has received funding from the European Research Council (ERC) under the European Unions Horizon 2020 research and innovation programme (grant agreement No 771005).

L. A. Gąsieniec et al. (Eds.): FCT 2019, LNCS 11651, pp. 3–17, 2019.
https://doi.org/10.1007/978-3-030-25027-0_1

This idea proved greatly useful for an interesting class of problems, so called fixed-template constraint satisfaction problems (CSPs), and eventually led to a full complexity classification result [17,30]. In a decomposition, suggested in [20] and proved in [24], Ψ assigns to a CSP a certain algebraic object that describes, informally, the high dimensional symmetries of the CSP. This basic insight of the so called *algebraic approach to CSPs* was later twice improved [7,16], giving us a chain

$$\text{CSPs} \xrightarrow{\Psi} \mathfrak{I}_1 \to \mathfrak{I}_2 \to \mathfrak{I}_3 \to \text{complexity classes.}$$

The basics of the algebraic theory can be adapted and applied in various generalizations and variants of the fixed-template CSPs, see surveys in [26]. One particularly exciting direction is a recently proposed significant generalization of CSPs, so called promise CPSs (PCSPs) [3,14]. This framework is substantially richer, both on the algorithmic and the hardness side, and a full complexity classification is wide open even in very restricted subclasses. On the other hand, the algebraic basics can be generalized from CSP to PCSP and, moreover, one of the results in [18] not only gives such a generalization but also provides an additional insight and simplifies the algebraic theory of CSPs.

The aim of this paper is to explain this result (here Theorem 6) and the development in CSPs leading to it (Theorems 1, 2 and 3). The most recent material comes from the conference papers [18] and [4], which will be merged and expanded in [5]. Very little preliminary knowledge is assumed but an interested reader may find an in depth introduction to the fixed-template CSP and its variants in [26].

2   CSP

Fur the purposes of this paper, we define a *finite relational structure* as a tuple $\mathbb{A} = (A; R_1, \ldots, R_n)$, where A is a finite set, called the *domain* of $\mathbb{A}$, and each R_i is a relation on A of some arity, that is, $R_i \subseteq A^{\mathrm{ar}(R_i)}$ where $\mathrm{ar}(R_i)$ is a natural number.

A *primitive positive formula (pp-formula)* over $\mathbb{A}$ is a formula that uses only existential quantification, conjunction, relations in $\mathbb{A}$, and the equality relation. We will work only with formulas in a prenex normal form.

Definition 1. *Fix a finite relational structure $\mathbb{A}$. The CSP over $\mathbb{A}$, written* $\text{CSP}(\mathbb{A})$, *is the problem of deciding whether a given pp-sentence over $\mathbb{A}$ is true.*
In this context, $\mathbb{A}$ is called the template *for* $\text{CSP}(\mathbb{A})$.

For example, if $\mathbb{A} = (A; R, S)$ and both R and S are binary, then an input to $\text{CSP}(\mathbb{A})$ is, e.g.,

$$(\exists x_1 \exists x_2 \ldots \exists x_5) \; R(x_1, x_3) \wedge S(x_5, x_2) \wedge R(x_3, x_3).$$

This sentence is true if there exists a *satisfying assignment*, that is, a mapping $f : \{x_1, \ldots, x_5\} \to A$ such that $(f(x_1), f(x_3)) \in R$, $(f(x_5), f(x_2)) \in S$, and

$(f(x_3), f(x_3)) \in R$. Each conjunct thus can be thought of as a constraint limiting f and the goal is to decide whether there is an assignment satisfying each constraint.

Clearly, CSP($\mathbb{A}$) is always in NP.

The CSP over $\mathbb{A}$ can be also defined as a search problem where the goal is to find a satisfying assignment when it exists. It has turned out that the search problem is no harder then the decision problem presented in Definition 1 [16].

2.1 Examples

Typical problems covered by the fixed-template CSP framework are satisfiability problems, (hyper)graph coloring problems, and equation solvability problems. Let us look at several examples. We use here the notation

$$E_k = \{0, 1, \ldots, k - 1\}.$$

Example 1. Let $3\mathbb{SAT} = (E_2; R_{000}, R_{001}, \ldots, R_{111})$, where

$$R_{abc} = E_2^3 \setminus \{(a, b, c)\} \quad \text{for all } a, b, c \in \{0, 1\}.$$

An input to CSP($3\mathbb{SAT}$) is, e.g.,

$$(\exists x_1 \exists x_2 \ldots) \; S_{001}(x_1, x_4, x_2) \wedge S_{110}(x_2, x_5, x_5) \wedge S_{000}(x_2, x_1, x_3) \wedge \ldots.$$

Observe that this sentence is true if and only if the propositional formula

$$(x_1 \vee x_4 \vee \neg x_2) \wedge (\neg x_2 \vee \neg x_5 \vee x_5) \wedge (x_2 \vee x_1 \vee x_3) \wedge \ldots$$

is satisfiable. Therefore CSP($3\mathbb{SAT}$) is essentially the same as the 3SAT problem, a well known NP-complete problem.

On the other hand, recall that the 2SAT problem, which is the CSP over $2\mathbb{SAT} = (E_2; R_{00}, R_{01}, R_{10}, R_{11})$, where $R_{ab} = E_2^2 \setminus \{(a, b)\}$, is in P.

Example 2. Let $\mathbb{K}_3 = (E_3; N_3)$, where N_3 is the binary inequality relation, i.e.,

$$N_3 = \{(a, b) \in E_3^2 : a \neq b\}.$$

An input to CSP($\mathbb{K}_3$) is, e.g.,

$$(\exists x_1 \ldots \exists x_5) \; N_3(x_1, x_2) \wedge N_3(x_1, x_3) \wedge N_3(x_1, x_4) \wedge N_3(x_2, x_3) \wedge N_3(x_2, x_4).$$

Here an input can be drawn as a graph – its vertices are the variables and vertices x, y are declared adjacent iff the input contains a conjunct $N_3(x, y)$ or $N_3(y, x)$. For example, the graph associated to the input above is the five vertex graph obtained by merging two triangles along an edge. Clearly, an input sentence is true if and only if the vertices of the associated graph can be colored by colors 0, 1, and 2 so that adjacent vertices receive different colors. Therefore CSP($\mathbb{K}_3$) is essentially the same as the 3-coloring problem for graphs, another well known NP-complete problem.

More generally, CSP($\mathbb{K}_k$) = (E_k, N_k), where N_k is the inequality relation on E_k, is NP-complete for $k \geq 3$ and in P for $k = 2$.

Example 3. Let $3\mathbb{NAE}_k = (E_k; 3NAE_k)$, where $3NAE_k$ is the ternary not-all-equal relation, i.e.,

$$3NAE_k = E_k^3 \setminus \{(a, a, a) : a \in E_k\}.$$

Taking the viewpoint of Example 1, the CSP over $3\mathbb{NAE}_2$ is the positive not-all-equal 3SAT, where one is given a 3SAT formula without negations and the aim is to decide whether there is an assignment such that, in every clause, not all variables get the same value. This problem is NP-complete [29].

From the graph theoretical viewpoint, CSP($3\mathbb{NAE}_k$) is the problem of deciding whether a given 3-uniform hypergraph[1] admits a coloring by k colors so that no hyperedge is monochromatic.

Example 4. Let $1\mathbb{IN}3 = (E_2; 1IN3)$, where

$$1IN3 = \{(1, 0, 0), (0, 1, 0), (1, 0, 0)\}.$$

The CSP over $1\mathbb{IN}3$ is the positive one-in-three SAT problem or, in other words, the problem of deciding whether a given 3-uniform hypergraph admits a coloring by colors 0 and 1 so that exactly one vertex in each hyperedge receives the color 1. This problem is, again, NP-complete [29].

Example 5. Let $3\mathbb{LIN}_5 = (E_5; L_{0000}, L_{0001}, \ldots, L_{4444})$, where

$$L_{abcd} = \{(x, y, z) \in E_5^3 : ax + by + cz = d \pmod 5\}.$$

An input, such as

$$(\exists x_1 \exists x_2 \ldots) \, L_{1234}(x_3, x_4, x_2) \wedge L_{4321}(x_5, x_1, x_3) \wedge \ldots$$

can be written as a system of linear equations over the 5-element field $\mathbb{Z}_5$, such as

$$1x_3 + 2x_4 + 3x_2 = 4, \; 4x_5 + 3x_1 + 2x_3 = 1, \; \ldots,$$

therefore CSP($3\mathbb{LIN}_5$) is essentially the same problem as deciding whether a system of linear equations over $\mathbb{Z}_5$ (with each equation containing 3 variables) has a solution. This problem is in P.

2.2   1st Step: Polymorphisms

The crucial concept for the algebraic approach to the CSP is a polymorphism, which is a homomorphism from a cartesian power of a structure to the structure:

Definition 2. *Let $\mathbb{A} = (A; R_1, \ldots, R_n)$ be a relational structure. A k-ary (total) function $f : A^k \to A$ is a polymorphism of $\mathbb{A}$ if it is compatible with every relation R_i, that is, for all tuples $\mathbf{r}_1, \ldots, \mathbf{r}_k \in R_i$, the tuple $f(\mathbf{r}_1, \ldots, \mathbf{r}_k)$ (where f is applied component-wise) is in R_i.*

By Pol($\mathbb{A}$) *we denote the set of all polymorphisms of $\mathbb{A}$.*

---

[1] Here we should rather say a hypergraph whose hyperedges have size at most 3 because of conjuncts of the form $3NAE_k(x, x, y)$ or $3NAE_k(x, x, x)$. Let us ignore this minor technical imprecision.

The compatibility condition is often stated as follows: for any $(\mathrm{ar}(R_i) \times k)$-matrix whose column vectors are in R_i, the vector obtained by applying f to its rows is in R_i as well.

Note that the unary polymorphisms of $\mathbb{A}$ are exactly the endomorphisms of $\mathbb{A}$. One often thinks of endomorphisms (or just automorphisms) as symmetries of the structure. In this sense, polymorphisms can be thought of as higher dimensional symmetries.

For any domain A and any $i \leq k$, the k-ary projection to the i-th coordinate, that is, the function $\pi_i^k : A^k \to A$ defined by

$$\pi_i^k(x_1, \ldots, x_n) = x_i,$$

is a polymorphism of every structure with domain A. These are the *trivial* polymorphisms. The following examples show some nontrivial polymorphisms.

Example 6. Consider the template 2SAT from Example 1. It is easy to verify that the ternary majority function $\mathrm{maj} : E_2^3 \to E_2$ given by

$$\mathrm{maj}(x, x, y) = \mathrm{maj}(x, y, x) = \mathrm{maj}(y, x, x) = x \quad \text{for all } x, y \in E_2$$

is a polymorphism of 2SAT.

In fact, whenever a relation $R \subseteq E_2^m$ is compatible with maj, it can be pp-defined (that is, defined by a pp-formula) from relations in 2SAT (see e.g. [25]). Now for any template $\mathbb{A} = (E_2; R_1, \ldots, R_n)$ with polymorphism maj, an input of CSP($\mathbb{A}$) can be easily rewritten to an equivalent input of CSP(2SAT) and therefore CSP($\mathbb{A}$) is in P.

Example 7. Consider the template 3LIN$_5$ from Example 5. Each relation in this structure is an affine subspace of $\mathbb{Z}_5^3$. Every affine subspace is closed under affine combinations, therefore, for every k and every $t_1, \ldots, t_k \in E_5$ such that $t_1 + \cdots + t_k = 1 \pmod 5$, the k-ary function $f_{t_1, \ldots, t_k} : E_5^k \to E_5$ defined by

$$f_{t_1, \ldots, t_k}(x_1, \ldots, x_k) = t_1 x_1 + \ldots, t_k x_k \pmod 5$$

is a polymorphism of 3LIN$_5$.

Conversely, every subset of A^m closed under affine combinations is an affine subspace of $\mathbb{Z}_5^m$. It follows that if every $f_{t_1, \ldots, t_k}$ is a polymorphism of $\mathbb{A} = (E_5; R_1, \ldots, R_n)$, then inputs to CSP($\mathbb{A}$) can be rewritten to systems of linear equations over $\mathbb{Z}_5$ and thus CSP($\mathbb{A}$) is in P.

The above examples also illustrate that polymorphisms influence the computational complexity. The first step of the algebraic approach was to realize that this is by no means a coincidence.

Theorem 1 ([24]). *The complexity of* CSP($\mathbb{A}$) *depends only on* Pol($\mathbb{A}$).

More precisely, if $\mathbb{A}$ *and* $\mathbb{B}$ *are finite relational structures and* Pol($\mathbb{A}$) $\subseteq$ Pol($\mathbb{B}$)*, then* CSP($\mathbb{B}$) *is (log-space) reducible to* CSP($\mathbb{A}$)*. In particular, if* Pol($\mathbb{A}$) = Pol($\mathbb{B}$)*, then* CSP($\mathbb{A}$) *and* CSP($\mathbb{B}$) *have the same complexity.*

Proof (sketch). If $\mathrm{Pol}(\mathbb{A}) \subseteq \mathrm{Pol}(\mathbb{B})$, then relations in $\mathbb{B}$ can be pp-defined from relations in $\mathbb{A}$ by a classical result in Universal Algebra [10,11,21]. This gives a reduction from $\mathrm{CSP}(\mathbb{B})$ to $\mathrm{CSP}(\mathbb{A})$.

Theorem 1 can be used as a tool for proving NP-hardness: when $\mathbb{A}$ has only trivial polymorphism (and has domain of size at least two), any CSP on the same domain can be reduced to $\mathrm{CSP}(\mathbb{A})$ and therefore $\mathrm{CSP}(\mathbb{A})$ is NP-complete. This NP-hardness criterion is not perfect, e.g., $\mathrm{CSP}(3\mathrm{NAE}_2)$ has a nontrivial endomorphism $x \mapsto 1 - x$.

2.3 2nd Step: Strong Maltsev Conditions

Theorem 1 shows that the set of polymorphisms determines the complexity of a CSP. What information do we really need to know about the polymorphisms to determine the complexity? It has turned out that it is sufficient to know which functional equations they solve. In the following definition we use a standard universal algebraic term for a functional equation, a strong Maltsev condition.

Definition 3. *A strong Maltsev condition over a set of function symbols Σ is a finite set of equations of the form $t = s$, where t and s are terms built from variables and symbols in Σ.*

Let $\mathcal{M}$ be a set of functions on a common domain. A strong Maltsev condition S is satisfied in $\mathcal{M}$ if the function symbols of Σ can be interpreted in $\mathcal{M}$ so that each equation in S is satisfied for every choice of variables.

Example 8. A strong Maltsev condition over $\Sigma = \{f, g, h\}$ (where f and g are binary symbols and h is ternary) is, e.g.,

$$f(g(f(x, y), y), z) = g(x, h(y, y, z))$$
$$f(x, y) = g(g(x, y), x).$$

This condition is satisfied in the set of all projections (on any domain) since, by interpreting f and g as π_1^2 and h as π_1^3, both equations are satisfied for every x, y, z in the domain – they are equal to x.

The strong Maltsev condition in the above example is not interesting for us since it is satisfied in every $\mathrm{Pol}(\mathbb{A})$. Such conditions are called trivial:

Definition 4. *A strong Maltsev condition is called* trivial *if it is satisfied in the set of all projections on a two-element set (equivalently, it is satisfied in $\mathrm{Pol}(\mathbb{A})$ for every $\mathbb{A}$).*

Two nontrivial Maltsev condition are shown in the following example.

Example 9. The strong Maltsev condition (over a single ternary symbol m)

$$m(x, x, y) = x$$
$$m(x, y, x) = x$$
$$m(y, x, x) = x$$

is nontrivial since each of the possible interpretations π_1^3, π_2^3, π_3^3 of m falsifies one of the equations. This condition is satisfied in $\mathrm{Pol}(2\mathbb{SAT})$ by interpreting m as the majority function, see Example 6.

The strong Maltsev condition

$$p(x, x, y) = y$$
$$p(y, x, x) = y$$

is also nontrivial. It is satisfied in $\mathrm{Pol}(3\mathbb{LIN}_5)$ by interpreting p as $x + 4y + z$ (mod 5).

In fact, if $\mathrm{Pol}(\mathbb{A})$ satisfies one of the strong Maltsev conditions in this example, then $\mathrm{CSP}(\mathbb{A})$ is in P (see e.g. [6]).

The following theorem is (a restatement of) the second crucial step of the algebraic approach.

Theorem 2 ([16], **see also** [9]). *The complexity of* $\mathrm{CSP}(\mathbb{A})$ *depends only on strong Maltsev conditions satisfied by* $\mathrm{Pol}(\mathbb{A})$.

More precisely, if $\mathbb{A}$ *and* $\mathbb{B}$ *are finite relational structures and each strong Maltsev condition satisfied in* $\mathrm{Pol}(\mathbb{A})$ *is satisfied in* $\mathrm{Pol}(\mathbb{B})$, *then* $\mathrm{CSP}(\mathbb{B})$ *is (log-space) reducible to* $\mathrm{CSP}(\mathbb{A})$. *In particular, if* $\mathrm{Pol}(\mathbb{A})$ *and* $\mathrm{Pol}(\mathbb{B})$ *satisfy the same strong Maltsev conditions, then* $\mathrm{CSP}(\mathbb{A})$ *and* $\mathrm{CSP}(\mathbb{B})$ *have the same complexity.*

Proof (sketch). The proof can be done in a similar way as for Theorem 1. Instead of pp-definitions one uses more general constructions called pp-interpretations and, on the algebraic side, the Birkhoff HSP theorem [8].

Theorem 2 gives us an improved tool for proving NP-hardness: if $\mathrm{Pol}(\mathbb{A})$ satisfies only trivial strong Maltsev conditions, then $\mathrm{CSP}(\mathbb{A})$ is NP-hard. This criterion is better, e.g., it can be applied to $\mathrm{CSP}(3\mathbb{NAE}_2)$, but still not perfect, e.g., it cannot be applied to the CSP over the disjoint union of two copies of $\mathbb{K}_3$.

2.4 3rd Step: Minor Conditions

Strong Maltsev conditions that appear naturally in the CSP theory or in Universal Algebra are often of an especially simple form, they involve no nesting of function symbols. The third step in the basics of the algebraic theory was to realize that this is also not a coincidence.

Definition 5. *A strong Maltsev condition is called a* minor condition *if each side of every equation contains exactly one occurrence of a function symbol.*

In other words, each equation in a strong Maltsev condition is of the form "symbol(variables) = symbol(variables)".

Example 10. The condition in Example 8 is not a minor condition since, e.g., the left-hand side of the first equation involves three occurrences of function symbols.

The conditions in Example 9 are not minor conditions either since the right-hand sides do not contain any function symbol. However, these conditions have close friends which are minor conditions. For instance, the friend of the second system is the minor condition

$$p(x, x, y) = p(y, y, y)$$
$$p(y, x, x) = p(y, y, y).$$

Note that this system is also satisfied in $Pol(3\mathbb{LIN}_5)$ by the same interpretation as in Example 9, that is, $x + 4y + z \pmod 5$.

The following theorem is a strengthening of Theorem 2. We give only the informal part of the statement, the precise formulation is analogous to Theorem 2.

Theorem 3 ([7]). *The complexity of* CSP($\mathbb{A}$) *(for finite $\mathbb{A}$) depends only on minor conditions satisfied by* Pol($\mathbb{A}$).

Proof (sketch). The proof again follows the same pattern by further generalizing pp-interpretations (to so called pp-constructions) and the Birkhoff HSP theorem.

2.5   Classification

Just like Theorems 1 and 2 give hardness criteria for CSPs, we get an improved sufficient condition for NP-hardness as a corollary of Theorem 3.

Corollary 1. *Let $\mathbb{A}$ be a finite relational structure which satisfies only trivial minor conditions. Then* CSP($\mathbb{A}$) *is NP-complete.*

Bulatov, Jeavons, and Krokhin have conjectured [16] that satisfying only trivial minor conditions is actually the only reason for hardness[2]. Intensive efforts to prove this conjecture, called the *tractability conjecture* or the *algebraic dichotomy conjecture*, have recently culminated in two independent proofs by Bulatov and Zhuk:

Theorem 4 ([17,30]). *If a finite relational structure $\mathbb{A}$ satisfies a nontrivial minor condition, then* CSP($\mathbb{A}$) *is in P.*

Thus we now have a complete classification result: every finite structure either satisfies a nontrivial minor condition and then its CSP is in P, or it does not and its CSP is NP-complete. The proofs of Bulatov and Zhuk are very complicated and it should be stressed out that the basic steps presented in this paper form only a tiny (but important) part of the theory.

In fact, the third step did not impact on the resolution of the tractability conjecture for CSP over finite domains at all. However, it turned out to be significant for some generalizations of the CSP, including the generalization that we discuss in the next section, the Promise CSP.

---

[2] Their conjecture is equivalent but was, of course, originally stated in a different language – the significance of minor conditions in CSPs was identified much later.

3 PCSP

Many fixed-template CSPs, such as finding a 3-coloring of a graph or finding a satisfying assignment to a 3SAT formula, are hard computational problems. There are two ways how to relax the requirement on the assignment in order to get a potentially simpler problem. The first one is to require a specified fraction of the constraints to be satisfied. For example, given a satisfiable 3SAT input, is it easier to find an assignment satisfying at least 90% of clauses? A celebrated result of Håstad [22], which strengthens the famous PCP Theorem [1,2], proves that the answer is negative – it is still an NP-complete problem. (Actually, any fraction greater than 7/8 gives rise to an NP-complete problem while the fraction 7/8 is achievable in polynomial time.)

The second type of relaxation, the one we consider in this paper, is to require that a specified weaker version of every constraint is satisfied. For example, we want to find a 100-coloring of a 3-colorable graph, or we want to find a valid $CSP(3NAE_2)$ assignment to a true input of $CSP(1IN3)$. This idea is formalized in the following definition.

Definition 6. *Let* $\mathbb{A} = (A; R_1^{\mathbb{A}}, R_2^{\mathbb{A}}, \ldots, R_n^{\mathbb{A}})$ *and* $\mathbb{B} = (B; R_1^{\mathbb{B}}, R_2^{\mathbb{B}}, \ldots, R_n^{\mathbb{B}})$ *be two similar finite relational structures (that is, $R^{\mathbb{A}}$ and $R^{\mathbb{B}}$ have the same arity for each i), and assume that there exists a homomorphism $\mathbb{A} \to \mathbb{B}$. Such a pair $(\mathbb{A}, \mathbb{B})$ is refered to as a* PCSP template.

The PCSP over $(\mathbb{A}, \mathbb{B})$, denoted $PCSP(\mathbb{A}, \mathbb{B})$, is the problem to distinguish, given a pp-sentence ϕ over the relational symbols $R_1, \ldots, R_n$, between the cases that ϕ is true in $\mathbb{A}$ (answer "Yes") and ϕ is not true in $\mathbb{B}$ (answer "No").

For example, consider $\mathbb{A} = (A; R^{\mathbb{A}}, S^{\mathbb{A}})$ and $\mathbb{B} = (B; R^{\mathbb{B}}, S^{\mathbb{B}})$, where all the relations are binary. An input to $PCSP(\mathbb{A}, \mathbb{B})$ is, e.g.,

$$(\exists x_1 \exists x_2 \ldots \exists x_5)\ R(x_1, x_3) \wedge S(x_5, x_2) \wedge R(x_3, x_3).$$

The algorithm should answer "Yes" if the sentence is true in $\mathbb{A}$, i.e., the following sentence is true

$$(\exists x_1 \exists x_2 \ldots \exists x_5)\ R^{\mathbb{A}}(x_1, x_3) \wedge S^{\mathbb{A}}(x_5, x_2) \wedge R^{\mathbb{A}}(x_3, x_3),$$

and the algorithm should answer "No" if the sentence is not true in $\mathbb{B}$. In case that neither of the cases takes place, we do not have any requirements on the algorithm. Alternatively, we can say that the algorithm is promised that the input satisfies either "Yes" or "No" and it is required to decide which of these two cases takes place.

Note that the assumption that $\mathbb{A} \to \mathbb{B}$ is necessary for the problem to make sense, otherwise, the "Yes" and "No" cases would not be disjoint. Also observe that $CSP(\mathbb{A})$ is the same problem as $PCSP(\mathbb{A}, \mathbb{A})$.

The search version of $PCSP(\mathbb{A}, \mathbb{B})$ is perhaps a bit more natural problem: the goal is to find a $\mathbb{B}$-satisfying assignment given an $\mathbb{A}$-satisfiable input. Unlike in the CSP, it is not known whether the search version can be harder than the decision version presented in Definition 6.

3.1 Examples

The examples below show that PCSPs are richer than CSP, both on the algorithmic and the hardness side.

Example 11. Recall the structure $\mathbb{K}_k$ from Example 2 consisting of the inequality relation on a k-element set. For $k \leq l$, the PCSP over $(\mathbb{K}_k, \mathbb{K}_l)$ is the problem to distinguish between k-colorable graphs and graphs that are not even l-colorable (or, in the search version, the problem to find an l-coloring of a k-colorable graph).

Unlike for the case $k = l$, the complexity of this problem for $3 \leq k < l$ is a notorious open question. It is conjectured that PCSP$(\mathbb{K}_k, \mathbb{K}_l)$ is NP-hard for every $k < l$, but this conjecture was confirmed only in special cases: for $l \leq 2k - 2$ [12] (e.g., 4-coloring a 3-colorable graph) and for a large enough k and $l \leq 2^{\Omega(k^{1/3})}$ [23]. The algebraic development discussed in the next subsection helped in improving the former result to $l \leq 2k - 1$ [18] (e.g., 5-coloring a 3-colorable graph).

Example 12. Recall the structure $3\mathrm{NAE}_k$ from Example 3 consisting of the ternary not-all-equal relation on a k-element set. For $k \leq l$, the PCSP over $(3\mathrm{NAE}_k, 3\mathrm{NAE}_l)$ is essentially the problem to distinguish between k-colorable 3-uniform hypergraphs and 3-uniform hypergraphs that are not even l-colorable.

This problem is NP-hard for every $2 \leq k \leq l$ [19], the proof uses strong tools, the PCP theorem and Lovász's theorem on the chromatic number of Kneser's graphs [27].

Example 13. Recall from Example 4 that $1\mathbb{IN}3$ denotes the structure on the domain E_2 with the ternary "one-in-three" relation $1IN3$. The PCSP over $(1\mathbb{IN}3, 3\mathrm{NAE}_2)$ is the problem to distinguish between 3-uniform hypergraphs, which admit a coloring by colors 0 and 1 so that exactly one vertex in each hyperedge receives the color 1, and 3-uniform hypergraphs that are not even 2-colorable.

This problem, even its search version, admits elegant polynomial time algorithms [14, 15] – one is based on solving linear equations over the integers, another one on linear programming. For this specific template, the algorithm can be further simplified as follows.

We are given a 3-uniform hypergraph, which admits a coloring by colors 0 and 1 so that $(x, y, z) \in 1IN3$ for every hyperedge $\{x, y, z\}$, and we want to find a 2-coloring. We create a system of linear equations *over the rationals* as follows: for each hyperedge $\{x, y, z\}$ we introduce the equation $x + y + z = 1$. By the assumption on the input hypergraph, this system is solvable in $\{0, 1\} \subseteq \mathbb{Q}$ (in fact, $\{0, 1\}$-solutions are the same as valid $1IN3$-assignments). Solving equations in $\{0, 1\}$ is hard, but it is possible to solve the system in $\mathbb{Q} \setminus \{1/3\}$ in polynomial time by a simple adjustment of Gaussian elimination. Now we assign 1 to a vertex x if $x > 1/3$ in our rational solution, and 0 otherwise. It is simple to see that we get a valid 2-coloring.

Interestingly, to solve $\mathrm{PCSP}(1\mathbb{IN}3, 3\mathbb{NAE}_2)$, the presented algorithm uses a CSP over an *infinite* structure, namely $(\mathbb{Q} \setminus \{1/3\}; R)$, where $R = \{(x, y, z) \in (\mathbb{Q} \setminus \{1/3\})^3 : x + y + z = 1\}$. In fact, the infinity is necessary for this PCSP, see [4] for a formal statement and a proof.

3.2   4th Step: Minor Conditions!

After the introduction of the PCSP framework, it has quickly turned out that both the notion of a polymorphism and Theorem 1 have straightforward generalizations.

Definition 7. *Let* $\mathbb{A} = (A; R_1^{\mathbb{A}}, \dots)$ *and* $\mathbb{B} = (B; R_1^{\mathbb{B}}, \dots)$ *be two similar relational structures. A k-ary (total) function* $f : A^k \to B$ *is a* polymorphism *of* $(\mathbb{A}, \mathbb{B})$ *if it is compatible with every pair* $(R_i^{\mathbb{A}}, R_i^{\mathbb{B}})$, *that is, for all tuples* $\mathbf{r}_1, \dots, \mathbf{r}_k \in R_i^{\mathbb{A}}$, *the tuple* $f(\mathbf{r}_1, \dots, \mathbf{r}_k)$ *is in* $R_i^{\mathbb{B}}$.

By $\mathrm{Pol}(\mathbb{A}, \mathbb{B})$ *we denote the set of all polymorphisms of* $(\mathbb{A}, \mathbb{B})$.

Example 14. For every k which is not disible by 3, the k-ary "1/3-threshold" function $f : E_2^k \to E_2$ defined by

$$f(x_1, \dots, x_k) = \begin{cases} 1 \text{ if } \sum x_i/k > 1/3 \\ 0 \text{ else} \end{cases}$$

is a polymorphism of the PCSP template $(1\mathbb{IN}3, 3\mathbb{NAE}_2)$ from Example 13. Any PCSP whose template (over the domains E_2 and E_2) admits all these polymorphisms is in P [14,15].

Theorem 5 ([13]). *The complexity of* $\mathrm{PCSP}(\mathbb{A}, \mathbb{B})$ *depends only on* $\mathrm{Pol}(\mathbb{A}, \mathbb{B})$.

Proof (sketch). Proof is similar to Theorem 1 using [28] instead of [10,11,21]. $\square$

Note that, in general, composition of polymorphisms is not even well-defined. Therefore the second step, considering strong Maltsev conditions satisfied by polymorphisms, does not make sense for PCSPs. However, minor conditions make perfect sense and they do capture the complexity of PCSPs, as proved in [18]. Furthermore, the paper [18] also provides an alternative proof by directly relating a PCSP to a computational problem concerning minor conditions!

Theorem 6 ([18]). *Let* $(\mathbb{A}, \mathbb{B})$ *be a PCSP template and* $\mathcal{M} = \mathrm{Pol}(\mathbb{A}, \mathbb{B})$. *The following computational problems are equivalent for every sufficiently large* N.

- $\mathrm{PCSP}(\mathbb{A}, \mathbb{B})$.
- *Distinguish, given a minor condition* $\mathbf{C}$ *whose function symbols have arity at most* N, *between the cases that* $\mathbf{C}$ *is trivial and* $\mathbf{C}$ *is not satisfied in* $\mathcal{M}$.

Proof (sketch). The reduction from $\mathrm{PCSP}(\mathbb{A}, \mathbb{B})$ to the second problem works as follows. Given an input to the PCSP we introduce one $|A|$-ary function symbol g_a for each variable a and one $|R^{\mathbb{A}}|$-ary function symbol f_C for each conjunct

$R(\dots)$. The way to build a minor condition is quite natural, for example, the input

$$(\exists a \exists b \exists c \exists d)\ R(c, a, b) \wedge R(a, d, c) \wedge \dots$$

to $\mathrm{PCSP}(1\mathbb{I}\mathbb{N}3, 3\mathbb{N}\mathbb{A}\mathbb{E}_2)$ is transformed to the minor condition

$$f_1(x_1, x_0, x_0) = g_c(x_0, x_1)$$
$$f_1(x_0, x_1, x_0) = g_a(x_0, x_1)$$
$$f_1(x_0, x_0, x_1) = g_b(x_0, x_1)$$

$$f_2(x_1, x_0, x_0) = g_a(x_0, x_1)$$
$$f_2(x_0, x_1, x_0) = g_d(x_0, x_1)$$
$$f_2(x_0, x_0, x_1) = g_c(x_0, x_1)$$

$$\dots$$

It is easy to see that a sentence that is true in $\mathbb{A}$ is transformed to a trivial minor condition. On the other hand, if the minor condition is satisfied in $\mathcal{M}$, say by the functions denoted $f_1', f_2', g_a', \dots$, then the mapping $a \mapsto g_a'(0, 1)$, $b \mapsto g_b'(0, 1)$, $\dots$ gives a $\mathbb{B}$-satisfying assignment of the sentence – this can be deduced from the fact that f''s and g' are polymorphisms.

The reduction in the other direction is based on the idea that the question "Is this minor condition satisfied by polymorphisms of $\mathbb{A}$?" can be interpreted as an input to $\mathrm{CSP}(\mathbb{A})$. The main ingredient is to look at functions as tuples (their tables); then "f is a polymorphism" translates to a conjunction, and equations can be simulated by merging variables.

Theorem 6 implies Theorem 3 (and its generalization to PCSPs) since the computational problem in the second item clearly only depends on minor conditions satisfied in $\mathcal{M}$. The proof sketched above

- is simple and does not (explicitly) use any other results, such as the correspondence between polymorphisms and pp-definitions used in Theorem 1 or the Birkhoff HSP theorem used in Theorem 2, and
- is based on constructions which have already appeared, in some form, in several contexts; in particular, the second item is related to important problems in approximation, versions of the Label Cover problem (see [5, 18]).

The theorem and its proof are simple, nevertheless, very useful. For example, the hardness of $\mathrm{PCSP}(\mathbb{K}_k, \mathbb{K}_{2k-1})$ mentioned in Example 11 was proved in [18] by showing that every minor condition satisfied in $\mathrm{Pol}(\mathbb{K}_k, \mathbb{K}_{2k-1})$ is satisfied in $\mathrm{Pol}(3\mathbb{N}\mathbb{A}\mathbb{E}_2, 3\mathbb{N}\mathbb{A}\mathbb{E}_l)$ (for some l) and then using the NP-hardness of $\mathrm{PCSP}(3\mathbb{N}\mathbb{A}\mathbb{E}_2, 3\mathbb{N}\mathbb{A}\mathbb{E}_l)$ proved in [19] (see Example 12).

4   Conclusion

The PCSP framework is much richer than the CSP framework; on the other hand, the basics of the algebraic theory generalize from CSP to PCSP, as shown in Theorem 6. Strikingly, the computational problems in Theorem 6 are (promise and restricted versions) of two "similar" problems:

(i) Given a structure $\mathfrak{A}$ and a first-order sentence ϕ over the same signature, decide whether $\mathfrak{A}$ satisfies ϕ.
(ii) Given a structure $\mathfrak{A}$ and a first-order sentence ϕ in a different signature, decide whether symbols in ϕ can be interpreted in $\mathfrak{A}$ so that $\mathfrak{A}$ satisfies ϕ.

Indeed, $\mathrm{CSP}(\mathbb{A})$ is the problem (i) with $\mathfrak{A}$ a fixed relational structure and ϕ a pp-sentence (and PCSP is a promise version of this problem), whereas a promise version of the problem (ii) restricted to a fixed $\mathfrak{A}$ of purely functional signature and universally quantified conjunctive first-order sentences ϕ is the second item in Theorem 6. Variants of problem (i) appear in many contexts throughout Computer Science. What about problem (ii)?

References

1. Arora, S., Lund, C., Motwani, R., Sudan, M., Szegedy, M.: Proof verification and the hardness of approximation problems. J. ACM **45**(3), 501–555 (1998). https://doi.org/10.1145/278298.278306
2. Arora, S., Safra, S.: Probabilistic checking of proofs: a new characterization of NP. J. ACM **45**(1), 70–122 (1998). https://doi.org/10.1145/273865.273901
3. Austrin, P., Guruswami, V., Håstad, J.: (2+ε)-sat is NP-hard. SIAM J. Comput. **46**(5), 1554–1573 (2017). https://doi.org/10.1137/15M1006507
4. Barto, L.: Promises make finite (constraint satisfaction) problems infinitary. In: LICS (2019, to appear)
5. Barto, L., Bulín, J., Krokhin, A.A., Opršal, J.: Algebraic approach to promise constraint satisfaction (2019, in preparation)
6. Barto, L., Krokhin, A., Willard, R.: Polymorphisms, and how to use them. In: Krokhin, A., Živný, S. (eds.) The Constraint Satisfaction Problem: Complexity and Approximability, Dagstuhl Follow-Ups, vol. 7, pp. 1–44. Schloss Dagstuhl-Leibniz-Zentrum fuer Informatik, Dagstuhl, Germany (2017). https://doi.org/10.4230/DFU.Vol7.15301.1
7. Barto, L., Opršal, J., Pinsker, M.: The wonderland of reflections. Isr. J. Math. **223**(1), 363–398 (2018). https://doi.org/10.1007/s11856-017-1621-9
8. Birkhoff, G.: On the structure of abstract algebras. Math. Proc. Camb. Philos. Soc. **31**(4), 433–454 (1935). https://doi.org/10.1017/S0305004100013463
9. Bodirsky, M.: Constraint satisfaction problems with infinite templates. In: Creignou, N., Kolaitis, P.G., Vollmer, H. (eds.) Complexity of Constraints. LNCS, vol. 5250, pp. 196–228. Springer, Heidelberg (2008). https://doi.org/10.1007/978-3-540-92800-3_8
10. Bodnarchuk, V.G., Kaluzhnin, L.A., Kotov, V.N., Romov, B.A.: Galois theory for post algebras. I. Cybernetics **5**(3), 243–252 (1969). https://doi.org/10.1007/BF01070906
11. Bodnarchuk, V.G., Kaluzhnin, L.A., Kotov, V.N., Romov, B.A.: Galois theory for Post algebras. II. Cybernetics **5**(5), 531–539 (1969)

12. Brakensiek, J., Guruswami, V.: New hardness results for graph and hypergraph colorings. In: Raz, R. (ed.) 31st Conference on Computational Complexity (CCC 2016). Leibniz International Proceedings in Informatics (LIPIcs), vol. 50, pp. 14:1–14:27. Schloss Dagstuhl-Leibniz-Zentrum fuer Informatik, Dagstuhl, Germany (2016). https://doi.org/10.4230/LIPIcs.CCC.2016.14

13. Brakensiek, J., Guruswami, V.: Promise constraint satisfaction: Algebraic structure and a symmetric boolean dichotomy. ECCC Report No. 183 (2016)

14. Brakensiek, J., Guruswami, V.: Promise constraint satisfaction: Structure theory and a symmetric boolean dichotomy. In: Proceedings of the Twenty-Ninth Annual ACM-SIAM Symposium on Discrete Algorithms, SODA 2018, pp. 1782–1801. Society for Industrial and Applied Mathematics, Philadelphia (2018). http://dl.acm.org/citation.cfm?id=3174304.3175422

15. Brakensiek, J., Guruswami, V.: An algorithmic blend of LPs and ring equations for promise CSPs. In: Proceedings of the Thirtieth Annual ACM-SIAM Symposium on Discrete Algorithms, SODA 2019, San Diego, California, USA, 6–9 January 2019, pp. 436–455 (2019). https://doi.org/10.1137/1.9781611975482.28

16. Bulatov, A., Jeavons, P., Krokhin, A.: Classifying the complexity of constraints using finite algebras. SIAM J. Comput. **34**(3), 720–742 (2005). https://doi.org/10.1137/S0097539700376676

17. Bulatov, A.A.: A dichotomy theorem for nonuniform CSPs. In: 2017 IEEE 58th Annual Symposium on Foundations of Computer Science (FOCS), pp. 319–330, October 2017. https://doi.org/10.1109/FOCS.2017.37

18. Bulín, J., Krokhin, A., Opršal, J.: Algebraic approach to promise constraint satisfaction. In: Proceedings of the 51st Annual ACM SIGACT Symposium on the Theory of Computing (STOC 2019). ACM, New York (2019). https://doi.org/10.1145/3313276.3316300

19. Dinur, I., Regev, O., Smyth, C.: The hardness of 3-uniform hypergraph coloring. Combinatorica **25**(5), 519–535 (2005). https://doi.org/10.1007/s00493-005-0032-4

20. Feder, T., Vardi, M.Y.: The computational structure of monotone monadic SNP and constraint satisfaction: a study through datalog and group theory. SIAM J. Comput. **28**(1), 57–104 (1998). https://doi.org/10.1137/S0097539794266766

21. Geiger, D.: Closed systems of functions and predicates. Pacific J. Math. **27**, 95–100 (1968). https://doi.org/10.2140/pjm.1968.27.95

22. Håstad, J.: Some optimal inapproximability results. J. ACM **48**(4), 798–859 (2001). https://doi.org/10.1145/502090.502098

23. Huang, S.: Improved hardness of approximating chromatic number. In: Raghavendra, P., Raskhodnikova, S., Jansen, K., Rolim, J.D.P. (eds.) APPROX/RANDOM-2013. LNCS, vol. 8096, pp. 233–243. Springer, Heidelberg (2013). https://doi.org/10.1007/978-3-642-40328-6_17

24. Jeavons, P.: On the algebraic structure of combinatorial problems. Theor. Comput. Sci. **200**(1–2), 185–204 (1998). https://doi.org/10.1016/S0304-3975(97)00230-2

25. Jeavons, P., Cohen, D., Gyssens, M.: Closure properties of constraints. J. ACM **44**(4), 527–548 (1997). https://doi.org/10.1145/263867.263489

26. Krokhin, A., Živný, S. (eds.): The Constraint Satisfaction Problem: Complexity and Approximability, Dagstuhl Follow-Ups, vol. 7. Schloss Dagstuhl - Leibniz-Zentrum für Informatik (2017)

27. Lovász, L.: Kneser's conjecture, chromatic number, and homotopy. J. Combin. Theory Ser. A **25**(3), 319–324 (1978). https://doi.org/10.1016/0097-3165(78)90022-5

28. Pippenger, N.: Galois theory for minors of finite functions. Discrete Math. **254**(1), 405–419 (2002). https://doi.org/10.1016/S0012-365X(01)00297-7

29. Schaefer, T.J.: The complexity of satisfiability problems. In: Proceedings of the Tenth Annual ACM Symposium on Theory of Computing, STOC 1978, pp. 216–226. ACM, New York (1978). https://doi.org/10.1145/800133.804350
30. Zhuk, D.: A proof of CSP dichotomy conjecture. In: 2017 IEEE 58th Annual Symposium on Foundations of Computer Science (FOCS), pp. 331–342, October 2017. https://doi.org/10.1109/FOCS.2017.38

Some Observations on Dynamic Random Walks and Network Renormalization

Bernard Chazelle[⊠]

Department of Computer Science, Princeton University, Princeton, NJ 08540, USA
chazelle@cs.princeton.edu

Abstract. We recently developed a general bifurcation analysis framework for establishing the periodicity of certain time-varying random walks. In this work, we look at the special case of lazy uniform-inflow random walks and show how a much simpler version of the argument can be used to resolve their analysis. We also revisit a renormalization technique for network sequences that we introduced earlier and we propose a few simplifications. This work can be viewed as a gentle introduction to *Markov influence systems*.

Keywords: Markov influence systems · Dynamic random walks · Network renormalization

1 Introduction

Markov chains have remarkably simple dynamics: they either mix toward a stationary distribution or oscillate periodically. The periodic regime can be easily ruled out by introducing self-loops; thus, from a dynamical-systems perspective, Markov chains are essentially trivial. Not so with time-varying Markov chains [1,4–12,14]. We recently introduced *Markov influence systems (MIS)* to model random walks in graphs whose transition probabilities and topologies change over time endogenously [3]. The presence of a feedback loop, through which the next graph is chosen as a function of the current distribution of the walk, plays a crucial role. Indeed, the dynamics ranges over the entire spectrum from fixed-point attraction to chaos. This stands in sharp contrast to not only classical Markov chains but also time-varying chains whose temporal changes are driven randomly [1].

The Research was sponsored by the Army Research Office and the Defense Advanced Research Projects Agency and was accomplished under Grant Number W911NF-17-1-0078. The views and conclusions contained in this document are those of the authors and should not be interpreted as representing the official policies, either expressed or implied, of the Army Research Office, the Defense Advanced Research Projects Agency, or the U.S. Government. The U.S. Government is authorized to reproduce and distribute reprints for Government purposes notwithstanding any copyright notation herein.

© Springer Nature Switzerland AG 2019
L. A. Gąsieniec et al. (Eds.): FCT 2019, LNCS 11651, pp. 18–28, 2019.
https://doi.org/10.1007/978-3-030-25027-0_2

We showed that, if the Markov chains used at each step are irreducible, then the *MIS* is almost surely asymptotically periodic [3]. We prove a similar result in the next section, but for a different family of random walks, called *uniform-inflow*. Though the proof borrows much of the architecture of our previous one, it is much simpler and can be viewed as a gentle introduction to Markov influence systems.

The main weakness of our bifurcation analysis is to impose topological restrictions on the graphs. As a step toward overcoming this limitation, we have developed a *renormalization* technique for graph sequences [3]. The motivation was to extend the standard classification of Markov chain states to the time-varying case. We revisit this technique in Sect. 3 and propose a number of useful simplifications.

2   Time-Varying Random Walks

Recall the definition of a Markov influence system [3]. Let $\mathbb{S}^{n-1}$ be the probability simplex $\{\, \mathbf{x} \in \mathbb{R}^n \mid \mathbf{x} \geq \mathbf{0},\, \|\mathbf{x}\|_1 = 1 \,\}$ and let $\mathcal{S}$ denote set of all n-by-n row-stochastic matrices. An *MIS* is a discrete-time dynamical system with phase space $\mathbb{S}^{n-1}$, which is defined by the map $f : \mathbf{x}^\top \mapsto f(\mathbf{x}) := \mathbf{x}^\top S(\mathbf{x})$, where $\mathbf{x} \in \mathbb{S}^{n-1}$ and S is a piecewise-constant function $\mathbb{S}^{n-1} \mapsto \mathcal{S}$ over the cells $\{C_k\}$ of a hyperplane arrangement $\mathcal{H}$ within $\mathbb{S}^{n-1}$ (Fig. 1); over the discontinuities $h \cap \mathbb{S}^{n-1}$ ($h \in \mathcal{H}$), we define f as the identity.[1]

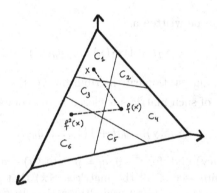

Fig. 1. The arrangement $\mathcal{H}$ consists of three hyperplanes. Each cell C_i in the simplex $\mathbb{S}^{n-1}$ is associated with a stochastic matrix defining the map f over it. The figure shows the first two iterates of $\mathbf{x}$ under f. The case $|\mathcal{H}| = 0$ corresponds to an ordinary random walk.

We focus our attention on *lazy uniform-inflow* random walks: each matrix $S(\mathbf{x})$ is associated with a probability distribution $(p_0(\mathbf{x}), \ldots, p_n(\mathbf{x})) \in \mathbb{S}^n$, such that $S(\mathbf{x})_{ij} = p_j(\mathbf{x}) + \delta_{ij} p_0(\mathbf{x})$ and δ_{ij} is the Kronecker delta. The cases $p_0(\mathbf{x}) =$

---

[1] The discontinuities can also be chosen to be real-algebraic varieties.

$0, 1$ are both trivial, so we may assume that $0 < p_0(\mathbf{x}) < 1$. Thus, any given $S(\mathbf{x})$ is the transition matrix of a lazy random walk. We state our main result:

Theorem 1. *Every orbit of a lazy uniform-inflow Markov influence system is almost surely asymptotically periodic.*

Note that lazy uniform-inflow random walks are not necessarily irreducible, so the theorem does not follow from [3].[2] In the remainder of this section, we discuss the meaning of the result and then we prove it. The *orbit* of $\mathbf{x} \in \mathbb{S}^{n-1}$ is the infinite sequence $(f^t(\mathbf{x}))_{t \geq 0}$ and its *itinerary* is the corresponding sequence of cells C_i's visited in the process. The orbit is *periodic* if $f^t(\mathbf{x}) = f^s(\mathbf{x})$ for any $s = t$ modulo a fixed integer. It is asymptotically periodic if it gets arbitrarily close to a periodic orbit over time. The discontinuities in the map f occur at the intersections of the simplex $\mathbb{S}^{n-1}$ with the hyperplanes $\{\mathbf{x} \in \mathbb{R}^n \mid \mathbf{a}_i^\top \mathbf{x} = 1\}$ of $\mathcal{H}$. The hyperplanes are perturbed into the form $\mathbf{a}_i^\top \mathbf{x} = 1 + \delta$, for $\delta \in \Omega = [-\omega, \omega]$ and $\omega > 0$. Assuming that $\mathcal{H}$ is in general position, ω can always be chosen small enough so that the perturbed arrangement remains topologically invariant over all $\delta \in \Omega$. Theorem 1 follows from the existence of a set of Lebesgue measure zero (coverable by a Cantor set of Hausdorff dimension less than one) such that, for any $\delta \in \Omega$ outside of it, there is a finite set of stable periodic orbits (ie, discrete limit cycles) such that every orbit is asymptotically attracted to one of them.

It is useful to begin with a few observations about the stochastic matrices involved in lazy uniform-inflow random walks:

1. The matrix $S(\mathbf{x})$ can be written as

$$p_0(\mathbf{x})I + \mathbf{1}\big(p_1(\mathbf{x}), \ldots, p_n(\mathbf{x})\big) \tag{1}$$

and it has the unique stationary distribution $\pi(\mathbf{x}) = \frac{1}{1-p_0(\mathbf{x})}\big(p_1(\mathbf{x}), \ldots, p_n(\mathbf{x})\big)$. The family of such matrices is closed under composition. Indeed,

$$S(\mathbf{x})S(\mathbf{y}) = q_0 I + \mathbf{1}\big(q_1, \ldots, q_n\big),$$

where $q_0 = p_0(\mathbf{x})p_0(\mathbf{y})$ and, for $i > 0$, $q_i = p_i(\mathbf{x})p_0(\mathbf{y}) + p_i(\mathbf{y})$.

2. Let $\mathcal{M}$ be the (finite) set of all the matrices $S(\mathbf{x})$ that arise in the definition of f. We just saw that $p_0(\mathbf{x})$ is multiplicative. In this case, it is equal to the coefficient of ergodicity of the matrix [13], which is defined as half the maximum ℓ_1-distance between any two of its rows. By our assumption, $\tau := \sup_\mathbf{x} p_0(\mathbf{x}) < 1$. Given $M_1, \ldots, M_k \in \mathcal{M}$, if π denotes the stationary distribution of $M_1 \cdots M_k$, then

$$\begin{cases} M_1 \cdots M_k = qI + (1-q)\mathbf{1}\pi^\top \\ \mathrm{diam}_{\ell_\infty}\big(\mathbb{S}^{n-1}M_1 \cdots M_k\big) = q \leq \tau^k. \end{cases} \tag{2}$$

---

[2] For example, the lazy random walk specified by the matrix $\left(\begin{smallmatrix} 1 & 0 \\ 0.5 & 0.5 \end{smallmatrix}\right)$ is not irreducible. Also, unlike in [3], we do not require the matrices to be rational.

Let D be the union of the discontinuities (defined by the intersection of the perturbed hyperplanes with the simplex), for some fixed $\delta \in \Omega$. Put $Z_t = \bigcup_{0 \leq k \leq t} f^{-k}(D)$ and $Z = \bigcup_{t \geq 0} Z_t$ and note that $Z_\nu = Z_{\nu-1}$ implies that $Z = Z_\nu$. Indeed, suppose that $Z_{t+1} \supset Z_t$ for $t \geq \nu$; then, $f^{t+1}(\mathbf{y}) \in D$ but $f^t(\mathbf{y}) \notin D$ for some $\mathbf{y} \in \mathbb{S}^{n-1}$; in other words, $f^\nu(\mathbf{x}) \in D$ but $f^{\nu-1}(\mathbf{x}) \notin D$ for $\mathbf{x} = f^{t-\nu+1}(\mathbf{y})$, which contradicts the equality $Z_\nu = Z_{\nu-1}$. The key to periodicity is to prove that ν is finite.[3]

Lemma 1. *There is a constant $c > 0$ such that, for any $\varepsilon > 0$, there exist an integer $\nu \leq c \log(1/\varepsilon)$ and a finite union K of intervals of total length at most ε such that $Z_\nu = Z_{\nu-1}$, for any $\delta \in \Omega \setminus K$.*

Proof of Theorem 1. The theorem can be shown to follow from Lemma 1 by using an argument from [3]. We reproduce the proof here for the sake of completeness. The polyhedral cells defined by the connected components of the complement of $Z = Z_\nu$ form the continuity pieces of $f^{\nu+1}$: by continuity, each one of them maps, under f, not simply to within a single cell of D but actually to within a single cell of Z itself.[4] This in turn implies the eventual periodicity of the symbolic dynamics. Once an itinerary becomes periodic at time t_o with period σ, the map f^t can be expressed locally by matrix powers. Indeed, divide $t - t_o$ by σ and let q be the quotient and r the remainder; then, locally, $f^t = g^q \circ f^{t_o+r}$, where g is specified by the stochastic matrix of a lazy, uniform-inflow random walk, which implies convergence to a periodic point. In fact, better than that, we know from (2) that the matrix corresponds to a random walk that mixes to a unique stationary distribution, so the attracting periodic orbits are stable and there are only a finite number of them.

To complete the proof, we apply Lemma 1 repeatedly, with $\varepsilon = 2^{-l}$ for $l = 1, 2, \ldots$ and denote by K_l be the corresponding union of "forbidden" intervals. Define $K^l = \bigcup_{j \geq l} K_j$ and $K^\infty = \bigcap_{l > 0} K^l$; then $\mathrm{Leb}(K^l) \leq 2^{1-l}$ and hence $\mathrm{Leb}(K^\infty) = 0$. Theorem 1 follows from the fact that any $\delta \in \Omega$ outside of K^∞ lies outside of K^l for some $l > 0$. $\square$

As in [3], we begin the proof of Lemma 1 with a discussion of the symbolic dynamics of the system. Given $\Delta \subseteq \Omega$, let L_Δ^t denote the set of t-long prefixes of any itinerary for any starting position $\mathbf{x} \in \mathbb{S}^{n-1}$ and any $\delta \in \Delta$. Fix $\rho > 0$ and define $\mathcal{D}_\rho = \big\{ [k\rho, (k+1)\rho] \cap \Omega \mid k \in \mathbb{Z} \big\}$.

Lemma 2. *There is a constant $b > 0$ such that, for any real $\rho > 0$ and any integer $T > 0$, there exist $t_\rho \leq b \log(1/\rho)$ and $V \subseteq \mathcal{D}_\rho$ of size at most b^T such that, for any $\Delta \in \mathcal{D}_\rho \setminus V$, any integer $t \geq t_\rho$, and any $\sigma \in L_\Delta^t$, we have $\big| \{ \sigma' \mid \sigma \cdot \sigma' \in L_\Delta^{t+T} \} \big| \leq b$.*

---

[3] The constants in this paper may depend on any of the input parameters, such as the dimension n, the number of hyperplanes, the hyperplane coefficients, and the matrix elements.

[4] Indeed, if that were not the case, then some $\mathbf{x}$ in a cell of Z, thus outside of Z, would be such that $f(\mathbf{x}) \in Z = Z_{\nu-1}$. It would follow that $f^k(\mathbf{x}) \in D$, for $k \leq \nu$; hence $\mathbf{x} \in Z_\nu = Z$, a contradiction.

Proof. Given $M_1, \ldots, M_k \in \mathcal{M}$, let $\varphi^k(\mathbf{x}) = \mathbf{x}^\top M_1 \cdots M_k$ for $\mathbf{x} \in \mathbb{R}^n$ and $k \leq T$; and let $h_\delta : \mathbf{a}^\top \mathbf{x} = 1 + \delta$ be some hyperplane in $\mathbb{R}^n$. Let $h_\Delta := \bigcup_{\delta \in \Delta} h_\delta$ and $X = \mathbf{x} + \rho[-1,1]^n$, for $\mathbf{x} \in \mathbb{S}^{n-1}$. We define an exclusion zone U outside of which the T iterated images of X can meet h_Δ at most once. This is a general position claim much stronger than the one we used in [3] and closer in spirit to a dimensionality argument for planar contractions from [2] that inspired our approach.

Claim A. *For some constant $d > 0$ (independent of ρ), there exists $U \subseteq \mathcal{D}_\rho$ of size at most dT^2 such that, for any $\Delta \in \mathcal{D}_\rho \setminus U$ and $\mathbf{x} \in \mathbb{S}^{n-1}$, there are at most one integer $k \leq T$ such that $\varphi^k(X) \cap h_\Delta \neq \emptyset$.*

Proof. The crux of the claim is that it holds for *any* probability distribution $\mathbf{x}$. We assume the existence of two integers $j < k \leq T$ such that $\varphi^i(X) \cap h_\Delta \neq \emptyset$, for $i = j, k$ and $\Delta \in \mathcal{D}_\rho$. We draw the consequences and then negate them in order to rule out the assumption: this, in turn, specifies the set U. The assumption implies the existence of δ and $\mathbf{x} \in \mathbb{S}^{n-1}$ such that $|(\mathbf{x} + \mathbf{u})^T M_1 \cdots M_j \mathbf{a} - (1 + \delta)| \leq d_o \rho$, with $\|\mathbf{u}\|_\infty \leq \rho$ and constant $d_o > 0$. Likewise, we have $|(\mathbf{x} + \mathbf{u}')^T M_1 \cdots M_k \mathbf{a} - (1 + \delta)| \leq d_o \rho$, with $\|\mathbf{u}'\|_\infty \leq \rho$. Writing $\mathbf{v} = \mathbf{x}^T M_1 \cdots M_j$, we have $|\mathbf{v}^T \mathbf{a} - (1 + \delta)| \leq d'_o \rho$ and $|\mathbf{v}^T M_{j+1} \cdots M_k \mathbf{a} - (1 + \delta)| \leq d'_o \rho$, for constant d'_o dependent on $\mathbf{a}$. By (1), $M_{j+1} \cdots M_k = q_0 I + \mathbf{1}(q_1, \ldots, q_n)$, for some $(q_0, \ldots, q_n) \in \mathbb{S}^n$. Since $\mathbf{v} \in \mathbb{S}^{n-1}$, it follows that

$$\left| q_0 \mathbf{v}^T \mathbf{a} + (q_1, \ldots, q_n) \mathbf{a} - (1 + \delta) \right| \leq d'_o \rho;$$

hence $|\delta + 1 - (q_1, \ldots, q_n)\mathbf{a}/(1 - q_0))| \leq 2d'_o \rho/(1 - \tau)$. To rule out the previous condition, we must keep δ outside of $O(1/(1 - \tau))$ intervals of $\mathcal{D}_\rho$. The claim follows from the fact that the number of products $M_{j+1} \cdots M_k$ is quadratic in T. □

To complete the proof of Lemma 2, we define V as the union of the sets U formed by applying Claim A to each one of the hyperplanes h_δ of $\mathcal{H}$ and every possible sequence of T matrices in $\mathcal{M}$; hence $|V| \leq b^T$ for constant $b > 0$. We fix $\Delta \in \mathcal{D}_\rho \setminus V$ and consider the (lifted) phase space $\mathbb{S} \times \Delta$ for the dynamical system induced by the map $f_\uparrow : (\mathbf{x}^\top, \delta) \mapsto (\mathbf{x}^\top S(\mathbf{x}), \delta)$. A continuity piece Υ_t for $f_\uparrow^t$ is a maximal polyhedron within $\mathbb{S}^{n-1} \times \Delta$ over which the t-th iterate of $f_\uparrow$ is linear.

Given any sequence $M_1, \ldots, M_k$ in $\mathcal{M}$, recall from (2) that $\mathrm{diam}_{\ell_\infty}(\mathbb{S}^{n-1} M_1 \cdots M_k) \leq \tau^k$. This implies the existence of an integer $t_\rho \leq b \log(1/\rho)$ (raising the previous constant b if necessary) such that, for any $t \geq t_\rho$, $f_\uparrow^t(\Upsilon_t) \subseteq (\mathbf{x} + \rho \mathbb{I}^n) \times \Delta$, for some $\mathbf{x} = \mathbf{x}(t, \Upsilon_t) \in \mathbb{S}^{n-1}$. Consider a nested sequence $\Upsilon_1 \supseteq \Upsilon_2 \supseteq \cdots$. Note that Υ_1 is a polyhedral cell within $\mathbb{S}^{n-1} \times \Delta$ and $f_\uparrow^k(\Upsilon_{k+1}) \subseteq f_\uparrow^k(\Upsilon_k)$. There is a *split* at k if $\Upsilon_{k+1} \subset \Upsilon_k$. Observe that, by Claim A, given any $t \geq t_\rho$, there are at most a constant number b_1 of splits between t and $t + T$ (at most one per hyperplane of $\mathcal{H}$). It follows that the number of nested sequences is bounded by the number of leaves in a tree of height T with at most b_1 nodes of degree greater than 1 along any path. Lemma 2 follows from the fact that no node has more than a constant number of children. □

Proof of Lemma 1. We use the notation of Lemma 2 and set T to a large enough constant. Fix $\varepsilon > 0$ and set $\rho = \frac{1}{2}\varepsilon/|V|$, and $\nu = t_\rho + kT$, where $k = T\log(1/\varepsilon)$. Since $t_\rho \leq b\log(1/\rho)$, note that $\nu = O(\log 1/\varepsilon)$. Let $P = M_1 \cdots M_\nu$, where $M_1, \ldots, M_\nu$ is the matrix sequence matching an element of L_Δ^ν, for $\Delta \in \mathcal{D}_\rho \backslash V$. By (2), $\operatorname{diam}_{\ell_\infty}(\mathbb{S}^{n-1} P) \leq \tau^\nu$, so there is a point $\mathbf{x}_P$ such that, given any point $\mathbf{y} \in \mathbb{S}^{n-1}$ whose ν-th iterate $f^\nu(\mathbf{y}) = \mathbf{z}^T$ is specified by $\mathbf{z}^T = \mathbf{y}^T P$, we have $\|\mathbf{x}_P - \mathbf{z}\|_\infty \leq \tau^\nu$. Given a discontinuity $h_\delta : \mathbf{a}_i^\top \mathbf{x} = 1 + \delta$ of the system, the point $\mathbf{z}$ lies on one side of h_δ if and only if $\mathbf{x}_P$ lies on the (relevant) side of some $h_{\delta'}$, for $|\delta' - \delta| \leq c_1 \tau^\nu$, for constant $c_1 > 0$. Thus, adding an interval of length $c_2 \tau^\nu$ to V, for constant c_2 large enough (independent of T), it is the case that, for any $h \in \mathcal{H}$, it holds that, for all $\mathbf{y} \in \mathbb{S}^{n-1}$, the ν-th iterates $f^\nu(\mathbf{y})$ specified by P all lie strictly on the same side of h_δ, for any $\delta \in \Delta$. We repeat this operation for every string L_Δ^ν and each one of the (at most) $1/\rho$ intervals $\Delta \in \mathcal{D}_\rho \backslash V$. This increases the length $\operatorname{Leb}(V)$ covered by V from its original $\rho|V| = \varepsilon/2$ to $\rho|V| + c_2 |L_\Delta^\nu| \tau^\nu / \rho \leq \varepsilon$. This last inequality follows from:

$$(\rho\varepsilon)^{-1} |L_\Delta^\nu| \leq (\rho\varepsilon)^{-1} c_3^{t_\rho} b^k \qquad [\text{ for constant } c_3 \text{ independent of } T]$$

$$\leq 2c_3^{t_\rho} b^{k+T} 4^{k/T} \qquad [\, 1/\rho\varepsilon \leq 2b^T/\varepsilon^2 \leq 2b^T 4^{k/T} \,]$$

$$\leq 2c_3^{b^2 k} b^{k+T} 4^{k/T} \qquad [\, t_\rho \leq b\log(1/\rho) \leq b^2 T \log(1/\varepsilon) \,]$$

$$\leq T^k \leq \tau^{-\nu}/(2c_2). \qquad [\text{ for } T \text{ large enough }]$$

Thus, for any $\delta \in \Omega$ outside a set of intervals covering a length at most ε, no $f^\nu(\mathbf{x})$ lies on a discontinuity. It follows that, for any such δ, we have $Z_\nu = Z_{\nu-1}$.
□

This completes the proof of Theorem 1.

3 Revisiting Network Sequence Renormalization

In [3], we proposed a mechanism for expressing an infinite sequence of networks as a hierarchy of graph clusters. The intention was to generalize to the time-varying case the standard classification of the states of a Markov chain. We review the main parts of this "renormalization" technique and propose a number of simplifications along the way. Our variant maintains the basic division of the renormalization process into temporal and topological parts, but it simplifies the overall procedure. For example, the new grammar includes only three productions, as opposed to four.

Throughout this discussion, a *digraph* is a directed graph with vertices in $[n] := \{1, \ldots, n\}$ and a self-loop attached to each vertex. Graphs and digraphs (words we use interchangeably) have no multiple edges. A *digraph sequence* $\mathbf{g} = (g_k)_{k>0}$ is an ordered (possibly infinite) list of digraphs over the same vertex set $[n]$. We define the product $g_i \times g_j$ as the digraph consisting of all the edges (x, y) with an edge (x, z) in g_i and another one (z, y) in g_j for at least one vertex z. The operation $\times$ is associative but not commutative; it corresponds roughly to matrix multiplication. The digraph $\prod_{\leq k} \mathbf{g} = g_1 \times \cdots \times g_k$ is called a *cumulant* and, for finite $\mathbf{g}$, we write $\prod \mathbf{g} = g_1 \times g_2 \times \cdots$

The cumulant links all the pairs of vertices that can be joined by a temporal walk of a given length. The mixing time of a random walk on a (fixed) graph depends on the speed at which information diffuses and, in particular, how quickly the cumulant becomes transitive. In the time-varying case, mixing is a more complicated proposition, but the emergence of transitive cumulants is still what guides the parsing process.

An edge (x, y) of a digraph g is *leading* if there is u such that (u, x) is an edge of g but (u, y) is not. The non-leading edges form a subgraph of g, which is denoted by $tf(g)$ and called the transitive front of g. For example, $tf(x \rightarrow y \rightarrow z)$ is the graph over x, y, z with the single edge $x \rightarrow y$ (and the three self-loops); on the other hand, the transitive front of a directed cycle over three or more vertices has no edges besides the self-loops. We denote by $cl(g)$ the transitive closure of g: it is the graph that includes an edge (x, y) for any two vertices x, y with a path from x to y. Note that $tf(g) \preceq g \preceq cl(g)$.

- An equivalent definition of the transitive front is that the edges of $tf(g)$ are precisely the pairs (i, j) such that $C_i \subseteq C_j$, where C_k denotes the set of vertices l such that (l, k) is an edge of g. Because each vertex has a self-loop, the inclusion $C_i \subseteq C_j$ implies that (i, j) is an edge of g. If g is transitive, then $tf(g) = g$. The set-inclusion definition of the transitive front shows that it is indeed transitive: ie, if (x, y) and (y, z) are edges, then so is (x, z). Given two graphs g, h over the same vertex set, we write $g \preceq h$ if all the edges of g are in h (with strict inclusion denoted by the symbol $\prec$). Because of the self-loops, $g, h \preceq g \times h$.

- A third characterization of $tf(g)$ is as the unique largest graph h over $[n]$ such that $g \times h = g$: we call this the *maximally-dense property* of the transitive front, and it is the motivation behind our use of the concept. Indeed, the failure of subsequent graphs to grow the cumulant implies a structural constraint on them. This is the sort of structure that parsing attempts to tease out.

A graph sequence $\mathbf{g} = (g_k)_{k>0}$ can be parsed into a rooted tree whose leaves are associated with $g_1, g_2, \ldots$ from left to right. The purpose of the parse tree is to track the creation of new temporal walks over time. This is based on the observation that, because of the self-loops, the cumulant $\prod_{\leq k} \mathbf{g}$ is monotonically nondecreasing with k (with all references to graph ordering being relative to $\preceq$). If the increase were strict at each step, then the parse tree would look like a fishbone. The cumulant cannot grow forever, obviously, and parsing is what tells us what to do when it reaches its maximum size. The underlying grammar consists of three productions: (1a) and (1b) renormalize the graph sequence along the time axis, while (2) creates the hierarchical clustering of the graphs in the sequence $\mathbf{g}$.

1. TEMPORAL RENORMALIZATION. We express the sequence $\mathbf{g}$ in terms of minimal subsequences with cumulants equal to $\prod \mathbf{g}$. There is a unique decomposition

$$\mathbf{g} = \mathbf{g}_1, g_{m_1}, \ldots, \mathbf{g}_k, g_{m_k}, \mathbf{g}_{k+1}$$

such that

(i) $\mathbf{g}_1 = g_1, \ldots, g_{m_1-1}$; $\mathbf{g}_i = g_{m_{i-1}+1}, \ldots, g_{m_i-1}$ $(1 < i \le k)$; and $\mathbf{g}_{k+1} = g_{m_k+1}, \ldots$.

(ii) $\left(\prod \mathbf{g}_i\right) \times g_{m_i} = \prod \mathbf{g}$, for any $i \le k$; and $\prod \mathbf{g}_i \prec \prod \mathbf{g}$, for any $i \le k+1$.

The two productions below create the *temporal parse tree*. Unless specified otherwise, the node corresponding to the sequence $\mathbf{g}$ is annotated by the transitive graph $cl(\prod \mathbf{g})$, called its *sketch*.

- *Transitivization*. Assume that $\prod \mathbf{g}$ is not transitive. We define $h = tf\left(\prod \mathbf{g}\right)$ and note that $h \prec \prod \mathbf{g}$. It follows from the maximally-dense property of the transitive front that $k = 1$. Indeed, $k > 1$ would imply that $\prod \mathbf{g} = \left(\prod \mathbf{g}_2\right) \times g_{m_2} \preceq tf\left\{\left(\prod \mathbf{g}_1\right) \times g_{m_1}\right\} = tf\left(\prod \mathbf{g}\right)$, which would contradict the non-transitivity of $\prod \mathbf{g}$. We have $\mathbf{g} = \mathbf{g}_1, g_{m_1}, \mathbf{g}_2$ and the production

$$\mathbf{g} \longrightarrow (\mathbf{g}_1)\, g_{m_1}\left((\mathbf{g}_2) \triangle h\right). \tag{1a}$$

In the parse tree, the node for $\mathbf{g}$ has three children: the first one serves as the root of the temporal parse subtree for $\mathbf{g}_1$; the second one is the leaf associated with the graph g_{m_1}; the third one is a *special* node annotated with the label $\triangle h$, which serves as the parent of the node rooting the parse subtree for $\mathbf{g}_2$. The purpose of annotating a special node with the label $\triangle h$ is to provide an intermediate approximation of $\prod \mathbf{g}_2$ that is strictly finer than the transitive closure. These coarse-grained approximations form the sketches. Note that special nodes have only one child.

- *Cumulant completion*. Assume that $\prod \mathbf{g}$ is transitive. We have the production

$$\mathbf{g} \longrightarrow (\mathbf{g}_1)\, g_{m_1} (\mathbf{g}_2)\, g_{m_2} \cdots (\mathbf{g}_k)\, g_{m_k} (\mathbf{g}_{k+1}). \tag{1b}$$

Note that the index k may be infinite and any of the subsequences $\mathbf{g}_i$ might be empty (for example, $\mathbf{g}_{k+1}$ if $k = \infty$).

2. TOPOLOGICAL RENORMALIZATION. Network renormalization exploits the fact that the information flowing across the system might get stuck in portions of the graph for some period of time: when this happens, we cluster the graphs using topological renormalization. Each nonspecial node v of the temporal parse tree is annotated by the sketch $cl(\prod \mathbf{g})$, where $\mathbf{g}$ is the graph sequence formed by the leaves of the subtree rooted at v. In this way, every path from the root of the temporal parse tree comes with a nested sequence of sketches $h_1 \succeq \cdots \succeq h_l$ (for both special and nonspecial nodes). Pick two consecutive ones, h_i, h_{i+1}: these are two transitive graphs whose strongly connected components, therefore, are cliques. Let $V_1, \ldots, V_a$ and $W_1, \ldots, W_b$ be the vertex sets of the cliques corresponding to h_i and h_{i+1}, respectively. Since h_{i+1} is a subgraph of h_i, it follows that each V_i is a disjoint union of the form $W_{i_1} \cup \cdots \cup W_{i_{s_i}}$.

- *Decoupling.* We decorate the temporal parse tree with additional trees connecting the sketches along its paths. These *topological parse trees* are formed by all the productions of the type:

$$V_i \longrightarrow W_{i_1} \cdots W_{i_{s_i}}. \tag{3}$$

A sketch at a node v of the temporal tree can be viewed as an acyclic digraph over cliques: its purpose is to place limits on the movement of the probability mass in any temporal random walk corresponding to the leaves of the subtree rooted at v. In particular, it indicates how decoupling might arise in the system over certain time intervals specified by the temporal parse tree.

The maximum depth of the temporal parse tree is $O(n^2)$ because each child's cumulant loses at least one edge from its parent's (or grandparent's) cumulant. To see why the quadratic bound is tight, consider a bipartite graph $V = L \cup R$, where $|L| = |R|$ and each pair from $L \times R$ is added one at a time as a bipartite graph with a single nonloop edge; the leftmost path of the parse tree is of quadratic length.

Left-to-Right Parsing. The temporal tree can be built on-line by scanning the graph sequence $\mathbf{g}$ with no need to back up. Let $\mathbf{g}'$ denote the sequence formed by appending the graph g to the end of the finite graph sequence $\mathbf{g}$. If $\mathbf{g}$ is empty, then the tree $T(\mathbf{g}')$ consists of a root with one child labeled g. If $\mathbf{g}$ is not empty and $\prod \mathbf{g} \prec \prod \mathbf{g}'$, the root of $T(\mathbf{g}')$ has one left child formed by the root of $T(\mathbf{g})$ as well as a right child (a leaf) labeled g. Assume now that $\mathbf{g}$ is not empty and that $\prod \mathbf{g} = \prod \mathbf{g}'$. Let v be the lowest internal nonspecial node on the rightmost path of $T(\mathbf{g})$ such that $c_v \times g = c_v$, where c_u denotes the product of the graphs associated with the leaves of the subtree rooted at node u of $T(\mathbf{g})$. Let w be the rightmost child of v; note that v and w always exist (the latter because v is internal). We explain how to form $T(\mathbf{g}')$ by editing $T(\mathbf{g})$.

1. *If c_v is transitive and w is a leaf.* Referring to (1b), v and w correspond to $\mathbf{g}$ and g_{m_k}, respectively, and $(\mathbf{g}_{k+1})$ is empty. If $g = c_v$, then $(\mathbf{g}_{k+1})$ remains empty while $g_{m_{k+1}} = g$ is created: accordingly, we attach a leaf to v as its new rightmost child and we label it g. On the other hand, if $g \prec c_v$, then $\mathbf{g}_{k+1}$ becomes the sequence consisting of g, so we attach a new rightmost child z to v and then a single leaf to z, which we label g, so as to form $(\mathbf{g}_{k+1})$.
2. *If c_v is transitive and w is not a leaf.* Again, referring to (1b), v and w correspond to $\mathbf{g}$ and the root of $(\mathbf{g}_{k+1})$, respectively. If $c_w \times g = c_v$, then $g = g_{m_{k+1}}$, so we attach a leaf to v as its new rightmost child and we label it g. On the other hand, if $c_w \times g \prec c_v$, then g is appended to the sequence $\mathbf{g}_{k+1}$. Because $c_w \prec c_w \times g$, we create a node z with w as its left child and, as its right child, a leaf labeled g: we attach z as the new rightmost child of v.
3. *If c_v is not transitive and w is a leaf.* Referring to (1a), v and w correspond to $\mathbf{g}$ and g_{m_1}, respectively, and $(\mathbf{g}_2)$ is empty. We know that $g \preceq \Delta tf(c_v) \prec c_v$. Accordingly, we give v a new rightmost child z, which we make into a special

node and annotate with the label $\Delta tf(c_v)$. We attach a leaf to z and label it g.

4. *If c_v is not transitive and w is not a leaf.* It follows then that w is a special node; let w' be its unique child. Referring to (1a), v and w correspond to $\mathbf{g}$ and the root of $(\mathbf{g_2})$, respectively. Because $c_w \prec c_w \times g \preceq \Delta tf(c_v) \prec c_v$, we create a node z with w' as its left child and, as its right child, a leaf labeled g: we attach z as the new unique child of the special node w.

Undirected Graphs. For our purposes, a graph is called undirected if any edge (x, y) with $x \neq y$ comes with its companion (y, x). Consider a sequence of undirected graphs over $[n]$. We begin with the observation that the cumulant of a sequence of undirected graphs might itself be directed; for example, the product $g_1 \times g_2 = (x \leftrightarrow y \quad z) \times (x \quad y \leftrightarrow z)$ has a directed edge from x to z but not from z to x. We can use undirectedness to strengthen the definition of the transitive front. Recall that $tf(g)$ is the unique maximal graph h such that $g \times h = g$. Its purpose is the following: if g is the current cumulant, the transitive front of g is intended to include any edge that might appear in subsequent graphs in the sequence without extending any path in g. Since, in the present case, the only edges considered for extension will be undirected, we might as well require that h itself (unlike g) should be undirected. In this way, we redefine the transitive front, now denoted by $utf(g)$, as the unique maximal *undirected* graph h such that $g \times h = g$. Its edge set includes all the pairs (i, j) such that $C_i = C_j$. Because of self-loops, the condition implies that (i, j) is an undirected edge of g. This forms an equivalence relation among the vertices, so that $utf(g)$ actually consists of disconnected, undirected cliques. To see the difference with the directed case, we take our previous example and note that $tf(g_1 \times g_2)$ has the edges $(x, y), (x, z), (y, z), (z, y)$ in addition to the self-loops, whereas $utf(g_1 \times g_2)$ has the single undirected edge (y, z) plus self-loops.

As observed in [3], the depth of the parse tree can still be as high as quadratic in n. Here is a variant of the construction. Given a clique C_k over k vertices $x_1, \ldots, x_k$ at time t, attach to it, at time $t + 1$, the undirected edge (x_1, y). The cumulant gains the undirected edge (x_1, y) and the directed edges (x_i, y) for $i = 2, \ldots, k$. At time $t + 2, \ldots, t + k$, visit each one of the $k - 1$ undirected edges (x_1, x_i) for $i > 1$, using single-edge undirected graphs with self-loops. Each such step will see the addition of a new directed edge (y, x_i) to the cumulant, until it becomes the undirected clique C_{k+1}. The quadratic lower bound on the tree depth follows immediately.

References

1. Avin, C., Koucký, M., Lotker, Z.: How to explore a fast-changing world (cover time of a simple random walk on evolving graphs). In: Aceto, L., Damgård, I., Goldberg, L.A., Halldórsson, M.M., Ingólfsdóttir, A., Walukiewicz, I. (eds.) ICALP 2008. LNCS, vol. 5125, pp. 121–132. Springer, Heidelberg (2008). https://doi.org/10.1007/978-3-540-70575-8_11
2. Bruin, H., Deane, J.H.B.: Piecewise contractions are asymptotically periodic. Proc. Am. Math. Soc. **137**(4), 1389–1395 (2009)

3. Chazelle, B.: Toward a theory of Markov influence systems and their renormalization. In: Proceedings of the 9th ITCS, pp. 58:1–58:18 (2018)
4. Condon, A., Hernek, D.: Random walks on colored graphs. Random Struct. Algorithms **5**, 285–303 (1994)
5. Denysyuk, O., Rodrigues, L.: Random walks on directed dynamic graphs. In: Proceedings of the 2nd International Workshop on Dynamic Networks: Algorithms and Security (DYNAS10), Bordeaux, France, July 2010. arXiv:1101.5944 (2011)
6. Denysyuk, O., Rodrigues, L.: Random walks on evolving graphs with recurring topologies. In: Kuhn, F. (ed.) DISC 2014. LNCS, vol. 8784, pp. 333–345. Springer, Heidelberg (2014). https://doi.org/10.1007/978-3-662-45174-8_23
7. Holme, P., Saramäki, J.: Temporal networks. Phys. Rep. **519**, 97–125 (2012)
8. Iacobelli, G., Figueiredo, D.R.: Edge-attractor random walks on dynamic networks. J. Complex Netw. **5**, 84–110 (2017)
9. Nguyen, G.H., Lee, J.B., Rossi, R.A., Ahmed, N., Koh, E., Kim, S.: Dynamic network embeddings: from random walks to temporal random walks. In: 2018 IEEE International Conference on Big Data, pp. 1085–1092 (2018)
10. Perra, N., Baronchelli, A., Mocanu, D., Gonçalves, B., Pastor-Satorras, R., Vespignani, A.: Random walks and search in time-varying networks. Phys. Rev. Lett. **109**, 238701 (2012)
11. Petit, J., Gueuning, M., Carletti, T., Lauwens, B., Lambiotte, R.: Random walk on temporal networks with lasting edges. Phys. Rev. E **98**, 052307 (2018)
12. Ramiro, V., Lochin, E., Sénac, P., Rakotoarivelo, T.: Temporal random walk as a lightweight communication infrastructure for opportunistic networks. In: 2014 IEEE 15th International Symposium (WoWMoM), June 2014
13. Seneta, E.: Non-Negative Matrices and Markov Chains, 2nd edn. Springer, New York (2006)
14. Starnini, M., Baronchelli, A., Barrat, A., Pastor-Satorras, R.: Random walks on temporal networks. Phys. Rev. E **85**(5), 056115–26 (2012)

Highly Succinct Dynamic Data Structures

Torben Hagerup[(⊠)] [ID]

Institut für Informatik, Universität Augsburg, 86135 Augsburg, Germany
hagerup@informatik.uni-augsburg.de

Abstract. It is attempted to elucidate a number of known dynamic data structures that come close to using the absolutely minimal amount of space, and their connections and ramifications are explored.

Keywords: Space efficiency · In-place chain technique · Constant-time initialization · Choice dictionaries · Arrays

1 Introduction

The study of space-efficient data structures has a long tradition. Many early such data structures are *static*, i.e., they support queries about the object represented, but do not allow changes to that object. A well-known example are (static) *rank-select structures* [4,5,8,18,25], which represent bit vectors and support queries for sums over given prefixes (*rank*) and for shortest prefixes with given sums (*select*). A data structure D capable of representing an arbitrary object drawn from a nonempty finite set $\mathcal{S}$, any two elements of which can be distinguished with queries supported by D, must store its state in at least $\lceil \log |\mathcal{S}| \rceil$ bits—we will use "log" throughout to denote the binary logarithm function $\log_2$. If $\mathcal{S}$ depends on one or more parameters, D is often said to be *succinct* if the number of bits occupied by D is $L + o(L)$, where $L = \lceil \log |\mathcal{S}| \rceil$, i.e., "the information-theoretic lower bound plus lower-order terms". This text offers a personal view of a small number of data structures that belong to the rarer species of *dynamic* space-efficient data structures, ones that support update operations, and that are also highly succinct. While no general definition of the latter term is proposed, here a highly succinct data structure can be taken to be one that occupies $L + (\log L)^{O(1)}$ bits, where L is defined as above.

The data structures surveyed here are all very basic and low-level. We first describe a data structure, due to Dodis, Pǎtraşcu and Thorup [6], that allows arrays of nonbinary values (e.g., ternary values drawn from $\{0, 1, 2\}$) to be realized on a usual binary computer with constant-time access, yet almost no space wasted. We then turn to the so-called *in-place chain technique* [22] and end with several of its applications, first to the constant-time initialization of arrays [22], then to the constant-time initialization of more general data structures [20], and finally to the realization of so-called *choice dictionaries* [11,12,14,21].

Our model of computation reflects what is available in a programming language like C on a present-day computer. The memory is divided into *words*,

© Springer Nature Switzerland AG 2019
L. A. Gąsieniec et al. (Eds.): FCT 2019, LNCS 11651, pp. 29–45, 2019.
https://doi.org/10.1007/978-3-030-25027-0_3

each word holds a bit pattern that may be interpreted as a nonnegative integer or a pointer to another word, and each instruction manipulates a constant number of words and is assumed to take constant time. Space-efficient algorithms and data structures need to "get at" the single bits in a word, for which reason we assume the availability not only of instructions for operations like integer multiplication and division, but also for bitwise Boolean operations and bit shifts by a variable number of bit positions. Technically, the model of computation is a word RAM [3,10]. We also assume that $\lfloor \log x \rfloor$ can be computed from $x \in \mathbb{N} = \{1, 2, \ldots\}$ in constant time, as this operation has the dedicated instruction bsr (bit scan reverse) in modern CPUs.

2   Collections of Nonbinary Values

Let A be a positive integer and suppose that we want to remember a value a of which we know only that it belongs to the set $[A] = \{0, \ldots, A-1\}$. Then we can store the $\lceil \log A \rceil$-bit binary representation of a, which allows us to inspect a and to replace it by another value drawn from $[A]$ in constant time. In some sense, nearly a whole bit may be wasted, for $A = 17$, e.g., but this is unavoidable—there is clearly no way to store a in fewer than $\lceil \log A \rceil$ bits.

If we want to store not one value a, but n values $a_1, \ldots, a_n \in [A]$, for some $n \in \mathbb{N}$, we can handle the values independently using the same method, but then the total number of bits used will be $n \lceil \log A \rceil$. Another possibility is to store (the binary representation of) the single number $\alpha = \sum_{i=0}^{n-1} a_{i+1} \cdot A^i$, i.e., the n-digit A-ary integer with digits $a_1, \ldots, a_n$ (in the order from least to most significant). We have $0 \leq \alpha < A^n$, so α can be represented in just $\lceil \log A^n \rceil = \lceil n \log A \rceil$ bits, which is optimal and may be almost n bits less than the $n \lceil \log A \rceil$ bits needed by the other method. The flip side is that it becomes harder to get at the digits "hidden" in α. For $i = 1, \ldots, n$, $a_i = \lfloor \alpha / A^{i-1} \rfloor \bmod A$, and unless we assume the availability of constant-time exponentiation—which we will not—constant time for reading and writing individual digits seems achievable only if we store the powers $A^0, \ldots, A^{n-1}$, which defeats the original goal of using little space. An even more serious problem with the approach is that unless n is quite small, it is not realistic to assume that integers almost as large as A^n can be manipulated in constant time. The rest of the section describes a simple and beautiful data structure, due to Dodis et al. [6], that gets around these problems and achieves constant time for reading and writing while using little more than $\lceil n \log A \rceil$ bits and keeping all integers polynomial in $n + A$.

2.1   Data Representation and Space Analysis

We will assume that the elements of the index set $V = \{1, \ldots, n\}$ are organized in a binary tree $T = (V, E)$, which we take to be an ordered intree. In a first instance, we generalize the problem by allowing a_v, for each node $v \in V$, to come from its own individual domain $[A_v]$, where $A_v \in \mathbb{N}$. Imagine a bottom-up computation in T in which each node $v \in V$ receives positive integers P_v^{L} and P_v^{R}

from its left and right child, respectively, also inputs A_v and produces positive integers B_v and P_v, of which P_v is sent to the parent of v, if any (see Fig. 1). If the left and/or right child of v is not present, (e.g., v may be a leaf in T), the missing P_v^L and/or P_v^R is given the default value 1. For $x \geq 1$, denote by $\llbracket x \rrbracket = 2^{\lceil \log x \rceil}$ the smallest power of 2 no smaller than x. The computation at v takes the following precise form:

$$B_v := \llbracket P_v^L P_v^R \sqrt{A_v} \rrbracket;$$
$$P_v := \lceil A_v / \lfloor B_v / (P_v^L P_v^R) \rfloor \rceil;$$

While these formulas may not be very intuitive, it is simple to verify that if the input values P_v^L, P_v^R and A_v are positive integers, then so are the output values B_v and P_v. Moreover, if $A = \max_{v \in V} A_v$, then $P_v \leq A$ and, by easy induction, $B_v \leq 2A^{2.5}$ for all $v \in V$.

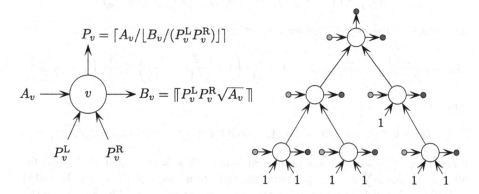

Fig. 1. The computation at a single node v (left) and in a complete tree T (right).

Let r be the root of T and write $\overline{P}$ for P_r. The plan is to represent $a_1, \ldots, a_n$ via integers $b_1, \ldots, b_n, \overline{p}$ (which are in turn stored in binary) with $b_v \in [B_v]$ for $v = 1, \ldots, n$ and $\overline{p} \in [\overline{P}]$. In order to investigate whether this is economical in terms of space, let us define R_v, for each node $v \in V$, as the sum of the logarithms of the outputs of v minus the sum of the logarithms of the inputs of v, i.e.,

$$R_v = \log B_v + \log P_v - \log P_v^L - \log P_v^R - \log A_v.$$

Intuitively, R_v is the "information loss" at the node v. We will show on the one hand that R_v is small for all $v \in V$ (under some assumptions) and on the other hand that $\sum_{v \in V} R_v$ is essentially the amount by which the space used overshoots the minimum space possible. The latter argument is more elegant: Because the "P contributions" to $\sum_{v \in V} R_v$ form a "tree-shaped telescoping sum" (one node's P (added) is another node's P^L or P^R (subtracted)),

$$\sum_{v \in V} R_v = \sum_{v \in V} \log B_v + \log \overline{P} - \sum_{v \in V} \log A_v.$$

Since B_v is a power of 2 for all $v \in V$, the number of bits used by our representation is exactly

$$\sum_{v \in V} \log B_v + \lceil \log \overline{P} \rceil = \left\lceil \sum_{v \in V} \log A_v + \sum_{v \in V} R_v \right\rceil. \tag{1}$$

On the other hand, $\lceil \sum_{v \in V} \log A_v \rceil = \lceil \log \prod_{v \in V} A_v \rceil$ is the minimum number of bits required to store $a_1, \ldots, a_n$. It is clearly of interest to bound R_v for all $v \in V$. To gain intuition, observe that if we were to compute P_v without rounding as $A_v/(B_v/(P_v^L P_v^R)) = A_v P_v^L P_v^R/B_v$, then R_v would be exactly zero. Thus the task at hand is to bound the effect of rounding in the formula $\lceil A_v/\lfloor B_v/(P_v^L P_v^R) \rfloor \rceil$.

Fix $v \in V$ and assume that $A_v \geq 4$. It is easy to see that with $Q_v = B_v/(P_v^L P_v^R)$, $\sqrt{A_v} \leq Q_v < 2\sqrt{A_v}$. Thus

$$\lfloor Q_v \rfloor \geq Q_v - 1 = Q_v(1 - 1/Q_v) \geq Q_v(1 - 1/\sqrt{A_v})$$

and, since $1/(1 - \epsilon) \leq 1 + 2\epsilon$ for $0 \leq \epsilon \leq 1/2$,

$$P_v = \left\lceil \frac{A_v}{\lfloor Q_v \rfloor} \right\rceil \leq \frac{A_v/Q_v}{1 - 1/\sqrt{A_v}} + 1 \leq \frac{A_v}{Q_v}\left(1 + \frac{2}{\sqrt{A_v}} + \frac{Q_v}{A_v}\right) \leq \frac{A_v}{Q_v}\left(1 + \frac{4}{\sqrt{A_v}}\right)$$

and

$$R_v = \log P_v + \log Q_v - \log A_v \leq \log(1 + 4/\sqrt{A_v}) \leq (4\log e)/\sqrt{A_v}.$$

Since $4\log e \leq \sqrt{34}$—a crude numerical bound—it follows that if $A_v \geq 34n^2$ for all $v \in V$, then $\sum_{v \in V} R_v \leq 1$ and the right-hand side of Eq. (1) is bounded by $\lceil \sum_{v \in V} A_v \rceil + 1$, i.e., our data structure uses at most one bit more than the absolute minimum. (Dodis et al. [6] argue that in the special case $A_1 = \cdots = A_n$, even the $+1$ can be squeezed out of the space bound.)

2.2 Reading and Writing

We still need to see how to represent $a_1, \ldots, a_n$ via $b_1, \ldots, b_n, \overline{p}$ and, above all, how to read and write an a_v stored in this manner. Write $\mathbb{N}_0 = \mathbb{N} \cup \{0\}$ and, for $p \in \mathbb{N}$, let $f_p : \mathbb{N}_0 \to \mathbb{N}_0 \times [p]$ be the function depicted in the left subfigure of Fig. 2, i.e., $f_p(z) = (\lfloor z/p \rfloor, z \bmod p)$ for all $z \in \mathbb{N}_0$. For all $p \in \mathbb{N}$, f_p is a bijection from $\mathbb{N}_0$ to $\mathbb{N}_0 \times [p]$, and its inverse $f_p^{-1} : \mathbb{N}_0 \times [p] \to \mathbb{N}_0$ is illustrated in the middle subfigure of Fig. 2. The function f_p writes its argument as two digits in a mixed-radix positional system whose least significant base is p, and f_p^{-1} recomposes the two digits to a single integer. In particular, for all $p, q \in \mathbb{N}$, f_p^{-1} maps $[q] \times [p]$ bijectively to $[pq]$.

Let G be the graph obtained from T by replacing each node $v \in V$ by the subgraph H_v shown in the right subfigure of Fig. 2. G can be viewed as a computational DAG with inputs, inner nodes that compute outgoing values from incoming values, and outputs, and we define the representation $(b_1, \ldots, b_n, \overline{p})$ to

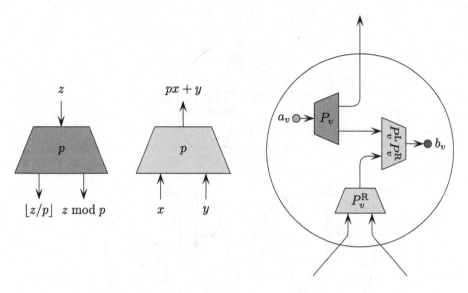

Fig. 2. The functions f_p (left) and f_p^{-1} (middle) and the graph H_v for a node v (right).

be the output obtained if the input $(a_1, \ldots, a_n)$ is fed into G (with the correspondence between input values and input nodes and output values of the form b_v and output nodes being as indicated in the right subfigure of Fig. 2 and $\overline{p}$ being output at the root). Note that missing inputs at nodes with fewer than two children are now taken to be 0, not 1.

A few things need to be checked: First, for every $v \in V$, the value sent from H_v to its parent belongs to $[P_v]$, and if the left and right inputs to H_v come from $[P_v^{\mathrm{L}}]$ and $[P_v^{\mathrm{R}}]$, respectively, then each of the two "f^{-1} boxes" in H_v, if labeled with a parameter p, receives a second argument (the y argument) that lies in $[p]$. All of this is immediate. Second, the value b_v computed indeed lies in $[B_v]$. To see this, note that the first argument of the "f^{-1} box" with parameter $P_v^{\mathrm{L}} P_v^{\mathrm{R}}$ is strictly smaller than $A_v/P_v \leq B_v/(P_v^{\mathrm{L}} P_v^{\mathrm{R}})$. Now comes the fun part.

The total degree of every vertex in G is bounded by three and the length of every path in G is bounded by four, as can easily be observed from Fig. 3. Of course, these properties also hold for the graph $\overleftarrow{G}$ obtained from G by replacing every edge (u, v) by the reverse (or antiparallel) edge (v, u). Because every inner node in G realizes an injective function, G's computation can be reversed. Recall that we are interested in $a_1, \ldots, a_n$, but actually store $b_1, \ldots, b_n, \overline{p}$. To recover $a_1, \ldots, a_n$ from $b_1, \ldots, b_n, \overline{p}$, we can simply replace every "f box" by an "f^{-1} box" with the same parameter, and vice versa, and let the computational DAG $\overleftarrow{G}$ do its job. Since $\overleftarrow{G}$ is of bounded degree and bounded maximum path length, each individual a_v, where $v \in V$, can be obtained in constant time. After a change to an input value, a_v, we simply recompute the output values that depend on a_v after using the method just described to recover the other input values on which these output values also depend. Again, this can happen in constant time.

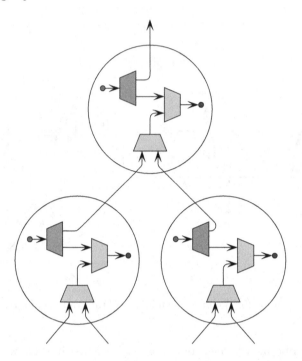

Fig. 3. A part of the computational DAG obtained by replacing all nodes by subgraphs.

We have paid careful attention to the space needed to store $b_1, \ldots, b_n, \bar{p}$ and showed this space to be within one bit of optimal if $A_v \geq 34n^2$ for all $v \in V$. Unfortunately, in order to be able to translate between the inputs and outputs of G as described above, we need additional information that is static in the sense that it depends only on the tree T and the parameters $A_1, \ldots, A_n$. First, navigation in T must be possible, i.e., given a node $v \in V$, we must be able to find the parent and the children of v, if any. Following Dodis et al. [6], we will assume that T has depth $\lfloor \log n \rfloor$ and the completely rigid structure known from Heapsort, i.e., the root of T is the node 1, the parent of each $v \in \{2, \ldots, n\}$ is $\lfloor v/2 \rfloor$, and every left child is even. Then the necessary navigation in T can happen in constant time.

Second, given a node $v \in V$, we must have access to (A_v, B_v, P_v), which we call the *type* of v. In order to keep the number of types small, we will assume that for $v \in V$, the integer A_v depends only on the depth of v in T (an alternative would be the height of v). Because of the special shape of T, the nodes on a given level in T (i.e., of a given depth) then have at most three different types, and it is easy (if somewhat tedious) to define a function $\tau : V \to \mathbb{N}_0$ with the following properties:

- For all $u, v \in V$, if $\tau(u) = \tau(v)$, then u and v have the same type.
- τ can be evaluated in constant time.

- For all $v \in V$, if the depth of v is d, then $\tau(v) \in \{3h - 2, 3h - 1, 3h\}$, where $h = \lfloor \log n \rfloor - d$.
- For all $i \in \mathbb{N}_0$, with $V_i = \{v \in V \mid \tau(v) = i\}$, the number $n_i = |V_i|$ can be computed in constant time, and there is a function ϕ_i, evaluable in constant time, that numbers the nodes in V_i consecutively, starting at 0, in an arbitrary order.

Third, given a node $v \in V$, we must be able to locate b_v. Let τ, n_i and ϕ_i (for $i \in \mathbb{N}_0$) be as above. For $i = 0, \ldots, 3\lfloor \log n \rfloor$, let m_i be the (common) value of $\log B_v$ for all $v \in V_i$ (arbitrary if $V_i = \emptyset$). For each $v \in V_i$, we can store b_v in the m_i bit positions starting in bit number $\sum_{j=0}^{i-1} n_j m_j + \phi_i(v) m_i$ in an array of $\sum_{j=0}^{3\lfloor \log n \rfloor} n_j m_j = \sum_{v \in V} \log B_v$ bits.

2.3 Results

The types of the inner nodes on one level in T and the relevant sums $s_i = \sum_{j=0}^{i-1} n_j m_j$ can be computed in constant time from the corresponding entities on the level below, if any. Only the calculation of $\sqrt{A_v}$ in the defining formula for B_v may appear problematic, but the analysis can compensate for a constant-factor error in the calculation, so an easy-to-obtain approximation to $\sqrt{A_v}$ suffices. (In contrast, P_v must be calculated very accurately.) For $v \in V$, the depth of v in T is $\lfloor \log v \rfloor$, and its height is at most $\log(n/v)$. If v is of height h, an access to a_v for reading or writing needs to know the types of nodes of height at most $h + 2$. One possibility is not to store the types of any nodes permanently, but to compute the types and the values s_i needed in a bottom-up computation as part of the access to a_v. This yields the following.

Theorem 1. *There is a data structure that can be initialized for arbitrary integers $n \geq 1$ and $C_0, \ldots, C_L \geq 34n^2$ in $O(\log n)$ time, where $L = \lfloor \log n \rfloor$, subsequently maintains a sequence $(a_1, \ldots, a_n)$, where $a_v \in \{0, \ldots, C_{\lfloor \log v \rfloor} - 1\}$ for $v = 1, \ldots, n$, occupies at most $\lceil \sum_{v \in V} \log C_{\lfloor \log v \rfloor} \rceil + 1$ bits plus the space needed to store $(n, C_0, \ldots, C_L)$, and supports reading and writing of a_v in $O(\log(2n/v))$ time, for $v = 1, \ldots, n$.*

The initialization time of the data structure D of the theorem is indicated as $O(\log n)$ because this is the time needed to access and stow away the parameters $C_0, \ldots, C_L$. If this is not necessary, e.g., because D is to be used only with $C_0 = \cdots = C_L$, the initialization time can be lowered to a constant. Note, however, that this stretches the definition of the space requirements of a data structure: Besides the space occupied by $(n, C_0, \ldots, C_L)$, D will never use more than $\lceil \sum_{v \in V} \log C_{\lfloor \log v \rfloor} \rceil + 1$ bits, but it can indicate its own size only after $\Theta(\log n)$ steps, which may make it difficult to pack D tightly among other data structures.

Although the average access time of Theorem 1 is a constant, where the average is taken over all $v \in \{1, \ldots, n\}$, we are more interested in having a constant worst-case time. In order to achieve this, we can precompute and store the types that occur in T in a table indexed via τ, as well as the sums s_i for $i = 1, \ldots, 3\lfloor \log n \rfloor$. Since a type can be stored in $O(\log C)$ bits, where $C = \max\{C_0, \ldots, C_L\}$, the number of bits needed by the table is $O(L(\log n + \log C)) = O(\log n \log(n + C))$. If $A_1 = \cdots = A_n = A \geq 2$, we can get rid of the annoying condition $A \geq 34n^2$ by returning to a simple idea explored in the beginning of the section. If $A < 34n^2$, then choose $\ell \in \mathbb{N}$ to make $A^\ell \geq 34n^2$, but $A^\ell = n^{O(1)}$, and partition the values $a_1, \ldots, a_n$, called *digits*, into groups of ℓ digits each, except that the first group has between ℓ and 2ℓ digits. For each group, combine the digits in the group to a single base-A integer of at most 2ℓ digits and maintain only these larger and sufficiently large integers in the tree-based data structure. If the first group is assigned to the root (as is natural), it will still be the case that the nodes on a given level in the tree have at most three distinct types.

Theorem 2. ([6]). *There is a data structure that can be initialized for arbitrary $n, A \in \mathbb{N}$ in $O(\log n)$ time, subsequently maintains a sequence $(a_1, \ldots, a_n)$, where $a_i \in \{0, \ldots, A - 1\}$ for $i = 1, \ldots, n$, occupies $n \log A + O(\log n \log(n + A))$ bits, and supports constant-time reading and writing of a_i, for $i = 1, \ldots, n$.*

3   The In-Place Chain Technique

Certain modern programming languages such as Java, VHDL and D stipulate that memory be initialized (e.g., cleared to zero) before it is allocated to application programs [9,17] or have this as the default behavior [2]. The initialization is carried out for security reasons—one program should not read what another program has written—and to ease debugging by eliminating one source of erratic program behavior. On the other hand, such initialization of large memory areas is rarely supported efficiently in hardware. In cases where it is to be expected that only a small part of the memory will actually be used, as in the context of adjacency matrices of sparse graphs, it is therefore of interest to simulate an initialized block of memory in uninitialized memory (memory that may contain arbitrary initial values). Here we can consider the block of memory to be an array of words, where a word is small enough to be manipulated in constant time but large enough to contain a pointer to another word in the block. Thus the task is to provide an *initializable array*, a data structure that supports the functionality of an array of words—reading and writing individual words in constant time—but allows all words in the array to be initialized (or reset) to zero in constant time. This problem has been considered at least since 1974 [1, Exercise 2.12] and investigated both theoretically and in practice in recent years [7,15,23,24]. The first really interesting new development, however, came in 2017 when Katoh and Goto [22] invented the so-called *in-place chain technique* and used it to develop an initializable array that leaves little to be desired. This section first describes their data structure and then gives an overview of later developments triggered by the advent of the in-place chain technique.

3.1 Initializable Arrays

Let us say that two (one-dimensional) arrays have the same *layout* if they have the same index set and elements of the same type. The folklore organization of an initializable array A, the one outlined in the hint to [1, Exercise 2.12], is illustrated in the left subfigure of Fig. 4, which shows the situation after the assignments $A[10] := $ J; $A[6] := $ F; $A[7] := $ G. The construction is based on the implicit observation that in order to realize an initializable array A, it suffices to maintain the set S of indices i for which the cell $A[i]$ has been written to at least once since the time when A was initialized or was last cleared. To see this, note that to simulate $A[i] := x$, it suffices to insert i in S and to store x in $V[i]$, where V is an array of the same layout as A, whereas $A[i]$ can be obtained as $V[i]$ if $i \in S$ and as zero if not. S is initially empty and must support only the two operations of insertion and membership test. With two bookkeeping arrays of the same layout as A, say T and F, the realization of S is easy: The elements of S are stored in no particular order in $T[1], T[2], \ldots, T[k]$, where $k = |S|$ is also remembered, and for each $i \in S$ the position in T of i is recorded in $F[i]$, so that membership in S can be tested in constant time.

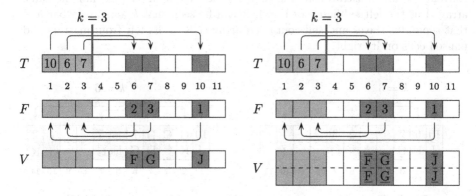

Fig. 4. The folklore initializable array with 1-word cells (left) and with 2-word cells (right). Slaves and masters are indicated in pale red and in blue, respectively. (Color figure online)

If $i \in S$ and $F[i] = j$, it is natural to view the cells $F[i]$ and $T[j]$ as pointing to each other. In this situation we will consider the triple $(T[i], F[i], V[i])$ as an entity, called a *master*, and $(T[j], F[j], V[j])$ as another entity, called the corresponding *slave*. Note that a master stores two words of data, namely in the arrays F and V, whereas a slave stores just one word, in T. The average number of words used, over a master and the corresponding slave, is 3/2, which is more than the number of words in A accommodated by the two in conjunction, namely 1. A small but crucial step towards obtaining a superior solution is to perceive A and V as composed of cells of two words each, rather than of single words, as suggested in the right subfigure of Fig. 4, which assumes that the data value F is now a pair $(\overline{F}, \underline{F})$ of two words, etc. If the first writing to a cell in A writes to only one word, we initialize the other word in the cell explicitly on the same occasion. Now a master stores three words, while a slave still stores just one word. The average is two, as is the number of words of A accommodated by the master and the slave, and we achieve "average for all" by relocating one word of data from the master to the slave. The word relocated should obviously not be chosen as the pointer to the slave, but otherwise the master can still access all its data in constant time.

This data organization opens up the prospect of storing the three arrays T, F and V together in a single array, which we continue to call V. According to our deliberations so far, a cell $V[i]$ of V would be either (if $i \leq k$) a slave or (if $i > k$) unused or a master. However, we must allow for one more possibility: If $i \in S$ for some $i \leq k$, $V[i]$ must store data of its own and cannot function as a slave. In this case we therefore store $A[i]$ directly in the cell $V[i]$ and call $V[i]$ *self-contained*. The only danger in this is that half of $V[i]$ might be misinterpreted as a pointer to a master $V[j]$. This is problematic only if $j > k$ and $V[j]$ points back to $V[i]$, but in that case $V[j]$ is unused, and we can break up the spurious master-slave pair by changing the pointer in $V[j]$ to let it point to $V[j]$ itself, say. The storage conventions are illustrated in Fig. 5, whose left subfigure represents the same array A as the left subfigure of Fig. 4. It can be seen that k acts as a "barrier" that separates slaves and self-contained array cells on its left from masters and unused cells on its right.

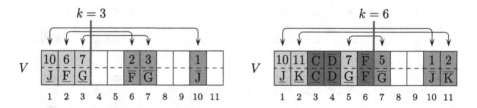

Fig. 5. The highly succinct initializable array of Katoh and Goto [22].

Now reading from A can happen in constant time, but writing to A must take care to preserve the storage invariants. Suppose that writing to A causes an index i to enter S and therefore increases $k = |S|$, from k_0 to $k_0 + 1$, say. If $i \leq k_0$, $V[i]$ must switch from being a slave to being self-contained, which leaves its former master without a slave. If $i > k_0 + 1$, $V[i]$ must switch from being unused to being a master, and again there is a master without a slave. The operation under consideration also moves the barrier to the right and across the cell $V[k_0 + 1]$, which changes the status of the latter. Assume that $i \neq k_0 + 1$. If $V[k_0+1]$ is unused before the operation, it becomes the required slave. Otherwise $V[k_0+1]$ changes from being a master to being self-contained, which sets free its former slave. In all situations, constant time suffices to pair up masters and slaves anew and to move the words relocated from masters with new slaves to these slaves. The case $i = k_0 + 1$ was omitted from consideration, but is particularly easy and left to the interested reader. The right subfigure of Fig. 5 shows the state arising from that of the left subfigure after the subsequent assignments $A[4] := D$; $A[3] := C$; $A[11] := K$.

As described so far, the data structure must store V, which is precisely as large as the initializable array A that it realizes, and k, which needs one additional word. While this is likely to be sufficiently succinct in every practical application, Katoh and Goto [22] describe a way to reduce the extra space from one word to just a single bit. The idea is to "hide" k in V as well. Recall that we increased the number ν of words per cell in A and V from 1 to 2 so that the 2ν words of a master and its corresponding slave could accommodate the ν words in a cell of A plus two pointers. If we increase ν still further, every cell in V to the right of the barrier (but not a self-contained cell to its left) can have $2\nu - \nu - 2 = \nu - 2$ words that are not used by the data structure. With $\nu = 3$, we can store k in the unused word in the rightmost cell of V, which is always available except in the "terminal state" in which the barrier has moved all the way to the right end of V. In the terminal state V has been fully initialized and coincides exactly with A, so that k is no longer needed. We still need to record in a single bit outside of V, however, whether the terminal state has been reached.

The number n of words in A and V is fixed once and for all when the data structure is initialized, and so far we have implicitly assumed n to be "known" (otherwise we cannot figure out where to look for k). If access to the data structure is provided through an outside pointer to its first word, a natural and common convention, we can get rid of this assumption by redefining k as $n - |S|$ and swapping the meanings of left and right. This moves k to the first cell of V, where it can be found with the external pointer. With $\nu = 4$, we can then also store n in the first cell. When the terminal state has been reached, however, n has disappeared and we can no longer catch out-of-range errors in accesses to A.

Theorem 3. ([22]). *For every array A, there is an initializable array of the same layout as A that occupies just one bit more than A.*

3.2   General Constant-Time Initialization

Suppose that we have two similar data structures with complementary good properties: D_1 is space-efficient, but needs a fairly substantial initialization time before it can process its first regular operation (i.e., operation other than the initialization)—the data structure of Theorem 2 is a case in point. D_2, on the other hand, can be initialized fast but is less space-efficient, or can maybe be used only for a limited time. We would like to combine the two into a data structure that can be initialized as fast as D_2, up to a constant factor, but has the space-efficiency of D_1. The in-place chain technique shows the way, as observed by Kammer and Sajenko [20]. This is one of the cases where describing the method is easier than formulating the precise conditions under which it can be applied, for which reason only an example will be provided.

We need to be more specific about D_1 and D_2. Suppose that they are both initialized with a parameter ξ drawn from a suitable set Ξ, that this takes $t_1(\xi)$ steps for D_1 and $t_2(\xi)$ steps for D_2 and that, following its initialization with some $\xi \in \Xi$, D_1 occupies at most $s_1(\xi)$ bits. Here t_1, t_2 and s_1 are all functions from Ξ to $\mathbb{N}$. In the case of the data structure of Theorem 2, e.g., it would be natural to let ξ be the tuple (n, A). Assume that, given $\xi \in \Xi$, $O(t_2(\xi))$ time suffices to compute a quantity $\widehat{s_1}(\xi) \in \mathbb{N}$ with $\widehat{s_1}(\xi) \leq s_1(\xi)$, but $t_1(\xi)w = o(\widehat{s_1}(\xi))$, where w is an upper bound with $w = \Omega(\log \widehat{s_1}(\xi))$ on the number of bits that may be occupied by an argument of an operation of D_1 and D_2. As for the latter condition, when f and g are arbitrary functions from Ξ to $\mathbb{N}$, we take the relation $f(\xi) = o(g(\xi))$ to mean that for some function $K : \mathbb{R}_+ \to \mathbb{N}$, where $\mathbb{R}_+$ is the set of positive real numbers, $f(\xi) \leq \epsilon g(\xi) + K(\epsilon)$ for all $\xi \in \Xi$ and all $\epsilon > 0$.

Concerning D_2, we will assume that the following holds for some function $T : \Xi \to \mathbb{N}$ with $t_1(\xi) = o(T(\xi))$: Following the initialization of both D_1 and D_2 with the same parameter $\xi \in \Xi$, every sequence Z of regular operations that can be executed by D_1 (in particular, is legal for D_1) in at most $T(\xi)$ steps can also be executed by D_2 and, moreover, during the execution of Z D_2 occupies at most $s_2(\xi)$ bits, where $s_2(\xi) = o(\widehat{s_1}(\xi))$, and executes every operation in at least t but $O(t)$ steps, where t is the number of steps taken by D_1 to execute the operation. If D_2 fails, we assume that it does so detectably.

Informally, the gist of the above can be formulated as follows: D_2 is strictly smaller than D_1 and can be used for longer than it takes to initialize D_1. Moreover, D_2 is as fast as D_1, up to a constant factor, but not faster than D_1. This already hints at the construction of a data structure D that can be initialized for an arbitrary parameter $\xi \in \Xi$ in $O(t_2(\xi))$ time and subsequently occupies at most $s_1(\xi) + 1$ bits and can execute every sequence of operations legal for D_1, with each operation taking as much time as in D_1, up to a constant factor. Let us now look at the details.

The first observation is that, spending one additional bit, we can implement the first $\widehat{s_1}(\xi)$ bits of D_1 as part of D in an initializable array of the type developed in Subsect. 3.1, called the *stage*, and with words of $\Omega(w) = \Omega(\log \widehat{s_1}(\xi))$ bits each (recall that a condition is that a word must be able to hold a pointer to a cell in the data structure). The salient property of our realization of the initializable array is that, provided that we choose $\nu > 2$, every cell to the right of the barrier (except possibly for the rightmost cell, which stores the position of the barrier) offers $\nu - 2 \geq 1$ words that are not used by the initializable array. Because the barrier begins at the leftmost end of the stage and moves right by at most one cell in each access to the initializable array, this gives us $\Omega(\widehat{s_1}(\xi))$ freely usable bits for $\Omega(\widehat{s_1}(\xi)/w)$ steps.

The operation of the combined data structure D is illustrated in Fig. 6. The horizontal axis corresponds to the layout of the data structure in memory. The first or leftmost $\widehat{s_1}(\xi) + 1$ bits of D contain the stage and, with it, a part of D_1. The subsequent $s_1(\xi) - \widehat{s_1}(\xi)$ bits contain the rest of D_1, but need not concern us here. The vertical axis of Fig. 6, from top to bottom, corresponds to the

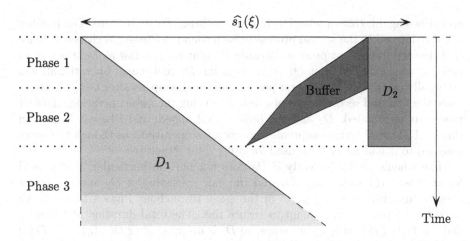

Fig. 6. A space-time diagram of the three phases in the initialization of D.

development over time. The figure is simplified in that it shows D_2 as occupying the last part of the stage. In reality, D_2 is stored only in the unused words of the corresponding cells, and the same is true of the buffer, whose function will be described shortly. Another simplification in the figure is that D_2 and the buffer can be stored next to each other, as shown, only if (a good approximation of) the size of D_2 can be computed beforehand. In general, D_2 and the buffer are stored in an interleaved fashion, i.e., alternately one word of D_2, then one word of the buffer, etc., and in reverse order, starting at the right end of the stage and growing towards the left. The data structure D_1 is shown as a triangle in the space-time diagram because the size of the leftmost part of the stage that no longer offers any unused words (i.e., is to the left of the barrier) grows at most linearly with time.

The initialization of D with a parameter $\xi \in \Xi$ computes $\widehat{s_1}(\xi)$ and initializes D_2 with parameter ξ. By assumption, this takes $O(t_2(\xi))$ time. The rest of the operation of D is divided into three phases. During Phase 1, two processes are run in an interleaved fashion, i.e., in sucessive *rounds*, where each round advances each process by a constant number of steps. The first process uses D_2 to execute the operations issued by a user of the data structure and also saves the type and the arguments of each operation in a FIFO queue, namely the buffer. The interleaved execution of the first process is slower than an execution directly by D_2 only by a constant factor, and D_2 in turn is slower than D_1 by another constant factor, so altogether the execution time of each operation is within a constant factor of its execution time on D_1, as required. The second process initializes D_1 with parameter ξ. This happens partly in the stage, using the initializable array, and partly to the right of the stage.

Phase 1 ends when the initialization of D_1 has been completed, at which point Phase 2 begins. In Phase 2 the first process continues as above, but the second process now iterates over the buffer to execute the sequence of regular operations again, this time in D_1, in order to bring D_1 up to date. The number of steps devoted to the second process in each round in Phase 2 is chosen as twice that devoted to the first process. Because D_1 is at least as fast as D_2, the second process, which had to wait $\Theta(t_1(\xi))$ steps for D_1 to become operational, will eventually catch up with the first process, and this happens after $O(t_1(\xi))$ steps. Once the first and second processes start executing the same operation, the first process is terminated, D_2 and the buffer are dropped, and Phase 3 begins. In this final phase D_1 reigns supreme and executes operations as though the space assigned to it had never been used for other data structures.

The scheme works correctly if D_2 does not fail—in particular, if it is used for at most $T(\xi)$ steps—and D_1 and the pair consisting of D_2 and the buffer operate in nonintersecting parts of the stage throughout Phases 1 and 2. All of our assumptions were set up to ensure this. The total duration of Phases 1 and 2 is $O(t_1(\xi)) = o(\widehat{s}_1(\xi)/w)$ steps, so D_2 is dropped after $O(t_1(\xi)) = o(T(\xi))$ steps. D_2 occupies $s_2(\xi) = o(\widehat{s}_1(\xi))$ bits, and the number of bits needed for the buffer at all times is $O(t_1(\xi)w) = o(\widehat{s}_1(\xi))$. As noted above, we have $\Omega(\widehat{s}_1(\xi))$ freely usable bits for $\Omega(\widehat{s}_1(\xi)/w)$ steps. Therefore, if the conditions formulated above are violated, we can detect this, $t_1(\xi)$ is bounded by a constant, and we can simply continue with D_1 as D.

We can use the scheme described above to reduce the initialization time of the data structure of Theorem 2 to a constant. Taking D_1 to be the data structure of the theorem, we need an alternative data structure D_2 with the same operations that can be initialized in constant time but may be less compact. A suitable choice is a depth-2 trie of degree $\Theta(\sqrt{n})$ and with the n array cells as its leaves, allocated lazily within an initializable array and slowed down by a constant factor to ensure that its operations are no faster than those of D_1. At any given time the nodes in the trie are precisely the ancestors of the leaves that have been written to, so that the space needed for the first t operations can be bounded by $O(t\sqrt{n}\log A)$. Taking ξ to be the pair (n, A), it is now trivial to verify that all conditions imposed in the description of the scheme are satisfied with $t_1(\xi) = O(\log n)$, $t_2(\xi) = O(1)$, $s_1(\xi) = n\log A + O(\log n \log(n + A))$ and, e.g., $\widehat{s}_1(\xi) = n\lfloor \log A\rfloor$, $w = \Theta(\log n + \log A)$, $T(\xi) = n^{1/4}$ and $s_2(\xi) = O(n^{3/4}\log A)$. Because D_1 is implemented in an initializable array, we can also assume, without any additional provisions, that the elements of the array are all initialized to 0.

Theorem 4. *There is a data structure that can be initialized in constant time with arbitrary $n, A \in \mathbb{N}$ and subsequently maintains a sequence $(a_1, \ldots, a_n) \in \{0, \ldots, A - 1\}^n$, initially $(0, \ldots, 0)$, occupies $n\log A + O(\log n \log(n + A))$ bits, and supports constant-time reading and writing of a_i, for $i = 1, \ldots, n$.*

3.3   Choice Dictionaries

In its most basic form, a *choice dictionary* is a data structure that is initialized with a positive integer n and subsequently maintains a subset S of $\{1, \ldots, n\}$ under the operations insertion, deletion, membership test and *choice*, which returns an (arbitrary) element of S if $S \neq \emptyset$, and 0 if $S = \emptyset$. From a different point of view, a choice dictionary is nothing but a bit vector with the additional operation "locate a 1". Choice dictionaries have numerous uses in space-efficient computing [14, 16, 19]. As observed by this author [11], it is not difficult to extend the initializable array of Subsect. 3.1 to turn it into a choice dictionary.

First, because the position of a nonzero bit within a nonzero word can be found with the operation $x \mapsto \lfloor \log x \rfloor$, the task at hand reduces to that of maintaining a sequence of words under inspection and update of individual bits (or words, for that matter) and a *choice* operation to locate a nonzero word, if any. Second, we change the conventions for the initializable array: What used to be a word written to at least once now will be a nonzero word. A nonzero word is then easy to find: There is such a word exactly if $k > 0$, and if so a nonzero word can be found either (if the cell $V[1]$ is self-contained) in $V[1]$ or (if $V[1]$ is a slave) in the data of the master corresponding to $V[1]$ (part of which data is relocated to $V[1]$). A new complication arises because a nonzero word may become zero again (whereas a word, once written to, cannot become uninitialized). If a cell that used to contain a nonzero word becomes all-zero again, it must change from being self-contained to being a slave (if it is to the left of the barrier) or from being a master to being "unused" (if it is to the right of the barrier)—again, an easy special case was left out of consideration. Simultaneously, the barrier moves left by one cell width, the cell that "jumps" the barrier must also change its status, and again masters and slaves can be paired up anew in constant time. Thus there is a choice dictionary that can be initialized with $S = \emptyset$ in constant time, uses $n + 1$ bits plus the space needed to store n, and executes every operation in constant time.

The basic choice dictionary considered above has been extended chiefly in two directions. First, rather than maintaining a sequence of n values drawn from $\{0, 1\}$, it is often necessary to maintain a sequence of n values, called *colors*, drawn from $\{0, \ldots, c - 1\}$ for some integer $c > 2$, with operations to inspect and change individual colors and to locate an occurrence, if any, of a given color. In this case one speaks of a *c-color* choice dictionary. Second, a very useful additional operation is the *robust iteration* over a given *color class* S_j, i.e., over the positions in $\{1, \ldots, n\}$ that hold a given color $j \in \{0, \ldots, c - 1\}$. The set S_j is allowed to change during the iteration, namely through interspersed color changes, and the iteration is robust if it enumerates all positions that hold the color j throughout the iteration but never enumerates a position more than once or when it does not hold the color j.

Going from the basic choice dictionary to a c-color choice dictionary can use the results of Sect. 2, but supporting *choice* efficiently in this situation needs additional ideas, as does the implementation of robust iteration. If we restrict attention to c-color choice dictionaries for which c is bounded by a constant, the current state of the art is approximately represented by the following results, where n denotes the number of colors maintained:

- A c-color choice dictionary that uses $n \log c + O(n/(\log n)^t)$ bits, for arbitrary fixed $t \in \mathbb{N}$, and executes robust iteration in constant time per element enumerated and every other operation in constant time [14].

- A c-color choice dictionary for which c is restricted to be a power of 2 and that uses $n \log c + O(\log n)$ bits and executes every operation except robust iteration in constant time [12, 21].
- A c-color choice dictionary that uses $n \log c + O(n^\epsilon)$ bits, for arbitrary fixed $\epsilon > 0$, and executes every operation except robust iteration in constant time [12].
- A c-color choice dictionary that uses $n \log c + O((\log n)^2)$ bits and executes every operation except robust iteration in $O(\log \log n)$ time [12].
- A c-color choice dictionary for $c = 3$ that uses $n \log c + O((\log n)^2)$ bits and executes almost robust iteration in a special case in constant amortized time per element enumerated and every operation other than robust iteration in constant amortized time [13]. The result generalizes to arbitrary values of c and to general fully robust iteration.

4   Open Problems

There are a number of ways in which one might hope to improve the results surveyed here. Theorem 1 allows the range $[A_v]$ associated with a node v in the tree T to depend on the depth of v in T, but it would be nice—for Theorems 1 and 2—if every node could pick its own individual range, as permitted in the initial part of the analysis. It is also an interesting question whether the extra space of Theorem 2, beyond the information-theoretic lower bound, can be brought down from $\Theta((\log n)^2)$ bits to $O(\log n)$ bits (for range sizes that are polynomial in n). Theorem 3 is optimal [15, 22], but a concise formulation of the construction described in Subsect. 3.2 and its limits would be desirable. As for c-color choice dictionaries, the current picture is complicated and unsatisfactory. It is to be hoped that a single unified data structure can be found that combines the advantages of all currently known c-color choice dictionaries.

References

1. Aho, A.V., Hopcroft, J.E., Ullman, J.D.: The Design and Analysis of Computer Algorithms. Addison-Wesley, Boston (1974)
2. Alexandrescu, A.: The D Programming Language. Addison-Wesley, Boston (2010)
3. Angluin, D., Valiant, L.G.: Fast probabilistic algorithms for Hamiltonian circuits and matchings. J. Comput. Syst. Sci. **18**(2), 155–193 (1979). https://doi.org/10.1016/0022-0000(79)90045-X
4. Baumann, T., Hagerup, T.: Rank-select indices without tears. In: Proceedings of the Algorithms and Data Structures Symposium (WADS 2019). LNCS. Springer, Cham (2019, to appear)
5. Clark, D.: Compact pat trees. Ph.D. thesis, University of Waterloo (1996)
6. Dodis, Y., Pătraşcu, M., Thorup, M.: Changing base without losing space. In: Proceedings of the 42nd ACM Symposium on Theory of Computing (STOC 2010), pp. 593–602. ACM (2010). https://doi.org/10.1145/1806689.1806771
7. Fredriksson, K., Kilpeläinen, P.: Practically efficient array initialization. J. Softw. Pract. Exper. **46**(4), 435–467 (2016). https://doi.org/10.1002/spe.2314

8. Golynski, A.: Optimal lower bounds for rank and select indexes. Theor. Comput. Sci. **387**(3), 348–359 (2007). https://doi.org/10.1016/j.tcs.2007.07.041

9. Gosling, J., Joy, B., Steele, G., Bracha, G., Buckley, A.: The Java Language Specification, Java SE 8 Edition. Oracle America (2015)

10. Hagerup, T.: Sorting and searching on the word RAM. In: Morvan, M., Meinel, C., Krob, D. (eds.) STACS 1998. LNCS, vol. 1373, pp. 366–398. Springer, Heidelberg (1998). https://doi.org/10.1007/BFb0028575

11. Hagerup, T.: An optimal choice dictionary. Computing Research Repository (CoRR) abs/1711.00808 [cs.DS] (2017)

12. Hagerup, T.: Small uncolored and colored choice dictionaries. Computing Research Repository (CoRR) abs/1809.07661 [cs.DS] (2018)

13. Hagerup, T.: Fast breadth-first search in still less space. In: Proceedings of the 45th Workshop on Graph-Theoretic Concepts in Computer Science (WG 2019). LNCS. Springer, Cham (2019, to appear)

14. Hagerup, T., Kammer, F.: Succinct choice dictionaries. Computing Research Repository (CoRR) abs/1604.06058 (2016)

15. Hagerup, T., Kammer, F.: On-the-fly array initialization in less space. In: Proceedings of the 28th International Symposium on Algorithms and Computation (ISAAC 2017). LIPIcs, vol. 92, pp. 44:1–44:12. Schloss Dagstuhl - Leibniz-Zentrum für Informatik (2017). https://doi.org/10.4230/LIPIcs.ISAAC.2017.44

16. Hagerup, T., Kammer, F., Laudahn, M.: Space-efficient Euler partition and bipartite edge coloring. Theor. Comput. Sci. **754**, 16–34 (2019). https://doi.org/10.1016/j.tcs.2018.01.008

17. IEC/IEEE International Standard; Behavioural languages – Part 1–1: VHDL Language Reference Manual. IEC 61691-1-1:2011(E) IEEE Std 1076-2008 (2011). https://doi.org/10.1109/IEEESTD.2011.5967868

18. Jacobson, G.: Succinct static data structures. Ph.D. thesis, Carnegie Mellon University (1988)

19. Kammer, F., Kratsch, D., Laudahn, M.: Space-efficient biconnected components and recognition of outerplanar graphs. Algorithmica **81**(3), 1180–1204 (2019). https://doi.org/10.1007/s00453-018-0464-z

20. Kammer, F., Sajenko, A.: Extra space during initialization of succinct data structures and dynamical initializable arrays. In: Proceedings of the 43rd International Symposium on Mathematical Foundations of Computer Science (MFCS 2018), pp. 65:1–65:16 (2018). https://doi.org/10.4230/LIPIcs.MFCS.2018.65

21. Kammer, F., Sajenko, A.: Simple 2^f-color choice dictionaries. In: Proceedings of the 29th International Symposium on Algorithms and Computation (ISAAC 2018). LIPIcs, vol. 123, pp. 66:1–66:12. Schloss Dagstuhl - Leibniz-Zentrum für Informatik (2018). https://doi.org/10.4230/LIPIcs.ISAAC.2018.66

22. Katoh, T., Goto, K.: In-place initializable arrays. Computing Research Repository (CoRR) abs/1709.08900 [cs.DS] (2017)

23. Loong, J.T.P., Nelson, J., Yu, H.: Fillable arrays with constant time operations and a single bit of redundancy. Computing Research Repository (CoRR) abs/1709.09574 (2017)

24. Navarro, G.: Spaces, trees, and colors: the algorithmic landscape of document retrieval on sequences. ACM Comput. Surv. **46**(4), 52:1–52:47 (2014). https://doi.org/10.1145/2535933

25. Raman, R., Raman, V., Satti, S.R.: Succinct indexable dictionaries with applications to encoding k-ary trees, prefix sums and multisets. ACM Trans. Algorithms **3**(4), 43:1–43:25 (2007)

Formal Methods

Winning Strategies for Streaming Rewriting Games

Christian Coester[1](✉), Thomas Schwentick[2](✉), and Martin Schuster[3]

[1] University of Oxford, Oxford, UK
christian.coester@cs.ox.ac.uk
[2] TU Dortmund, Dortmund, Germany
thomas.schwentick@tu-dortmund.de
[3] University of Edinburgh, Edinburgh, UK

Abstract. Context-free games on strings are two-player rewriting games based on a set of production rules and a regular target language. In each round, the first player selects a position of the current string; then the second player replaces the symbol at that position according to one of the production rules. The first player wins as soon as the current string belongs to the target language. In this paper the one-pass setting for context-free games is studied, where the knowledge of the first player is incomplete: She selects positions in a left-to-right fashion and only sees the current symbol and the symbols from previous rounds. The paper studies conditions under which dominant and undominated strategies exist for the first player, and when they can be chosen from restricted types of strategies that can be computed efficiently.

Keywords: Context-free games · Rewriting games · Streaming · Incomplete information · strategies

1 Introduction

Context-free games on strings are rewriting games based on a set of *production rules* and a regular *target language*. They are played by two players, *Juliet* and *Romeo*, and consist of several rounds. In each round, first Juliet selects a position of the current string; then Romeo replaces the symbol at that position according to one of the production rules. Juliet wins as soon as the current string belongs to the target language. Context-free games were introduced by Muscholl, Schwentick and Segoufin [14] as an abstraction of *Active XML*.

Active XML (AXML) is a framework that extends XML by "active nodes". In AXML documents, some of the data is given explicitly while other parts are given by means of embedded calls to web services [11]. These embedded calls can be invoked to materialise more data. As an example (adapted from [11,14]), consider a document for the web page of a local newspaper. The document may

The first author is supported by EPSRC Award 1652110.

L. A. Gąsieniec et al. (Eds.): FCT 2019, LNCS 11651, pp. 49–63, 2019.
https://doi.org/10.1007/978-3-030-25027-0_4

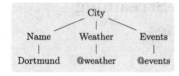

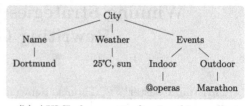

(a) AXML document before invoking calls

(b) AXML document after invoking calls

Fig. 1. An AXML document before and after the invocation of service calls

contain some explicit data, such as the name of the city, whereas information about the weather and local events is given by means of calls to a weather forecast service and an events service (see Fig. 1a). By invoking these calls, the data is materialised, i.e. replaced by concrete weather and events data (Fig. 1b). The data returned by the service call may contain further service calls.

It might not be necessary to invoke all possible service calls. In the example of Fig. 1, data about the weather might be relevant only if there are outdoor events and otherwise it does not need to be materialised. The choice which data needs to be materialised by the sender and the receiver may be influenced by considerations about performance, capabilities, security and functionalities and can be specified, for instance, by a DTD [11]. An overview about AXML is given in [1].

The question whether a document can be rewritten so that it satisfies the specification then basically translates to the *winning problem* for context-free games: given a game and a string[1], does Juliet have a winning strategy? In general, this problem is undecidable, however it becomes decidable if Juliet has to follow a *left-to-right*-strategy [14]. With such a strategy, Juliet basically traverses the string from left to right and decides, for each symbol, whether to play *Read* (keep the symbol and go to the next symbol) or *Call* (let Romeo replace the symbol).

With applications in mind where the AXML document comes as a data stream, Abiteboul, Milo and Benjelloun [3] initiated the study of a further strategy restriction, called *one-pass strategies*: Juliet still has to process the string from left-to-right, but now she does not even see the remaining part of the string, beyond the current symbol.

Due to the lack of knowledge of Juliet, one-pass strategies are more difficult to analyse and have less desirable properties than left-to-right strategies. For instance, in the sandbox game with one replacement rule $a \rightarrow b$ and the target language $\{ab, bc\}$, Juliet has a winning strategy that wins on the word ab (Read the initial a) and one that wins on ac (Call the initial a), but none that wins on both [3]. This example shows that even for some extremely simple games and input strings, there is no *dominant* strategy for Juliet, i.e., a strategy that wins

---
[1] The restriction to strings instead of trees was justified in [11].

on all words on which she has a winning strategy at all. However, both mentioned strategies are optimal in the sense that they can not be strictly improved; we call such strategies *undominated*.[2]

Since our focus will be on one-pass strategies, we will often just say strategy for short. We consider several restricted types of strategies. A strategy is *forgetful* if it does not need to remember all decisions it made, but only the (prefix of the) current string. Abiteboul et al. [3] also introduced *regular strategies*, a simple type of one-pass strategies defined by a finite state automaton, and therefore efficiently computable. Some strategies that are both regular and forgetful are of a particularly simple form such that they computed by an automaton that is derived from the minimal automaton for the target language. We call this most restricted type *strongly regular*. We refer to Sect. 2 for precise definitions of these notions.

We study the following questions.

- Under which circumstances does an undominated one-pass strategy exist?
- When can a dominant or undominated strategy even be chosen from one of the restricted types (regular, forgetful or strongly regular)?

This paper is based on the Master's thesis of the first author, supervised by the other two authors [8]. The thesis contains further results, investigating also the computational complexity of some related problems.

1.1 Our Results

Regarding the existence of undominated strategies, we show that the situation is much better than for dominant strategies. Although it remains unclear whether undominated strategies exist for *all* context-free games, we identify important classes of games in which they do exist. More precisely, we give a semantical restriction, the bounded depth property, that guarantees existence of undominated strategies, and a fairly natural syntactical restriction, prefix-freeness[3], that guarantees existence of an undominated strategy which can even be chosen regular. For two other families of restricted games, non-recursive games (i.e., the production rules do not allow recursive generation of a symbol from itself) and games with finite target language, we show that if they have a dominant strategy, then even a strongly regular one.

Theorem 1. *(a) Every game with the bounded depth property has an undominated strategy.*

(b) Every prefix-free game has a regular undominated strategy.

(c) Every game with a finite target language that has a dominant strategy has a strongly regular dominant strategy.

---

[2] In [3], dominant and undominated strategies were called optimum and optimal respectively.

[3] As explained later, every game can be transformed into a very similar prefix-free game.

(d) Every non-recursive game with a dominant strategy has a strongly regular dominant strategy.

Games with a finite target language as well as non-recursive games also have the bounded depth property, so these two classes also have an undominated strategy by part (a) (cf. [8; Theorem 3.6]).

We complement the implication results of Theorem 1(c) and (d) by negative results, showing that these implications do not generalise to arbitrary games or to undominated strategies. More strongly, the following theorem states that even much weaker implications do not hold in general.

Theorem 2. *(a) There exists a game G_1 with a regular dominant strategy but no forgetful dominant strategy.*

(b) There exists a game G_2 with a forgetful regular dominant strategy but no strongly regular dominant strategy.

(c) The statements (a) and (b) hold also when "dominant" is replaced by "undominated". In this case, G_1 and G_2 can even be chosen as non-recursive games with a finite target language.

1.2   Related Work

Further background about AXML is given in [1, 2, 11]. Context-free games were introduced in [13], which is the conference paper corresponding to [14]. The article studies the decidability and complexity of deciding whether a winning unrestricted or left-to-right strategy exists for a word in the general case and several restricted cases. More recently, Holík et al. gave a new algorithm for determining the winner of a left-to-right context-free game and determining a winning strategy [10]. One-pass strategies and (forgetful) regular strategies were introduced in [3]. The complexity of deciding, for a given context-free game, whether Juliet has a winning left-to-right strategy for every word for which she has a winning unrestricted strategy is studied in [5]. Extended settings of context-free games with nested words (resembling the tree structure of (A)XML documents) are examined in [15, 12].

The article [14] also showed a tight connection between context-free games and pushdown games [17, 7, 16]. In [4, 6], variants of these pushdown games in settings of imperfect information are studied.

1.3   Organization

In Sect. 2 we provide several definitions and some basic lemmas. We prove Theorem 1(a) in Sect. 3. In the two subsequent sections, we sketch the proofs of parts (b), (c) and (d). We also explain our motivation for prefix-freeness in Sect. 4, which we consider the most relevant case and with the most interesting proof. Section 6 contains proofs of Theorem 2(a) and (b). Most proof details of Theorem 1 as well as the proof of Theorem 2(c) can be found in the full version of our paper [9].

2   Preliminaries

We denote the set of strings over an alphabet Σ by Σ^* and the set of non-empty strings by Σ^+. Σ^k denotes the set of strings of length k and $\Sigma^{\leq k}$ the set of strings of length at most k.

A *nondeterministic finite automaton (NFA)* is a tuple $\mathcal{A} = (Q, \Sigma, \delta, s, F)$, where Q is the set of states, Σ the alphabet, $\delta \subseteq Q \times \Sigma \times Q$ the transition relation, $s \in Q$ the initial state and $F \subseteq Q$ the set of accepting states. A *run* on a string $w = w_1 \cdots w_n$ is a sequence $q_0, \ldots, q_n$ of states such that $q_0 = s$ and, for each $i \leq n$, $(q_{i-1}, w_i, q_i) \in \delta$. A run is *accepting* if $q_n \in F$. A word w is in the language $L(\mathcal{A})$ of $\mathcal{A}$ if $\mathcal{A}$ has an accepting run on w. If $\mathcal{A}$ is deterministic, i.e., for each p and a, there is exactly one state q such that $(p, a, q) \in \delta$, then we consider δ as *transition function* $Q \times \Sigma \to Q$ and also use the *extended transition function* $\delta^* : Q \times \Sigma^* \to Q$, as usual.

Context-Free Games. A *context-free game*, or a *game* for short, is a tuple $G = (\Sigma, R, T)$ consisting of a finite alphabet Σ, a minimal[4] DFA $T = (Q, \Sigma, \delta, s, F)$, and a binary relation $R \subseteq \Sigma \times \Sigma^+$ such that for each $a \in \Sigma$, the *replacement language* $L_a \overset{\text{def}}{=} \{v \in \Sigma^+ \mid (a, v) \in R\}$ of a is regular. We call $L(T)$ the *target language* of G. By $\Sigma_f = \{a \in \Sigma \mid \exists v \in \Sigma^+ : (a, v) \in R\}$ we denote the set of *function symbols*, i.e. the symbols occurring as the left hand side of a rule. The languages L_a are usually represented by regular expressions R_a for each $a \in \Sigma_f$, and we specify R often by expressions of the form $a \to R_a$. We note that the definition of context-free games assures $\epsilon \notin L_a$.

The semantics of context-free games formalises the intuition given in the introduction. In a configuration, we summarise the information about a current situation of a play together with some information about the history of the play. For the latter, let $\widehat{\Sigma_f} = \{\widehat{a} \mid a \in \Sigma_f\}$ be a disjoint copy of the set Σ_f of function symbols, and let $\overline{\Sigma} = \Sigma \,\dot{\cup}\, \widehat{\Sigma_f}$. A *configuration* is a tuple $(\alpha, u) \in \overline{\Sigma}^* \times \Sigma^*$. If u is non-empty, i.e. $u = av$ for $a \in \Sigma$ and $v \in \Sigma^*$, then we also denote this configuration by (α, a, v), consisting of a *history string* α, a *current symbol* $a \in \Sigma$ and a *remaining string* $v \in \Sigma^*$. We denote the set of all (syntactically) possible configurations by $\mathcal{K}$. Intuitively, if the ith symbol of the history string is $b \in \Sigma$ then this shall denote that Juliet's ith move was to read the symbol b, and if it is $\widehat{b} \in \widehat{\Sigma_f}$ then this shall denote that Juliet's ith move was to call b. The remaining string is the string of symbols that have not been revealed to Juliet yet. By $\natural : \overline{\Sigma}^* \to \Sigma^*$ we denote[5] the homomorphism which deletes all symbols from $\widehat{\Sigma_f}$ and is the identity on Σ. We call $\delta^*(s, \natural\alpha)$ the *T-state* of the configuration (α, u).

A play is a sequence of configurations, connected by moves. In one move at a configuration (α, a, v) Juliet can either "read" a or "call" a. In the latter case, Romeo can replace a by a string from L_a. More formally, a *play* of a game

---

[4] The assumption that T is minimal will be convenient at times.
[5] We usually omit brackets and write, e.g., $\natural\alpha\beta$ for $\natural(\alpha\beta)$.

is a finite or infinite sequence $\Pi = (K_0, K_1, K_2, \ldots)$ of configurations with the following properties:

(a) The *initial configuration* is of the form $K_0 = (\epsilon, w)$, where $w \in \Sigma^*$ is called the *input word*.
(b) If $K_n = (\alpha, a, v)$, then either $K_{n+1} = (\alpha a, v)$ or $K_{n+1} = (\alpha \widehat{a}, xv)$ with $x \in L_a$. In the former case we say that Juliet plays a *Read* move, otherwise she plays a *Call* move and Romeo replies by x.
(c) If $K_n = (\alpha, \epsilon)$, then K_n is the last configuration of the sequence. Its history string α is called the *final history string* of Π. Its *final string* is $\natural \alpha$.

A play is *winning for Juliet* (and *losing for Romeo*) if it is finite and its final string is in the target language $L(T)$. A play is *losing for Juliet* (and *winning for Romeo*) if it is finite and its final string is not in $L(T)$. An infinite play is neither winning nor losing for any player.

Strategies. As mentioned in the introduction, we are interested in so-called one-pass strategies for Juliet, where Juliet's decisions do not depend on any symbols of the remaining string beyond the current symbol.

A *one-pass strategy for Juliet* is a map $\sigma \colon \overline{\Sigma}^* \times \Sigma_f \to \{Call, Read\}$, where the argument corresponds to the first two components of a configuration. A *strategy for Romeo* is a map $\tau \colon \overline{\Sigma}^* \times \Sigma_f \to \Sigma^+$ where $\tau(\alpha, a) \in L_a$ for each $(\alpha, a) \in \overline{\Sigma}^* \times \Sigma_f$.[6] We generally denote strategies for Juliet by $\sigma, \sigma', \sigma_1, \ldots$ and Romeo strategies by $\tau, \tau', \tau_1, \ldots$. We often just use the term *strategy* to refer to a one-pass strategy for Juliet.

The *play of σ and τ on w*, denoted $\Pi(\sigma, \tau, w)$, is the (unique) play $(K_0, K_1, \ldots)$ with input word w satisfying that

– if $K_n = (\alpha, a, v)$ and $\sigma(\alpha, a) = Read$, then $K_{n+1} = (\alpha a, v)$,
– if $K_n = (\alpha, a, v)$ and $\sigma(\alpha, a) = Call$, then $K_{n+1} = (\alpha \widehat{a}, \tau(\alpha, a)v)$.

The *depth* of a finite play is its maximum nesting depth of *Call* moves. E.g., if Romeo replaces some symbol a by a string u and Juliet calls a symbol in u, the nesting depth of this latter *Call* move is 2.

A strategy σ is *terminating* if each of its plays is finite. The *depth* of σ is the supremum of depths of plays of σ. Note that each strategy with finite depth is terminating. The converse, however, is not true and it is easy to construct counter-examples of a game and a strategy σ where each play of σ has finite depth but depths are arbitrarily large.

A strategy σ *wins* on a string $w \in \Sigma^*$ if every play of σ on w is winning (for Juliet). By $W(\sigma) = W_G(\sigma)$ we denote the set of words on which σ wins in G. In contrast, σ *loses* on w if there exists a losing play of σ on w. Note that σ neither wins nor loses on w if there exists an infinite play of σ on w but no losing play of σ on w.

---

[6] Even though we think of Romeo as an omniscient adversary, it is not necessary to provide the remaining string as an argument to τ: The remaining string is uniquely determined by the input word and his own and Juliet's previous moves.

A strategy σ *dominates* a strategy σ' if $W(\sigma') \subseteq W(\sigma)$. A strategy σ is *dominant* if it dominates every other (one-pass) strategy. It is *undominated* if there is no strategy σ' with $W(\sigma) \subsetneq W(\sigma')$.

In the proofs of Theorem 1(a) and (b), we will actually show a slightly stronger form of optimality than "undominated", which we call weakly dominant. To define it, fix some total order $<$ of the alphabet Σ. We order strings by *shortlex order*, i.e. for two strings $v, w \in \Sigma^*$ we define $v <_{\mathsf{sl}} w$ if $|v| < |w|$ or if $|v| = |w|$ and v precedes w in the lexicographical order. We extend this to a total order $\leq_{\mathsf{sl}}$ on sets of words as follows. Let $V, W \subseteq \Sigma^*$ be two sets with $V \neq W$. Their order is determined by the minimal string w (with respect to shortlex order $\leq_{\mathsf{sl}}$) that is contained in only one of the two sets. If $w \in W$, then $V <_{\mathsf{sl}} W$; otherwise $W <_{\mathsf{sl}} V$. We observe that if $V \subsetneq W$ then $V <_{\mathsf{sl}} W$. A strategy σ is *weakly dominant* if, for every strategy σ' it holds $W(\sigma') \leq_{\mathsf{sl}} W(\sigma)$. Thus, a weakly dominant strategy can be seen as a best undominated strategy with respect to $\leq_{\mathsf{sl}}$.

The following lemma is convenient as it will often allow us to assume, without loss of generality, that a given strategy is terminating.

Lemma 1. *Each strategy is dominated by a terminating strategy.*

Proof. Given a strategy σ, we construct a terminating strategy σ' with $W(\sigma) \subseteq W(\sigma')$ as follows.

Consider some $\alpha \in \overline{\Sigma}^*$ and $a \in \Sigma_f$. If there exists a play Π of σ that contains a configuration (α, av) for some $v \in \Sigma^*$ such that no later configuration of the form $(\alpha\alpha', v)$ occurs in Π, then let $\sigma'(\alpha, a) = Read$ and $\sigma'(\alpha a\beta, b) = Read$ for each $\beta \in \overline{\Sigma}^*$ and $b \in \Sigma_f$. For all elements of the domain for which σ' is not already defined by this, we define σ' like σ. Clearly, σ' is terminating and $W(\sigma) \subseteq W(\sigma')$. $\square$

Restricted Strategy Types. A strategy σ is *regular* if the set L of strings αa with $\sigma(\alpha, a) = Call$ is regular. In this case, a DFA $\mathcal{A}$ for L is called a *strategy automaton* for $\sigma = \sigma_{\mathcal{A}}$. A strategy is *forgetful* if its decisions are independent of symbols from $\overline{\Sigma}_f$ in the history string, i.e. if $\sigma(\alpha, a) = \sigma(\beta, a)$ whenever $\natural\alpha = \natural\beta$. Clearly, if σ is regular *and* forgetful, then $L' \stackrel{\text{def}}{=} \{\natural\alpha a \mid \sigma(\alpha, a) = Call\}$ is regular. A DFA $\mathcal{A}$ for L' is also called a strategy automaton, and we write $\sigma_{\mathcal{A}} = \sigma$ again.

We are particularly interested in the special case of regular forgetful strategies where Juliet's decisions depend only on the current T-state and the current symbol. More precisely, if $T = (Q, \Sigma, \delta, s, F)$ is the target automaton and the strategy automaton is of the form $\mathcal{A} = (Q \cup \{Call\}, \Sigma, \delta_{\mathcal{A}}, s, \{Call\})$ with $\delta_{\mathcal{A}}(q, a) \in \{\delta(q, a), Call\}$, for each q and a, then $\sigma_{\mathcal{A}}$ is called *strongly regular*.

Classes of Games. A game G has the *bounded depth property* if there exists a sequence $(B_k)_{k \in \mathbb{N}_0} \subseteq \mathbb{N}$ such that for each one-pass strategy σ for G and each $k \in \mathbb{N}$ there exists a one-pass strategy σ_k that wins on each $w \in W(\sigma) \cap \Sigma^{\leq k}$ with plays of depth at most $B_{|w|}$. Roughly speaking, the bounded depth property means the following: When input words are restricted to a finite set, Juliet can choose a strategy whose depth is bounded on words from her winning set, without losing strategic power.

A game is *prefix-free* if each replacement language L_a is prefix-free, that is, there are no $u, v \in L_a$ where u is a proper prefix of v.

A game is *non-recursive* if no symbol can be derived from itself by a sequence of rules, i.e. there do not exist $a_0, \ldots, a_n \in \Sigma_f$, $n \geq 1$, such that $a_0 = a_n$ and for each $k = 1, \ldots, n$ there exists a word in $L_{a_{k-1}}$ containing a_k.

Convergence of Strategies. A concept used in the proofs of parts (a) and (d) is the convergence of a sequence of one-pass strategies. A sequence $(\sigma_k)_{k \in \mathbb{N}}$ of strategies *converges to* a strategy σ if for each $n \in \mathbb{N}$ there exists $k_0 \in \mathbb{N}$ such that for each $k \geq k_0$ and $(\alpha, a) \in \overline{\Sigma}^{\leq n} \times \Sigma$ it holds that $\sigma(\alpha, a) = \sigma_k(\alpha, a)$.

Lemma 2. *Let G be a game and $(\sigma_k)_{k \in \mathbb{N}}$ be a sequence of one-pass strategies that converges to some one-pass strategy σ. Let $L_1 \subseteq L_2 \subseteq \cdots$ be an infinite sequence of languages such that, for every k, $L_k \subseteq W(\sigma_k)$ and let $L \overset{def}{=} \bigcup_{k \in \mathbb{N}} L_k$.*

Then σ does not lose on any word $w \in L$.

It should be noted, however, that σ might fail to win on some of these words due to infinite plays.

Proof of Lemma 2. Towards a contradiction, suppose that σ loses on a word $w \in L$. Then there exists a strategy of Romeo with which he wins the (finite) play $\Pi = \Pi(\sigma, \tau, w) = (K_0, \ldots, K_n)$. Let $k_0 \in \mathbb{N}$ be such that for each $k \geq k_0$ and $(\alpha, a) \in \overline{\Sigma}^{\leq n} \times \Sigma$ it holds that $\sigma(\alpha, a) = \sigma_k(\alpha, a)$. Let furthermore k_1 be such that $w \in L_{k_1}$ and let $k \overset{def}{=} \max(k_0, k_1)$. Then Π is also a play of σ_k on w. But then σ_k loses on $w \in L_{k_1} \subseteq L_k$, the desired contradiction.   □

3   Games with the Bounded Depth Property

In this section, we show that each game with the bounded depth property admits a weakly dominant strategy, implying Theorem 1(a). Since non-recursive games trivially have the bounded depth property, and also games with a finite target language (cf. [8; Lemma 3.5]) and prefix-free games (cf. Sect. 4) have the bounded depth property, it follows that any such game has an undominated strategy. In fact, all of these games satisfy the stronger version of the bounded depth property where $B_k = B$ does not depend on k.

For a strategy σ and some $i \geq 0$, we denote by $\sigma\big|_i$ the restriction of σ to the first i rounds of the game. Thus $\sigma\big|_i$ is a mapping $\sigma\big|_i : \overline{\Sigma}^{\leq i-1} \times \Sigma \to \{Call, Read\}$ and $\sigma\big|_0$ is the mapping with empty domain.

Proof of Theorem 1(a). Let G be a context-free game with the bounded depth property and let $(B_k)_{k \in \mathbb{N}_0} \subseteq \mathbb{N}$ be its sequence of depth bounds.

We first define a language L which will serve as the winning set of the weakly dominant strategy that will be constructed below.

The definition of L is by induction. For each $k \geq 0$, we define a set $L_k \subseteq \Sigma^{\leq k}$ such that $L_k \subseteq L_{k+1}$, and finally let $L \stackrel{\text{def}}{=} \bigcup_k L_k$.

Let $L_0 = \{\epsilon\}$ if ϵ is in the target language of G, and $L_0 = \emptyset$ otherwise.

For $k \geq 0$, we define L_{k+1} as the maximal set with respect to $\leq_{\mathsf{sl}}$ of the form $W(\sigma) \cap \Sigma^{\leq k+1}$ for some strategy σ with $L_k \subseteq W(\sigma)$. It is easy to see that the following two properties hold by construction.

(1) For each k, there is a strategy σ such that $L_k \subseteq W(\sigma)$.
(2) There is no strategy σ with $L <_{\mathsf{sl}} W(\sigma)$.

Thanks to property (2), it suffices to construct a strategy $\hat{\sigma}$ with $L \subseteq W(\hat{\sigma})$.

For each $k \geq 0$, let S_k be the set of strategies σ with $L_k \subseteq W(\sigma)$ for which each play on a word $w \in L_k$ has depth at most $B_{|w|}$. Because of property (1) and since G has the bounded depth property, we have $S_k \neq \emptyset$ for every $k \geq 0$.

We will construct mappings $\rho_k \colon \overline{\Sigma}^{\leq k-1} \times \Sigma \to \{Call, Read\}$ such that for every $k \geq 0$,

- ρ_{k+1} extends ρ_k; more precisely: $\rho_{k+1}\big|_k = \rho_k$, and
- for each $\ell \geq k$ there exists $\sigma_\ell^k \in S_\ell$ with $\rho_k = \sigma_\ell^k\big|_k$.

Let ρ_0 be the mapping with empty domain. Fix k such that $\rho_0, \ldots, \rho_k$ are defined and have the stated properties. Since there are only finitely many mappings $\overline{\Sigma}^{\leq k} \times \Sigma \to \{Call, Read\}$, one of them has to occur infinitely often within $\big(\sigma_\ell^k\big|_{k+1}\big)_{\ell \geq k}$. Let ρ_{k+1} be such a mapping. For $\ell' \geq k+1$ we can choose $\sigma_{\ell'}^{k+1} = \sigma_\ell^k$ for some $\ell \geq \ell'$ with $\rho_{k+1} = \sigma_\ell^k\big|_{k+1}$. This defines a sequence $\rho_0, \rho_1, \ldots$ with the properties above. Let $\hat{\sigma}$ be the strategy that is uniquely determined by $\hat{\sigma}\big|_k = \rho_k$, for every k. Clearly $(\rho_k)_{k \in \mathbb{N}}$ converges[7] to $\hat{\sigma}$.

Thanks to Lemma 2 it suffices to show that $\hat{\sigma}$ terminates on L. Let thus $w \in L$ and τ be a Romeo strategy. We show that the depth of $\Pi \stackrel{\text{def}}{=} \Pi(\hat{\sigma}, \tau, w)$ is at most $B_{|w|}$. Otherwise let k be such that in the kth round Juliet does a $Call$ move of nesting depth $B_{|w|} + 1$. However, $\hat{\sigma}\big|_k = \rho_k = \sigma_\ell^k\big|_k$, where $\ell = \max\{k, |w|\}$, and $\sigma_\ell^k \in S_\ell$ has depth at most $B_{|w|}$ on $w \in L_\ell$, a contradiction. Therefore the depth of Π is at most $B_{|w|}$ and by König's Lemma Π is thus finite, completing the proof. □

4   Prefix-Free Games

Prefix-freeness appears as a realistic constraint for a practical (Active XML) setting since it can be easily enforced by suffixing each replacement string with a special end-of-file symbol: *Every game* $G = (\Sigma, R, T)$ *can be transformed into a prefix-free game* $G' = (\Sigma', R', T')$ *by letting* $\Sigma' = \Sigma \,\dot{\cup}\, \{\$\}$ *for some new*

---

[7] Since the ρ_k are only partially defined, one might consider the strategies σ_k that result from the ρ_k which take the value $Call$ whenever ρ_k is undefined.

symbol $\$ \notin \Sigma$ that shall denote the end of replacement strings, and further letting $R_a' = R_a\$$ for each $a \in \Sigma_f$ to enforce that replacement words end with $\$$, and adding a loop transition for the symbol $\$$ to each state of T (accomplishing that the symbol $\$$ is "ignored" by the target language). Another special case of prefix-free games, which is similar to the *one-pass with size* setting discussed in [3], are games where the alphabet Σ contains (besides other symbols) numbers $1, \ldots, N$ for some $N \in \mathbb{N}$ and all replacement strings are of the form nx where $x \in \Sigma^+$ and $n = |x|$. Our result for prefix-free games also easily transfers to the setting where the input word is revealed to Juliet in a one-pass fashion, but Romeo's replacement words are revealed immediately.

We sketch the proof of Theorem 1(b) in the following.

A context-free game on a string $w = a_1 \cdots a_n$ can be viewed as a sequence of n games on the single symbols $a_1, \ldots, a_n$. Intuitively, in prefix-free games Juliet has the benefit to know when a subgame on some symbol a_i has ended and when the next subgame starts.

This allows us to view strategies of Juliet in a hierarchical way: they consist of a top-level strategy that chooses, whenever a subgame on some a_i starts, a strategy for this subgame. This choice may take the current history string into account. We will use this view to proceed in an inductive fashion: we establish that there are automata for the subgame strategies and then combine these automata with suitable automata for a "top-level" strategy.

It turns out that the choice of the top-level strategy boils down to an "online word problem" for NFAs which we introduce and study first.

The Online Word Problem for NFAs. In the online-version of the word problem for an NFA $\mathcal{N}$, denoted ONLINENFA($\mathcal{N}$), the single player gets to know the symbols of a word one by one, and always needs to decide which transition $\mathcal{N}$ should take before the next symbol is revealed. We only consider the case that $\mathcal{N} = (Q, \Sigma, \delta, s, F)$ has at least one transition for each symbol from each state. Formally, a strategy is a map $\rho: \Sigma^* \to Q$ such that $\rho(\epsilon) = s$ and $(\rho(w), a, \rho(wa)) \in \delta$ for each $w \in \Sigma^*$ and $a \in \Sigma$.

Given a strategy ρ for ONLINENFA($\mathcal{N}$), we denote by $W_\mathcal{N}(\rho)$ the *winning set* of words that are accepted by $\mathcal{N}$ if the player follows ρ. A strategy ρ is *weakly dominant* if, for every strategy ρ', it holds $W_\mathcal{N}(\rho') \leq_{\mathsf{sl}} W_\mathcal{N}(\rho)$.

We are interested in strategies that can be computed by automata. A particularly simple such strategy for ONLINENFA($\mathcal{N}$) can be obtained by transforming $\mathcal{N}$ into a DFA $\mathcal{D}$ by removing transitions. The associated strategy $\rho_\mathcal{D}$ is the one that only uses the transitions of $\mathcal{D}$. We prove that $\mathcal{D}$ can be chosen such that $\rho_\mathcal{D}$ is weakly dominant.

Lemma 3. *For each NFA $\mathcal{N}$, there exists a DFA $\mathcal{D}$ obtained by removing transitions from $\mathcal{N}$ such that $\rho_\mathcal{D}$ is a weakly dominant strategy for* ONLINENFA($\mathcal{N}$).

Game Composition and Game Effects. Let in the following, $G = (\Sigma, R, T)$ be a prefix-free game with $T = (Q, \Sigma, \delta, s, F)$. Let furthermore, for every $a \in \Sigma$,

$\mathcal{A}_a = (Q_a, \Sigma, \delta_a, s_a, \{f_a\})$ be a minimal DFA for the replacement language L_a of a. Since L_a is prefix-free, $\mathcal{A}_a$ has a unique accepting state f_a.

For a strategy σ (for Juliet or Romeo) and a string $\alpha \in \overline{\Sigma}^*$, we define the *substrategy* σ^α of σ by $\sigma^\alpha(\beta, a) = \sigma(\alpha\beta, a)$.

In the following, $states(q, w, \sigma)$ denotes the set of T-states that can be reached at the end of a play of σ on w if the initial state of T were q. More precisely, it is the set of states of the form $\delta^*(q, \natural\alpha)$ where α is a final history string of a play of σ on w.

An *effect triple* (p, a, S) consists of a state $p \in Q$, a symbol $a \in \Sigma$ and a set $S \subseteq Q$. We say that (p, a, S) is an effect triple of σ if $states(p, a, \sigma) \subseteq S$. That is, starting from p and processing a according to σ, one is guaranteed to reach a state in S. We call (p, a, S) *trivial* if $\delta(p, a) \in S$, i.e. if it is an effect triple of a strategy that plays Read on a. The *single-symbol effect* $\mathsf{sse}(\sigma)$ of a strategy σ is the set of all its effect triples. Finally, we define the *effect set* $E(\sigma)$ of a strategy σ as $E(\sigma) \stackrel{\text{def}}{=} \bigcup_{\alpha \in \overline{\Sigma}^*} \mathsf{sse}(\sigma^\alpha)$. That is, $E(\sigma)$ contains all effect triples that are induced by substrategies of σ.

For a set E of effect triples, consider the NFA $\mathcal{N}_E = (\mathcal{P}(Q), \Sigma, \delta_E, \{s\}, \mathcal{P}(F))$, where δ_E is defined as follows. For sets $S, S' \subseteq Q$ and $a \in \Sigma$, $(S, a, S') \in \delta_E$ if, for each $p \in S$, there is some $S'' \subseteq S'$ such that $(p, a, S'') \in E$.

Proposition 1. *Let $G = (\Sigma, R, T)$ be a prefix-free game, E a set of effect triples, and σ a terminating strategy such that $E(\sigma) \subseteq E$. Then there is a strategy ρ for* ONLINENFA$(\mathcal{N}_E)$ *such that $W_G(\sigma) = W_{\mathcal{N}_E}(\rho)$.*

Proof. It is straightforward to verify that $\rho(w) \stackrel{\text{def}}{=} states(s, w, \sigma)$ yields a well-defined strategy ρ for ONLINENFA$(\mathcal{N}_E)$. The proposition follows, since $w \in W_{\mathcal{N}_E}(\rho)$ if and only if $\rho(w) \subseteq F$, and $w \in W_G(\sigma)$ if and only if $states(s, w, \sigma) \subseteq F$. □

We say that a strategy automaton $\mathcal{A} = (Q_\mathcal{A}, \overline{\Sigma}, \delta_\mathcal{A}, s_\mathcal{A}, F_\mathcal{A})$ is (p, a, S)-*inducing* if $\sigma_\mathcal{A}$ is terminating, $a \in \Sigma_f$, and the following conditions hold.

- For each $u \in L_a$, $states(p, u, \sigma_\mathcal{A}) \subseteq S$.
- There are disjoint subsets $Q_{\mathcal{A},q} \subseteq Q_\mathcal{A}$, for $q \in S$, such that for every play of σ on some $u \in L_a$ with final history string α, it holds $\delta_\mathcal{A}^*(s_\mathcal{A}, \alpha) \in Q_{\mathcal{A},q} \Leftrightarrow \delta^*(p, \natural\alpha) = q$.
  Furthermore, there is no proper prefix β of α for which $\delta_\mathcal{A}^*(s_\mathcal{A}, \beta) \in Q_{\mathcal{A},r}$, for any r.

Proposition 2. *Let $G = (\Sigma, R, T)$ be a prefix-free game and E a set of effect triples such that for each non-trivial $t \in E$, there exists a t-inducing strategy automaton $\mathcal{A}_t$. Then there is a strategy automaton $\mathcal{A}$ for G such that, for each strategy ρ for* ONLINENFA$(\mathcal{N}_E)$, $W_{\mathcal{N}_E}(\rho) \leq_{sl} W_G(\sigma_\mathcal{A})$.

A crucial ingredient is the following proposition, which will allow us to restrict our attention to strategies of finite depth. An almost identical proof can also be used to show that prefix-free games have the bounded depth property, and we could use this and Theorem 1(a) to deduce immediately that they have weakly dominant strategies. However, we are aiming for the stronger result that they have *regular* weakly dominant strategies.

Proposition 3. *In prefix-free games, each effect triple of a terminating strategy is also an effect triple of a strategy of bounded depth.*

Proof Idea. Let E denote the set of effect triples of bounded depth strategies. Consider an effect triple $(p, a, S) \notin E$ of a strategy $\hat{\sigma}$. If the effect triples of $\hat{\sigma}$'s substrategies on the symbols of a's replacement word were all in E, then these substrategies could be replaced so as to obtain a bounded depth strategy σ for (p, a, S), contradicting $(p, a, S) \notin E$. Thus, for each $(p, a, S) \notin E$, Romeo can force a configuration in the play against $\hat{\sigma}$ on a where the effect triple of the substrategy is again not in E. But Romeo can do this repeatedly, so $\hat{\sigma}$ is not terminating.   □

The finite depth allows us to construct t-inducing strategy automata by induction on the depth of a strategy with effect triple t.

Proposition 4. *Let $G = (\Sigma, R, T)$ be a prefix-free game and (p, a, S) a non-trivial effect triple of some terminating strategy σ. Then there exists a (p, a, S)-inducing strategy automaton.*

Now we are ready to prove Theorem 1(b).

Proof of Theorem 1(b). Let $G = (\Sigma, R, T)$ be prefix-free. Let E be the set of effect triples of terminating strategies. By Proposition 4 there is a t-inducing automaton for each non-trivial $t \in E$. Let σ_A be the regular strategy as guaranteed by Proposition 2. We show that σ_A is weakly dominant.

To this end, let σ be any terminating strategy for G (cf. Lemma 1). Since $E(\sigma) \subseteq E$, Proposition 1 guarantees a strategy ρ for $\text{ONLINENFA}(\mathcal{N}_E)$ such that $W_G(\sigma) = W_{\mathcal{N}_E}(\rho)$. By Proposition 2, $W_{\mathcal{N}_E}(\rho) \leq_{\text{sl}} W_G(\sigma_A)$, and therefore, altogether $W_G(\sigma) \leq_{\text{sl}} W_G(\sigma_A)$ as required.   □

5   Strongly Regular Dominant Strategies

In this section, we sketch the main ideas of the proof of Theorem 1(c) and (d).

We say that a strategy σ has a (q, a)-*conflict*, for $q \in Q$ and $a \in \Sigma_f$, if there are configurations (α_1, a, u_1) and (α_2, a, u_2) in plays on words from $W(\sigma)$ such that $\delta^*(s, \natural\alpha_1) = \delta^*(s, \natural\alpha_2) = q$ and $\sigma(\alpha_1, a) \neq \sigma(\alpha_2, a)$. If σ has no conflicts, then changing it to a strongly regular strategy requires modification only on configurations that do not occur in plays on words from $W(\sigma)$.

Proof Idea of Theorem 1(c). We show that a dominant strategy σ with some conflicts can be transformed into a dominant strategy σ' with less conflicts. To do so, we find a configuration $(v, a, *)$ with $\delta^*(s, v) = q$ that has no (q, a)-conflict with any later configuration of the same play. The existence of such a strategy can be concluded from the fact that T is acyclic, hence the only way for a configuration to conflict with a later configuration is if all intermediate moves are *Call*. A strategy without (q, a)-conflict can be obtained by "copying" the substrategy starting from $(v, a, *)$ to any conflicting configuration.   □

Proof Idea of Theorem 1(d). We inductively construct a sequence $(\sigma_1, \sigma_2, \dots)$ of dominant strategies that is either finite and ends with a strategy that has no conflict, or is infinite and converges to such a strategy. In the convergent case, since each strategy in a non-recursive game is terminating, the limit strategy is also dominant by Lemma 2. To construct a strategy σ_{k+1}, the idea is to modify σ_k so as to shift the earliest conflict to a later time in the future.   □

6   Negative Results

Finally, we provide proofs for Theorem 2(a) and (b).

Proof of Theorem 2(a). The game $G_1 = (\Sigma, R, T)$ with $\Sigma = \{a\}$, the only replacement rule being $a \to aa$ and $L(T) = \{a^k \mid k \geq 2\}$ has the stated property. The strategy plays Call exactly if it has not seen any symbol $\hat{a}$. Since this strategy wins on every word, it is dominant. However, a forgetful strategy that plays Call on the first symbol a is bound to play Call forever, and therefore does not win on any word. On the other hand, a forgetful strategy that plays Read on the first symbol a does not win on the word a and is therefore not dominant, either.   □

Proof of Theorem 2(b). Let $G_2 = (\Sigma, R, T)$ with $\Sigma = \{a, b, c, d\}$, rule set R given by

$$a \to b$$
$$c \to ac$$
$$d \to bad$$

and the target language automaton $T = (Q, \Sigma, \delta_T, q_0, F)$ depicted in Fig. 2a. We claim that the regular forgetful strategy σ_A based on the automaton A shown in Fig. 2b fulfils $W(\sigma_A) = \Sigma^*$ and is thus dominant. Indeed, by induction on the length of w, the following is easy to show: for each input word w, the strategy σ_A yields a terminating play with a final string u such that: $\delta_T(q_0, u) = \delta_A(q_0, u)$, if u does not end with a, and $\delta_T(q_0, u) = q_0$ and $\delta_A(q_0, u) = q_0'$, otherwise.

However, for a strategy automaton $B = (Q \cup \{Call\}, \Sigma, \delta_B, q_0, \{Call\})$ of a strongly regular strategy σ_B, it holds that $W(\sigma_B) \subsetneq \Sigma^*$, and thus no such σ_B is dominant. For a proof of this claim, we can assume that $\delta_B(q_0, c) = \delta_B(q_1, a) = \delta_B(q_1, d) = Call$ since otherwise σ_B would lose on c, ba or bd. If $\delta_B(q_0, a) = q_0$, then the play of σ_B on ac is infinite. On the other hand, if $\delta_B(q_0, a) = Call$, then the play of σ_B on ad is infinite.   □

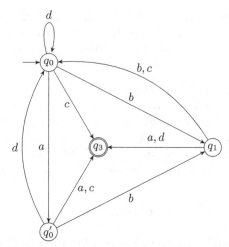

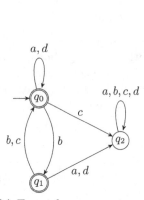

(a) Target language automaton T

(b) Strategy automaton $\mathcal{A}$; missing transitions lead into a non-accepting sink state (not shown).

Fig. 2. Automata used in the proof of Theorem 2(b)

7 Conclusion and Open Questions

It remains unclear whether undominated strategies always exist. Our main positive result is that prefix-free replacement languages allow for an undominated strategy that is regular. Indeed, prefix-freeness seems to be a realistic solution in practice, because it can be achieved easily and with almost no overhead by suffixing replacement words with an end-of-file symbol. Also non-recursive rules or a finite target language lead to many positive properties, in particular because they have the bounded depth property. While finiteness of the target language may seem like a strong restriction, one particular instance of it is if there is a single target document that has to be reached [14]. Restrictions that bound the number of recursive replacements also seem plausible in practise.

It is actually conceivable that each game has the bounded depth property (even with a constant bounding sequence) and therefore undominated strategies. Another question is whether every game with an undominated strategy also has a regular one.

References

1. Abiteboul, S., Benjelloun, O., Milo, T.: The active XML project: an overview. VLDB J. **17**(5), 1019–1040 (2008)
2. Abiteboul, S., Bonifati, A., Cobéna, G., Manolescu, I., Milo, T.: Dynamic XML documents with distribution and replication. In: Proceedings of the 2003 ACM SIGMOD International Conference on Management of Data, pp. 527–538. ACM (2003)

3. Abiteboul, S., Milo, T., Benjelloun, O.: Regular rewriting of active XML and unambiguity. In: PODS, pp. 295–303. ACM (2005)
4. Aminof, B., Legay, A., Murano, A., Serre, O., Vardi, M.Y.: Pushdown module checking with imperfect information. Inf. Comput. **223**, 1–17 (2013). https://doi.org/10.1016/j.ic.2012.11.005
5. Björklund, H., Schuster, M., Schwentick, T., Kulbatzki, J.: On optimum left-to-right strategies for active context-free games. In: Joint 2013 EDBT/ICDT Conferences, ICDT 2013 Proceedings, pp. 105–116 (2013)
6. Bozzelli, L.: New results on pushdown module checking with imperfect information. In: Proceedings of Second International Symposium on Games, Automata, Logics and Formal Verification, GandALF 2011, Minori, Italy, 15–17 June 2011, pp. 162–177 (2011). https://doi.org/10.4204/EPTCS.54.12
7. Cachat, T.: Symbolic strategy synthesis for games on pushdown graphs. In: Widmayer, P., Eidenbenz, S., Triguero, F., Morales, R., Conejo, R., Hennessy, M. (eds.) ICALP 2002. LNCS, vol. 2380, pp. 704–715. Springer, Heidelberg (2002). https://doi.org/10.1007/3-540-45465-9_60
8. Coester, C.: One-pass strategies for context-free games. Master's thesis, TU Dortmund (2015). https://doi.org/10.17877/DE290R-16466
9. Coester, C., Schwentick, T., Schuster, M.: Streaming rewriting games: winning strategies and complexity. CoRR (2018). http://arxiv.org/abs/1804.10292
10. Holík, L., Meyer, R., Muskalla, S.: Summaries for context-free games. In: 36th IARCS Annual Conference on Foundations of Software Technology and Theoretical Computer Science, FSTTCS 2016, Chennai, India, 13–15 December 2016, pp. 41:1–41:16 (2016). https://doi.org/10.4230/LIPIcs.FSTTCS.2016.41
11. Milo, T., Abiteboul, S., Amann, B., Benjelloun, O., Ngoc, F.D.: Exchanging intensional XML data. ACM Trans. Database Syst. **30**(1), 1–40 (2005)
12. Schuster, M.: Transducer-based rewriting games for active XML. In: 41st International Symposium on Mathematical Foundations of Computer Science, MFCS 2016, Kraków, Poland, 22–26 August 2016, pp. 83:1–83:13 (2016). https://doi.org/10.4230/LIPIcs.MFCS.2016.83
13. Muscholl, A., Schwentick, T., Segoufin, L.: Active context-free games. In: Diekert, V., Habib, M. (eds.) STACS 2004. LNCS, vol. 2996, pp. 452–464. Springer, Heidelberg (2004). https://doi.org/10.1007/978-3-540-24749-4_40
14. Muscholl, A., Schwentick, T., Segoufin, L.: Active context-free games. Theory Comput. Syst. **39**(1), 237–276 (2006)
15. Schuster, M., Schwentick, T.: Games for active XML revisited. In: 18th International Conference on Database Theory, ICDT 2015, pp. 60–75 (2015)
16. Serre, O.: Note on winning positions on pushdown games with [omega]-regular conditions. Inf. Process. Lett. **85**(6), 285–291 (2003). https://doi.org/10.1016/S0020-0190(02)00445-3
17. Walukiewicz, I.: Pushdown processes: games and model-checking. Inf. Comput. **164**(2), 234–263 (2001). https://doi.org/10.1006/inco.2000.2894

Nominal Syntax with Atom Substitutions: Matching, Unification, Rewriting

Jesús Domínguez and Maribel Fernández[(✉)]

Department of Informatics, King's College London, London, UK
{jesus.dominguez,maribel.fernandez}@kcl.ac.uk

Abstract. Unification and matching algorithms are essential components of logic and functional programming languages and theorem provers. Nominal extensions have been developed to deal with syntax involving binding operators: nominal unification takes into account α-equivalence; however, it does not take into account non-capturing substitutions, which are not primitive in nominal syntax. We consider an extension of nominal syntax with non-capturing substitutions and show that matching is decidable and finitary but unification is undecidable. We provide a matching algorithm and characterise problems for which matching is unitary, giving rise to expressive and efficient rewriting systems.

Keywords: Nominal syntax · Non-capturing substitution ·
Rewriting · Unification

1 Introduction

Nominal syntax is a generalisation of first-order syntax that deals with variable binding using atom permutations and freshness constraints (see [17,28]). Nominal syntax uses two kinds of variables: atoms $a, b, \ldots$, which can be abstracted but not substituted ($[a]t$ means that a is abstracted in t), and meta-variables $X, Y, \ldots$, called simply variables, which may be decorated with atom permutations. Unification of nominal terms (i.e., modulo α-equivalence) is *decidable and unitary* [28]. Efficient algorithms exist that solve nominal unification problems in polynomial time [5,7,21]. Nominal matching (a form of unification where only one of the terms can be instantiated) can be solved in linear time [6].

Nominal unification and matching have applications in logic and functional languages [2,8,25,26] and automated reasoning [10,11,13,16,23,27] among others. However, nominal terms do not provide a built-in form of substitution for atoms that would permit direct definitions of systems such as the λ-calculus. Instead, atom substitution has to be defined explicitly, by rewrite rules or equations [14,15], as in the following system, where (explicit) substitutions are sugared to $t\{a \mapsto t'\}$ and $a \# t$ means that a is not free in t.

L. A. Gąsieniec et al. (Eds.): FCT 2019, LNCS 11651, pp. 64–79, 2019.
https://doi.org/10.1007/978-3-030-25027-0_5

$$\begin{array}{lll}
\text{(Beta)} & \mathsf{app}(\lambda[a]X, X') & \to X\{a \mapsto X'\} \\
(\sigma_{\mathsf{var}}) & a\{a \mapsto X\} & \to X \\
(\sigma_{\epsilon}) & a \mathbin{\#} Y \vdash Y\{a \mapsto X\} & \to Y \\
(\sigma_{\mathsf{app}}) & \mathsf{app}(X, X')\{a \mapsto Y\} & \to \mathsf{app}(X\{a \mapsto Y\}, X'\{a \mapsto Y\}) \\
(\sigma_{\mathsf{lam}}) & b \mathbin{\#} Y \vdash (\lambda[b]X)\{a \mapsto Y\} & \to \lambda[b](X\{a \mapsto Y\})
\end{array}$$

An extension of nominal syntax with a primitive capture-avoiding atom substitution, which avoids the need to introduce explicit substitution rules, was presented in [12]; however, its rewriting theory was not developed. Here we show that unification in this extended syntax is undecidable in general but matching remains decidable (albeit no longer unitary) and the rewriting relation can be effectively computed. The undecidability result is obtained by reducing Hilbert's tenth problem to extended nominal unification, inspired by Goldfarb's proof of undecidability of second-order unification [18]. Our main contributions are an algorithm that computes complete sets of solutions for solvable matching problems, and a characterisation of a wide class of problems for which matching is unitary, inducing a well-behaved rewriting relation. This class includes the Beta and Eta reduction rules of the λ-calculus (we give details in Sect. 5). These results open the way for the development of expressive reasoning frameworks based on nominal syntax.

Related Work. Our syntax for extended nominal terms is inspired by [12], where a dependent type system for extended terms is presented. Matching was used in [12] to type-check terms given a set of declarations for function symbols. It was noted that restrictions were needed to ensure unitary matching, however, no matching algorithm was provided. Capture-avoiding atom substitution was previously studied in the context of nominal algebra by Gabbay and Mathijssen [15,16], but its unification theory was not considered.

In [13], a nominal reduction system for the λ-calculus is given, with an explicit atom-substitution operation defined by a set of rewrite rules. The extended nominal syntax proposed here reduces the verbosity of such systems by internalising capture-avoiding substitutions.

Efficient nominal unification algorithms were developed by Calvès and Fernández [4,5] and Levy and Villaret [21]. Both approaches were later unified by Calvès [3]. Kumar and Norrish [19] also studied efficient forms of nominal unification. Cheney [9] proved that a more general version of nominal unification, called equivariant unification, is NP-complete.

We followed Goldfarb's methodology [18] to prove the undecidability of nominal unification extended with atom substitutions. Goldfarb [18] proved that second-order unification is undecidable by reducing Hilbert's tenth problem to a second-order unification problem. An alternative undecidability proof for second-order unification by a direct encoding of the Halting problem is given by Levy and Veanes [20], which could also be adapted to our language.

2 Background

Fix countably infinite, pairwise disjoint sets of **atoms**, $a, b, c, \ldots \in \mathcal{A}$; **variables**, $X, Y, Z, \ldots \in \mathcal{X}$; and **term-formers** $f, g, \ldots \in \mathcal{F}$. A **permutation** π is a bijection on a finite subset of $\mathcal{A}$ called **support** of π, $Support(\pi)$. A **swapping** $(a \; b)$ is a particular case where a maps to b, b maps to a and all other atoms c map to themselves. We follow the **permutative convention** [15, Convention 2.3] for atoms throughout the paper, i.e., atoms a, b, c range permutatively over $\mathcal{A}$ so that they are always distinct unless stated otherwise. **Atom substitutions** ϕ, or just **a-substitutions**, are mappings with finite **domain** from atoms to terms, i.e., the set of atoms such that $\phi(a) \neq a$, written $Dom(\phi)$, is finite. Permutations π, a-substitutions ϕ, and **terms with atom substitutions** s, t, or just **(extended) terms**, are generated by the following grammar.

Definition 1 (Syntax).

$$\pi ::= \mathsf{Id} \mid \pi(a \; b) \quad \phi ::= \mathsf{Id} \mid [a \mapsto s]\phi \quad s, t ::= a \mid \phi\hat{\;}\pi{\cdot}X \mid [a]s \mid fs \mid (s_1, \ldots, s_n)$$

The final Id is usually omitted from permutations and a-substitutions. Write π^{-1} for the **inverse** of π, e.g., if $\pi = (a \; b)(b \; c)$ then $\pi(c) = a$ and $c = \pi^{-1}(a)$. A-substitutions are *simultaneous* bindings, abbreviated as $[a_1 \mapsto s_1; \ldots; a_n \mapsto s_n]$ where atoms a_i are pairwise distinct. Write $\phi^{-a_1, \ldots, a_n}$ for the a-substitution ϕ with domain restricted to $Dom(\phi) \setminus \{a_1, \ldots, a_n\}$. $Img(\phi)$ denotes the set of terms $\{\phi(a) \mid a \in Dom(\phi)\}$. Term constructors as given in Definition 1 are called respectively **atoms**, **moderated variables**, **abstractions**, **function applications** (where $f()$ is denoted as f) and **tuples** ($n \geq 0$). A moderated variable $\phi\hat{\;}\pi{\cdot}X$ comprises a variable X, and **suspended** permutation π and a-substitution ϕ. As in first-order syntax, variables denote unknown parts of the term, but here they are decorated with permutations and atom-substitutions, that will act when the variable is instantiated, as shown below. We abbreviate $\mathsf{Id}\hat{\;}\pi{\cdot}X$ (resp. $\phi\hat{\;}\mathsf{Id}{\cdot}X$) as $\pi{\cdot}X$ (resp. $\phi{\cdot}X$) and $\mathsf{Id}\hat{\;}\mathsf{Id}{\cdot}X$ as X if there is no ambiguity.

Permutations **act** on terms and a-substitutions; $\circ$ denotes composition:

$$\pi{\cdot}a \triangleq \pi(a) \quad \pi{\cdot}[a]t \triangleq [\pi(a)]\pi{\cdot}t \quad \pi{\cdot}ft \triangleq f\pi{\cdot}t \quad \pi{\cdot}(t_1, \ldots, t_n) \triangleq (\pi{\cdot}t_1, \ldots, \pi{\cdot}t_n)$$

$$\pi{\cdot}(\phi\hat{\;}\pi'{\cdot}X) \triangleq (\pi{\cdot}\phi)\hat{\;}(\pi \circ \pi'){\cdot}X \text{ where } \pi{\cdot}\mathsf{Id} \triangleq \mathsf{Id}, \quad \pi{\cdot}([a \mapsto t]\phi) \triangleq [\pi(a) \mapsto \pi{\cdot}t](\pi{\cdot}\phi)$$

Write $V(t)$ for the set of variable symbols appearing in a term t and $A(t)$ for the set of atoms in t; this includes atoms in the domain and image of a-substitutions and atoms in the support of permutations.

A **position** p, q is a string of positive integers denoting a path in the abstract syntax tree of a term. The set of **positions** of a term s, $Pos(s)$, is defined inductively as usual [1] with an additional case for a moderated variable:

$$Pos([a_1 \mapsto t_1; \cdots; a_n \mapsto t_n]\hat{\;}\pi{\cdot}X) \triangleq \{\epsilon\} \cup \bigcup_{i=1}^{n} \{i \cdot p \mid p \in Pos(t_i)\}.$$

An arbitrary ordering (e.g., lexicographic) is chosen when defining the positioning of the terms in the image of suspended a-substitutions. Since we are dealing with simultaneous a-substitutions, the choice of ordering does not matter. The size of a term t, $|t|$, is the cardinality of $\mathcal{P}os(t)$. Call $t|_p$ the **subterm** of t at position p. If $p \in \mathcal{P}os(t)$, then $t[s]_p$ denotes the term obtained from t by replacing its subterm at position p by the term s and $(\cdots (s[t_1]_{p_1}) \cdots)[t_n]_{p_n}$ is abbreviated as $s[t_1 \cdots t_n]_{p_1 \cdots p_n}$.

Example 1. Let $\mathtt{map}, \mathtt{cons}$ and $\mathtt{nil}$ be term-formers; $\mathtt{map}([a]F, \mathtt{cons}(H, \mathtt{nil}))$ is a term and so is t defined as $\mathtt{cons}([a \mapsto H] \cdot F, \mathtt{map}([b]F, \mathtt{nil}))$. $A(t) = \{a, b\}, V(t) = \{F, H\}$. $\mathcal{P}os(t) = \{\epsilon, 1, 11, 111, 12, 121, 1211, 12111, 1212\}$ so that, $t|_{111} = H$ and $t|_{1212} = \mathtt{nil}$, for instance. See [12,13,28] for more examples.

Call $a \# t$ a **freshness constraint**. Let $\Delta, \nabla, \ldots$ range over finite sets of **primitive constraints** of the form $a \# X$; call such sets **freshness contexts**. Call $s \approx_\alpha t$ an **α-equivalence constraint**. Write $\nabla \vdash a \# t$ and $\nabla \vdash s \approx_\alpha t$, called **freshness** and **α-equivalence judgements** respectively, when a derivation exists using the syntax-directed rules from Definition 2 where, for a-substitutions ϕ, ϕ' and permutations π, π', $\mathcal{D}om(\phi) \cup \mathcal{D}om(\phi')$ is abbreviated as $\mathcal{D}omP(\phi, \phi')$ and $Support(\pi) \cup Support(\pi')$ as $SupportP(\pi, \pi')$. We write $a, b \# t$ (resp. $a\#s, t$) instead of $a \# t, b \# t$ (resp. $a \# s, a \# t$), and abbreviate $\varnothing \vdash s \approx_\alpha t$ as $s \approx_\alpha t$.

Definition 2 (Freshness and α-equivalence judgements).

$$\frac{}{\nabla \vdash a \# b}\ (\#\mathbf{ab}) \qquad \frac{}{\nabla \vdash a \# [a]s}\ (\#[\mathbf{a}]) \qquad \frac{\nabla \vdash a \# s}{\nabla \vdash a \# [b]s}\ (\#[\mathbf{b}]) \qquad \frac{\nabla \vdash a \# s}{\nabla \vdash a \# fs}\ (\#\mathbf{f})$$

$$\frac{\bigwedge_{b \in \mathcal{D}om(\phi) \cup \{a\}} (\nabla \vdash a \# \phi(b) \vee (\pi^{-1}(b) \# X \in \nabla))}{\nabla \vdash a \# \phi\hat{}\pi \cdot X}\ (\#\mathbf{X}) \qquad \frac{\nabla \vdash a \# s_1 \cdots \nabla \vdash a \# s_n}{\nabla \vdash a \# (s_1, \ldots, s_n)}\ (\#\mathbf{tupl})$$

$$\frac{}{\nabla \vdash a \approx_\alpha a}\ (\approx_\alpha \mathbf{a}) \qquad \frac{\nabla \vdash s \approx_\alpha t}{\nabla \vdash [a]s \approx_\alpha [a]t}\ (\approx_\alpha[\mathbf{a}]) \qquad \frac{\nabla \vdash (b\ a) \cdot s \approx_\alpha t \quad \nabla \vdash b \# s}{\nabla \vdash [a]s \approx_\alpha [b]t}\ (\approx_\alpha[\mathbf{b}])$$

$$\frac{\nabla \vdash s \approx_\alpha t}{\nabla \vdash fs \approx_\alpha ft}\ (\approx_\alpha \mathbf{f}) \qquad \frac{\nabla \vdash s_1 \approx_\alpha t_1 \cdots \nabla \vdash s_n \approx_\alpha t_n}{\nabla \vdash (s_1, \ldots, s_n) \approx_\alpha (t_1, \ldots, t_n)}\ (\approx_\alpha \mathbf{tupl})$$

$$\frac{\bigwedge_{a \in (\mathcal{D}omP(\phi, \phi') \cup SupportP(\pi, \pi'))} (\nabla \vdash \phi(\pi(a)) \approx_\alpha \phi'(\pi'(a)) \vee (a \# X \in \nabla))}{\nabla \vdash \phi\hat{}\pi \cdot X \approx_\alpha \phi'\hat{}\pi' \cdot X}\ (\approx_\alpha \mathbf{X})$$

The most interesting rules are (#**x**) and ($\approx_\alpha$**x**). The first one specifies that a is fresh in $\phi\hat{}\,\pi\cdot X$ if it is fresh in the image by ϕ of any atom that could occur in an instance of $\pi\cdot X$. The second ensures that the atom actions produce the same effect for any valid instance of X, in other words, any atom that could be affected by the atom actions suspended in X is either affected in the same way on both sides of the equality constraint, or it must be fresh in X. The relation $\approx_\alpha$ is indeed an equivalence relation [12].

Example 2. We can derive $a\,\#\,[b \mapsto Y]\hat{}\,(a\ c)\cdot X$ from $\nabla_1 = \{a\,\#\,Y, c\,\#\,X\}$ or from $\nabla_2 = \{b\,\#\,X, c\,\#\,X\}$ using rule (#**x**), and $[c \mapsto (a\ b)\cdot Y]\cdot X \approx_\alpha [b \mapsto Y]\cdot(b\ c)\cdot X$ from $\nabla_1 = \{b\,\#\,X, a\,\#\,Y, b\,\#\,Y\}$ or $\nabla_2 = \{b\,\#\,X, c\,\#\,X\}$ using rule ($\approx_\alpha$**x**). In contrast, using standard (non-extended) nominal syntax, for each derivable constraint there exists a unique least freshness context entailing it [28].

The action of an a-substitution ϕ on a term t relies on a freshness context ∇ and therefore is defined over **terms-in-context**, written $\nabla \vdash t$, or simply $\vdash t$ if $\nabla = \varnothing$. Below we abbreviate $(\nabla \vdash t)\phi$ as $\nabla \vdash t\phi$.

Definition 3 (A-substitution action).

$$\nabla \vdash a\phi \triangleq \nabla \vdash \phi(a) \quad \nabla \vdash (ft)\phi \triangleq \nabla \vdash ft\phi$$

$$\nabla \vdash (t_1, \ldots, t_n)\phi \triangleq \nabla \vdash (t_1\phi, \ldots, t_n\phi)$$

$$\nabla \vdash (\phi'\hat{}\,\pi\cdot X)\phi \triangleq \nabla \vdash (\phi' \bullet \phi)\hat{}\,\pi\cdot X \quad where \ \bullet \ denotes \ composition$$

$$\nabla \vdash ([a]t)\phi \triangleq \nabla \vdash [b]((a\ b)\cdot t)\phi^{-b} \quad where \ \nabla \vdash b\,\#\,t, \mathcal{I}mg(\phi)$$

A-substitutions work uniformly on α-equivalence classes of terms, that is, the choice of b in Definition 3 is irrelevant [12]. Capture-avoidance is guaranteed by selecting an α-equivalent representative of $\nabla \vdash [a]t$, i.e., $\nabla \vdash [b](a\ b)\cdot t$, with fresh b. There exists always some $b \in (\mathcal{A} \setminus (A(t) \cup A(\mathcal{I}mg(\phi))))$ such that $\nabla \vdash b\,\#\,t, \mathcal{I}mg(\phi)$, assuming primitive constraints $b\,\#\,X$ in ∇ for each X in $(V(t) \cup V(\mathcal{I}mg(\phi)))$, since variables have finite support [24]. We assume ∇ is large enough (in practice, it can be augmented whenever required). This approach is also taken in [8,12,13] and tacitly assumed in the rest of the paper.

Variable substitutions $\sigma, \theta, \ldots$, or just **v-substitutions**, are mappings from variables to terms, with finite **domain** $\mathcal{D}om(\sigma)$. They are generated by the grammar: $\sigma, \theta ::= \mathsf{Id} \mid [X \mapsto s]\sigma$ where Id is commonly omitted, and interpreted as *simultaneous* bindings, abbreviated $[X_1 \mapsto s_1; \ldots; X_n \mapsto s_n]$ where variables X_i are pairwise distinct. The application of a v-substitution θ to a moderated variable $\phi\hat{}\,\pi\cdot X$ induces the action of ϕ on the term $\pi\cdot\theta(X)$. The action of v-substitutions, σ, on terms, t, written $t\sigma$, is also parameterised by freshness contexts but left implicit in Definition 4. Given v-substitution σ and freshness contexts ∇, Δ, we write $\Delta \vdash \nabla\sigma$ to denote $\Delta \vdash a\,\#\,\sigma(X)$ for each $a\,\#\,X \in \nabla$.

Definition 4 (V-substitution action).

$$a\sigma \triangleq a \quad ([a]t)\sigma \triangleq [a]t\sigma \quad (ft)\sigma \triangleq ft\sigma \quad (t_1, \ldots, t_n)\sigma \triangleq (t_1\sigma, \ldots, t_n\sigma)$$

$$(\phi\hat{}\,\pi\cdot X)\sigma \triangleq (\pi\cdot\sigma(X))(\phi\sigma) \quad where \ \mathsf{Id}\sigma \triangleq \mathsf{Id} \ and \ ([a \mapsto s]\phi)\sigma \triangleq [a \mapsto s\sigma](\phi\sigma)$$

Permutations and a-substitutions commute: $\nabla \vdash \pi \cdot (s\phi) \approx_\alpha (\pi \cdot s)(\pi \cdot \phi)$ and $\nabla \vdash (\pi \cdot s)\phi \approx_\alpha \pi \cdot (s(\pi^{-1} \cdot \phi))$. Also, v-substitutions commute with permutations, $\nabla \vdash \pi \cdot (s\sigma) \approx_\alpha (\pi \cdot s)\sigma$, and a-substitutions, $\nabla \vdash (s\sigma)\phi\sigma \approx_\alpha (s\phi)\sigma$.

3   Unification, Matching and Rewriting

Definition 5. *Let C range over freshness and α-equality constraints. A **unification problem** $\mathbf{P}$ is a finite set of such constraints, where α-equivalence constraints are written as **unification constraints** $s\ _?\approx_? t$. A **solution** to $\mathbf{P}$ is a pair $(\mathbf{F}, \sigma)$ of a non-empty collection $\mathbf{F}$ of freshness contexts and a v-substitution σ such that $\Delta \vdash C\sigma$ for each $\Delta \in \mathbf{F}$ and $C \in \mathbf{P}$.*

*Write $\mathcal{U}(\mathbf{P})$ for the **set of all solutions** of $\mathbf{P}$. $(\mathbf{F}, \sigma) \in \mathcal{U}(\mathbf{P})$ is **more general** than $(\mathbf{F}', \sigma') \in \mathcal{U}(\mathbf{P})$, written $(\mathbf{F}, \sigma) \leq (\mathbf{F}', \sigma')$, if for each $\Delta' \in \mathbf{F}'$ there exists $\Delta \in \mathbf{F}$ and a v-substitution θ such that $\Delta' \vdash X(\sigma \bullet \theta) \approx_\alpha X\sigma'$ for all X and $\Delta' \vdash \Delta\theta$. If there is no $(\mathbf{F}', \sigma') \in \mathcal{U}(\mathbf{P})$ such that $(\mathbf{F}', \sigma') < (\mathbf{F}, \sigma)$ then $(\mathbf{F}, \sigma)$ is a **principal** or **most general solution**.*

The unification problem $\{[a \mapsto c] \cdot X\ _?\approx_? c\}$ has principal solutions $(\{\varnothing\}, [X \mapsto a])$ and $(\{\varnothing\}, [X \mapsto c])$. In fact, the unification theory of extended nominal terms is *infinitary*. We give an example after defining complete sets of solutions. Note that solutions of unification problems use collections of contexts, since there may be several independent contexts that solve a constraint, as shown in Example 2.

Definition 6. *Call W a **complete set of solutions** for $\mathbf{P}$ if $W \subseteq \mathcal{U}(\mathbf{P})$; $\forall (\mathbf{F}, \theta) \in W, \mathcal{D}om(\theta) \subseteq V(\mathbf{P})$; and $\forall (\mathbf{F}, \sigma) \in \mathcal{U}(\mathbf{P}), \exists (\mathbf{F}', \theta) \in W : (\mathbf{F}', \theta) \leq (\mathbf{F}, \sigma)$. W is a **complete set of most general solutions** if each element is principal.*

The unification problem $\{[c \mapsto f(a, b)] \cdot X\ _?\approx_? f(a, [c \mapsto b] \cdot X)\}$ has an infinite number of principal solutions of the form $(\{\varnothing\}, \sigma_n)$ where $\sigma_n = [X \mapsto f(a, f(a, \ldots, f(a, c) \cdots))]$ and n is the number of occurrences of function symbol f and atom a in $\sigma_n(X)$. In particular, $\sigma_0 = [X \mapsto c]$, $\sigma_1 = [X \mapsto f(a, c)]$ and $\sigma_2 = [X \mapsto f(a, f(a, c))]$.

A **matching constraint** is a unification constraint $s\ _?\approx_? t$ where only variables in s may be instantiated; occurrences of variables in t are seen as constants. We sometimes write $s\ _?\approx t$ to emphasise that we are dealing with matching, and refer to s as the **pattern** and t as the **matched term**. Matching plays an important role in rewriting: given a set of rewriting rules, the nominal rewriting relation is generated by solving pattern-matching problems as defined below.

Definition 7. *A **matching problem** $\mathbf{P}$ is a set of matching constraints $s_i\ _?\approx t_i$ such that $(\bigcup_i V(s_i)) \cap (\bigcup_i V(t_i)) = \varnothing$. We denote $\bigcup_i V(s_i)$ by $V_{LHS}(\mathbf{P})$ and $\bigcup_i V(t_i)$ by $V_{RHS}(\mathbf{P})$. A **pattern matching problem** consists of a pair of terms-in-context, written $(\nabla \vdash l)\ _?\approx (\Delta \vdash t)$ such that $V(\nabla \vdash l) \cap V(\Delta \vdash t) = \varnothing$.*

A solution to a pattern matching problem $(\nabla \vdash l)\ _?\approx (\Delta \vdash s)$ is a v-substitution σ, such that there exists $\mathbf{F}$ such that $(\mathbf{F}, \sigma)$ is a solution to $\{l\ _?\approx s\} \cup \nabla$, and $\Delta \vdash \nabla_i$ for some $\nabla_i \in \mathbf{F}$.

Definition 8. *An **extended nominal rewrite rule**, or just **rewrite rule**, is a tuple, written $R = (\nabla \vdash l \to r)$, where ∇ is a freshness context, and l, r are extended nominal terms such that $V(\nabla, r) \subseteq V(l)$. We write $l \to r$ for $\varnothing \vdash l \to r$.*

Example 3. – $\mathtt{par(out(a,b), in(a,[c]P))} \to [\mathtt{c} \mapsto \mathtt{b}]\cdot \mathtt{P}$ is a rewrite rule , representing communication in the π calculus.
– $\mathtt{a \# X \vdash X \to lam([a]app(X,a))}$ is the η-expansion rule of the λ-calculus. The β and η reduction rules are:

$$(\beta)\ \mathtt{app(lam([a]X), Y)} \to [\mathtt{a} \mapsto \mathtt{Y}]\cdot \mathtt{X}$$
$$(\eta)\ \mathtt{a \# X \vdash lam([a]app(X,a))} \to \mathtt{X}$$

Using standard nominal rules, four additional rules are needed to define explicit substitution (see the Introduction and [13]).
– The higher-order function *map* (see Example 1) is defined by rules:

$$\mathtt{map([a]F, nil)} \to \mathtt{nil}$$
$$\mathtt{map([a]F, cons(H,T))} \to \mathtt{cons([a} \mapsto \mathtt{H]\cdot F, map([a]F, T))}$$

To generate the rewrite relation, terms in rewrite rules are considered up to renaming of variables and atoms (metalevel equivariance [13,24]), denoted t^π.

Definition 9. *A rewrite system $\mathcal{R}$ induces a **rewrite step** $\Delta \vdash s \xrightarrow{R} t$ if there exists $(\nabla \vdash l \to r) \in \mathcal{R}$, $p \in \mathcal{P}os(s)$ and a permutation π such that the pattern-matching problem $(\nabla^\pi \vdash l^\pi)\ _? \approx (\Delta \vdash s|_p)$ has solution θ, and $\Delta \vdash s[r^\pi \theta]_p \approx_\alpha t$:*

$$\frac{\Delta \vdash \{\nabla^\pi \theta, \quad l^\pi \theta \approx_\alpha s|_p, \quad s[r^\pi \theta]_p \approx_\alpha t\}}{\Delta \vdash s \xrightarrow{R} t}\ (\to\mathbf{Rew})$$

*The **(multi-step) rewrite relation** $\Delta \vdash_{\mathcal{R}} s \to t$ is the reflexive, transitive closure of the one-step rewrite relation.*

Example 4. The term-in-context $\vdash \mathtt{app(lam([a]lam([b]app(a,b))), b)}$ rewrites to a normal form in one step with the rule (β) (see Example 3), at position ϵ with permutation Id and v-substitution $\theta = [\mathtt{X} \mapsto \mathtt{lam([b]app(a,b))}; \mathtt{Y} \mapsto \mathtt{b}]$ as follows,

$$\vdash \mathtt{app(lam([a]lam([b]app(a,b))), b)} \to_{\langle(\beta),\epsilon,\mathsf{Id},\theta\rangle} \mathtt{lam([c]app(b,c))}$$

Capture of the unabstracted atom $\mathtt{b}$ has been avoided by the internal machinery of the extended nominal framework implementing a-substitution. By relegating the semantics of capture-avoiding substitution to the metal-level, where they are managed by our formalism, we have reduced the set of rewrite rules necessary to provide a nominal representation of the rewrite system at hand. The same reduction requires several steps using explicit substitution rules [13].

4 Solving Matching Problems

A sound and complete matching algorithm can be built by converting the set of derivation rules given in Definition 2 into a simplification system. This algorithm can then be used to implement rewriting, and also to check *closedness* of terms and rewrite rules (see [13]). Principal solutions are not unique in general but matching is unitary for a restricted but practically useful class of problems (cf. Theorem 2).

In a matching problem $\mathbf{P} = s_1\, ?{\approx}\, t,\ldots, s_n\, ?{\approx}\, t_n$, variables on the right-hand side of matching constraints are treated as constants. Hence, without loss of generality, we assume $a \# X$ for any $X \in V_{RHS}(\mathbf{P})$ and $a \in \mathcal{A}$.

Although, initially, the sets of variables in left- and right-hand sides of matching constraints are disjoint, this property is not preserved during the process of solving matching problems (due to variable instantiations). Thus, given a matching problem $\mathbf{P}_0$ to solve, we start by computing the set $V_{RHS}(\mathbf{P}_0)$ of variables that should not be instantiated.

The following auxiliary functions Cap and Ψ are used in the matching algorithm to handle constraints where the pattern is a moderated variable: To solve a constraint of the form $\phi\hat{\ }\pi{\cdot}X\ ?{\approx}_?\ t$ where $t \neq \phi'\hat{\ }\pi'{\cdot}X$, one checks if some subterm $t|_p$ of t is contained in the image of ϕ, that is, $[\pi(a) \mapsto t|_p] \in \phi$. In order to find such position p and subterm $t|_p$, the matching algorithm generates *cap constraints* of the form $(t[a_1 \cdots a_n]_{p_1 \cdots p_n})\phi\ ?{\approx}_?\ t$, where $p_i \in \mathcal{P}os(t)$ and $a_i \in \mathcal{D}om(\phi)$, using the function Cap defined below.

Definition 10 (*Cap* terms). *Let t be a term, $\mathbb{A}$ a finite set of atoms.*
$$Cap(t, \mathbb{A}) = \{t[a_1 \cdots a_n]_{p_1 \cdots p_n} \mid n \in Nat, a_i \in \mathbb{A}, p_i \in \mathcal{P}os(t), 1 \leq i \leq n\}.$$

Thus, $Cap(t, \mathbb{A})$ returns the set of all the terms obtained by replacing subterms of t with atoms from $\mathbb{A}$. Note that $Cap(t, \mathbb{A})$ also includes the term t.

Example 5. $Cap(cons([a \mapsto H]{\cdot}F, T), \{b, c\}) = \{b, c, cons\ b, cons\ c, cons(b, b),$
$cons(b, c), cons(c, b), cons(c, c), cons(b, T), cons(c, T), cons([a \mapsto b]{\cdot}F, b),$
$cons([a \mapsto c]{\cdot}F, b), cons([a \mapsto b]{\cdot}F, c), cons([a \mapsto c]{\cdot}F, c), cons([a \mapsto b]{\cdot}F, T),$
$cons([a \mapsto c]{\cdot}F, T), cons([a \mapsto H]{\cdot}F, b), ([a \mapsto H]{\cdot}F, c), cons([a \mapsto H]{\cdot}F, T)\}.$

The function Ψ is used to handle constraints of the form $\phi\hat{\ }\pi{\cdot}X\ ?{\approx}_?\ \phi'\hat{\ }\pi'{\cdot}X$ or $a \# \phi\hat{\ }\pi{\cdot}X$, i.e., Ψ deals with the premises of rules $(\approx_\alpha \mathbf{X})$ and $(\# \mathbf{X})$ (see Definition 2).

Definition 11 (Function Ψ). *Let s and t be either two moderated variables $\phi\hat{\ }\pi{\cdot}X$ and $\phi'\hat{\ }\pi'{\cdot}X$, or an atom a and a moderated variable $\phi\hat{\ }\pi{\cdot}X$. Let $\mathbf{P}$ be a matching problem, $\mathbb{A}$ a finite set of atoms and b an atom in $\mathbb{A}$. Then, $\Psi(s, t)^{\mathbb{A}} = \Psi'(s, t, \varnothing)^{\mathbb{A}}$ where Ψ' computes a set of problems (i.e., a collection of sets of constraints) as follows: $\Psi'(s, t, \mathbf{P})^{\mathbb{A}} \triangleq$*

$$
\begin{cases}
\bullet\ \{\mathbf{P}\} & \text{if } \mathbb{A} = \varnothing \\
\bullet\ \Psi'(s, t, \mathbf{P} \cup \{s \# \phi(b)\})^{\mathbb{A} \setminus \{b\}} \cup \Psi'(s, t, \mathbf{P} \cup \{\pi^{-1}(b) \# X\})^{\mathbb{A} \setminus \{b\}} & \text{if } s = a, t = \phi\hat{\ }\pi{\cdot}X \\
\bullet\ \Psi'(s, t, \mathbf{P} \cup \{\phi(\pi(b))\ ?{\approx}_?\ \phi'(\pi'(b))\})^{\mathbb{A} \setminus \{b\}}\ \cup \Psi'(s, t, \mathbf{P} \cup \{b \# X\})^{\mathbb{A} \setminus \{b\}} \\
\hfill \text{if } s = \phi\hat{\ }\pi{\cdot}X, t = \phi'\hat{\ }\pi'{\cdot}X
\end{cases}
$$

Ψ' deals with one atom from the given finite set $\mathbb{A}$ in each recursive call (thus ensuring termination); the order in which elements of $\mathbb{A}$ are considered is irrelevant since $\Psi'(s,t,\mathbf{P})^{\mathbb{A}}$ is a collection of sets. Hence Ψ' is indeed a function.

Freshness constraints of form $a \# a$ are **inconsistent**. A matching constraint $s\,_?\approx_?\,t$ is **clashing** when s,t have different term constructors at the root except if s is a moderated variable $\phi\hat{}\pi{\cdot}X$ and $X \notin V_{RHS}(\mathbf{P})$. For example, if $V_{RHS}(\mathbf{P}) = \{Y\}$ then $a\,_?\approx_?\,(a\ b){\cdot}Y$, $f\,a\,_?\approx_?\,g\,a$ and $[a \mapsto b]{\cdot}Y\,_?\approx_?\,[b]a$ are clashing but $[a \mapsto b]{\cdot}X\,_?\approx_?\,f(c,b)$ and $[a \mapsto b]{\cdot}Y\,_?\approx_?\,[b \mapsto a]{\cdot}Y$ are not. Clashing and inconsistent constraints are not derivable; failure rules ($\bot$) will be specified to deal with them.

Definition 12 (Matching steps). *Let $\mathbf{P}_0$ be a matching problem and $\mathbb{X} = V_{RHS}(\mathbf{P}_0)$. $\mathcal{P}, \mathcal{Q}$ denote sets of pairs $(\mathbf{P}, \theta)$, where $\mathbf{P}$ is a unification problem and θ a v-substitution. Write $\mathcal{P} \xRightarrow{\ \ \mathbb{X}\ \ }_{?\approx} \mathcal{Q}$ (resp. $\mathcal{P} \Longrightarrow_{\#} \mathcal{Q}$), if $\mathcal{Q}$ is obtained from $\mathcal{P}$ by application of one matching (resp. freshness) reduction rule below. As usual, $\Longrightarrow^{*}_{?\approx}$ (resp. $\Longrightarrow^{*}_{\#}$) denotes reflexive transitive closure; arrow subindices are omitted if there is no ambiguity.*

$(_?\approx\bot)^{1}$ $\qquad (\{s\,_?\approx_?\,t\} \cup \mathbf{P}, \theta) \xRightarrow{\ \mathbb{X}\ }_{?\approx} \varnothing \quad$ *if clashing*

$(_?\approx\equiv)$ $\qquad (\{t\,_?\approx_?\,t\} \cup \mathbf{P}, \theta) \xRightarrow{\ \mathbb{X}\ }_{?\approx} (\mathbf{P}, \theta)$

$(_?\approx\mathbf{f})$ $\qquad (\{fs\,_?\approx_?\,ft\} \cup \mathbf{P}, \theta) \xRightarrow{\ \mathbb{X}\ }_{?\approx} (\{s\,_?\approx_?\,t\} \cup \mathbf{P}, \theta)$

$(_?\approx[\mathbf{a}])$ $\qquad (\{[a]s\,_?\approx_?\,[a]t\} \cup \mathbf{P}, \theta) \xRightarrow{\ \mathbb{X}\ }_{?\approx} (\{s\,_?\approx_?\,t\} \cup \mathbf{P}, \theta)$

$(_?\approx[\mathbf{b}])$ $\qquad (\{[a]s\,_?\approx_?\,[b]t\} \cup \mathbf{P}, \theta) \xRightarrow{\ \mathbb{X}\ }_{?\approx} (\{(b\ a){\cdot}s\,_?\approx_?\,t, b \# s\} \cup \mathbf{P}, \theta)$

$(_?\approx\mathbf{tupl})$ $\ (\{(s_1, \ldots, s_n)\,_?\approx_?\,(t_1, \ldots, t_n)\} \cup \mathbf{P}, \theta) \xRightarrow{\ \mathbb{X}\ }_{?\approx} (\{s_1\,_?\approx_?\,t_1, \ldots, s_n\,_?\approx_?\,t_n\} \cup \mathbf{P}, \theta)$

$(_?\approx\mathbf{X})^{1}$ $\ (\{\phi\hat{}\pi{\cdot}X\,_?\approx_?\,\phi'\hat{}\pi'{\cdot}X\} \cup \mathbf{P}, \theta) \xRightarrow{\ \mathbb{X}\ }_{?\approx} \bigcup_{\mathbf{P}' \in \Psi(\phi\hat{}\pi{\cdot}X, \phi'\hat{}\pi'{\cdot}X)^{\mathbb{A}}} \{(\mathbf{P}' \cup \mathbf{P}, \theta)\}$

$\qquad \qquad \qquad$ *where* $\mathbb{A} = (\mathcal{S}upport P(\pi, \pi') \cup \mathcal{D}om P(\phi, \phi'))$

$(_?\approx\mathbf{Inst})^{1}$ $\qquad (\{\phi\hat{}\pi{\cdot}X\,_?\approx_?\,t\} \cup \mathbf{P}, \theta) \xRightarrow{\ \mathbb{X}\ }_{?\approx} \bigcup_{s \in \mathcal{C}ap(t, \mathcal{D}om(\phi))} \{(\{s(\phi\theta')\,_?\approx_?\,t\} \cup \mathbf{P}\theta', \theta \bullet \theta')\}$

$\qquad \qquad \qquad$ *if* $t \neq \phi'\hat{}\pi'{\cdot}Y$ $(Y \in \mathcal{X})$, $(X \notin \mathbb{X})$, $\theta' = [X \mapsto \pi^{-1}{\cdot}s]$

$(_?\approx\mathbf{XY})^{1}$ $\ (\{\phi\hat{}\pi{\cdot}X\,_?\approx_?\,\phi'\hat{}\pi'{\cdot}Y\} \cup \mathbf{P}, \theta) \xRightarrow{\ \mathbb{X}\ }_{?\approx} \bigcup_{s \in (\mathcal{C}ap(\phi'\hat{}\pi'{\cdot}Y, \mathcal{D}om(\phi)) \cup \{\pi'{\cdot}Y\})} \{(\{s(\phi\theta')\,_?\approx_?\,\phi'\hat{}\pi'{\cdot}Y\} \cup \mathbf{P}\theta', \theta \bullet \theta')\}$

$\qquad \qquad \qquad \qquad \qquad$ *if* $(X \notin \mathbb{X})$ *and* $\theta' = [X \mapsto \pi^{-1}{\cdot}s]$

$(\#\bot)$ $\qquad \qquad \{a \# a\} \cup \mathbf{P} \Longrightarrow_{\#} \bot$

$(\#\mathbf{ab})$ $\qquad \qquad \{a \# b\} \cup \mathbf{P} \Longrightarrow_{\#} \mathbf{P}$

$(\#[\mathbf{a}])$ $\qquad \quad \{a \# [a]s\} \cup \mathbf{P} \Longrightarrow_{\#} \mathbf{P}$

$(\#[\mathbf{b}])$ $\qquad \quad \{a \# [b]s\} \cup \mathbf{P} \Longrightarrow_{\#} \{a \# s\} \cup \mathbf{P}$

$(\#\mathbf{f})$ $\qquad \quad \{a \# fs\} \cup \mathbf{P} \Longrightarrow_{\#} \{a \# s\} \cup \mathbf{P}$

$(\#\mathbf{tupl})\ \{a \# (s_1, \ldots, s_n)\} \cup \mathbf{P} \Longrightarrow_{\#} \{a \# s_1, \ldots, a \# s_n\} \cup \mathbf{P}$

$(\#\mathbf{X})^{1}$ $\qquad \{a \# \phi\hat{}\pi{\cdot}X\} \cup \mathbf{P} \Longrightarrow_{\#} \bigcup_{\mathbf{P}' \in \Psi(a, \phi\hat{}\pi{\cdot}X)^{\mathbb{A}}} \{\mathbf{P}' \cup \mathbf{P}\} \quad (where\ \mathbb{A} = \mathcal{D}om(\phi) \cup \{a\})$

$\qquad \qquad \qquad \qquad \qquad \qquad \qquad \qquad$ *if* $\phi \neq \mathsf{Id} \wedge \pi \neq \mathsf{Id}$

Rule $(_?\approx\equiv)$ has priority; it is an optimisation to reduce trivial matching constraints in one step, subsuming rule $(\approx_\alpha\mathbf{a})$ (Definition 2). The right-hand side of rule $(_?\approx\bot)$ is the empty set since this pair cannot produce solutions (but other pairs in the problem could, so we do not use $\bot$). Rules $(_?\approx\mathbf{Inst})$ and $(_?\approx\mathbf{XY})$

---

[1] In this rule, the right-hand side is a set; we assume a flattening step is performed after each application of the rule (to avoid nested sets).

are **instantiating rules**. Note that the matching steps in Definition 12 provide an algorithmic presentation of Definition 2, where instantiating rules have been added and the symbol $\approx_\alpha$ has been replaced by $_?\approx_?$ to represent the constraints to be solved.

Termination of the simplification process follows from the fact that the instantiating rules decrease the number of variables in the problem. For any other rule, an interpretation based on the multiset of sizes of the constraints in the problem can be shown to be strictly decreasing using the multiset extension of the standard ordering on natural numbers, $\leq_{mul}$. Confluence under the imposed strategy then follows by Newman's Lemma, since there are only trivial overlaps. Hence normal forms are unique.

Remark 1 (Matching algorithm). The algorithm has *two phases.*

Input: Assume $\mathbf{P}_0$ is the given matching problem, where $\mathbb{X} = V_{RHS}(\mathbf{P}_0)$.

Phase 1 ($\Longrightarrow_{?\approx}$-Normalisation): $\{(\mathbf{P}_0, \mathsf{Id})\} \overset{\mathbb{X}}{\underset{?\approx}{\Longrightarrow}}^* \langle \mathbf{P}_0 \rangle_{\mathtt{nf}_{?\approx}}$ where $\langle \mathbf{P}_0 \rangle_{\mathtt{nf}_{?\approx}}$ is the normal form of $\{(\mathbf{P}_0, \mathsf{Id})\}$ by application of $\Longrightarrow_{?\approx}$.

Phase 2 ($\Longrightarrow_{\#}$-Normalisation): $\forall (\mathbf{P}_i, \theta_i) \in \langle \mathbf{P}_0 \rangle_{\mathtt{nf}_{?\approx}}$, compute $\{\mathbf{P}_i\} \Longrightarrow_{\#}^* \langle \mathbf{P}_i \rangle_{\mathtt{nf}_{\#}}$ where $\langle \mathbf{P}_i \rangle_{\mathtt{nf}_{\#}}$ is the normal form of the set of freshness constraints $\mathbf{P}_i$ by application of $\Longrightarrow_{\#}$.

Output: $\langle \mathbf{P}_0 \rangle_{out} = \{ (\langle \mathbf{P}_i \rangle_{\mathtt{nf}_{\#}} \setminus \bot, \theta_i) \mid (\mathbf{P}_i, \theta_i) \in \langle \mathbf{P}_0 \rangle_{\mathtt{nf}_{?\approx}}, \langle \mathbf{P}_i \rangle_{\mathtt{nf}_{\#}} \neq \bot \}$.

Informally, **Phase 1** reduces the matching problem until no matching constraints are left, resolving into a set of pairs $(\{C_{ij}\}, \theta_i)(i, j \in Nat)$ where each C_{ij} is a (possibly empty) set of freshness constraints and θ_i a v-substitution. Then, **Phase 2** reduces each C_{ij} into freshness contexts C'_{ij}, discarding along the way any set C_{ij} containing inconsistent freshness constraints. Finally, the remaining pairs (C'_{ij}, θ_i) in the set are solutions to the initial matching problem $\mathbf{P}_0$. If no pairs are left, i.e., all sets of freshness constraints have been discarded, then the matching problem is unsolvable.

Example 6. The matching problem $\mathbf{P} = (\{[a \mapsto Y] \cdot X \ _?\approx [a \mapsto b] \cdot Z\})$ has principal solutions $(\{a \# Z\}, [X \mapsto Z]), (\varnothing, [X \mapsto Z; Y \mapsto b]), (\varnothing, [X \mapsto [a \mapsto b] \cdot Z]), (\{a \# Z\}, [X \mapsto a; Y \mapsto Z]), (\varnothing, [X \mapsto a; Y \mapsto [a \mapsto b] \cdot Z])$ computed by the algorithm as follows. Below, the affected parts of each reduction are highlighted and outer brackets in singleton collections of freshness contexts are omitted for readability. Here $V_{RHS}(\mathbf{P}) = \{Z\}$.

$$\{ (\{[a \mapsto Y] \cdot X \,_?\approx_? [a \mapsto b] \cdot Z\}, \mathsf{Id}) \}$$

$$\overset{X \notin \{Z\}}{\Longrightarrow}_{(_?\approx \mathbf{XY})} \{(\{Y \,_?\approx_? [a \mapsto b] \cdot Z\}, \mathsf{Id} \bullet [X \mapsto a])\}$$
$$\cup\{(\{[a \mapsto Y] \cdot Z \,_?\approx_? [a \mapsto b] \cdot Z\}, \mathsf{Id} \bullet [X \mapsto Z])\}$$
$$\cup\{ (\{[a \mapsto b] \cdot Z \,_?\approx_? [a \mapsto b] \cdot Z\}, \mathsf{Id} \bullet [X \mapsto [a \mapsto b] \cdot Z]) \}$$
$$where \; Cap([a \mapsto b] \cdot Z, \{a\}) = \{a, Z, [a \mapsto b] \cdot Z\}$$

$$\Longrightarrow_{(_?\approx \equiv)} \{ (\{Y \,_?\approx_? [a \mapsto b] \cdot Z\}, [X \mapsto a]) \} \cup$$
$$\{(\{[a \mapsto Y] \cdot Z \,_?\approx_? [a \mapsto b] \cdot Z\}, [X \mapsto Z])\} \cup \{(\varnothing, [X \mapsto [a \mapsto b] \cdot Z])\}$$

$$\overset{Y \notin \{Z\}}{\Longrightarrow}_{(_?\approx \mathbf{XY})} \{ (\{[a \mapsto b] \cdot Z \,_?\approx_? [a \mapsto b] \cdot Z\}, [X \mapsto a] \bullet [Y \mapsto [a \mapsto b] \cdot Z]) \}$$
$$\cup\{(\{Z \,_?\approx_? [a \mapsto b] \cdot Z\}, [X \mapsto a] \bullet [Y \mapsto Z])\} \cup$$
$$\{(\{[a \mapsto Y] \cdot Z \,_?\approx_? [a \mapsto b] \cdot Z\}, [X \mapsto Z])\} \cup \{(\varnothing, [X \mapsto [a \mapsto b] \cdot Z])\}$$
$$where \; Cap([a \mapsto b] \cdot Z, \varnothing) = \{[a \mapsto b] \cdot Z\}$$

$$\Longrightarrow_{(_?\approx \equiv)} \{(\varnothing, [X \mapsto a] \bullet [Y \mapsto [a \mapsto b] \cdot Z])\}$$
$$\cup\{ (\{Z \,_?\approx_? [a \mapsto b] \cdot Z\}, [X \mapsto a] \bullet [Y \mapsto Z]) \} \cup$$
$$\{(\{[a \mapsto Y] \cdot Z \,_?\approx_? [a \mapsto b] \cdot Z\}, [X \mapsto Z])\} \cup \{(\varnothing, [X \mapsto [a \mapsto b] \cdot Z])\}$$

$$\Longrightarrow_{(_?\approx \mathbf{X})} \{(\varnothing, [X \mapsto a] \bullet [Y \mapsto [a \mapsto b] \cdot Z])\}$$
$$\cup\{(\{a \,\#\, Z\}, [X \mapsto a] \bullet [Y \mapsto Z])\}$$
$$\cup\{ (\{a \,_?\approx_? b\}, [X \mapsto a] \bullet [Y \mapsto Z]) \}$$
$$\cup\{ (\{[a \mapsto Y] \cdot Z \,_?\approx_? [a \mapsto b] \cdot Z\}, [X \mapsto Z]) \}$$
$$\cup\{(\varnothing, [X \mapsto [a \mapsto b] \cdot Z])\}$$
$$where \; \Psi(Z, [a \mapsto b] \cdot Z)^{\{a\}} = \{\{a \,_?\approx_? b\}, \{a \,\#\, Z\}\}$$

$$\Longrightarrow_{(_?\approx \perp)} \Longrightarrow_{(_?\approx \mathbf{X})} \{(\varnothing, [X \mapsto a] \bullet [Y \mapsto [a \mapsto b] \cdot Z])\}$$
$$\cup\{(\{a \,\#\, Z\}, [X \mapsto a] \bullet [Y \mapsto Z])\} \cup \{(\{a \,\#\, Z\}, [X \mapsto Z])\}$$
$$\cup\{ (\{Y \,_?\approx_? b\}, [X \mapsto Z]) \} \cup \{(\varnothing, [X \mapsto [a \mapsto b] \cdot Z])\}$$
$$where \; \Psi([a \mapsto Y] \cdot Z, [a \mapsto b] \cdot Z)^{\{a\}} = \{\{a \,\#\, Z\}, \{Y \,_?\approx_? b\}\}$$

$$\overset{Y \notin \{Z\}}{\Longrightarrow}_{(_?\approx \mathbf{Inst})} \Longrightarrow_{(_?\approx \equiv)} \{(\varnothing, [X \mapsto a] \bullet [Y \mapsto [a \mapsto b] \cdot Z])\}$$
$$\cup\{(\{a \,\#\, Z\}, [X \mapsto a] \bullet [Y \mapsto Z])\} \cup \{(\{a \,\#\, Z\}, [X \mapsto Z])\}$$
$$\cup\{(\varnothing, [X \mapsto Z] \bullet [Y \mapsto b])\} \cup \{(\varnothing, [X \mapsto [a \mapsto b] \cdot Z])\}.$$

Phase 2 is trivial and thus omitted.

As a consequence of the termination and confluence properties, the relation $\Longrightarrow$ defines a function from matching problems to their unique normal form. Write $\langle \mathbf{P} \rangle_{out}$ for the normal form of $\{(\mathbf{P}, \mathsf{Id})\}$. $\langle \mathbf{P} \rangle_{out}$ may contain solutions $(\mathbf{F}, \sigma), (\mathbf{F}', \sigma')$ with α-equivalent substitutions but different collections of freshness contexts. For instance, $(\varnothing, [X \mapsto a; Y \mapsto [a \mapsto b] \cdot Z])$ and $(\{a \,\#\, Z\}, [X \mapsto a; Y \mapsto Z])$ in Example 6 could be merged as $(\varnothing, [X \mapsto a; Y \mapsto [a \mapsto b] \cdot Z])$.

Definition 13 (Merging solutions). *Let W be a set of solutions, such that there are two different elements $(\mathbf{F}, \sigma)$ and $(\mathbf{F}', \sigma')$ in W satisfying $\forall \Delta \in \mathbf{F}.\Delta \vdash \sigma \approx_\alpha \sigma'$. This pair of solutions can be replaced with a single solution as follows:*

$$([\mathbf{W_1}]) \qquad (\mathbf{F}, \sigma), (\mathbf{F}', \sigma') \qquad \Longrightarrow_{[W]} \qquad (\mathbf{F}' \cup \mathbf{F}, \sigma')$$

Further, if $(\mathbf{F}, \sigma)$ *contains the empty set as one of the freshness contexts in* $\mathbf{F}$, *then any other freshness context in* $\mathbf{F}$ *is redundant and can be discarded:*

$$([\mathbf{W_2}]) \quad (\mathbf{F}, \sigma) \quad \Longrightarrow_{[W]} \quad (\{\varnothing\}, \sigma) \quad if \; \mathbf{F} \neq \{\varnothing\}, \; \varnothing \in \mathbf{F}$$

Write $[W]$ *for the normal form of* W *by the rules above.* $\langle \mathbf{P} \rangle_{sol}$ *denotes the normal form by* $([\mathbf{W_1}])$ *and* $([\mathbf{W_2}])$ *of* $\langle \mathbf{P} \rangle_{out}$, *that is:* $\langle \mathbf{P} \rangle_{sol} = [\langle \mathbf{P} \rangle_{out}]$.

Example 7 (Merging solutions). By application of rules $[W_1]$ and $[W_2]$ to the solution set W from Example 6, solution $(\varnothing, [X \mapsto a; Y \mapsto [a \mapsto b]{\cdot}Z])$ replaces in W the pair $(\varnothing, [X \mapsto a; Y \mapsto [a \mapsto b]{\cdot}Z])$, $(\{a \# Z\}, [X \mapsto a; Y \mapsto Z])$. Similarly, $(\varnothing, [X \mapsto [a \mapsto b]{\cdot}Z])$ replaces the pair $(\varnothing, [X \mapsto [a \mapsto b]{\cdot}Z])$, $(\{a \# Z\}, [X \mapsto Z])$.

Theorem 1 (Soundness and completeness). $\langle \mathbf{P} \rangle_{sol} \subseteq \mathcal{U}(\mathbf{P})$ *(soundness);*
$\quad \forall (\mathbf{F}, \sigma) \in \mathcal{U}(\mathbf{P}), \; \exists (\mathbf{F}_i, \theta_i) \in \langle \mathbf{P} \rangle_{sol}$ *such that* $(\mathbf{F}_i, \theta_i) \leq (\mathbf{F}, \sigma)$ *(completeness).*

5 Unitary Matching for Simple Problems

When using matching to generate, for instance, rewrite steps for a given nominal rewriting rule, it is useful to have a unique most general matching solution. Below we characterise a class of matching constraints, which we call **simple**, for which matching is unitary. The idea is to require each variable symbol in a pattern to have at least one occurrence with trivial suspended a-substitution (Id) and not in a suspension (see below). Constraints whose pattern is a moderated variable with non-trivial a-substitutions will be **postponed**.

Moderated variables occurring in suspended a-substitutions will be called **suspended (variable) occurrences**, and the others will be called **fixed (variable) occurrences**. For instance, in the term $([a \mapsto Z]{\cdot}X, [a \mapsto b]{\cdot}Y)$, both $[a \mapsto Z]{\cdot}X$ and $[a \mapsto b]{\cdot}Y$, are fixed, but Z is a suspended occurrence since it occurs in the image of the a-substitution suspended over X. Write $V_f(t)$ for the subset of $V(t)$ such that each variable *has at least one fixed occurrence with trivial a-substitutions*. The set $V_f(t)$ will play an important role in the characterisation of unitary matching problems.

Definition 14. *A term* s *is* **simple** *if* $V(s) \subseteq V_f(s)$, *that is, for each variable* $X \in V(s)$ *there is one or more fixed occurrences of the form* $\mathsf{Id}{\hat{}}\pi{\cdot}X$.[2]

A **simple matching constraint** *is a matching constraint* $s \mathbin{?}\approx t$ *such that* s *is a simple term and* $V(s) \cap V(t) = \varnothing$. *A* **simple matching problem** *is a problem as specified in Definition 7 where* $(\ldots, s_i, \ldots) \mathbin{?}\approx (\ldots, t_i, \ldots)$ *is simple.*[3]

---

[2] This means that each variable has an occurrence that does not involve a-substitution.

[3] We use a constraint $(\ldots, s_i, \ldots) \mathbin{?}\approx (\ldots, t_i, \ldots)$ in order to ensure that all the variables that can be instantiated have an occurrence that does not involve a-substitution somewhere in the problem.

Example 8. The constraints $\mathrm{cons}([a \mapsto Y]{\cdot}X, \mathrm{map}([b]X, Y))_? \approx \mathrm{cons}(H, \mathrm{nil})$ and $\mathrm{map}([a]X, \mathrm{cons}(Y, \mathrm{nil}))_? \approx \mathrm{map}([a \mapsto H]{\cdot}F, \mathrm{cons}(H, \mathrm{nil}))$ are simple but the constraint $\mathrm{map}([a \mapsto Y]{\cdot}X, X)_? \approx \mathrm{map}([a \mapsto \mathrm{nil}]{\cdot}F, F)$ is not; the latter does not have a simple pattern term, there is no fixed occurrence of Y.

Given a simple matching problem $\mathbf{P}$, postponed constraints (of the form $\phi\hat{\ }\pi{\cdot}X \,_? \approx t$ where $t \neq \phi'\hat{\ }\pi'{\cdot}X$, $\phi \neq \mathsf{Id}$ and $X \notin \{X \mid s \,_? \approx t \in \mathbf{P}, X \in V(t)\}$) are delayed until an instantiation for X is readily available. The definition of simple constraint (Definition 14) ensures such instantiation exists. The matching rule $(_? \approx \mathbf{XY})$ is not included in the simple-matching algorithm and rule $(_? \approx \mathbf{Inst})$ is adapted following the standard instantiating rule (see [28, Fig. 3]) as follows.

Definition 15 (Simple-matching algorithm). *Let* $\mathbf{P}$ *be a simple matching problem,* $\mathbb{X} = V_{RHS}(\mathbf{P})$ *and assume* $X \notin \mathbb{X}$. *Take the rule set of Definition 12, discard rule* $(_? \approx \mathbf{XY})$ *and replace rule* $(_? \approx \mathbf{Inst})$ *with:*

$$(_? \approx \sigma) \; (\{\pi{\cdot}X \,_? \approx_? t\} \cup \mathbf{P}, \theta) \xRightarrow{\mathbb{X}}{}_{? \approx} (\mathbf{P}[X \mapsto \pi^{-1}{\cdot}t], \theta \bullet [X \mapsto \pi^{-1}{\cdot}t])$$

The **simple-matching algorithm** *follows the two-phase reduction strategy described in Remark 1, using the modified rule set where rule* $(_? \approx \sigma)$ *has the highest priority along with rule* $(_? \approx \equiv)$, *rule* $(_? \approx \mathbf{X})$ *has the lowest priority and all other rules have equal priority. Let* $\langle \mathbf{P} \rangle_{\mathrm{nf}_? \approx}$ *be the normal form of* $\mathbf{P}$ *with respect to the set of updated rules.*

The priority imposed on rule $(_? \approx \sigma)$ forces the generation of v-substitutions as soon as possible, whilst by giving lowest precedence to the rule $(_? \approx \mathbf{X})$, we ensure it is simply checking α-equality (as specified by $(\approx_\alpha \mathbf{X})$) since no variables are left to be instantiated. As a result, each distinct solution $(\mathbf{F}, \sigma)$ from the solution set W shares the same unifier, σ, and by application of the merging rule to W in the final part of the algorithm, the solution set is reduced to $[W] = \{(\bigcup \mathbf{F}, \sigma)\}$. We formalise this claim in Theorem 2. Write $\mathtt{Match}(\mathbf{P}, V_{RHS}(\mathbf{P}))$ for the normal form of the matching problem $\mathbf{P}$ by the simple-matching algorithm (Definition 15). Then, $\langle \mathbf{P} \rangle_{sol_? \approx}$ is the result of applying function $[\cdot]$ from Definition 13 to $\mathtt{Match}(\mathbf{P}, V_{RHS}(\mathbf{P}))$. The following theorem is the main result of this section.

Theorem 2 (Normal form of a simple problem). *Given a simple matching problem* $\mathbf{P}$, *either* $\langle \mathbf{P} \rangle_{sol_? \approx} = [\mathtt{Match}(\mathbf{P}, V_{RHS}(\mathbf{P}))] = \{(\mathbf{F}, \theta)\}$ *and* $(\mathbf{F}, \theta)$ *is a solution for* $\mathbf{P}$, *or* $\langle \mathbf{P} \rangle_{sol_? \approx} = \varnothing$ *and* $\mathbf{P}$ *has no solution.*

Example 9. The rewriting rules in Example 3 have simple terms as patterns and the rewrite relation generated uses only simple matching: indeed, all the terms used in left-hand sides of rewrite rules are standard nominal terms (without atom substitutions), only the matched terms may have a-substitutions.

A-substitutions are used in the right-hand side of rules in Example 3 to implement function application in a direct way (avoiding the introduction of an additional set of rewrite rules to define non-capturing atom substitution as in standard nominal rewriting systems).

6   Undecidability of Extended Nominal Unification

To prove the undecidability of extended nominal unification, we encode Hilbert's tenth problem, proved undecidable in [22]. The main idea is to build unification problems for which *ground unifiers* simulate addition or multiplication. Then, one can represent Diophantine equations. To simplify the encoding, we consider a restricted language.

Definition 16 (Terms in L). L-*terms are generated from a triple* $(\mathcal{A}, \mathcal{X}, \mathcal{F}_L)$ *of pairwise disjoint sets, where* $\mathcal{F}_L$ *is empty and* $\mathcal{X}, \mathcal{A}$ *are countable sets of variables and atoms respectively (as described in Sect. 2), using the grammar given in Definition 1 without abstraction terms.*

Our representation of natural numbers is inspired by Goldfarb numbers [18], which are themselves inspired by Church numerals. In L, the natural number n is written: $\overline{n}_a c = (a, (a, \ldots (a, c)))$ with n occurrences of a and a single occurrence of c, where $a, c \in \mathcal{A}$. L-terms of this form, which we call **L-Goldfarb numbers**, are exactly those that solve extended nominal unification problems of the form

$$\{(a, [c \mapsto a]{\cdot}F) \; {}_?{\approx}_? \; [c \mapsto (a, a)]{\cdot}F\}. \tag{1}$$

Example 10 (L-Goldfarb numbers). The number 0 is represented as $\overline{0}_a c$, that is, c; the number 1 is represented as $\overline{1}_a c = (a, c)$; 3 is represented as $\overline{3}_a c = (a, (a, (a, c)))$. The term $\overline{2}_a(\overline{1}_a a)$ is in L but is not an L-Goldfarb number (it does not solve Eq. 1). Note also that, $\overline{2}_a(\overline{1}_a a) = (a, (a, (a, a))) = \overline{2 + 1}_a a$.

To simulate addition, we adapt Church's λ-term $add = \lambda n.\lambda m.\lambda x.n(m(x))$: we use a constraint $[c \mapsto X_i]{\cdot}X_j \; {}_?{\approx}_? \; X_k$. To simulate multiplication we use nested a-substitutions. Undecidability of extended nominal unification follows from Lemmas 1 and 2.

Lemma 1 (Addition).   *For all* $m, n, p \geq 0$, *there exists a ground unifier* θ *for the unification problem* $\{[c \mapsto X_i]{\cdot}X_j \; {}_?{\approx}_? \; X_k\}$ *such that* $\{[X_i \mapsto \overline{n}_a c; X_j \mapsto \overline{m}_a c; X_k \mapsto \overline{p}_a c]\} \subseteq \theta$ *if and only if* $p = m + n$.

Lemma 2 (Multiplication).   *Let* $\mathbf{P}^{\times} = \{s_1 \; {}_?{\approx}_? \; s_2, s_3 \; {}_?{\approx}_? \; s_4\}$ *where*
$s_1 = [c_1 \mapsto a; c_2 \mapsto b; c_3 \mapsto (((c \mapsto a]{\cdot}X_k, [c \mapsto b]{\cdot}X_j), a)]{\cdot}G$,
$s_2 = ((a, b), [c_1 \mapsto [c \mapsto a]{\cdot}X_i; c_2 \mapsto \overline{1}_a b; c_3 \mapsto a]{\cdot}G)$,
$s_3 = [c_1 \mapsto b; c_2 \mapsto a; c_3 \mapsto (((c \mapsto b]{\cdot}X_k, [c \mapsto a]{\cdot}X_j), b)]{\cdot}G$,
$s_4 = ((b, a), [c_1 \mapsto [c \mapsto b]{\cdot}X_i; c_2 \mapsto \overline{1}_a a; c_3 \mapsto b]{\cdot}G)$.
    For all $m, n, p \geq 0$, *there is a ground unifier* θ *for* $\mathbf{P}^{\times}$ *such that* $\sigma = [X_i \mapsto \overline{m}_a c; X_j \mapsto \overline{n}_a c; X_k \mapsto \overline{p}_a c]$ *and* $\sigma \subset \theta$ *if and only if* $p = m \times n$.

Theorem 3.   *There is an effective reduction of Hilbert's tenth problem to nominal unification of* L-*terms. Therefore unification of extended terms is undecidable.*

7   Conclusion

The matching algorithm provided in this paper induces a notion of rewriting that avoids the need to introduce extra rules to encode non-capturing substitutions. In future work, we will analyse the relationship between higher-order matching/unification and the corresponding problems in our language. The study of the complexity of the algorithms and the development of efficient implementations using graph representations of terms will also be subject of future research.

Acknowledgements. We thank James Cheney, Elliot Fairweather and Jordi Levy for helpful comments and useful pointers.

References

1. Baader, F., Nipkow, T.: Term Rewriting and All That. Cambridge University Press, New York (1998)
2. Byrd, W., Friedman, D.: α-kanren: a fresh name in nominal logic programming. In: Proceedings of the 2007 Workshop on Scheme and Functional Programming, pp. 79–90. Université Laval Technical Report DIUL-RT-0701 (2007)
3. Calvès, C.: Unifying nominal unification. In: RTA-13. LIPIcs, vol. 21, pp. 143–157 (2013). http://drops.dagstuhl.de/opus/volltexte/2013/4059
4. Calvès, C., Fernández, M.: Nominal matching and alpha-equivalence. In: Hodges, W., de Queiroz, R. (eds.) WoLLIC 2008. LNCS (LNAI), vol. 5110, pp. 111–122. Springer, Heidelberg (2008). https://doi.org/10.1007/978-3-540-69937-8_11
5. Calvès, C., Fernández, M.: A polynomial nominal unification algorithm. Theor. Comput. Sci. **403**(2–3), 285–306 (2008)
6. Calvès, C., Fernández, M.: Matching and alpha-equivalence check for nominal terms. J. Comput. Syst. Sci. **76**(5), 283–301 (2009). Special issue: Selected papers from WOLLIC 2008
7. Calvès, C., Fernández, M.: The first-order nominal link. In: Alpuente, M. (ed.) LOPSTR 2010. LNCS, vol. 6564, pp. 234–248. Springer, Heidelberg (2011). https://doi.org/10.1007/978-3-642-20551-4_15
8. Cheney, J., Urban, C.: αProlog: a logic programming language with names, binding and α-equivalence. In: Demoen, B., Lifschitz, V. (eds.) ICLP 2004. LNCS, vol. 3132, pp. 269–283. Springer, Heidelberg (2004). https://doi.org/10.1007/978-3-540-27775-0_19
9. Cheney, J.: The complexity of equivariant unification. In: Díaz, J., Karhumäki, J., Lepistö, A., Sannella, D. (eds.) ICALP 2004. LNCS, vol. 3142, pp. 332–344. Springer, Heidelberg (2004). https://doi.org/10.1007/978-3-540-27836-8_30
10. Clouston, R.A., Pitts, A.M.: Nominal equational logic. In: Cardelli, L., Fiore, M., Winskel, G. (eds.) Computation, Meaning and Logic. Articles dedicated to Gordon Plotkin, ENTCS, vol. 1496. Elsevier (2007)
11. Dowek, G., Gabbay, M., Mulligan, D.: Permissive nominal terms and their unification: an infinite, co-infinite approach to nominal techniques. In: IGPL 2010, vol. 18, no. 6, pp. 769–822 (2010)
12. Fairweather, E., Fernández, M., Szasz, N., Tasistro, A.: Dependent types for nominal terms with atom substitutions. In: TLCA 2015. LIPIcs, vol. 38, pp. 180–195 (2015). http://drops.dagstuhl.de/opus/volltexte/2015/5163

13. Fernández, M., Gabbay, M.: Nominal rewriting. Inf. Comput. **205**(6), 917–965 (2007). https://doi.org/10.1016/j.ic.2006.12.002
14. Fernández, M., Gabbay, M.J.: Closed nominal rewriting and efficiently computable nominal algebra equality. In: Proceedings 5th International Workshop on Logical Frameworks and Meta-Languages: Theory and Practice, LFMTP 2010, Edinburgh, UK, 14 July 2010, pp. 37–51 (2010). https://doi.org/10.4204/EPTCS.34.5
15. Gabbay, M.J., Mathijssen, A.: Capture-avoiding substitution as a nominal algebra. Formal Aspects Comput. **20**(4–5), 451–479 (2008). https://doi.org/10.1007/s00165-007-0056-1
16. Gabbay, M.J., Mathijssen, A.: Nominal universal algebra: equational logic with names and binding. J. Logic Comput. **19**(6), 1455–1508 (2009)
17. Gabbay, M.J., Pitts, A.M.: A new approach to abstract syntax with variable binding. Formal Aspects Comput. **13**(3–5), 341–363 (2001)
18. Goldfarb, W.D.: The undecidability of the second-order unification problem. Theor. Comput. Sci. **13**(2), 225–230 (1981)
19. Kumar, R., Norrish, M.: (Nominal) unification by recursive descent with triangular substitutions. In: Kaufmann, M., Paulson, L.C. (eds.) ITP 2010. LNCS, vol. 6172, pp. 51–66. Springer, Heidelberg (2010). https://doi.org/10.1007/978-3-642-14052-5_6
20. Levy, J., Veanes, M.: On the undecidability of second-order unification. Inf. Comput. **159**(1–2), 125–150 (2000). https://doi.org/10.1006/inco.2000.2877
21. Levy, J., Villaret, M.: An efficient nominal unification algorithm. In: RTA 2010, pp. 209–226 (2010). https://doi.org/10.4230/LIPIcs.RTA.2010.209
22. Matiyasevich, Y.V.: Enumerable sets are Diophantine (in Russian). Soviet Math. Doklady **191**(2), 279–282 (1970)
23. Near, J.P., Byrd, W.E., Friedman, D.P.: αleanTAP: a declarative theorem prover for first-order classical logic. In: Garcia de la Banda, M., Pontelli, E. (eds.) ICLP 2008. LNCS, vol. 5366, pp. 238–252. Springer, Heidelberg (2008). https://doi.org/10.1007/978-3-540-89982-2_26
24. Pitts, A.M.: Nominal logic, a first order theory of names and binding. Inf. Comput. **186**(2), 165–193 (2003). https://doi.org/10.1016/S0890-5401(03)00138-X
25. Pitts, A.M., Gabbay, M.J.: A metalanguage for programming with bound names modulo renaming. In: Backhouse, R., Oliveira, J.N. (eds.) MPC 2000. LNCS, vol. 1837, pp. 230–255. Springer, Heidelberg (2000). https://doi.org/10.1007/10722010_15
26. Shinwell, M.R., Pitts, A.M., Gabbay, M.J.: FreshML: programming with binders made simple. In: SIGPLAN 2003, vol. 38, no. 9, pp. 263–274 (2003)
27. Suzuki, T., Kikuchi, K., Aoto, T., Toyama, Y.: Confluence of orthogonal nominal rewriting systems revisited. In: Fernández, M. (ed.) 26th International Conference on Rewriting Techniques and Applications, RTA 2015, Warsaw, Poland, 29 June to 1 July 2015. LIPIcs, vol. 36, pp. 301–317. Schloss Dagstuhl - Leibniz-Zentrum fuer Informatik (2015). https://doi.org/10.4230/LIPIcs.RTA.2015.301
28. Urban, C., Pitts, A.M., Gabbay, M.: Nominal unification. Theor. Comput. Sci. **323**(1–3), 473–497 (2004)

Two Characterizations of Finite-State Dimension

Alexander Kozachinskiy[1,4(✉)] and Alexander Shen[2,3]

[1] National Research University Higher School of Economics, Moscow, Russia
`kozlach@mail.ru`
[2] LIRMM CNRS & University of Montpellier, Montpellier, France
`alexander.shen@lirmm.fr`
[3] IITP RAS, Moscow, Russia
[4] Lomonosov Moscow State University, Moscow, Russia
`https://www.lirmm.fr/~ashen`

Abstract. In this paper we provide two equivalent characterizations of the notion of finite-state dimension introduced by Dai, Lathrop, Lutz and Mayordomo [7]. One of them uses Shannon's entropy of non-aligned blocks and generalizes old results of Pillai [12] and Niven – Zuckerman [11]. The second characterizes finite-state dimension in terms of super-additive functions that satisfy some calibration condition (in particular, superadditive upper bounds for Kolmogorov complexity). The use of superadditive bounds allows us to prove a general sufficient condition for normality that easily implies old results of Champernowne [5], Besicovitch [1], Copeland and Erdös [6], and also a recent result of Calude, Staiger and Stephan [4].

Keywords: Finite-state dimension ·
Superadditive complexity functions · Normal sequences

1 Introduction

The notion of finite-state dimension of a bit sequence was introduced by Dai et al. [7] using finite-state gales. Later Bourke et al. [2] characterized the finite-state dimension in terms of Shannon entropies of aligned bit blocks (a prefix of the sequence is split into k-bit blocks for some k, and a random variable "uniformly chosen block" is considered).

In this paper we provide two new characterizations (equivalent definitions) of this notion. First (Sect. 2) we extend old results of Niven – Zuckerman [11] and Pillai [12] to the case of arbitrary finite-state dimension. These results were proven for normal sequences, i.e., sequences of finite-state dimension 1, and new

A. Shen—On leave from IITP RAS.
Supported by RaCAF ANR-15-CE40-0016-01 grant. The article was prepared within the framework of the HSE University Basic Research Program and funded by the Russian Academic Excellence Project '5-100'.

© Springer Nature Switzerland AG 2019
L. A. Gąsieniec et al. (Eds.): FCT 2019, LNCS 11651, pp. 80–94, 2019.
https://doi.org/10.1007/978-3-030-25027-0_6

tools (including Shearer-type inequality for entropies) are needed for the case of arbitrary finite-state dimension. Namely, we show (Theorem 1) that one can equivalently define the finite-state dimension using *non-aligned* blocks. For that, for a given n we consider a random variable "uniformly chosen k-bit factor" of the n-bit prefix of the sequence, take the lim inf of its Shannon entropy as $n \to \infty$, divide this lim inf by k and then take infimum (or limit) over k. We also provide examples showing that this equivalence works only in the limit ($k \to \infty$), not for blocks of fixed size.

The second characterization of finite-state dimension is given in Sect. 3. It does not use finite-state machines or entropies at all. We consider non-negative *superadditive* functions on bit strings, i.e., functions F such that $F(uv) \geqslant F(u) + F(v)$ for all u and v. Additionally we require some calibration property saying that F cannot be too small on too many inputs. Given a sequence $\alpha = \alpha_0\alpha_1\alpha_2 \ldots$, we consider $\liminf_n F(\alpha_0\alpha_1 \ldots \alpha_{n-1})/n$. We prove that the finite-state dimension of α is the infimum of these quantities taken over all F that satisfy our requirements.

The first example of a normal sequence was given by Champernowne [5]. It was the sequence $0\,1\,10\,11\,100\,101\,110\,111\,1000\,1001 \ldots$ (concatenation of integers $0, 1, 2, 3, \ldots$ written in binary[1]). Later a more general class of examples was suggested by Copeland and Erdős [6]. In Sect. 4, using superadditive functions, we prove a general sufficient condition for normality (=finite-state dimension 1) for a sequence that is a concatenation of some finite strings x_1, x_2, x_3, etc. This sufficient condition is formulated in terms of Kolmogorov complexity of x_i: the average Kolmogorov complexity of strings $x_1, \ldots, x_k$ should have the same asymptotic growth as the average length of these strings (under some technical conditions; see the exact statement of Theorem 4). In [3] Calude, Salomaa and Roblot introduced the notion of automatic complexity and asked whether this notion can be used to characterize normality. This question was answered negatively in [4]. We give an alternative proof of this result using our sufficient condition for normality.

The notion of automatic complexity that can be used to characterize normality and finite-state dimension (and was the starting point for us) was introduced in [13]. A self-contained exposition, including the results of the current paper and other results about finite-state dimension, automatic complexity, finite-state a priori probability and martingales, as well as applications of these notions, will be included in the arxiv version of [13].

2   Non-aligned Entropies

Consider a sequence $\alpha = \alpha_0\alpha_1\alpha_2 \ldots$, and some positive integer k. We can split the sequence α into k-bit consecutive non-overlapping blocks (aligned version), or consider all k-bit substrings of α (non-aligned version, see below the exact definition). Then we consider limit frequencies of these blocks. In this way we

---

[1] In fact, Champernowne spoke about decimal notation and sequences of digits, but this does not make a big difference.

get some distribution on the set $\{0,1\}^k$ of all k-bit blocks. We want to define the finite-state dimension of α as the limit of the normalized (i.e., divided by k) Shannon entropy of this distribution when k goes to infinity.

However, we should be more careful since these limit frequencies may not exist. Here is the exact definition. For every N take the first N blocks of length k and choose one of them uniformly at random. In this way we obtain a random variable taking values in $\{0,1\}^k$. Consider the Shannon entropy of this random variable (for the definition of Shannon entropy of a random variable see, e.g., [14, Chap. 7]). This can be done in an aligned (a) and non-aligned (na) settings, so we get two quantities: $H_{k,N}^a(\alpha) = H(\alpha_{kI} \ldots \alpha_{kI+k-1})$, $H_{k,N}^{na}(\alpha) = H(\alpha_I \ldots \alpha_{I+k-1})$, where $I \in \{0,\ldots,N-1\}$ (the block number) is chosen uniformly at random, and H denotes the Shannon entropy of the corresponding random variable.

Then we apply the $\liminf_N$ as $N \to \infty$ and let $H_k^a(\alpha) = \liminf_{N\to\infty} H_{k,N}^a(\alpha)$ and $H_k^{na}(\alpha) = \liminf_{N\to\infty} H_{k,N}^{na}(\alpha)$. The following result says that both quantities $H_k^a(\alpha)$ and $H_k^{na}(\alpha)$, divided by the block length k, converge to the same value as $k \to \infty$, and this value can also be defined as $\inf_k H_k(\alpha)/k$ (both in aligned and non-aligned versions).

Theorem 1. *For every bit sequence α we have*

$$\lim_k \frac{H_k^a(\alpha)}{k} = \inf_k \frac{H_k^a(\alpha)}{k} = \lim_k \frac{H_k^{na}(\alpha)}{k} = \inf_k \frac{H_k^{na}(\alpha)}{k}.$$

This common value is called the *finite-state dimension* of α and denoted by FSD(α). The original definition of finite-state dimension [7] was different, and the equivalence between it and the aligned version of the definition given above was shown in [2]. The equivalence between non-aligned and aligned versions seems to be new.

To prove this result, it is enough to prove two symmetric lemmas. The first one guarantees that if $H_k^a(\alpha)/k$ is small (less than some threshold) for some k, then $H_K^{na}(\alpha)/K$ is also small (less than the same threshold) for all sufficiently large K; the second says the same with aligned and non-aligned versions exchanged.

Lemma 1. *For every α, every k, every $K \geqslant k$:* $\frac{H_K^{na}(\alpha)}{K} \leqslant \frac{H_k^a(\alpha)}{k} + O\left(\frac{k}{K}\right).$

Lemma 2. *For every α, every k, every $K \geqslant k$:* $\frac{H_K^a(\alpha)}{K} \leqslant \frac{H_k^{na}(\alpha)}{k} + O\left(\frac{k}{K}\right).$

This two lemmas easily imply Theorem 1 by taking $\limsup_{K\to\infty}$ and then $\inf_k$ of both sides of both inequalities. So it remains to prove them.

Proof (of Lemma 1). Fix some sequence α, and consider some integer N. Take $I \in \{0,1\ldots,N-1\}$ uniformly at random and consider a random variable

$$\xi = \alpha_I \ldots \alpha_{I+K-1}$$

whose values are K-bit strings. By definition, the entropy of ξ is $H_{K,N}^{na}(\alpha)$. Let us look at aligned k-bit blocks covered by the block ξ (i.e., the aligned k-bit blocks inside $I \ldots I+K-1$). The exact number of these blocks may vary depending on I, but there are at least $m = \lfloor K/k \rfloor - 1$ of them (if there were only $m - 1$ complete blocks, plus maybe two incomplete blocks, then the total length would be at most $k(m - 1) + 2k - 2 = km + k - 2$, but we have $K/k \geqslant m + 1$, i.e., $K \geqslant km + k$). We number the first m blocks from left to right and get m random variables $\xi_1, \ldots, \xi_m$ (defined on the same space $\{0, \ldots, N - 1\}$). For example, ξ_1 is the leftmost aligned k-bit block of α in the interval $I \ldots I + K - 1$. To reconstruct the value of ξ when all ξ_i are known, we need to specify the prefix and suffix of ξ that are not covered by ξ_i (including their lengths). This requires $O(k)$ bits of information, so

$$H_{K,N}^{\mathrm{na}}(\alpha) = H(\xi) \leqslant H(\xi_1) + \ldots + H(\xi_m) + O(k).$$

We will show that for each $s \in \{1, \ldots, m\}$ the distribution of the random variable ξ_s is close to the uniform distribution over the first $\lfloor N/k \rfloor$ aligned k-bit blocks of α. The standard way to measure how close are two distributions on the same set A is to measure the *statistical distance* between them, defined as

$$\delta(P, Q) = \frac{1}{2} \sum_{a \in A} |P(a) - Q(a)|.$$

We claim that (for each $s \in \{1, 2, \ldots, m\}$) the statistical distance between the distribution of ξ_s and the uniform distribution on the first $\lfloor N/k \rfloor$ aligned blocks converges to 0 as $N \to \infty$. First, let us note that for a fixed aligned block its probability to become s-th aligned block inside a random nonaligned block is exactly k/N (there are k possible positions for a random non-aligned block when this happens). The only exception to this rule are aligned blocks that are near the endpoints, and we have at most $O(K/k)$ of them. When we choose a random aligned block, the probability to choose some position is exactly $1/\lfloor N/k \rfloor$, so we get some difference due to rounding. It is easy to see that the impact of both factors on the statistical distance converges to 0 as $N \to \infty$. Indeed, the number of the boundary blocks is $O(K/k)$, and the bound does not depend on N, while the probability of each block (in both distributions) converges to zero.[2] Also, since $m = N/k$ and $m' = \lfloor N/k \rfloor$ differ at most by 1, the difference between $1/m$ and $1/m'$ is of order $1/m^2$, and converges to 0 even if multiplied by m (the number of blocks is about m).

Now we use the continuity (more precisely, the uniform continuity) of the entropy function and note that all $m = \lfloor N/k \rfloor - 1$ random variables in the right hand side are close to the uniform distribution on first $\lfloor N/k \rfloor$ aligned blocks (the statistical distance converges to 0), so

$$\liminf_{N \to \infty} H_{K,N}^{\mathrm{na}}(\alpha) \leqslant (\lfloor K/k \rfloor - 1) \liminf_{N \to \infty} H_{k, \lfloor N/k \rfloor}^{\mathrm{a}}(\alpha) + O(k),$$

and dividing by K we get the statement of Lemma 1. $\square$

---

[2] More precisely, we should speak not about the probability of a given block, since the same k-bit block may appear in several positions, but about the probability of its appearance in a given position. Formally speaking, we use the following obvious fact: if we apply some function to two random variables, the statistical difference between them may only decrease. Here the function forgets the position of a block.

Proof (of Lemma 2). Take $I \in \{0, 1 \ldots, N - 1\}$ uniformly at random. We need an upper bound for $H^{\mathrm{a}}_{K,N}(\alpha)$, i.e., for $H(\alpha_{KI} \ldots \alpha_{KI+K-1})$. For that we use Shearer's inequality (see, e.g., [14, Sect. 7.2 and Chap. 10]). In general, this inequality can be formulated as follows. Consider a finite family of arbitrary random variables $\eta_0, \ldots, \eta_{m-1}$ indexed by integers in $\{0, \ldots, m - 1\}$. For every $U \subset \{0, \ldots, m - 1\}$ consider the tuple η_U of all η_u where $u \in U$. If a family of subsets $U_0, \ldots, U_{s-1} \subset \{0, \ldots, m - 1\}$ covers each element of U at least r times, then

$$H(\eta_U) \leqslant \tfrac{1}{r} \left(H(\eta_{U_0}) + \ldots + H(\eta_{U_{s-1}}) \right).$$

In our case we have K variables $\eta_0, \ldots \eta_{K-1}$ that are individual bits of a K-bit block $\alpha_{KI} \ldots \alpha_{KI+K-1}$ (for random I), i.e., $\eta_0 = \alpha_{KI}$, $\eta_1 = \alpha_{KI+1}$, etc. The set U contains all indices $0, \ldots, K - 1$, and the sets U_i contains k indices $i, i + 1, \ldots, i + k - 1$ (where operations are performed modulo K, so there are U_i that combine the prefix and suffix of a random K-bit block). Each η_i is covered k times due to this cyclic arrangement. In other words, the variable η_{U_i} is a substring of the random string $\eta_U = \alpha_{KI} \ldots \alpha_{KI+K-1}$ that starts from ith position and wraps around if there is not enough bits. There are $k - 1$ tuples of this "wrap-around" type (block of length k may cross the boundary in $k - 1$ ways). These tuples are not convenient for our analysis, so we just bound their entropy by k. In this way we obtain the following upper bound:

$$H^{\mathrm{a}}_{K,N}(\alpha) = H(\alpha_{KI} \ldots \alpha_{KI+K-1}) \leqslant$$

$$\leqslant \frac{1}{k} \left(\sum_{s=0}^{K-k} H(\alpha_{KI+s} \ldots \alpha_{KI+s+k-1}) + (k-1)k \right).$$

Adding $k - 1$ terms (replacing the wrap-around terms by some other entropies), we increase the right hand side:

$$H^{\mathrm{a}}_{K,N}(\alpha) \leqslant \frac{1}{k} \left(\sum_{s=0}^{K-1} H(\alpha_{KI+s} \ldots \alpha_{KI+s+k-1}) + (k-1)k \right).$$

Let us look at the variable $\alpha_{KI+s} \ldots \alpha_{KI+s+k-1}$ in the right hand side for some fixed s. It has the same distribution as the random non-aligned k-bit block $\alpha_J \ldots \alpha_{J+k-1}$ for uniformly chosen J in $\{0, \ldots, NK - 1\}$ conditional on the event "$J \bmod K = s$":

$$H(\alpha_{KI+s} \ldots \alpha_{KI+s+k-1}) = H(\alpha_J \ldots \alpha_{J+k-1} \,|\, J \bmod K = s).$$

The average of these K entropies (for $s = 0, \ldots, K - 1$) is the conditional entropy $H(\alpha_J \ldots \alpha_{J+k-1} \,|\, J \bmod K)$ that does not exceed the unconditional entropy. So we get

$$H^{\mathrm{a}}_{K,N}(\alpha) \leqslant \frac{1}{k} \left(K \cdot H^{\mathrm{na}}_{k,KN}(\alpha) + (k-1)k \right).$$

By taking the $\liminf$ as $N \to \infty$ we obtain

$$\frac{H^{\mathrm{a}}_K(\alpha)}{K} = \liminf_{N \to \infty} \frac{H^{\mathrm{a}}_{K,N}(\alpha)}{K} \leqslant \liminf_{N \to \infty} \frac{H^{\mathrm{na}}_{k,KN}(\alpha)}{k} + O\left(\frac{k}{K}\right).$$

However, the lim inf in the right hand side is taken over multiples of K and we want it to be over all indices. Formally, it remains to show that

$$\liminf_{N \to \infty} \frac{H^{\mathrm{na}}_{k,KN}(\alpha)}{k} = \liminf_{N \to \infty} \frac{H^{\mathrm{na}}_{k,N}(\alpha)}{k}$$

as the latter is by definition equal to $H^{\mathrm{na}}_k(\alpha)/k$. Indeed, the statistical distance between distributions on the first KN (non-aligned) blocks and the distribution on the first $KN + r$ blocks (where r the remainder modulo K) tends to zero since the first distribution is the second one conditioned on the event whose probability converges to 1 (i.e., the event "the randomly chosen block is not among the r last ones" whose probability is $KN/(KN + r)$).   □

As we have mentioned, this result implies that non-aligned and aligned versions of normality (uniform distribution on non-aligned and aligned blocks) are equivalent. However, note the asymptotic nature of this argument: to prove that the distribution of (say) non-aligned k-bit blocks is uniform, it is not enough to know that aligned k-bit blocks have uniform distribution; we need to know that the distribution of K-bit blocks is uniform for arbitrarily large values of K. This is unavoidable, as the following result shows.

Theorem 2.

(a) *For all k there exists an infinite sequence α such that $H^{\mathrm{na}}_2(\alpha) < 2$ and $H^a_i(\alpha) = i$ for all $i \leqslant k$.*

(b) *For all k there exists an infinite sequence α such that $H^a_2(\alpha) < 2$ and $H^{\mathrm{na}}_i(\alpha) = i$ for all $i \leqslant k$.*

Proof. (a) Consider all k-bit strings. It is easy to arrange them in some order $B_0, B_1, \ldots$ such that the last bit of B_i is the same as the first bit of B_{i+1}, for all i, and the last bit of the last block is the same as the first bit of the first block. For example, consider (for every $x \in \{0,1\}^{k-2}$) four k-bit strings $0x0, 0x1, 1x1, 1x0$ and concatenate these 2^{k-2} quadruples in arbitrary order.

Then consider a periodic sequence with period $B_0 B_1 \ldots B_{2^k-1}$. Obviously all aligned k-bit blocks have the same frequency, so $H^a_k(\alpha) = k$. However, for non-aligned bit blocks of length 2 we have two cases: this pair can be completely inside some B_i, or be on the boundary between blocks. The pairs of the first type are balanced (since we have all possible k-bit blocks), but the boundary pairs could be only 00 or 11 due to our construction. So the non-aligned frequency of these two blocks is $1/4 + \Omega(1/k)$, and for two other blocks we have $1/4 - \Omega(1/k)$, so $H^{\mathrm{na}}_2(\alpha) < 2$.

However, in this construction we do not necessarily have that $H^a_i(\alpha) = i$ for $i < k$. But this is easy to fix. Note that $H^a_k(\alpha) = k$ implies $H^a_i(\alpha) = i$ whenever i is a divisor of k. So we can just use the same construction with blocks of length $k!$ instead of k.

(b) Now let us consider a sequence constructed in the same way, but blocks $B_0, B_1, \ldots, B_{2^k-1}$ go in the lexicographical ordering. First let us note that all k-bit blocks have the same *non-aligned* frequencies in the periodic sequence with period $B_0 B_1 \dots B_{2^k-1}$. (For aligned k-blocks it was obvious, but the non-aligned case needs some proof.) Indeed, consider some k-bit string U; we need to show that it appears exactly k times in the (looped) sequence $B_0 B_1 \dots B_{2^k-1}$. In fact, it appears exactly once for each position modulo k. For example, it appears once among the blocks B_i. Why the same it true for some other position $s \bmod k$ where the $k - s$ first bits of U appear as a suffix of B_{i-1} and the last s bits of U appear as a prefix of B_i? Note that $(k - s)$-bit suffixes of $B_0, B_1, B_2, \dots$ form a cycle modulo 2^{k-s}, so the first $k - s$ bits of U uniquely determine the *last* $k - s$ bits of B_i, whereas the first s bits of B_i are just written in the s-bit suffix of U.

This implies that non-aligned frequencies for all k-bit blocks are the same. Therefore, they are the same also for all smaller values of k. In particular, we can assume for the rest that k is odd.

Now let us consider *aligned* blocks of size 2. We will show that aligned frequency of the block 10 in the sequence $B_0 B_1 \dots B_{2^k-1}$ is $1/4 - \Omega(1/k)$. Since k is odd (see above), when we cut our sequence into blocks of size 2, there are "border" blocks that cross the boundaries between B_i and B_{i+1}, and other non-border blocks. Each second boundary is crossed (between B_0 and B_1, then B_2 and B_3, and so on), so the border blocks *all have the first bit* 0. In particular, 10 never appears on such positions. This creates discrepancy of order $1/k$ for 10, and we should check that it is not compensated by non-boundary blocks. In the blocks B_i with even i we delete that last bit and cut the rest into bit pairs. After deleting the last bit we have all possible $(k - 1)$-bit strings, so no discrepancy arises here. In the blocks B_i with odd i we delete the first bit, and then cut the rest into bit pairs. In the last pair the last bit is 1 (since i is odd), so once again we never have 10 here, as required (the other positions are balanced).   □

3   Superadditive Complexity Measures

The finite-state dimension is a scaled-down version of effective Hausdorff dimension [8]. The effective Hausdorff dimension of a sequence $\alpha = \alpha_0 \alpha_1 \dots$ can be equivalently defined as the $\liminf \mathrm{C}(\alpha_0 \dots \alpha_{N-1})/N$, where C stands for the Kolmogorov complexity function [9,10]. We use here plain complexity, but prefix, a priori or monotone complexity (see, e.g., [14, Chap. 6]) will work as well, since they all differ only by $O(\log n)$ for n-bit strings (see, e.g., [14] for more details about Kolmogorov complexity and effective dimension). It is natural to look for a similar characterization of finite-state dimension in terms of compressibility. Such a characterization was given in [7, Sect. 7]. However, it did not use a complexity notion that can replace C in the definition of effective Hausdorff dimension, using finite-state compressors instead. A suitable complexity notion was introduced in [13], and it indeed gives the desired characterization. We may also use superadditive upper bounds for Kolmogorov complexity. In this extended abstract we present only a version that does not mention Kolmogorov complexity or finite-state machines at all.

Consider a non-negative function F defined on strings. Recall that F is *super-additive* if $F(xy) \geqslant F(x) + F(y)$ for all x and y. We call F *calibrated* if for every

n the sum $\sum 2^{-F(x)}$ taken over all strings x of length n does not exceed some constant (not depending on n).

Theorem 3. *Let $\alpha = \alpha_0\alpha_1\alpha_2\ldots$ be an infinite bit sequence. Then*

$$\mathrm{FSD}(\alpha) = \inf_F \left(\liminf_{N\to\infty} \frac{F(\alpha_0\ldots\alpha_{N-1})}{N} \right),$$

where the infimum is taken over all superadditive calibrated $F\colon \{0,1\}^* \to [0,+\infty)$.

Proof. We start with an upper bound for the finite-state dimension. Let F be a superadditive calibrated function. We need to show that

$$\mathrm{FSD}(\alpha) \leqslant \liminf_{N\to\infty} \frac{F(\alpha_0\ldots\alpha_{N-1})}{N}.$$

Since $\mathrm{FSD}(\alpha)$ can be defined as $\lim_k H_k^{\mathrm{a}}(\alpha)/k$, it is enough to prove that

$$H_k^{\mathrm{a}}(\alpha)/k \leqslant \liminf_{N\to\infty} \frac{F(\alpha_0\ldots\alpha_{N-1})}{N} + O(1/k) \qquad (*)$$

for all k. Fix some $k \in \mathbb{N}$. We can split $\alpha_0\ldots\alpha_{N-1}$ into $M = \lfloor N/k \rfloor$ aligned k-bit blocks $b_1,\ldots,b_M$ and a tail of length less than k. Since F is superadditive, its value of $\alpha_0\ldots\alpha_{N-1}$ is at least the sum of its values on blocks $b_1,\ldots,b_M$ (plus the value on the tail; it is non-negative and we ignore it). So we need a lower bound for the sum $F(b_1) + \ldots + F(b_M)$.

How do we get such a bound? We know that the sum of $2^{-F(b)}$ (taken over all blocks b of length k) is bounded by some constant c that does not depend on k. Assume first for simplicity that this constant is 1 and all values of F are integers. Then there exists a prefix-free code for all k-bit blocks where every block b has code of length at most $F(b)$. Then the sum $F(b_1) + \ldots + F(b_M)$, divided by M, is an average code length for the distribution with entropy $H_{k,M}^{\mathrm{a}}(\alpha)$, therefore

$$F(b_1) + \ldots + F(b_M) \geqslant M H_{k,M}^{\mathrm{a}}(\alpha),$$

and

$$F(\alpha_0\ldots\alpha_{N-1}) \geqslant \lfloor N/k \rfloor H_{k,\lfloor N/k\rfloor}^{\mathrm{a}}(\alpha).$$

Now, dividing both sides by N and taking the $\liminf$, we get the desired inequality $(*)$ even without $O(1/k)$ term. This term appears when we recall that the sum of $2^{-F(b)}$ over all blocks of length k is bounded by a constant (instead of 1) and that the values of F are not necessary integers. To rescue the argument, we need to add some constant to F and perform rounding that adds a constant term to the average code length bound. We get

$$F(b_1) + \ldots + F(b_M) \geqslant M(H_{k,M}^{\mathrm{a}}(\alpha) - O(1))$$

and

$$F(\alpha_0\ldots\alpha_{N-1}) \geqslant \lfloor N/k \rfloor (H_{k,\lfloor N/k\rfloor}^{\mathrm{a}}(\alpha) - O(1)).$$

Dividing by N, we get a correction of order $O(1/k)$, as claimed.

For the other direction, we need to assume that $H^{\mathrm{a}}_k(\alpha)/k$ is small (less than some threshold) for some k and construct a calibrated superadditive function F such that $\liminf F(\alpha_0 \ldots \alpha_{N-1})/N$ is small (does not exceed the same threshold). For that, we need some general method to construct superadditive calibrated functions. This method is a finite-state version of the a priori complexity notion from algorithmic information theory [14, Sect. 5.3]. Here it is.

Consider a finite set S of vertices (states). Assume that each vertex has two outgoing edges labeled by $(0, p_0)$ and $(1, p_1)$, where p_0 and p_1 are some nonnegative reals such that $p_0 + p_1 = 1$. Then we may consider a probabilistic process: being in state s, the machine emits 0 (with probability p_0) or 1 (with probability p_1), and changes state following the corresponding edge. In addition to such a labeled graph G, fix some state $s \in S$ as an initial state. Then we get a probabilistic algorithm that emits bits, and the corresponding measure $P_{G,s}$ on the space of bit sequences. Let $P_{G,s}(u)$ be the probability of the event "starting from s, the process emits a bit sequence with prefix u". For each k the sum of $P_{G,s}(u)$ over all strings u of length k is exactly 1, so the function $u \mapsto -\log_2 P_{G,s}(u)$ is calibrated. However, it may not be superadditive. To get superadditivity, we take the maximum probability over all initial states s.

Lemma 3. *Let G be a labeled graph of the described type, and all probabilities on labels are positive.[3] Then the function $F_G(u) = -\log \max_{s \in S} P_{G,s}(u)$ is calibrated and superadditive.*

Proof (of Lemma 3). (Calibration) Since $\max_{s \in S}$ does not exceed $\sum_{s \in S}$, we conclude that the sum of $2^{-F_G(u)}$ over all strings of given length does not exceed the number of states.

(Superadditivity) We need to prove that

$$\max_{s \in S} P_{G,s}(uv) \leqslant \max_{s \in S} P_{G,s}(u) \cdot \max_{s' \in S} P_{G,s'}(v).$$

We need an upper bound for $P_{G,s}(uv)$ for each s. Indeed, the probability of emitting uv starting from s is equal to the product of the probability of emitting u, starting from s, and the conditional probability of emitting v if u was emitted before. The first probability is $P_{G,s}(u)$ (and does not exceed the maximal value taken over all s). The second probability is $P_{G,s'}(v)$, where s' is the state s' after emitting u. Lemma 3 is proven.   □

Now assume that $H^{\mathrm{a}}_k(\alpha)/k$ (for some k) is less than some threshold β. This means that there exists a sequence of prefixes of α such that the entropies of corresponding aligned distributions on $\{0,1\}^k$ converge to some number less than βk. Compactness arguments show that we may assume that the corresponding distributions on $\{0,1\}^k$ converge to some distribution Q whose entropy $H(Q)$ is less that βk. Assume for now that all blocks have positive Q-probabilities. Consider a probabilistic process that generates a concatenation of independent k-bit strings each having distribution Q. To generate one string according to Q, we

---

[3] This is a technical condition needed to avoid infinities in the logarithms.

generate its bits sequentially, with corresponding conditional probabilities. So the state is the sequence of bits that are already generated; the states form a tree. Finally, generating the last (kth) bit of this string, we return to the initial state (the root of this tree) and are ready to generate new independent strings with the same distribution.

If G is the labeled graph constructed in this way, all labels are positive (recall that we assume that all Q-values are positive). If s is the root, then $P_{G,s}(b_0 \ldots b_{m-1}) = Q(b_0) \cdot \ldots \cdot Q(b_{m-1})$ for arbitrary k-bit blocks $b_0, \ldots, b_{m-1}$. Now let $b_0 b_1 \ldots b_{m-1}$ be the prefix of α from the subsequence of prefixes where the corresponding distributions converge to Q. If $F(u)$ is defined as $- \log P_{G,s}(u)$, then $F(b_0 \ldots b_{m-1}) = \sum_{i=0}^{m-1} (- \log Q(b_i))$. Recall that the frequencies of all k-bit blocks among $b_0, \ldots, b_{m-1}$ converge to Q. Therefore,

$$F(b_0 \ldots b_{m-1}) = (H(Q) + o(1))m < \beta km$$

for sufficiently large m such that the prefix $b_0 \ldots b_{m-1}$ belongs to the subsequence. Dividing both sides by the length km, we get $\liminf_N F(\alpha_0 \ldots \alpha_{N-1})/N \leqslant \beta$. The only problem is that $F(u)$ may not be superadditive, but we can replace it by a smaller superadditive calibrated function $- \log P_G(u)$ (taking the maximum of probabilities over all states).

This ends the proof for the case when Q is everywhere positive. If not, we may consider another distribution Q' that is close to Q but has all positive probabilities. Then $F(b_0 \ldots b_{m-1})$ will be bigger, and the increase is Kullback – Leibler divergence between Q and Q'. So we just need to make this divergences less than $\beta k - H(Q)$.

Theorem 3 is proven.   □

4   Sufficient Condition for Normality

Assume that some non-empty strings $x_1, x_2, \ldots$ are given, and consider the infinite sequence $\varkappa = x_1 x_2 \ldots$ obtained by their concatenation. The following theorem provides some conditions that guarantee that $\varkappa$ is a normal sequence.

Theorem 4. *Let L_n be the average length of the first n strings, i.e., $L_n = (|x_1| + \ldots + |x_n|)/n$. Let C_n be the average Kolmogorov complexity of the same strings, i.e., $C_n = (\mathrm{C}(x_1) + \ldots + \mathrm{C}(x_n))/n$. Assume that $|x_n|/(|x_1| + \ldots + |x_{n-1}|) \to 0$ and $L_n \to \infty$ as $n \to \infty$. If $C_n/L_n \to 1$ as $n \to \infty$, then $\varkappa = x_1 x_2 \ldots$ is normal. In general, $\mathrm{FSD}(\varkappa) \geqslant \liminf\limits_{n \to \infty} C_n/L_n$.*

Recall that normal sequences can be defined as sequences of finite-state dimension 1.

For example, in the Champernowne sequence the string x_n is the binary representation of n. It is easy to check all three conditions (the latter one uses that the average Kolmogorov complexity of k-bit strings is $k - O(1)$).

This theorem and its proof require some notions and results from algorithmic information theory (all needed information can be found, e.g., in [14]): the notion of plain Kolmogorov complexity $C(x)$ in used in its statement, the notion of a priori complexity (the logarithm of the continuous a priori probability) is used in the proof. However, this theorem has a corollary that can be formulated without Kolmogorov complexity. For that we consider a random variable i uniformly distributed in $\{1, \ldots, n\}$, random variable x_i whose value are binary strings, and replace C_n by the entropy H_n of this variable. (If all x_i are different, this entropy is $\log n$.) Again, if $H_n/L_n \to 1$, then $\varkappa$ is normal, and $\mathrm{FSD}(\varkappa) \geqslant \liminf H_n/L_n$ in the general case. To derive this corollary, we note that the difference between a priori and prefix complexity is negligible (logarithmic compared to length, see below the comparison between a priori and plain complexities), and prefix complexity provides a prefix-free code for the random variable x_i (with random i), so the average length of the code is at least the Shannon entropy of this variable.

Proof (of Theorem 4). To prove this result, we need to recall the proof of Theorem 3 and note that we can restrict the $\inf_F$ in the right hand side to functions F that are computable upper bounds for the a priori complexity up to $O(1)$ precision (see [14, Sect. 5.1] for the definition). Indeed, in the proof we have constructed a distribution on the Cantor space (product of distribution Q on k-bit blocks). If Q were computable, then all the transition probabilities in the graph G we constructed would be computable, and $P_{G,s}$ would be a computable measure on the Cantor space for each s, therefore its negative logarithm would be an upper bound for a priori complexity (up to $O(1)$ precision), and the same is true for the minimum over (finitely many) states s.

However, we may not assume that Q is computable: it is the limit distribution in a sequence of prefixes and may be arbitrary. Still (see the discussion above) we may always choose Q' that is close to Q, is computable (even rational) and has non-zero probabilities.

Therefore it remains to show that for every F that is a superadditive upper bound for a priori complexity, the liminf of $F(u)/|u|$, where u is a prefix of $\varkappa$, is at least $\liminf_n C_n/L_n$. If u ends on the block boundary, i.e., if $u = x_1 \ldots x_n$ for some n, then

$$F(u) = F(x_1 \ldots x_n) \geqslant F(x_1) + \ldots + F(x_n) \geqslant \mathrm{KA}(x_1) + \ldots + \mathrm{KA}(x_n) - O(n),$$

where KA is a priori complexity (we use superadditivity of F and recall that F is an upper bound for KA up to $O(1)$ additive term). Assume for a while that we have plain complexity C in this inequality. Then we may continue and write $F(u) \geqslant C(x_1) + \ldots + C(x_n) - O(n) = nC_n - O(n)$ and $|u| = nL_n$, so $F(u)/|u| \geqslant C_n/L_n - O(1/L_n)$, and the last term is $o(1)$, since $L_n \to \infty$ as $n \to \infty$.

Now we should consider u that do not end on the block boundary. We can delete the last incomplete block and get slightly shorter u'. For this u' we use the same bound as before, and due to the superadditivity it works as a bound for u. However, we have $|u|$ in the denominator, not $|u'|$. This does not change the lim inf, since we assume that $|x_n| = o(|x_1| + \ldots + |x_{n-1}|)$, so the length of the incomplete block is negligible compared to the total length of previous complete blocks, and the correction factor converges to 1.

Finally, the difference between plain and a priori complexity is $O(\log m)$ for strings of length m. Therefore, we get a bound (for prefixes $u = x_1 \ldots x_n$)

$$\frac{F(u)}{|u|} \geqslant \frac{KA(x_1) + \ldots + KA(x_n) - O(n)}{|x_1| + \ldots + |x_n|} \geqslant$$
$$\geqslant \frac{C(x_1) + \ldots + C(x_n) - O(\log|x_1| + \ldots + \log|x_n|) - O(n)}{|x_1| + \ldots + |x_n|}.$$

Both O-terms do not change the limit; we have already discussed this for $O(n)$ (recall that n is small compared to the total length, since $L_n \to \infty$), and the convexity of logarithm (Cauchy inequality) allows us to write

$$\frac{\log|x_1| + \ldots + \log|x_n|}{|x_1| + \ldots + |x_n|} \leqslant \frac{n \cdot \log\left(|x_1|/n + \ldots + |x_n|/n\right)}{|x_1| + \ldots + |x_n|} = \frac{\log L_n}{L_n} \to 0.$$

Theorem 4 is proven.   □

As we have noted, this sufficient condition implies the normality of the Champernowne number [5]. It is also easy to see that Copeland – Erdös criterion [6] can be derived in the same way. In this result some integers are skipped, but in such a way that the bit length of the ith remaining integer is still $(1 + o(1)) \log i$, and the sufficient condition can be still applied. More work is needed to derive the result of Besicovitch [1] saying that concatenated binary representations of perfect squares form a normal number. For this example x_m is a binary representation of m^2, has length about $2 \log m$ and complexity about m, so we get only the lower bound $1/2$ for its finite-state dimension from Theorem 4. To prove normality, we should split the string x_m into two halves of the same length $x_m = y_m z_m$. It is easy to see that the most significant half of m^2 determines m almost uniquely, so the complexity of y_m is close to the complexity of m. For z_m it is not the case: if m has j trailing zeros in the binary representation, then m^2 has $2j$ trailing zeros and its complexity decreases at least by $j - O(1)$ compared to the complexity of m. A simple analysis shows that this estimate is exact, and since the average number of trailing zeros in a random s-bit string is $O(1)$, we get the required bound.

Now let us give more details. Let z_m be the suffix of x_m of length $\lfloor \log_2 m \rfloor + 1$, i.e., the length of z_m is exactly the length of the binary representation of m, and let $y_m \in \{0,1\}^*$ be the corresponding prefix, i.e., $x_m = y_m z_m$. Note that the length of y_m is $\log_2 m + O(1)$. Therefore, the average length of $y_1, z_1, \ldots, y_m, z_m$ is $\log_2 m + O(1)$, and it remains to show that the average Kolmogorov complexity of these strings is $\log m \cdot (1 - o(1))$. We will do this by showing that the average of conditional complexities $C(i|y_i), C(i|z_i)$ over $i \in \{1, \ldots, m\}$ is $O(\log \log m)$. Since we already know that the average of $C(i)$ over $i \in \{1, \ldots, m\}$ is $\log_2 m + O(1)$, this would give the desired bound. Indeed, this follows from the chain rule:

$$C(y_i) \geqslant C(i) - C(i|y_i) - O(\log \log m), \qquad C(z_i) \geqslant C(i) - C(i|z_i) - O(\log \log m).$$

For the first part we will show not only that the average of $C(i|y_i)$ is at most $O(\log \log m)$, but that the same is true for *every* i. Indeed, assume that you know y_i and the length of the binary representation of i (let us denote this quantity by k). Then there is at most $O(1)$ different j of length k such that $y_j = y_i$. Indeed, the difference between i^2 and j^2 is $|i^2 - j^2| = \Omega(|i-j| \cdot 2^k)$. On the other hand, by definition we have that $i^2 = 2^k y_i + z_i, j^2 = 2^k y_i + z_j$, which means that that difference between i^2 and j^2 is $|z_i - z_j| = O(2^k)$. Therefore, if for the k-bit number j we have $y_j = y_i$, then j differs from i only by some constant. We need only to specify the length of the binary representation of i, using $O(\log \log m)$ bits.

As we mentioned earlier, we need a more complicated argument to show that the average of $C(i|z_i)$ is $O(\log \log m)$. The reason is that it is true only for averages: there are some i such that $C(i|z_i)$ is of order $\log m$. We have to show somehow that the number of "bad" i is negligible. To do so we need the following technical lemma.

Lemma 4. *Let $t(n)$ denote the largest natural number d such that n is divisible by 2^d (i.e., $t(n)$ is the number of trailing zeros in the binary representation of n). Then for every $a \in \mathbb{N}$ the number of $x \in \{0, 1, \ldots, 2^k - 1\}$ such that $x^2 \equiv a^2$ (mod 2^k) is at most $O(2^{t(a)})$.*

Proof. Indeed, assume that a has z trailing zeros and $x^2 = a^2$ (mod 2^k) for some $x \in \{0, 1 \ldots, 2^k - 1\}$. Then $x^2 - a^2 = (x - a)(x + a)$ is a multiple of 2^k, therefore $x - a$ is a multiple of 2^u and $x + a$ is a multiple of 2^v for some u, v such that $u + v = k$. Then $2a = (x + a) - (x - a)$ is a multiple of $2^{\min(u,v)}$, so $\min(u, v) \leqslant z - 1$. Then $\max(u, v) \geqslant k - z - 1$, so one of $x - a$ and $x + a$ is a multiple of 2^{k-z-1}, and each case contributes at most $2^{z+1} = O(2^z)$ solutions for the equation $x^2 = a^2$ (mod 2^k). $\square$

This lemma implies that $C(i|z_i) = O(t(i) + \log \log m)$. Indeed, assume that z_i and the length of the binary representation of i (denoted by k in the sequel) are given. Suppose that j is a k-bit number satisfying $z_j = z_i$. Then, as $i^2 = 2^k \cdot y_i + z_i, j^2 = 2^k \cdot y_j + z_i$, the difference between i^2 and j^2 is the multiple of 2^k. By Lemma 4 the number of such j is $O(2^{t(a)})$, i.e., specifying one of them requires $t(a) + O(1)$ bits.

As the average of $t(i)$ is $O(1)$, this gives the required bound for the average value of $C(i|z_i)$.

Calude, Salomaa and Roblot [3, Sect. 6] define a version of automatic complexity in the following way. A deterministic transducer (finite automaton that reads an input string and at each step produces some number of output bits) maps a description string to a string to be described, and the complexity of y is measured as the minimal sum of the sizes of the transducer and the input string needed to produce y; the minimum is taken over all pairs (transducer, input string) producing y. The size of the transducer is measured via some encoding, so the complexity function depends on the choice of this encoding. "It will be interesting to check whether finite-state random strings are Borel normal" [3, p. 5677]. Since normality is defined for infinite sequences, one probably should interpret this question in the following way: is it true that normal infinite sequences can be characterized as sequences whose prefixes have finite-state complexity close to length?

It turns out [4] that this is only a sufficient condition, not a criterion. More precisely, there is a normal sequence such that finite-state complexity of its first n bits is $o(n)$. This example is also an easy consequence of Theorem 4. Indeed, let us denote the complexity defined in [3] by $\mathrm{CSR}(x)$. It depends on the choice of the encoding for transducers, but the following theorem is true for every encoding, so we assume that some encoding is fixed and omit it in the notation.

Theorem 5 ([4]).

(a) *If a sequence $\alpha = a_0 a_1 \ldots$ is not normal, then there exists some $c < 1$ such that the $\mathrm{CSR}(a_0 \ldots a_{n-1}) < cn$ for infinitely many n.*
(b) *$\liminf \mathrm{CSR}(b_0 \ldots b_{n-1})/n = 0$ for some normal sequence $\beta = b_0 b_1 \ldots$*

Proof. The first part of the statement can be proven using Shannon coding in the same way as in [13]. For the second part we construct an example of a normal sequence using Champernowne's idea and Theorem 4. The sequence will have the form $\beta = (B_1)^{n_1} (B_2)^{n_2} \ldots$; here B_i is the concatenation of all strings of length i (say, in lexicographical ordering, but this does not matter), and n_i is a fast growing sequence of integers.

To choose n_i, let us note first that for a periodic sequence (of the form XY^∞) the CSR-complexity of its prefixes of the form XY^k is $o(\text{length})$. Indeed, we may consider a transducer that first outputs X, then outputs Y for each input bit 1. So $\mathrm{CSR}(XY^m) = m + O(1)$, and the compression ratio is about $1/|Y|$. To get $o(\text{length})$, we use Y^c for some constant c as a period to improve the compression.

Now consider the complexity/length ratio for the prefixes of β if the sequence n_i grows fast enough. Indeed, assume that $n_1, n_2, \ldots, n_k$ are already chosen and we now choose the value of n_{k+1}. We may use the bound explained in the previous paragraph and let $X = (B_1)^{n_1} \ldots (B_k)^{n_k}$ and $Y = B_{k+1}$. For sufficiently large n_{k+1} we get arbitrarily small complexity/length ratio. (Note that good compression is guaranteed only for some prefixes; when increasing k, we need to switch to another transducer, and we know nothing about the length of its encoding.)

It remains to apply Theorem 4 to show that for some fast growing sequence $n_1, n_2, \ldots$ the sequence β is normal. We apply the criterion by splitting B_k into pieces of length k (so all strings of length k appear once in this decomposition of B_k). We already know that the average Kolmogorov complexity of the pieces in B_k is $k - O(1)$ (and the length of all pieces is k). This is enough to satisfy the conditions of Theorem 4 if $x_1 \ldots x_n$ ends on the boundary of the block B_k. But this is not guaranteed; in general we need also to consider the last incomplete group of blocks that form a prefix of some B_k. The total length of these blocks is bounded by $|B_k|$, i.e., by $k2^k$. We need this group to be short compared to the rest, and this will be guaranteed if n_{k-1} (the lower bound for the length of the previous part) is much bigger than $k2^k$. And we assume that n_k grow very fast, so this condition is easy to satisfy. Theorem 5 is proven. $\square$

References

1. Besicovitch, A.: The asymptotic distribution of the numerals in the decimal representation of the squares of the natural numbers. Math. Z. **39**(1), 146–156 (1935). https://doi.org/10.1007/BF01201350
2. Bourke, C., Hitchcock, J.M., Vinodchandran, N.: Entropy rates and finite-state dimension. Theor. Comput. Sci. **349**(3), 392–406 (2005). https://doi.org/10.1016/j.tcs.2005.09.040
3. Calude, C.S., Salomaa, K., Roblot, T.K.: Finite state complexity. Theor. Comput. Sci. **412**(41), 5668–5677 (2011). https://doi.org/10.1016/j.tcs.2011.06.021
4. Calude, C.S., Staiger, L., Stephan, F.: Finite state incompressible infinite sequences. Inf. Comput. **247**, 23–36 (2016). https://doi.org/10.1016/j.ic.2015.11.003
5. Champernowne, D.G.: The construction of decimals normal in the scale of ten. J. London Math. Soc. **1**(4), 254–260 (1933). https://doi.org/10.1112/jlms/s1-8.4.254
6. Copeland, A.H., Erdös, P.: Note on normal numbers. Bull. Am. Math. Soc. **52**(10), 857–860 (1946). https://doi.org/10.1090/S0002-9904-1946-08657-7
7. Dai, J.J., Lathrop, J.I., Lutz, J.H., Mayordomo, E.: Finite-state dimension. Theor. Comput. Sci. **310**(1–3), 1–33 (2004). https://doi.org/10.1016/S0304-3975(03)00244-5
8. Lutz, J.H.: Dimension in complexity classes. SIAM J. Comput. **32**(5), 1236–1259 (2003). https://doi.org/10.1137/S0097539701417723
9. Lutz, J.H.: The dimensions of individual strings and sequences. Inf. Comput. **187**(1), 49–79 (2003). https://doi.org/10.1016/S0890-5401(03)00187-1
10. Mayordomo, E.: A Kolmogorov complexity characterization of constructive Hausdorff dimension. Inf. Process. Lett. **84**(1), 1–3 (2002). https://doi.org/10.1016/S0020-0190(02)00343-5
11. Niven, I., Zuckerman, H., et al.: On the definition of normal numbers. Pac. J. Math. **1**(1), 103–109 (1951). https://doi.org/10.2140/pjm.1951.1.103
12. Pillai, S.: On normal numbers. Proc. Indian Acad. Sci. Sect. A **12**(2), 179–184 (1940). https://doi.org/10.1007/BF03173913
13. Shen, A.: Automatic kolmogorov complexity and normality revisited. In: Klasing, R., Zeitoun, M. (eds.) FCT 2017. LNCS, vol. 10472, pp. 418–430. Springer, Heidelberg (2017). https://doi.org/10.1007/978-3-662-55751-8_33
14. Shen, A., Uspensky, V.A., Vereshchagin, N.: Kolmogorov Complexity and Algorithmic Randomness, vol. 220. American Mathematical Society (2017). https://doi.org/10.1090/surv/220

Largest Common Prefix
of a Regular Tree Language

Markus Lohrey[1]([⊠]) and Sebastian Maneth[2]

[1] University of Siegen, Siegen, Germany
lohrey@eti.uni-siegen.de
[2] University of Bremen, Bremen, Germany
maneth@uni-bremen.de

Abstract. A family of tree automata of size n is presented such that the size of the largest common prefix (lcp) tree of all accepted trees is exponential in n. Moreover, it is shown that this prefix tree is not compressible via DAGs (directed acyclic graphs) or tree straight-line programs. We also show that determining whether or not the lcp trees of two given tree automata are equal is coNP-complete; the result holds even for deterministic bottom-up tree automata accepting finite tree languages. These results are in sharp contrast to the case of context-free string grammars.

1 Introduction

For a given language L one can define the *largest common prefix* of L as the longest string which is a prefix of every word in L. This definition can be extended to tree languages in a natural way. One motivation to compute the largest common prefix of a set of strings or trees is the so called earliest normal form, which has been studied for string transducers [1,8] and tree transducers [3]. The existence of an earliest normal form has several important consequences. For instance, the transducer can in a simple further step be made canonical, which allows deciding equivalence and gives rise to Gold-style learning algorithms [5,9]. Intuitively, an earliest transducer produces its output "as early as possible". In order to compute the earliest form of a given transducer, one has to consider all possible inputs (for a certain set of states), and has to determine if the corresponding outputs have a non-empty common prefix; if so, then the transducer is *not* earliest, because this common prefix is independent of the input and hence should have been produced before. The questions arise how large such common prefixes can possibly be, and whether or not they can be compressed.

In this paper we address these questions in a general setting where the trees of which the common prefix is computed are given by a finite tree automaton. We present a family of tree automata of size $\Theta(n)$ such that their largest common prefixes (lcps) are of size exponential in n and are essentially incompressible via common tree compression methods such as DAGs (directed acyclic graphs) or tree straight-line programs [4,6]. Recently it has been shown that for a given

© Springer Nature Switzerland AG 2019
L. A. Gąsieniec et al. (Eds.): FCT 2019, LNCS 11651, pp. 95–108, 2019.
https://doi.org/10.1007/978-3-030-25027-0_7

context-free string grammar, a representation of the largest common prefix can be computed in polynomial time [7].

Whenever above we mention "tree automaton", we always mean "nondeterministic (top-down or bottom-up) tree automaton". Let us now consider the case of *deterministic* tree automata. It is known that in the deterministic case, *top-down* automata are strictly less expressive than bottom-up tree automata; in fact, they are so weak that they cannot even recognize finite tree languages such as $\{f(a, f(a, a)), f(f(a, a), a)\}$ (here, we use the standard term representation for trees; see Sect. 2). It turns out that the largest common prefix of the trees recognized by a deterministic top-down tree automaton can be computed by a simple (top-down) procedure. Moreover, the resulting lcps are compressible via DAGs, and the procedure can produce in linear time a DAG of the lcp. In contrast, for deterministic bottom-up tree automata (which are equally expressive as nondeterministic tree automata), such a procedure is not possible. Surprisingly, for such automata similar results can be proven as for nondeterministic automata, e.g., a family of automata of size $\Theta(n^2)$ can be defined, such that the size of their lcp is exponential in n. Technically, one ingredient of both families of automata (nondeterministic and deterministic ones) is the well-known fact that an automaton needs exponentially many states in order to recognize strings where the n-th last symbol carries a specific label.

We then address a second important problem for largest common prefixes given by tree automata, namely to determine whether or not the largest common prefixes of two given tree automata coincide. Note that when constructing an earliest canonical ("minimal") transducer, we need to determine whether two given states are equivalent; for this to hold, several lcps must be checked for equality. The following question arises: what is the precise complexity of checking equality of the lcps of two given tree automata? In this paper, we prove that this problem is **coNP**-complete using a reduction from the complement of 3-SAT.

2   Preliminaries

We assume that the reader is familiar with words and finite automata on words. A language $L \subseteq \{0, 1\}^*$ is a *right-ideal* if $L = L\{0, 1\}^*$. A set $S \subseteq \{0, 1\}^*$ is *prefix-closed*, if $uv \in S$ implies that $u \in S$ for all $u, v \in \{0, 1\}^*$. Note that L is a right-ideal if and only if $\{0, 1\}^* \setminus L$ is prefix-closed.

A DFA (deterministic finite automaton) over a finite alphabet Γ is a 5-tuple $A = (Q, \Gamma, q_0, F, \delta)$, where Q is the finite set of states, $q_0 \in Q$ is the initial state, $F \subseteq Q$ is the set of final states, and $\delta : Q \times \Gamma \to Q$ is the transition mapping. The language $L(A)$ accepted by A is defined in the usual way. For an NFA (nondeterministic finite automaton) we have a set $I \subseteq Q$ of initial states and the transition function δ maps from $Q \times \Gamma$ to 2^Q (the powerset of Q).

We consider finite *binary trees* that are unlabeled, rooted, and ordered. The latter means that there is an order on the children of a node. Moreover, every node is either a leaf or has exactly two children. We will use two equivalent formalizations of such trees. We can view them as formal expressions over the

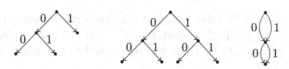

Fig. 1. Trees $t_0 = f(f(a, a), a)$ (left), $t_1 = f(f(a, a), f(a, a))$ (midle) and the minimal DAG of t_1 (right).

set of function symbols $\{f, a\}$, where f gets two arguments and a is a constant-symbol (i.e., gets no arguments). The set of all such expressions is denoted by T_2 and is inductively defined by the following conditions: $a \in T_2$ and if $t_1, t_2 \in T_2$ then also $f(t_1, t_2) \in T_2$. Trees from T_2 are binary trees, where each leaf is labeled with a and every internal node is labeled with f. Obviously, the labeling bears no information, and trees from T_2 can be identified with unlabeled binary trees. For instance, the expression $f(f(a, a), a)$ represents the binary tree t_0 from Fig. 1. Alternatively, we can specify a binary tree by a path language. A *path language* P is a finite non-empty subset of $\{0, 1\}^*$ such that

- P is prefix-closed and
- for every $w \in \{0, 1\}^*$, $w0 \in P$ if and only if $w1 \in P$.

A binary tree $t \in T_2$ can be uniquely represented by a path language $P(t)$, and vice versa. Formally, we define $P(t)$ inductively as follows:

- $P(a) = \{\epsilon\}$
- $P(f(t_1, t_2)) = \{\epsilon\} \cup \{iw \mid i \in \{0, 1\}, w \in P(t_i)\}$.

For instance, for the binary tree t_0 from Fig. 1 we have $P(t_0) = \{\epsilon, 0, 1, 00, 01\}$. The root of a tree corresponds to the empty word ϵ, $u0$ denotes the left child of u, and $u1$ denotes the right child of u. The leaves of a tree t correspond to those words in $P(t)$ that are maximal with respect to the prefix relation. The *depth* of $t \in T_2$ can be defined as the maximal length of a word in $P(t)$. Note that the intersection of an arbitrary number of path languages is again a path language.

A *nondeterministic top-down tree automaton* (NTTA for short) is a 4-tuple $B = (Q, I, F, \delta)$, where Q is a finite set of states, $I \subseteq Q$ with $I \neq \emptyset$ is the set of initial states, $F \subseteq Q$ is the set of final states, and $\delta : Q \to 2^{Q^2}$ is the transition function (here and in the following we view elements of Q^2 as words of length two over the alphabet Q). A run of B on a tree t is a mapping $\rho : P(t) \to Q$ such that:

- If $v \in P(t)$ is a leaf of t, then $\rho(v) \in F$.
- If $v, v0, v1 \in P(t)$ with $\rho(v) = p$, $\rho(v_0) = p_0$ and $\rho(v_1) = p_1$ then $p_0 p_1 \in \delta(p)$.

For $q \in Q$, we let $T(B, q)$ denote the set of all trees t for which there exists a run ρ of B such that $\rho(\varepsilon) = q$. Finally we define $T(B) = \bigcup_{q \in I} T(B, q)$ as the tree language accepted by B.

An NTTA $B = (Q, I, F, \delta)$ is called *productive* if $T(B, q) \neq \emptyset$ for every $q \in Q$. From a given NTTA B with $T(B) \neq \emptyset$ one can construct in polynomial time an equivalent productive NTTA B'. One first computes in polynomial time the set $P = \{p \in Q \mid T(B, p) \neq \emptyset\}$. Note that $F \subseteq P$. Then B' is obtained from B by removing all states from $Q \setminus P$. To do this, one also has to replace every set $\delta(q)$ $(q \in P)$ by $\delta(q) \cap P^2$.

A *deterministic top-down tree automaton* (DTTA for short) is a 4-tuple $B = (Q, q_0, F, \delta)$, where Q is a finite set of states, $q_0 \in Q$ is the initial state, $F \subseteq Q$ is the set of final states, and $\delta : Q \to Q^2$ is the transition function. We can identify this 4-tuple with the NTTA $(Q, \{q_0\}, F, \delta')$ where $\delta'(q) = \{\delta(q)\}$. This allows us the transfer all definitions from NTTAs to DTTAs.

Finally a *deterministic bottom-up tree automaton* (DBTA for short) is an NTTA $B = (Q, I, F, \delta)$ such that $|F| = 1$ and for every $q_1 q_2 \in Q^2$ there is at most one $q \in Q$ such that $q_1 q_2 \in \delta(q)$. In other words, the sets $\delta(q)$ $(q \in Q)$ are pairwise disjoint. This allows defining a partially defined inverse δ^{-1} of δ by $\delta^{-1}(q_1 q_2) = q$ if $q_1 q_2 \in \delta(q)$. For every tree t there is at most one run of B on t and this run ρ can be constructed bottom-up by first setting $\rho(u) = q_f$ for every leaf u of t, where q_f is the unique state in F. Then, for all $v, v0, v1 \in P(t)$ such that $\rho(v0)$ and $\rho(v1)$ have been already defined, one sets $\rho(v) = \delta^{-1}(\rho(v0)\rho(v1))$.

It is well known that for every NTTA there exists an equivalent DBTA accepting the same tree language. On the other hand, there exist NTTAs which do not have an equivalent DTTA; see [2] for examples.

The *minimal DAG* for a tree $t \in T_2$ is obtained by keeping for every subtree s of t exactly one isomorphic copy to which all tree edges that point to occurrences of s are redirected. The size of the minimal DAG of t (measured in number of nodes) is exactly the number of pairwise non-isomorphic subtrees of t.

Lemma 1. *Let t be a tree. The following statements are equivalent:*

1. *The minimal DAG for the tree t has n nodes.*
2. *The minimal DFA for the path language $P(t)$ has $n + 1$ states.*

The proof of the lemma is straightforward. Consider for instance the tree $t_1 = f(f(a, a), f(a, a))$ from Fig. 1. Its minimal DAG is shown in Fig. 1 on the right. It yields a DFA for $P(t_1)$ by taking the root node as the initial state, all othern nodes as final states and adding a failure state (note that a DFAs has a totally defined transition mapping according to our definition).

Largest Common Prefix Tree. Consider a non-empty tree language $L \subseteq T_2$. The *largest common prefix* $\mathrm{lcp}(L)$ of L is the unique binary tree t such that $P(t) = \bigcap_{t \in L} P(t)$. For instance, for $L = \{f(f(a, a), a), f(a, a)\}$ we obtain $\mathrm{lcp}(L) = f(a, a)$.

Lemma 2. *Assume that B is an NTTA with n states and such that $T(B) \neq \emptyset$. Then every word $w \in P(\mathrm{lcp}(T(B))) = \bigcap_{t \in T(B)} P(t)$ has length at most $n - 1$, i.e., the depth of $\mathrm{lcp}(T(B))$ is at most $n - 1$.*

Proof. It well-known that B must accept a tree t of depth at most $n - 1$; see e.g. [2, Corollary 1.2.3]. Hence, $|w| \leq n - 1$ for every word $w \in P(t)$. This implies the statement of the lemma. $\qquad\square$

It is straightforward extend all the notions from this section to labelled binary trees. A Σ-labelled binary tree can be defined as a pair (P, λ) where $P \subseteq \{0, 1\}^*$ is a path language and $\lambda : P \to \Sigma$ is the labelling function. Given a set L of Σ-labelled binary trees, one can define its lcp as the unique tree (P, λ) where P is the largest (with respect to inclusion) path language such that for all $(P', \lambda') \in L$ we have: $P \subseteq P'$ and $\lambda(u) = \lambda'(u)$ for all $u \in P$. All results in this paper also hold for Σ-labelled binary trees. Since the focus of this paper is on lower bounds, we decided to restrict our considerations to unlabelled trees.

3   From NTTAs to DFAs

In this and the next section we establish a correspondence between largest common prefix trees of regular tree languages and finite automata (on words).

Let $B = (Q, I, F, \delta)$ be a productive NTTA. We extend $\delta : Q \to 2^{Q^2}$ to $\hat{\delta} : 2^Q \to 2^{Q^2}$ by setting $\hat{\delta}(Q') = \bigcup_{p \in Q'} \delta(p) \subseteq Q^2$ for $Q' \subseteq Q$. For a state pair $p_0 p_1 \in Q^2$ and $i \in \{0, 1\}$ we define the projection $\pi_i(p_0 p_1) = p_i$. For a set $S \subseteq Q^2$ and $i \in \{0, 1\}$ we define $\pi_i(S) = \{\pi_i(pq) \mid pq \in S\}$.

We fix a fresh state $q_f \notin Q$ and define a DFA B^s (s for string) by

$$B^s = (2^Q \setminus \{\emptyset\} \uplus \{q_f\}, \{0, 1\}, I, 2^Q \setminus \{\emptyset\}, \delta^s)$$

($\uplus$ denotes disjoint union) where for all $Q' \subseteq Q$ with $Q' \neq \emptyset$ and $i \in \{0, 1\}$ we set

$$\delta^s(Q', i) = \begin{cases} \pi_i(\hat{\delta}(Q')) & \text{if } Q' \cap F = \emptyset \\ q_f & \text{if } Q' \cap F \neq \emptyset. \end{cases}$$

Moreover, $\delta^s(q_f, 0) = \delta^s(q_f, 1) = q_f$. The state q_f is called the failure state of B^s. Note that if $Q' \cap F = \emptyset$ for $Q' \neq \emptyset$, then the productivity of B implies that $\hat{\delta}(Q') \neq \emptyset$. In particular, we have $\pi_i(\hat{\delta}(Q')) \neq \emptyset$ if $Q' \neq \emptyset$, which implies that δ^s is well-defined.

Lemma 3. *Let B be a productive NTTA. Then*

$$L(B^s) = P(\text{lcp}(T(B))) = \bigcap_{t \in T(B)} P(t).$$

Proof. Consider a word $w = a_1 a_2 \cdots a_n$ with $a_1, \ldots, a_n \in \{0, 1\}$. Let us first assume that $w \in L(B^s)$ and let $t \in T(B)$. We have to show that $w \in P(t)$. In order to get a contradiction, assume that $w \notin P(t)$. Let v be a longest prefix of w that belongs to $P(t)$. Since $\varepsilon \in P(t)$, v is well-defined. Clearly, v is a proper prefix of w and v is a leaf of t. Thus, we can write v as $v = a_1 a_2 \cdots a_k$ for $k < n$. Fix a run ρ of B on t such that $\rho(\varepsilon) \in I$. Let $q_i = \rho(a_1 \cdots a_i)$ for $0 \leq i \leq k$. Since v is a leaf of t we have $q_k \in F$. Since $w \in L(B^s)$ there exists a path

$$I = Q_0 \xrightarrow{a_1} Q_1 \xrightarrow{a_2} Q_2 \xrightarrow{a_3} \cdots \xrightarrow{a_n} Q_n,$$

where, for $0 \leq i \leq n$, $Q_i \subseteq Q$ and $Q_i \neq \emptyset$, and for $0 \leq i \leq n-1$, $Q_i \cap F = \emptyset$ and $Q_{i+1} = \pi_{a_i}(\hat{\delta}(Q_i))$. The latter point implies by induction on i that $q_i \in Q_i$ for $0 \leq i \leq k$. Since $Q_k \cap F = \emptyset$, we must have $q_k \notin F$, which is a contradiction.

Now assume that $w \notin L(B^s)$. Hence, the unique run of B^s on w ends in the failure state q_f. Thus, there must exist a proper prefix $v = a_1 \cdots a_k$ of w such that $k < n$ and the run of B^s on w has the form

$$I = Q_0 \xrightarrow{a_1} Q_1 \xrightarrow{a_2} \cdots \xrightarrow{a_k} Q_k \xrightarrow{a_{k+1}} q_f \xrightarrow{a_{k+2}} \cdots \xrightarrow{a_n} q_f.$$

where $Q_i \subseteq Q$ and $Q_i \neq \emptyset$ for $0 \leq i \leq k$, $Q_i \cap F = \emptyset$ and $Q_{i+1} = \pi_{a_i}(\hat{\delta}(Q_i))$ for $0 \leq i \leq k-1$, and $Q_k \cap F \neq \emptyset$. Let $q_k \in Q_k \cap F$.

We have to construct a tree $t \in T(B)$ such that $w \notin P(t)$. For this we choose states $q_i \in Q_i$ for $0 \leq i \leq k$. The state $q_k \in Q_k \cap F$ has already been chosen in the last paragraph. Assume that $q_{i+1} \in Q_{i+1}$ has been defined for some $0 \leq i \leq k-1$. To define q_i note that $q_{i+1} \in \pi_{a_i}(\hat{\delta}(Q_i))$. Hence, there exist states $p \in Q_i$ and $q'_{i+1} \in Q$ such that the following holds: if $a_i = 0$ then $q_{i+1}q'_{i+1} \in \delta(p)$ and if $a_i = 1$ then $q'_{i+1}q_{i+1} \in \delta(p)$. We set $q_i = p$. By the productivity of B there exist trees $t'_i \in T(B, q'_i)$ for $1 \leq i \leq k$. Moreover, since $q_k \in Q_k \cap F$, the one-node tree a belongs to $T(B, q_k)$. From the trees $t'_1, \ldots, t'_k, a$ we can now construct a tree $t \in T(B)$ such that $v = a_1 \cdots a_k$ is a leaf of t (and hence $w \notin P(t)$). For instance, if $v = 1^k$ then we take $t = f(t'_1, f(t'_2, f(t'_3, \cdots f(t'_k, a) \cdots)))$. For the general case, we define trees $t_0, t_1, \ldots, t_k$ inductively as follows:

- $t_k = a$,
- $t_i = f(t_{i+1}, t'_{i+1})$ if $0 \leq i \leq k-1$ and $a_{i+1} = 0$, and
- $t_i = f(t'_{i+1}, t_{i+1})$ if $0 \leq i \leq k-1$ and $a_{i+1} = 1$.

Finally, let $t = t_0$. Then t has the desired properties.   □

Note that the size of the above DFA B^s is exponential in the size of B. In the case where we start with a DTTA, we can easily modify the above construction in order to construct in linear time a DFA (of linear size). Hence, let us redefine for a DTTA $B = (Q, q_0, F, \delta)$ the DFA $B^s = (Q \uplus \{q_f\}, \{0, 1\}, q_0, Q, \delta^s)$ by setting for all $q \in Q$ and $i \in \{0, 1\}$:

$$\delta^s(q, i) = \begin{cases} \pi_i(\delta(q)) & \text{if } q \notin F \\ q_f & \text{if } q \in F. \end{cases}$$

Moreover, $\delta^s(q_f, 0) = \delta^s(q_f, 1) = q_f$. The proof of the following lemma is similar as the proof for Lemma 3.

Lemma 4. *Let B be a DTTA with $T(B) \neq \emptyset$. Then*

$$L(B^s) = P(\mathrm{lcp}(T(B))) = \bigcap_{t \in T(B)} P(t).$$

4 From NFAs to NTTAs

We now consider NFAs that generate languages L over $\{0,1\}$ such that the complement of L is a finite path language. An NFA $A = (Q, \{0,1\}, I, F, \delta)$ is *well-behaved*, if there are two different states $q_e, q_f \in Q$ such that

1. $F = \{q_f\}$ and $q_f \notin I$,
2. $\delta(q, a) \neq \emptyset$ for all $q \in Q$ and all $a \in \{0,1\}$,
3. $q_f \notin \delta(q, a)$ for all $q \in Q \setminus \{q_e, q_f\}$ and all $a \in \{0,1\}$,
4. $\delta(q_e, 0) = \delta(q_e, 1) = \delta(q_f, 0) = \delta(q_f, 1) = \{q_f\}$,
5. the NFA obtained from A by removing the state q_f is acyclic, and
6. all states are reachable from I.

In a well-behaved NFA A every path of length at least $|Q| - 1$ that starts in a state $q \neq q_f$ must visit q_e (this follows from points 2 and 5) . Moreover, the complement $\{0,1\}^* \setminus L(A)$ is a path language.

From a well-behaved NFA $A = (Q, \{0,1\}, I, \{q_f\}, \delta)$ we construct the NTTA $A^t = (Q \setminus \{q_f\}, I, \{q_e\}, \delta^t)$ (t for tree) with

- $\delta^t(q) = \{q_1 q_2 \mid q_1 \in \delta(q, 0), q_2 \in \delta(q, 1)\}$ for $q \in Q \setminus \{q_e, q_f\}$, and
- $\delta^t(q_e) = \emptyset$.

Note that for every well-behaved NFA A, the NTTA A^t is productive.

Lemma 5. *Let A be a well-behaved NFA. Then $P(\mathrm{lcp}(T(A^t))) = \{0,1\}^* \setminus L(A)$.*

Proof. Let $A = (Q, \{0,1\}, I, F, \delta)$. We first assume that $w \in L(A)$ and show that $w \notin P(\mathrm{lcp}(T(A^t)))$. For this, we have to prove that there exists a tree $t \in T(A^t)$ such that $w \notin P(t)$. Since $w \in L(A)$ we can write $w = uv$ with $v \neq \epsilon$ such that in A there exists a u-labeled path from $q_0 \in I$ to q_e. Let us write this path as

$$q_0 \xrightarrow{a_1} q_1 \xrightarrow{a_2} q_2 \xrightarrow{a_3} \cdots \xrightarrow{a_n} q_n = q_e,$$

where $u = a_1 a_2 \cdots a_n$ and $a_1, \ldots, a_n \in \{0,1\}$. Note that $q_0, \ldots, q_{n-1} \in Q \setminus \{q_e, q_f\}$. For $1 \leq i \leq n$ let us choose any state $q_i' \in \delta(q_{i-1}, \bar{a}_i)$ (where $\bar{0} = 1$ and $\bar{1} = 0$). Such a state q_i' must exist since A is well-behaved. Moreover, choose for every $1 \leq i \leq n$ a tree $t_i \in T(A^t, q_i')$. Finally let t be the unique tree with

$$P(t) = \{u\} \cup \bigcup_{i=1}^n \{a_1 \cdots a_{i-1} \bar{a}_i u' \mid u' \in P(t_i)\}.$$

From the construction of A^t it follows that $t \in T(A^t)$. Moreover, since $w = uv$ with $v \neq \varepsilon$ we get $w \notin P(t)$. This concludes the first part of the proof.

Now assume that $w \notin P(\mathrm{lcp}(T(A^t)))$. We have to show that $w \in L(A)$. Since $w \notin P(\mathrm{lcp}(T(A^t)))$, there exists $t \in T(A^t)$ such that $w \notin P(t)$. We can factorize $w = uv$ with $v \neq \varepsilon$, where $u = a_1 \cdots a_n$ is the longest prefix of w with $u \in P(t)$. Hence, u leads in the tree t to a leaf. Since $t \in T(A^t)$, there exists a run ρ of A^t on t such that $\rho(\epsilon) \in I$. Let $q_i = \rho(a_1 \cdots a_n)$ for $0 \leq i \leq n$. Since u leads to a leaf of t we must have $q_n = q_e$. Then

$$q_0 \xrightarrow{a_1} q_1 \xrightarrow{a_2} q_2 \xrightarrow{a_3} \cdots \xrightarrow{a_n} q_n = q_e$$

is a u-labeled path in A from $q_0 \in I$ to q_e. Since $v \neq \varepsilon$ we get $w = uv \in L(A)$. $\square$

5 Incompressibility of Largest Common Prefix Trees

5.1 Incompressibility by DAGs

In this section we present our first main result, which shows that there is a family of tree automata such that the size of the minimal DAG of the corresponding largest common prefix tree is exponential in the automata size.

For $n \geq 1$ we consider the following language L_n:

$$L_n = \{0, 1\}^{2n+3} \{0, 1\}^* \cup \bigcup_{i=0}^{n-1} \left(\{0, 1\}^i 0 \{0, 1\}^n 0 \{0, 1\}^+ \right).$$

Let us first establish that the complement $V_n = \{0, 1\}^* \setminus L_n$ is a path language. Since L_n is a right ideal, the complement V_n is prefix closed. Since all words of length at least $2n + 3$ belong to L_n, the language V_n is finite. Finally, $w0 \in \{0, 1\}^{2n+3}\{0, 1\}^*$ iff $|w0| \geq 2n+3$ iff $|w1| \geq 2n+3$ iff $w1 \in \{0, 1\}^{2n+3}\{0, 1\}^*$ and $w0 \in \{0, 1\}^i 0 \{0, 1\}^n 0 \{0, 1\}^+$ iff $w1 \in \{0, 1\}^i 0 \{0, 1\}^n 0 \{0, 1\}^+$. Hence, $w0 \in L_n$ if and only if $w1 \in L_n$, and the same property must hold for the complement V_n of L_n. Thus, V_n is a path language.

Lemma 6. *The minimal DFA A for V_n has at least 2^n states.*

Proof. Let $A = (Q, \{0, 1\}, \delta, q_0, F)$. Consider the extension $\delta : Q \times \{0, 1\}^* \to Q$ with $\delta(q, \epsilon) = q$ and $\delta(q, ua) = \delta(\delta(q, u), a)$ for $u \in \{0, 1\}^*$, $a \in \{0, 1\}$. We claim that $\delta(q_0, u) \neq \delta(q_0, v)$ for every $u, v \in \{0, 1\}^n$ with $u \neq v$, which implies that A has at least 2^n states (and hence size at least 2^n). Assume by contradiction that $\delta(q_0, u) = \delta(q_0, v)$ for some $u, v \in \{0, 1\}^n$ with $u \neq v$. We can write u and v as $u = x0y$ and $v = x1z$ (or vice versa) for some $x, y, z \in \{0, 1\}^*$. Note that $0 \leq |x| \leq n - 1$ and $|y| = |z|$. We define the words $u' = x0y1^{n-|y|}01 = u1^{n-|y|}01$ and $v' = x1z1^{n-|z|}01 = v1^{n-|y|}01$. Since $\delta(q_0, u) = \delta(q_0, v)$ we have $\delta(q_0, u') = \delta(q_0, v')$. It should be clear that $u' \in L_n = \{0, 1\}^* \setminus V_n$. Hence, in order to get a contradiction, it suffices to show $v' \notin L_n$. First, note that $|v'| = 2n - |y| + 2 \leq 2n + 2$. This implies that if $v' \in L_n$, then it must belong to $\{0, 1\}^i 0 \{0, 1\}^n 0 \{0, 1\}^+$ for some $0 \leq i \leq n - 1$. But the word $v' = x1z1^{n-|z|}01$ contains no factor from $0\{0, 1\}^n 0$ (note that $x1z$ has length n and hence cannot contain such a factor). □

Figure 2 shows a well-behaved NFA A_n with $\Theta(n)$ states for the language L_n. The NTTA $B_n := A_n^t$ has $\Theta(n)$ states as well and satisfies $P(\text{lcp}(T(B_n))) = \{0, 1\}^* \setminus L(A_n) = V_n$ by Lemma 5. From Lemmas 1 and 6 it follows that the minimal DAG for $\text{lcp}(T(A_n^t))$ has at least $2^n - 1$ nodes. We have shown:

Theorem 1. *For every n there is a NTTA B_n with $\Theta(n)$ states such that the minimal DAG for the tree $\text{lcp}(T(B_n))$ has at least $2^n - 1$ nodes.*

The bound $2^n - 1$ in Theorem 1 is optimal up to constant factors in the exponent: If B is an NTTA with n states then by Lemma 2, $\text{lcp}(T(B))$ has depth at most $n - 1$ and hence at most $2^n - 1$ nodes. A variation of the above construction yields a slightly weaker lower bound for DBTAs:

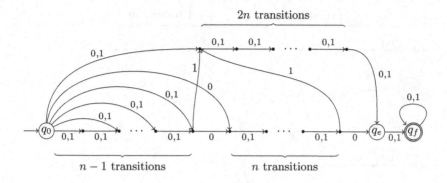

Fig. 2. The well-behaved NFA A_n recognizing the language L_n

Theorem 2. *For every n there is a DBTA B'_n with $\Theta(n^2)$ states such that the minimal DAG for the tree $\mathrm{lcp}(T(B'_n))$ has at least $2^n - 1$ nodes.*

Proof. Consider the NFA $A'_n = (Q, \{0,1\}, I, \{q_f\}, \delta)$ from Fig. 3. It is a well-behaved NFA that recognizes the language L_n as well. The transition function δ can be viewed as a mapping $\delta : Q \times \{0,1\} \to Q$ (nondeterminism only comes from the fact that there are several initial states). Moreover, δ has the property that for all states $p, q \in Q \setminus \{q_f\}$, if $p \neq q$ then $\delta(p,0)\delta(p,1) \neq \delta(q,0)\delta(q,1)$. Together with the fact that q_e is the unique final state of the NTTA A'^t, this implies that A'^t is a DBTA. $\square$

5.2   Incompressibility by Tree Straight-Line Programs

So far we considered the compression of trees by DAGs. Let us now consider the more general formalism of tree straight-line programs (TSLPs) [4,6].[1] As explained in Sect. 2 we consider binary trees as expressions over the leaf symbol a and the binary symbol f. A TSLP is a 4-tuple $\mathcal{G} = (V_0, V_1, \rho, S)$ where V_0 and V_1 are finite disjoint sets of variables, $S \in V_0$ is the start nonterminal, and ρ is a function that assigns to each variable A a formal expression (the right-hand side of A) such that one of the following conditions holds:

(a) $A \in V_0$ and $\rho(A) = a$,
(b) $A, B, C \in V_0$ and $\rho(A) = f(B, C)$,
(c) $A, C \in V_0$, $B \in V_1$ and $\rho(A) = B(C)$,
(d) $A, B, C \in V_1$ and $\rho(A) = B(C)$,
(e) $A \in V_1$, $B \in V_0$ and $\rho(A) = f(B, x)$,
(f) $A \in V_1$, $B \in V_0$ and $\rho(A) = f(x, B)$.

---

[1] We define here monadic TSLPs in normal form [6] which makes no difference with respect to succinctness; see [6].

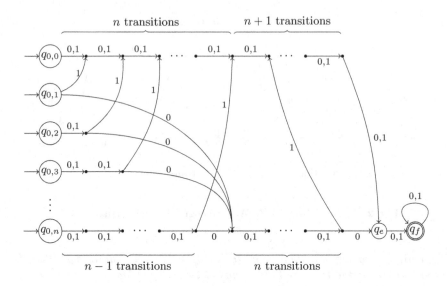

Fig. 3. The well-behaved NFA A'_n recognizing L_n from the proof of Theorem 2.

We require that the binary relation $E(\mathcal{G}) = \{(B, A) \mid B$ occurs in $\rho(A)\}$ is acyclic. We can therefore define a partial order $\leq_\mathcal{G}$ as the reflexive transitive closure of $E(\mathcal{G})$. The idea is that with the above rules, every variable $A \in V_0$ evaluates to a unique binary tree $\llbracket A \rrbracket_\mathcal{G}$, whereas every variable $A \in V_1$ evaluates to a unique binary tree $\llbracket A \rrbracket_\mathcal{G}$ with a marked leaf. This marked leaf is denoted by the special symbol x. For instance, $f(f(a, x), f(a, a))$ would be such a tree. We let $T_{2,x}$ denote the set of all such trees. For $s \in T_{2,x}$ and $t \in T_{2,x} \cup T_2$ we let $s[t]$ denote the result of replacing in s the unique occurrence of x by t. For instance, for $s = f(f(a, x), f(a, a))$ and $t = f(a, x)$ we have $s[t] = f(f(a, f(a, x)), f(a, a))$. Here are the formal inductive rules for the evaluation of variables. In all cases $t_B := \llbracket B \rrbracket_\mathcal{G}$ and $t_C := \llbracket C \rrbracket_\mathcal{G}$ are already defined by induction.

- if $A \in V_0$ and $\rho(A) = a$, then $\llbracket A \rrbracket_\mathcal{G} = a$,
- if $A, B, C \in V_0$ and $\rho(A) = f(B, C)$, then $\llbracket A \rrbracket_\mathcal{G} = f(t_B, t_C)$,
- if $A, C \in V_0$, $B \in V_1$ and $\rho(A) = B(C)$, then $\llbracket A \rrbracket_\mathcal{G} = t_B[t_C]$,
- if $A, B, C \in V_1$ and $\rho(A) = B(C)$, then $\llbracket A \rrbracket_\mathcal{G} = t_B[t_C]$,
- if $A \in V_1$, $B \in V_0$ and $\rho(A) = f(B, x)$, then $\llbracket A \rrbracket_\mathcal{G} = f(t_B, x)$,
- if $A \in V_1$, $B \in V_0$ and $\rho(A) = f(x, B)$, then $\llbracket A \rrbracket_\mathcal{G} = f(x, t_B)$.

Finally, we define $\llbracket \mathcal{G} \rrbracket = \llbracket S \rrbracket_\mathcal{G} \in T_2$. Readers that are familiar with the notion of context-free tree grammars will notice that a TSLP is a context-free tree grammar that produces a unique tree. A DAG corresponds to a TSLP where only variables of the above types (a) and (b) are present. In contrast to DAGs, TSLPs can also compress deep narrow trees, such as caterpillar trees, for example.

Lemma 7. *Let $\mathcal{G} = (V_0, V_1, \rho, S)$ be a TSLP with $t = \llbracket \mathcal{G} \rrbracket$ and let d be the depth of t. Then the minimal DAG for t has at most $|V_0| \cdot d$ nodes.*

Proof. We count the number of pairwise non-isomorphic subtrees of t. Consider a specific subtree $s \in T_2$ of t. By walking down from the start variable $S \in V_0$ we can determine the smallest variable A (with respect to $\leq_{\mathcal{G}}$) such that s is a subtree of $[\![A]\!]_{\mathcal{G}}$. Let us consider the cases (a)–(f) for the right-hand side $\rho(A)$.

If (a) or (b) holds, then we must have $s = [\![A]\!]_{\mathcal{G}}$. The cases (e) and (f) cannot occur (in both cases s would be a subtree of $[\![B]\!]_{\mathcal{G}}$). Similarly, (d) cannot occur since s would be a subtree of either $[\![B]\!]_{\mathcal{G}}$ or $[\![C]\!]_{\mathcal{G}}$. Finally in case (c), since s is neither a subtree of $[\![B]\!]_{\mathcal{G}}$ nor $[\![C]\!]_{\mathcal{G}}$, the subtree s must be rooted at one of the nodes on the path leading from the root of $[\![A]\!]_{\mathcal{G}}$ to the position of the symbol x in $[\![B]\!]_{\mathcal{G}}$ (excluding the position of x). There are at most d such nodes. It follows that $[\![\mathcal{G}]\!]$ contains at most $|V_0| \cdot d$ different subtrees. $\quad\square$

Theorem 3. *For every n there is an NTTA B_n with $\Theta(n)$ states such that the smallest TSLP for the tree $\mathrm{lcp}(T(B_n))$ has $\Omega(2^n/n)$ variables. Moreover, for every n there is a DBTA B'_n with $\Theta(n^2)$ states such that the smallest TSLP for the tree $\mathrm{lcp}(T(B'_n))$ has $\Omega(2^n/n)$ variables.*

Proof. We take the tree automata families from Theorems 1 and 2, respectively. Assume that $\mathcal{G} = (V_0, V_1, \rho, S)$ is a TSLP for the tree $\mathrm{lcp}(T(B_n))$ from Theorem 1. The minimal DAG for $\mathrm{lcp}(T(B_n))$ has at least $2^n - 1$ nodes. Recall that $P(\mathrm{lcp}(T(B_n))) = V_n = \{0,1\}^* \setminus L_n$. Since L_n contains all word of length at least $2n+3$, the path language V_n contains only words of length at most $2n+2$. Thus, the depth of the tree $\mathrm{lcp}(T(B_n))$ is at most $2n + 2$. With Lemma 7 it follows that the smallest TSLP for $\mathrm{lcp}(T(B_n))$ has at least $(2^n - 1)/(2n + 2)$ variables. For DBTAs one can argue analogously using Theorem 2. $\quad\square$

The upper bound $\Omega(2^n/n)$ for NTTAs in Theorem 3 cannot be improved much: As remarked before, if an NTTA B has n states then the tree $\mathrm{lcp}(T(B))$ has at most 2^n nodes. By [4], $\mathrm{lcp}(T(B))$ has a TSLP with $\mathcal{O}(2^n/n)$ variables.

6 Checking Equality of Largest Common Prefixes

We now deal with the problem of checking whether to tree languages yield the same lcp (or whether one lcp is contained in the other lcp). For DTTAs this is possible in polynomial time, whereas the problem becomes **coNP**-complete for DBTAs.

Theorem 4. *The problem of checking $P(\mathrm{lcp}(T(B_1))) \subseteq P(\mathrm{lcp}(T(B_2)))$ for two given DTTAs B_1 and B_2 can be solved in polynomial time.*

Proof. We compute the DFAs B_1^s and B_2^s from Sect. 3. Since B_1 and B_2 are DTTAs, these DFAs can be computed in polynomial time. By Lemma 4 we have $P(\mathrm{lcp}(T(B_1))) \subseteq P(\mathrm{lcp}(T(B_2)))$ if and only if $L(B_1^s) \subseteq L(B_2^s)$. The theorem follows because inclusion of DFAs can be checked in polynomial time. $\quad\square$

Theorem 5. *The problem of checking $P(\mathrm{lcp}(T(B_1))) \subseteq P(\mathrm{lcp}(T(B_2)))$ for two given NTTAs B_1 and B_2 belongs to **coNP**.*

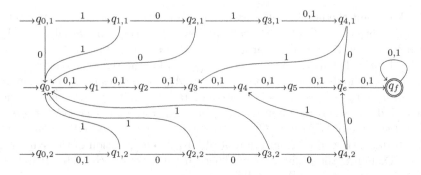

Fig. 4. The construction from the proof of Theorem 6 for the 3-SAT formula $C = C_1 \wedge C_2$ with $C_1 = (\neg x_1 \vee x_2 \vee \neg x_3)$ and $C_2 = (x_2 \vee x_3 \vee x_4)$ (so $n = 4$ and $m = 2$).

Proof. We show that there exists a nondeterministic polynomial time machine that checks whether there exists $u \in P(\mathrm{lcp}(T(B_1)))$ with $u \notin P(\mathrm{lcp}(T(B_2)))$. W.l.o.g. we can assume that B_1 and B_2 are productive. Let m be the number of states of B_1. By Lemma 2 we know that $P(\mathrm{lcp}(T(B_1)))$ only contains words of length at most $m - 1$. Hence, we can nondeterministically guess a word u of length at most $m - 1$ and then verify whether $u \in P(\mathrm{lcp}(T(B_1)))$ and $u \notin P(\mathrm{lcp}(T(B_2)))$. For this we use the DFAs B_1^s and B_2^s from Lemma 3 and check whether $u \in L(B_1^s)$ and $u \notin L(B_2^s)$. For this, we do not have to construct the DFAs B_1^s and B_2^s explicitly (they have exponential size); it suffices to run B_1^s and B_2^s on the fly on the word u (recall that u has polynomial length). $\square$

Theorem 6. *The problem of checking* $\mathrm{lcp}(T(B_1)) = \mathrm{lcp}(T(B_2))$ *for two given NTTAs B_1 and B_2 is* **coNP**-*complete. The* **coNP** *lower bound already holds for the case that B_1 and B_2 are DBTAs.*

Proof. Since **coNP** is closed under intersection, we obtain the upper bound from Theorem 5. Let us now show **coNP**-hardness for DBTAs by a reduction from the complement of 3-SAT. Consider a 3-SAT formula $C = \bigwedge_{i=1}^{m} C_i$ where every C_i is a disjunction of three literals (possibly negated variables). Let $x_1, \ldots, x_n$ be the variables that occur in C. W.l.o.g. we can assume that $n \geq m$ (we can add dummy variables if necessary) and that there is no clause C_i and variable x_j such that x_j and $\neg x_j$ both belong to C_i. Given a bit string $w = a_1 a_2 \cdots a_n$ with $a_i \in \{0, 1\}$ we write $w \models C_i$ (resp., $w \models C$) if C_i (resp., C) becomes true when every variable x_i gets the truth value a_i.

We first construct an (incomplete) acyclic DFA A_i for the language $\{w0 \mid w \in \{0, 1\}^n, w \not\models C_i\}$. The states of A_i are $q_{0,i}, q_{1,i}, \ldots, q_{n,i}, q_{n+1,i}$, $q_{0,i}$ is the initial state, $q_{n+1,i}$ is the final state, and the transitions are defined as follows, where $1 \leq j \leq n$:

- $q_{j-1,i} \xrightarrow{0} q_{j,i}$ if x_j belongs to C_i,
- $q_{j-1,i} \xrightarrow{1} q_{j,i}$ if $\neg x_j$ belongs to C_i,

- $q_{j-1,i} \xrightarrow{0,1} q_{j,i}$ if neither x_j nor $\neg x_j$ belongs to C_i,
- $q_{n,i} \xrightarrow{0} q_{n+1,i}$.

By taking the disjoint union of the DFAs A_i, we obtain an NFA A with

$$L(A) = \bigcup_{i=1}^{n} L(A_i)$$

$$= \bigcup_{i=1}^{n} \{w0 \mid w \in \{0,1\}^n, w \not\models C_i\}$$

$$= \{w0 \mid w \in \{0,1\}^n, w \not\models C\}.$$

Hence, we have $L(A) = \{0,1\}^n 0$ if and only if C is not satisfiable. Note that the initial states of A are the states $q_{0,1}, \ldots, q_{0,m}$.

We finally construct a well-behaved NFA A_1 from A as follows (an example is shown in Fig. 4):

- Merge the final states $q_{n+1,i}$ ($1 \le i \le m$) into a single non-final state q_e.
- Add states $q_0, q_1, \ldots, q_{n+1}, q_f$, where q_0 is an initial state (hence, the initial states of A_1 are $q_0, q_{0,1}, \ldots, q_{0,m}$) and q_f is the unique final state of A_1.
- Add the transitions $q_j \xrightarrow{0,1} q_{j+1}$ for $0 \le j \le n$, $q_{n+1} \xrightarrow{0,1} q_e \xrightarrow{0,1} q_f \xrightarrow{0,1} q_f$.
- If some state $q_{j-1,i}$ ($1 \le i \le m$, $1 \le j \le n$) has no outgoing a-transition for $a \in \{0,1\}$ (this happens if $a = 0$ and $\neg x_j$ belongs to C_i or $a = 1$ and x_j belongs to C_i) then add the transition $q_{j-1,i} \xrightarrow{a} q_0$ to A_1.
- For every $1 \le i \le m$ we add a 1-transition from $q_{n,i}$ to one of the states $q_0, \ldots, q_{n+1}$ in such a way that no two such 1-transitions enter the same state. Since $m \le n + 2$, this is possible.

The automaton A_1 satisfies $L(A_1) = L(A)\{0,1\}\{0,1\}^* \cup \{0,1\}^{n+3}\{0,1\}^*$ and is is well-behaved. The transition function δ of A_1 can be viewed as a mapping $\delta : Q \times \{0,1\} \to Q$ (nondeterminism only comes from the fact that there are several initial states). Moreover, δ has the property that for all states $p, q \in Q \setminus \{q_f\}$, if $p \ne q$ then $\delta(p,0)\delta(p,1) \ne \delta(q,0)\delta(q,1)$. Together with the fact that q_e is the unique final state of the NTTA A_1^t, this implies that A_1^t is a DBTA.

It is straightforward to construct a well-behaved NFA A_2 such that $L(A_2) = \{0,1\}^n 0\{0,1\}\{0,1\}^* \cup \{0,1\}^{n+3}\{0,1\}^*$ and A_2^t is a DBTA (one can make the above construction with an unsatisfiable 3-SAT formula). We get the following equivalences:

$$C \text{ is unsatisfiable} \Leftrightarrow L(A) = \{0,1\}^n 0$$

$$\Leftrightarrow L(A_1) = \{0,1\}^n 0\{0,1\}\{0,1\}^* \cup \{0,1\}^{n+3}\{0,1\}^*$$

$$\Leftrightarrow L(A_1) = L(A_2)$$

$$\Leftrightarrow \{0,1\}^* \setminus L(A_1) = \{0,1\}^* \setminus L(A_2)$$

$$\Leftrightarrow P(\mathrm{lcp}(T(A_1^t))) = P(\mathrm{lcp}(T(A_2^t)))$$

$$\Leftrightarrow \mathrm{lcp}(T(A_1^t)) = \mathrm{lcp}(T(A_2^t)).$$

This concludes the proof of the theorem. $\square$

References

1. Choffrut, C.: Minimizing subsequential transducers: a survey. Theor. Comput. Sci. **292**(1), 131–143 (2003)
2. Comon, H., et al.: Tree automata techniques and applications (2007). http://tata. gforge.inria.fr/
3. Engelfriet, J., Maneth, S., Seidl, H.: Deciding equivalence of top-down XML transformations in polynomial time. J. Comput. Syst. Sci. **75**(5), 271–286 (2009)
4. Ganardi, M., Hucke, D., Jez, A., Lohrey, M., Noeth, E.: Constructing small tree grammars and small circuits for formulas. J. Comput. Syst. Sci. **86**, 136–158 (2017)
5. Lemay, A., Maneth, S., Niehren, J.: A learning algorithm for top-down XML transformations. In: Proceedings of PODS 2010, pp. 285–296. ACM (2010)
6. Lohrey, M., Maneth, S., Schmidt-Schauß, M.: Parameter reduction and automata evaluation for grammar-compressed trees. J. Comput. Syst. Sci. **78**(5), 1651–1669 (2012)
7. Luttenberger, M., Palenta, R., Seidl, H.: Computing the longest common prefix of a context-free language in polynomial time. In: Proceedings of STACS 2018. LIPIcs, vol. 96, pp. 48:1–48:13. Schloss Dagstuhl - Leibniz-Zentrum für Informatik (2018)
8. Mohri, M.: Minimization algorithms for sequential transducers. Theor. Comput. Sci. **234**(1–2), 177–201 (2000)
9. Oncina, J., García, P., Vidal, E.: Learning subsequential transducers for pattern recognition interpretation tasks. IEEE Trans. Pattern Anal. Mach. Intell. **15**(5), 448–458 (1993)

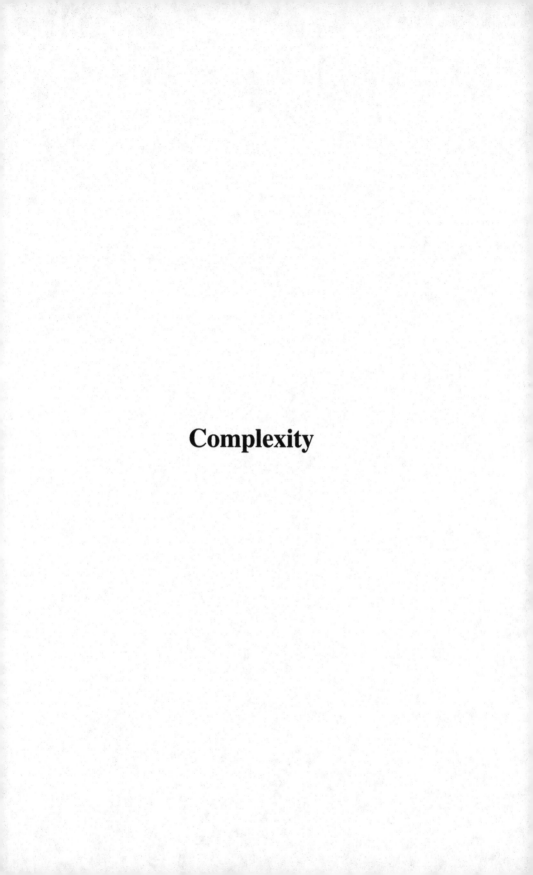

Complexity

On Weisfeiler-Leman Invariance: Subgraph Counts and Related Graph Properties

Vikraman Arvind[1], Frank Fuhlbrück[2]([⊠]) [ID], Johannes Köbler[2], and Oleg Verbitsky[2]

[1] The Institute of Mathematical Sciences (HBNI), Chennai, India
arvind@imsc.res.in
[2] Humboldt-Universität zu Berlin, Unter den Linden 6, 10099 Berlin, Germany
{fuhlbfra,koebler,verbitsky}@informatik.hu-berlin.de

Abstract. The k-dimensional Weisfeiler-Leman algorithm (k-WL) is a fruitful approach to the Graph Isomorphism problem. 2-WL corresponds to the original algorithm suggested by Weisfeiler and Leman over 50 years ago. 1-WL is the classical color refinement routine. Indistinguishability by k-WL is an equivalence relation on graphs that is of fundamental importance for isomorphism testing, descriptive complexity theory, and graph similarity testing which is also of some relevance in artificial intelligence. Focusing on dimensions $k = 1, 2$, we investigate subgraph patterns whose counts are k-WL invariant, and whose occurrence is k-WL invariant. We achieve a complete description of all such patterns for dimension $k = 1$ and considerably extend the previous results known for $k = 2$.

Keywords: Isomorphism and similarity of graphs ·
Weisfeiler-Leman algorithm · Subgraph counts

1 Introduction

Color refinement is a classical procedure widely used in isomorphism testing and other areas. It initially colors each vertex of an input graph by its degree and refines the vertex coloring in rounds, taking into account the colors appearing in the neighborhood of each vertex. This simple and efficient procedure successfully canonizes almost all graphs in linear time [5]. Combined with *individualization*, it is the basis of the most successful practical algorithms for the graph isomorphism problem; see [27] for an overview and historical comments.

The first published work on color refinement dates back at least to 1965 (Morgan [29]). In 1968 Weisfeiler and Leman [37] gave a procedure that assigns colors to *pairs* of vertices of the input graph $G = (V, E)$. The initial colors are *edge*, *nonedge*, and *loop*. The coloring is refined in rounds by assigning a new

O. Verbitsky was supported by DFG grant KO 1053/8–1. He is on leave from the IAPMM, Lviv, Ukraine.

L. A. Gąsieniec et al. (Eds.): FCT 2019, LNCS 11651, pp. 111–125, 2019.
https://doi.org/10.1007/978-3-030-25027-0_8

color to each pair $(u, v) \in V \times V$ based on the color types of the 2-walks uwv, as w ranges V. The procedure terminates when the color partition of $V \times V$ *stabilizes*. I.e., the color partition does not refine any further. The output coloring is an isomorphism invariant of the input graph. It yields an edge-colored complete directed graph with certain highly regular properties. This object, known as a *coherent configuration*, has independently been discovered in other contexts in statistics (Bose [8]) and algebra (Higman [23]).

A natural extension of this idea, due to Babai (see [4,9]), is to iteratively classify k-tuples of vertices. This is the *k-dimensional Weisfeiler-Leman procedure*, abbreviated as k-WL. Thus, 2-WL is the original Weisfeiler-Leman algorithm [37], and 1-WL is color refinement. The running time of k-WL is $n^{O(k)}$, where n denotes the number of vertices in an input graph. Cai, Fürer, and Immerman [9] have shown infinitely many pairs of nonisomorphic graphs (G_i, H_i) such that k-WL fails to distinguish between them for any $k = o(n)$. Nevertheless, the Weisfeiler-Leman procedure, as an essential component in isomorphism testing, can hardly be overestimated. A constant dimension often suffices to solve the isomorphism problem for important graph classes. A striking result here (Grohe [21]) is that for any graph class excluding a fixed minor (like bounded genus or bounded treewidth graphs) isomorphism can be tested using k-WL for a constant k that only depends on the excluded minor. Moreover, Babai's quasipolynomial-time algorithm [4] for general graph isomorphism crucially uses k-WL for logarithmic k.

We call two graphs G and H *k-WL-equivalent* and write $G \equiv_{k\text{-WL}} H$ if they are indistinguishable by k-WL; formal definitions are given in Sects. 2 ($k = 1$) and 3 ($k \geq 2$). It follows from the Cai-Fürer-Immerman result [9] that for any k the $\equiv_{k\text{-WL}}$-equivalence is coarser than the isomorphism relation on graphs.

Definition 1. *A graph property (that is, an isomorphism-invariant family of graphs) $\mathcal{P}$ is k-WL-invariant if for any pair of graphs G and H:*

$$G \in \mathcal{P} \text{ and } G \equiv_{k\text{-WL}} H \text{ implies } H \in \mathcal{P}.$$

In particular, a graph parameter π is k-WL-invariant if $\pi(G) = \pi(H)$ whenever $G \equiv_{k\text{-WL}} H$.

The broad question of interest in this paper is which graph properties (and graph parameters) are k-WL-invariant for a specified k. The motivation for it comes from various areas. Understanding the power of k-WL, even for small values of k, is important for both isomorphism testing and graph similarity testing. For example, the largest eigenvalues of 1-WL-equivalent graphs are equal [34]. Moreover, 2-WL-equivalent graphs are cospectral [13,17]. Consequently, by Kirchhoff's theorem, 2-WL-equivalent graphs have the same number of spanning trees. Also the 2-WL-invariance of certain metric graph parameters such as diameter is easy to show. Fürer [18] recently asked which basic combinatorial parameters are 2-WL-invariant. While it is readily seen that 2-WL-equivalence preserves the number of 3-cycles, Fürer pointed out, among other interesting observations, that also the number of s-cycles is 2-WL-invariant for each $s \leq 6$. More recently, Dell, Grohe, and Rattan [14] characterized k-WL-equivalence in terms of homomorphism profiles. Specifically, they show that $G \equiv_{k\text{-WL}} H$ if and only if the number of homomorphisms from F to G equals the number of homomorphisms from F to H for all graphs F of treewidth at most k.

As a heuristic for graph similarity testing, the Weisfeiler-Leman procedure has been applied in artificial intelligence; see [35] for an 1-WL-based application and [30] for a multidimensional version. It is noteworthy that 1-WL turns out to be exactly as powerful as graph neural networks [31]. Comparing subgraph frequencies is also widely used for testing graph similarity and detecting structure of large real-life graphs; see, e.g., [19,20,28,36]. For example, just knowing the number of triangles is valuable information about a social network; see, e.g., [22]. Important structural information can also be found from the number of paths of length 2 and from the degree distribution, i.e., the statistics of star subgraphs; see [32]. This poses a natural question on how much the two approaches—one based on k-WL-equivalence and the other on subgraph statistics—are related to each other.

Finally, k-WL-equivalence is of fundamental importance for finite and algorithmic model theory. A graph property $\mathcal{P}$ is k-WL-invariant precisely when $\mathcal{P}$ is definable in the $(k + 1)$-variable infinitary counting logic. If $\mathcal{P}$ is not $\equiv_{k\text{-WL}}$-invariant for any k then, in fact, $\mathcal{P}$ is not definable in fixed-point logic with counting (FPC); see, e.g., the survey [12]. A systematic study of k-WL-invariant constraint satisfaction problems was undertaken by Atserias, Bulatov, and Dawar [3].

Our Results. Let F be a fixed pattern graph and G be any given graph. We investigate the k-WL-invariance of: (a) the property that G contains F as a subgraph, and (b) the number of subgraphs of G isomorphic to F. We use $\text{sub}(F, \cdot)$ to denote the subgraph count function. Thus, $\text{sub}(F, G)$ denotes the number of subgraphs of G isomorphic to F.

Definition 2. *Let $\mathcal{C}(k)$ denote the class of all pattern graphs F for which the subgraph count $\text{sub}(F, \cdot)$ is $\equiv_{k\text{-WL}}$-invariant. Furthermore, $\mathcal{R}(k)$ consists of all pattern graphs F such that the property of a graph containing F as a subgraph is $\equiv_{k\text{-WL}}$-invariant.*

In a sense, the graph classes $\mathcal{C}(k)$ and $\mathcal{R}(k)$ correspond to algorithmic *counting* and *recognition* problems, respectively.

Note that $\mathcal{C}(k) \subseteq \mathcal{R}(k)$. We use this notation to state some consequences of prior work. The k-WL-equivalence characterization [14], stated above, can be used to show that $\mathcal{C}(k)$ contains every F such that all homomorphic images of F have treewidth no more than k. We say that such an F has *homomorphism-hereditary treewidth* at most k; see Sect. 3 for details. The striking result of Anderson et al. [1], showing the expressibility of the matching number in FPC, implies that there is some k such that $\mathcal{R}(k)$ contains *all* matching graphs sK_2, where sK_2 denotes the disjoint union of s edges. On the other hand, there is no

k such that $\mathcal{R}(k)$ contains all cycle graphs C_s. This readily follows from Dawar's result [11,12] that graph hamiltonicity is not $\equiv_{k\text{-WL}}$-invariant for any k.

Our results are as follows.

Complete Description of $\mathcal{C}(1)$ and $\mathcal{R}(1)$ (Invariance Under Color Refinement). We prove that, up to adding isolated vertices, $\mathcal{C}(1)$ consists of all star graphs $K_{1,s}$ and the 2-matching graph $2K_2$. Hence, $\mathcal{C}(1)$ contains exactly the pattern graphs of homomorphism-hereditary treewidth equal to 1. An interesting consequence is that, for every $F \in \mathcal{C}(1)$, the subgraph count $\text{sub}(F,G)$ is determined just by the degree sequence of a graph G. We obtain a complete description of $\mathcal{R}(1)$ by proving that this class consists of the graphs in $\mathcal{C}(1)$ and three forests $P_3 + P_2$, $P_3 + 2P_2$, and $2P_3$, where P_s denotes the path graph on s vertices.

Case Study for $\mathcal{C}(2)$ and $\mathcal{R}(2)$ (Invariance Under the Original Weisfeiler-Leman Algorithm). An explicit characterization of $\mathcal{C}(2)$ and $\mathcal{R}(2)$ appears challenging. Indeed, it is not a priori clear whether testing membership in these graph classes is possible in polynomial time. While it is unknown whether $\mathcal{C}(2)$ consists exactly of graphs with homomorphism-hereditary treewidth bounded by 2, we prove that this is indeed the case for some standard graph sequences. These results are related to questions that have been discussed in the literature.

- Beezer and Farrell [6] proved that the first five coefficients of the matching polynomial of a strongly regular graph are determined by its parameters.[1] I.e., if G and H are strongly regular graphs with the same parameters, then $\text{sub}(sK_2, G) = \text{sub}(sK_2, H)$ for $s \leq 5$. We prove that $sK_2 \in \mathcal{C}(2)$ if and only if $s \leq 5$. It follows that the Beezer-Farrell result extends to 2-WL-equivalent graphs. I.e., if G and H are any two 2-WL-equivalent graphs, then the first five coefficients of their matching polynomials coincide. Moreover, this result is tight and cannot be extended to a larger s. Note that strongly regular graphs with the same parameters are the simplest example of 2-WL-equivalent graphs.
- Fürer [18] proved that $C_s \in \mathcal{C}(2)$ for $3 \leq s \leq 6$ and $C_s \notin \mathcal{C}(2)$ for $8 \leq s \leq 16$. We close the gap and show that C_7 is the largest cycle graph in $\mathcal{C}(2)$. We also prove that $\mathcal{C}(2)$ contains $P_1, \ldots, P_7$ and no other path graphs.
  The result on cycles admits the following generalization. First, we observe that the girth $g(G)$ of a graph G is a 2-WL-invariant parameter. Then, we prove that if $G \equiv_{2\text{-WL}} H$, then $\text{sub}(C_s, G) = \text{sub}(C_s, H)$ for each $3 \leq s \leq 2\,g(G)+1$. Neither the factor of 2, nor the additive term of 1 can here be improved.

Characterization of $\mathcal{R}(2)$ appears to be still harder. Fürer [18] has shown that $\mathcal{R}(2)$ does not contain K_4 (the complete graph with 4 vertices). Building on that, we show that $\mathcal{R}(2)$ also does not contain any graph F with a unique 4-clique. In view of this result, it is natural to conjecture that $\mathcal{R}(2)$ does not contain any graph of clique number more than 3.

Due to space restrictions some proofs are only available in the full version of this article, see [2].

---

[1] The result of [6] is actually stronger and applies even to distance-regular graphs.

Notation. The *girth* $g(G)$ is the minimum length of a cycle in G. If G is acyclic, then $g(G) = \infty$. We denote the vertex set of G by $V(G)$ and the edge set by $E(G)$. Furthermore, $v(G) = |V(G)|$ and $e(G) = |E(G)|$. The set of vertices adjacent to a vertex $u \in V(G)$ forms its neighborhood $N(u)$. The vertex-disjoint union of graphs G and H is denoted by $G+H$. Furthermore, we write mG for the disjoint union of m copies of G. We use the standard notation K_n for complete graphs, P_n for paths, and C_n for cycles on n vertices. Furthermore, $K_{s,t}$ denotes the complete bipartite graph whose vertex classes have s and t vertices. Likewise, $K_{1,1} = K_2 = P_2$, $K_{1,2} = P_3$, $C_3 = K_3$ etc.

2   Color Refinement Invariance

Given a graph G, the *color-refinement* algorithm (abbreviated as 1-WL) iteratively computes a sequence of colorings C^i of $V(G)$. The initial coloring C^0 is monochromatic, that is $C^0(u)$ is the same for all vertices u. Then,

$$C^{i+1}(u) = \left(C^i(u), \{\!\{\, C^i(a) : a \in N(u)\}\!\}\right), \tag{1}$$

where $\{\!\{\ldots\}\!\}$ denotes a multiset (i.e., the multiplicity of each element counts).

If ϕ is an isomorphism from G to H, then a straightforward inductive argument shows that $C^i(u) = C^i(\phi(u))$ for each vertex u of G. This readily implies that, if graphs G and H are isomorphic, then

$$\{\!\{\, C^i(u) : u \in V(G)\}\!\} = \{\!\{\, C^i(v) : v \in V(H)\}\!\} \tag{2}$$

for all $i \geq 0$. We write $G \equiv_{\text{1-WL}} H$ exactly when this condition is met.

A direct consequence of the definition is the following.

Lemma 3. *If $A \equiv_{\text{1-WL}} B$ and $A' \equiv_{\text{1-WL}} B'$, then $A + A' \equiv_{\text{1-WL}} B + B'$.*

1-WL *distinguishes* graphs G and H if $G \not\equiv_{\text{1-WL}} H$. In fact, the algorithm does not need to check (2) for infinitely many i: If Equality (2) is false for some i then it is false for $i = n$, where n denotes the number of vertices in each of the graphs.

Let F be a graph and s be a positive integer. Note that F belongs to $\mathcal{C}(k)$ or $\mathcal{R}(k)$ if and only if the graph $F + sK_1$ belongs to this class. Therefore, we will ignore isolated vertices.

Theorem 4. *Up to adding isolated vertices, the two classes $\mathcal{C}(1)$ and $\mathcal{R}(1)$ are formed by the following graphs.*

1. *$\mathcal{C}(1)$ consists of the star graphs $K_{1,s}$ for all $s \geq 1$ and the 2-matching graph $2K_2$.*
2. *$\mathcal{R}(1)$ consists of the graphs in $\mathcal{C}(1)$ and the following three forests:*

$$P_3 + P_2, \ P_3 + 2P_2, \ and \ 2P_3. \tag{3}$$

The proof occupies the rest of this section.

Membership in $\mathcal{C}(1)$. If two graphs are indistinguishable by color refinement, they have the same degree sequence. Notice that

$$\text{sub}(K_{1,s}, G) = \sum_{v \in V(G)} \binom{\deg v}{s},$$

where $\deg v$ denotes the degree of a vertex v. This equality shows that $K_{1,s} \in \mathcal{C}(1)$. Since any two edges constitute either $2K_2$ or $K_{1,2}$, we have

$$\text{sub}(2K_2, G) = \binom{e(G)}{2} - \text{sub}(K_{1,2}, G). \tag{4}$$

As $e(G) = \frac{1}{2}\sum_{v \in V(G)} \deg v$, it follows that $2K_2 \in \mathcal{C}(1)$.

The equality (4) has been reported earlier; see, e.g., [16, Lemma 1] and the comments therein.

Non-membership in $\mathcal{C}(1)$. To prove that a graph F is not in $\mathcal{C}(1)$, we need to exhibit 1-WL-equivalent graphs G and H such that $\text{sub}(F, G) \neq \text{sub}(F, H)$. For each of the 3 forests in (3) we can easily find witnesses G and H that are regular graphs with the same number of vertices and of the same degree. Specifically, $\text{sub}(P_3 + P_2, C_6) = 12$, while $\text{sub}(P_3 + P_2, 2C_3) = 18$; $\text{sub}(2P_3, C_6) = 3$, while $\text{sub}(2P_3, 2C_3) = 9$; and $\text{sub}(P_3 + 2P_2, C_7) = 7$, while $\text{sub}(P_3 + 2P_2, C_4 + C_3) = 6$. The non-membership of all other graphs in $\mathcal{C}(1)$ follows from their non-membership in $\mathcal{R}(1)$, which will be proved in the corresponding subsection below.

Membership in $\mathcal{R}(1)$. We call a graph H *amenable* if color refinement distinguishes H from any other nonisomorphic graph G. For each of the three forests F in (3), we are able to explicitly describe the class $Forb(F)$ of F-free graphs. Based on this description, we can show that, with just a few exceptions, every F-free graph is amenable.

Lemma 5.

1. *Every $(P_3 + P_2)$- or $2P_3$-free graph H is amenable.*
2. *Every $(P_3 + 2P_2)$-free graph H is amenable unless $H = 2C_3$ or $H = C_6$.*

Proving that $F \in \mathcal{R}(1)$ means proving the following:

$$G \equiv_{1\text{-WL}} H \quad \& \quad H \in Forb(F) \implies G \in Forb(F). \tag{5}$$

This implication is trivial whenever H is an amenable graph because then $G \cong H$. By Part 1 of Lemma 5, we immediately conclude that the graphs $P_3 + P_2$ and $2P_3$ are in $\mathcal{R}(1)$. Part 2 ensures (5) for all $(P_3 + 2P_2)$-free graphs except $2C_3$ and C_6. However, the implication (5) also holds for each exceptional graph $H \in \{2C_3, C_6\}$ for the following simple reason. Since H has 6 vertices, any 1-WL-indistinguishable graph G must have also 6 vertices and hence cannot contain a $P_3 + 2P_2$ subgraph.

The proof of Lemma 5 relies on an explicit description of the class of F-free graphs for each $F \in \{P_3 + P_2, 2P_3, P_3 + 2P_2\}$, which requires a detailed combinatorial analysis; see Appendix of the full version [2].

Non-membership in $\mathcal{R}(1)$. We first show that $\mathcal{R}(1)$ can contain only forests of stars.

Lemma 6 (see Bollobás [7, Corollary 2.19] or Wormald [38, Theorem 2.5]). *Let $d, g \geq 3$ be fixed, and dn be even. Let $\mathcal{G}_{n,d}$ denote a random d-regular graph on n vertices. Then the probability that $\mathcal{G}_{n,d}$ has girth g converges to a non-zero limit as n grows large.*

Lemma 7. *$\mathcal{R}(1)$ can contain only acyclic graphs.*

Proof. Assume that a graph F has a cycle of length m. We show that it is not in $\mathcal{R}(1)$. Let $d = v(F) - 1$. By Lemma 6 there is a d-regular graph X of girth strictly more than m. Then F does not appear as a subgraph in $H = (d+1)X$ but clearly does in $G = v(X) K_{d+1}$. It remains to notice that G and H are both d-regular with the same number of vertices. □

Lemma 8. *$\mathcal{R}(1)$ can contain only forests of stars.*

Proof. Suppose that $F \in \mathcal{R}(1)$. By Lemma 7, F is a forest. In order to prove that every connected component of F is a star, it is sufficient and necessary to prove that F does not contain P_4 as a subgraph. Assume, to the contrary, that F has P_4-subgraphs.

Let T be a connected component of F containing P_4. Consider a diametral path $v_1 v_2 v_3 \ldots v_d$ in T, where $d \geq 4$. Note that v_1 is a leaf. Let T' be obtained from T by identifying the vertices v_1 and v_4. Thus, T' is a unicyclic graph, where the vertices v_2, v_3, and $v_4 = v_1$ form a cycle C_3. Obviously, $v(T') < v(T)$.

Consider now the graph $H_T = 2T'$. Identify one component of H_T with T' and fix an isomorphism α from this to the other component of H_T. Let G_T be obtained from H_T by removing the edges $v_2 v_4$ and $\alpha(v_2)\alpha(v_4)$ and adding instead the new edges $v_2\alpha(v_4)$ and $v_4\alpha(v_2)$. Note that, by construction, $V(T') \subset V(H_T) = V(G_T)$. Note that G_T contains a subgraph isomorphic to T. We now prove that

$$G_T \equiv_{1\text{-WL}} H_T. \tag{6}$$

Indeed, define a map $\phi : V(G_T) \to V(T')$ by $\phi(u) = \phi(\alpha(u)) = x$ for each $u \in V(T') \subset V(G_T)$. Note that ϕ is a *covering map* from G_T to T'. That is, ϕ is a surjective homomorphism whose restriction to the neighborhood of each vertex of G_T is surjective. A straightforward inductive argument shows that ϕ preserves the coloring produced by 1-WL. More precisely, $C^i(\phi(u)) = C^i(u)$ for all i, where C^i is defined by (1). Thus, the multiset $\{\!\!\{\, C^i(u) : u \in V(G_T)\,\}\!\!\}$ is obtained from the multiset $\{\!\!\{\, C^i(u) : u \in V(T')\,\}\!\!\}$ by doubling the multiplicity of each color. Since H_T consists of two disjoint copies of T', it readily follows that G_T and H_T are indistinguishable by 1-WL, which yields (6).

If a connected component T of F does not contain P_4, we set $G_T = H_T = 2T$. The equivalence (6) is true also in this case. Define $G = \sum_T G_T$ and $H = \sum_T H_T$

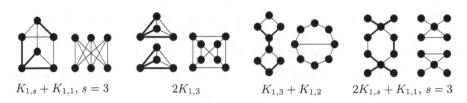

$K_{1,s} + K_{1,1},\ s = 3$ $2K_{1,3}$ $K_{1,3} + K_{1,2}$ $2K_{1,s} + K_{1,1},\ s = 3$

Fig. 1. G/H-certificates for each basic star forest F.

where the disjoint union is taken over all connected components T of F. Then $G \equiv_{\text{1-WL}} H$ by Lemma 3. Since each G_T contains a subgraph isomorphic to T, the graph G contains a subgraph isomorphic to F. On the other hand, H does not contain any subgraph isomorphic to F. To see this, let F_0 be a non-star component of F with maximum number of vertices. Then H cannot contain even F_0 because every non-star component of H has fewer vertices than F_0. This contradicts the assumption that $F \in \mathcal{R}(1)$.   □

Lemma 8 reduces our task to showing that star forests not listed in Theorem 4 are not in $\mathcal{R}(1)$. I.e., $\mathcal{R}(1)$ contains only the star forests listed below:

$$K_{1,s}\ (s \geq 1),\ 2K_{1,1},\ K_{1,2} + K_{1,1},\ 2K_{1,2},\ K_{1,2} + 2K_{1,1} \tag{7}$$

Our proof has the following structure. First, we directly argue the *non-membership* in $\mathcal{R}(1)$ for a few *basic* star forests. Then we obtain two *derivation rules*, based on some closure properties of $\mathcal{R}(1)$. Finally, in order to show that $F \notin \mathcal{R}(1)$ for some star forest F, we apply the derivation rules to F to derive one of the excluded basic star forests.

Lemma 9 (Basic star forests). *None of the star forests $K_{1,s} + K_{1,1}$ for any $s \geq 3$, $K_{1,3} + K_{1,2}$, $2K_{1,3}$, and $2K_{1,s} + K_{1,1}$ for any $s \geq 1$ belongs to $\mathcal{R}(1)$.*

Proof. For each basic star forest F listed in the lemma, we provide 1-WL-indistinguishable graphs G and H such that G contains F as a subgraph while H does not; see also Fig. 1.

$K_{1,s} + K_{1,1}$, $s \geq 3$: $H = K_{s,s}$ and G is obtained from $2K_s$ by adding a perfect matching between the two K_s parts.

$2K_{1,3}$: $G = 2K_4$ and H is the *Wagner graph* (or *4-Möbius ladder*).

$K_{1,3} + K_{1,2}$: G is obtained from $2C_4$ by adding an edge between the two C_4 parts, and H is obtained from C_8 by adding an edge between two antipodal vertices of the 8-cycle in H.

$2K_{1,s} + K_{1,1}$, $s \geq 1$: Both graphs G and H are obtained from $2K_{1,s+1}$ by adding two edges e. Let a and b be two leaves of the fist copy of $K_{1,s+1}$, and let a' and b' be two leaves of the other copy of $K_{1,s+1}$. Then G additionally contains two edges aa' and bb', whereas H additionally contains two edges ab and $a'b'$.   □

Lemma 10 (Derivation rules).

1. If $K_{1,i_1} + \ldots + K_{1,i_s} + K_{1,i_{s+1}} \in \mathcal{R}(1)$, then $K_{1,i_1} + \ldots + K_{1,i_s} \in \mathcal{R}(1)$.
2. If $K_{1,i_1+1} + \ldots + K_{1,i_s+1} \in \mathcal{R}(1)$, then $K_{1,i_1} + \ldots + K_{1,i_s} \in \mathcal{R}(1)$.

Proof. 1. Suppose that $K_{1,i_1} + \ldots + K_{1,i_s} \notin \mathcal{R}(1)$. Let G and H be two graphs witnessing this. That is, $G \equiv_{1\text{-WL}} H$ and G contains this star forest while H does not. Then the graphs $G + K_{1,i_{s+1}}$ and $H + K_{1,i_{s+1}}$, which are 1-WL-indistinguishable by Lemma 3, witness that $K_{1,i_1} + \ldots + K_{1,i_s} + K_{1,i_{s+1}} \notin \mathcal{R}(1)$.

2. Suppose that $K_{1,i_1} + \ldots + K_{1,i_s} \notin \mathcal{R}(1)$ and this is witnessed by G and H. Given a graph X, let X' be the graph obtained by attaching a new degree-1 vertex x' to each vertex x of X (thus, $v(X') = 2v(X)$). Then the graphs G' and H' witness that $K_{1,i_1+1} + \ldots + K_{1,i_s+1} \notin \mathcal{R}(1)$. Indeed, it is easy to see that X contains $K_{1,i_1} + \ldots + K_{1,i_s}$ if and only if X' contains $K_{1,i_1+1} + \ldots + K_{1,i_s+1}$ as a subgraph. The equivalence $G' \equiv_{1\text{-WL}} H'$ follows from the equivalence $G \equiv_{1\text{-WL}} H$. $\square$

Now, let F be a star forest not listed in (7). Assume that $F \in \mathcal{R}(1)$. Lemma 10 provides us with two derivations rules:

- if a star forest X is in $\mathcal{R}(1)$, then the result of removing one connected component from X is also in $\mathcal{R}(1)$;
- if a star forest X is in $\mathcal{R}(1)$, then the result of cutting off one leaf in each connected component of X is also in $\mathcal{R}(1)$.

Note that, applying these derivation rules, F can be reduced to one of the basic star forests. By Lemma 9, we get a contradiction, which completes the proof of Theorem 4.

3   Weisfeiler-Leman Invariance

The original algorithm described by Weisfeiler and Leman in [37], which is nowadays more often referred to as the *2-dimensional Weisfeiler-Leman algorithm*, operates on the Cartesian square V^2 of the vertex set of an input graph G. Initially it assigns each pair $(u, v) \in V^2$ one of three colors, namely *edge* if u and v are adjacent, *nonedge* if $u \neq v$ and u and v are non-adjacent, and *loop* if $u = v$. Denote this coloring by C^0. The coloring of V^2 is then refined step by step. The coloring after the i-th refinement step is denoted by C^i and is computed as

$$C^i(u, v) = C^{i-1}(u, v) \mid \{\!\{ C^{i-1}(u, w) \mid C^{i-1}(w, v) \}\!\}_{w \in V}, \tag{8}$$

where $\{\!\{ \}\!\}$ denotes the multiset and $\mid$ denotes the string concatenation (an appropriate encoding is assumed).

The k-dimensional version of the algorithm, k-WL, operates on V^k. The initial coloring of a tuple $(u_1, \ldots, u_k)$ encodes its equality type and the isomorphism type of the subgraph of G induced by the vertices $u_1, \ldots, u_k$. The color refinement is performed similarly to (8). We write $\mathrm{WL}_k^r(G, u_1, \ldots, u_k)$ to denote the

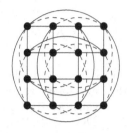
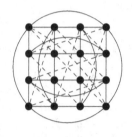

n	$\mathrm{sub}(P_n, G)$	$\mathrm{sub}(P_n, H)$
8	275616	274560
9	880128	877440
10	2506752	2512512
11	6239232	6283392
12	13189248	13293696
13	22631040	22754688
14	29376000	29457408
15	25532928	25560576
16	11197440	11115264

Fig. 2. The 4×4 rook's graph G and the Shrikhande graph H (some edges are dashed just to ensure readability). The table shows the counts of P_n, $8 \le n \le 16$, in G and H.

color of the tuple $(u_1, \dots, u_k)$ produced by the k-dimensional Weisfeiler-Leman algorithm after performing r refinement steps. The length of $\mathrm{WL}_k^r(G, u_1, \dots, u_k)$ grows exponentially as r increases, which is remedied by renaming the tuple colors after each step and retaining the corresponding color substitution tables. However, in our analysis of the algorithm we will use $\mathrm{WL}_k^r(G, u_1, \dots, u_k)$ in its literal, iteratively defined meaning.

Let $\mathrm{WL}_k^r(G) = \{\!\{\,\mathrm{WL}_k^r(G, \bar{u}) : \bar{u} \in V^k\}\!\}$ denote the color palette observed on the input graph G after r refinement rounds. We say that the k-dimensional Weisfeiler-Leman algorithm *distinguishes* graphs G and H if $\mathrm{WL}_k^r(G) \ne \mathrm{WL}_k^r(H)$ after some number of rounds r. The standard color stabilization argument shows that if n-vertex graphs G and H are distinguishable by k-WL, then they are distinguished within n^k refinement rounds. If this does not happen, we say that G and H are *k-WL-equivalent* and write $G \equiv_{k\text{-WL}} H$.

Obviously, isomorphic graphs are k-WL-equivalent for every k. Recall also that any two strongly regular graphs with the same parameters are 2-WL-equivalent. The smallest pair of non-isomorphic strongly regular graphs with the same parameters consists of the 4×4-rook's graph and the Shrikhande graph (see Fig. 2). The 2-WL-equivalence of these graphs will be used several times below.

Theorem 11 (Dell, Grohe, and Rattan [14]). *Let* $\mathrm{hom}(F, G)$ *denote the number of homomorphisms from a graph F to a graph G. For each F of treewidth k, the homomorphism count* $\mathrm{hom}(F, \cdot)$ *is* $\equiv_{k\text{-WL}}$*-invariant.*

Definition 12. *We define the* homomorphism-hereditary treewidth *of a graph F, denoted by* $\mathrm{htw}(F)$*, to be the maximum treewidth* $\mathrm{tw}(F')$ *over all homomorphic images F' of F.*

The following result follows directly from Theorem 11 and the fact established by Lovász [26, Sect. 5.2.3] that the subgraph count $\mathrm{sub}(F, G)$ is expressible as a function of the homomorphism counts $\mathrm{hom}(F', G)$ where F' ranges over homomorphic images of F (see also [10], where algorithmic consequences of this relationship are explored).

Corollary 13. $\mathcal{C}(k)$ *contains all* F *with* $htw(F) \leq k$.

It is easy to see that $htw(F) = 1$ if and only if F is a star graph or the matching graph $2K_2$ (up to adding isolated vertices). Thus, Theorem 4 implies that $\mathcal{C}(1)$ consists exactly of the pattern graphs F with $htw(F) = 1$. We now characterize the class of graphs F with $htw(F) \leq 2$.

Given a graph G and a partition P of the vertex set $V(G)$, we define the *quotient graph* G/P as follows. The vertices of G/P are the elements of P, and $X \in P$ and $Y \in P$ are adjacent in G/P if and only if $X \neq Y$ and there are vertices $x \in X$ and $y \in Y$ adjacent in G.

Lemma 14. $htw(F) > 2$ *if and only if there is a partition* P *of* $V(F)$ *such that* $F/P \cong K_4$.

Proof. Let us make two basic observations. First, H is a homomorphic image of G if and only if there is a partition P of $V(G)$ into independent sets such that $H \cong G/P$. Second, H is a minor of G if and only there is a partition P of $V(G)$ such that the graph $G[X]$ is connected for every $X \in P$ and H is isomorphic to a subgraph of G/P.

These observations imply the following fact, which is more general than stated in the lemma. Let $\mathcal{S}_k$ be the set of the minimal forbidden minors for the class of graphs with treewidth at most k. Note that, since the last class of graphs is minor-closed, $\mathcal{S}_k$ exists and is finite by the Robertson–Seymour theorem. Then $htw(F) > k$ if and only if $V(F)$ admits a partition P such that G/P contains a subgraph isomorphic to a graph in $\mathcal{S}_k$.

The lemma now follows from the well-known fact [15, Chapter 12] that $\mathcal{S}_2 = \{K_4\}$. Note that, if F/P contains K_4 as a subgraph, then $V(F)$ admits a partition P' such that F/P' is itself isomorphic to K_4 as the superfluous nodes of F/P can be merged. $\qquad\square$

Whether or not $htw(F) \leq 2$ is a necessary condition for the membership of F in $\mathcal{C}(2)$, is open. We now show the equivalence of $F \in \mathcal{C}(2)$ and $htw(F) \leq 2$ for several standard graph sequences.

Theorem 15. $\mathcal{C}(2)$ *contains*

1. $K_2, 2K_2, 3K_2, 4K_2, 5K_2$ *and no other matching graphs;*
2. $C_3, \ldots, C_7$ *and no other cycle graphs;*
3. $P_1, \ldots, P_7$ *and no other path graphs.*

Theorem 15 is related to some questions that have earlier been discussed in the literature. Beezer and Farrell [6] proved that the first five coefficients of the matching polynomial of a strongly regular graph are determined by its parameters. In other terms, if G and H are strongly regular graphs with the same parameters (in fact, even distance-regular graphs with the same intersection array), then $\text{sub}(sK_2, G) = \text{sub}(sK_2, H)$ for $s \leq 5$. Part 1 of Theorem 15 implies that this is true in a much more general situation, namely when G and H are arbitrary 2-WL-equivalent graphs. Moreover, this cannot be extended to larger s.

Fürer [18] classified all C_s for $s \leq 16$, except the C_7, with respect to membership in $\mathcal{C}(2)$. Part 2 of Theorem 15 fills this gap and also shows that the positive result for C_7 is optimal.

Part 2 of Theorem 15 can be generalized as follows. Recall that $g(G)$ denotes the girth of a graph G.

Theorem 16. *Suppose that $G \equiv_{2\text{-WL}} H$. Then*

1. $g(G) = g(H)$.
2. $\mathrm{sub}(C_s, G) = \mathrm{sub}(C_s, H)$ *for each* $3 \leq s \leq 2\,g(G) + 1$.

Proof. 1. The proof uses the logical characterization of the $\equiv_{k\text{-WL}}$-equivalence in [9]. According to this characterization, $G \equiv_{k\text{-WL}} H$ if and only if G and H satisfy the same sentences in the first-order $(k+1)$-variable logic with counting quantifiers $\exists^{\geq t}$, where an expression $\exists^{\geq t} x\, \Phi(x)$ for any integer t means that there are at least t vertices x with property $\Phi(x)$.

Assume that $g(G) < g(H)$ and show that then $G \not\equiv_{2\text{-WL}} H$. It is enough to show that G and H are distinguishable in 3-variable logic with counting quantifiers.

Case 1: $g(G)$ is odd. In this case, G and H are distinguishable even in the standard 3-variable logic (with quantifiers $\exists$ and $\forall$ only). As it is well known [24], two graphs G and H are distinguishable in first-order k-variable logic if and only if Spoiler has a winning strategy in the *k-pebble Ehrenfeucht-Fraïssé game* on G and H. In the 3-pebble game, the players *Spoiler* and *Duplicator* have equal sets of 3 pebbles $\{a, b, c\}$. In each round, Spoiler takes a pebble and puts it on a vertex in G or in H; then Duplicator has to put her copy of this pebble on a vertex of the other graph. Duplicator's objective is to ensure that the pebbling determines a partial isomorphism between G and H after each round; when she fails, she immediately loses.

Spoiler wins the game as follows. Let C be a cycle of length $g(G)$ in G. In the first three rounds, Spoiler pebbles a 3-path along C by his pebbles a, b, and c in this order. Then, keeping the pebble a fixed, Spoiler moves the pebbles b and c, in turns, around C so that the two pebbled vertices are always adjacent. In the end, there arises a pebbled acb-path, which is impossible in H.

Case 2: $g(G)$ is even. Let $g(G) = 2m$. Consider the following statement in the 3-variable logic with counting quantifiers:

$$\exists x \exists y\, \big(dist(x,y) = m \wedge \exists^{\geq 2} z (z \sim y \wedge dist(z,x) = m-1) \big),$$

where $dist(x,y) = m$ is a 3-variable formula expressing the fact that the distance between vertices x and y is equal to m. This statement is true on G and false on H.

2. The proof of this part is based on the result by Dell et al. (Theorem 11) and the Lovász result [26, Sect. 5.2.3] on the expressibility of $\mathrm{sub}(F, G)$ through the homomorphism counts $\mathrm{hom}(F', G)$ for homomorphic images F' of F. By these results, it suffices to prove that, if $s \leq 2\,g(G) + 1$ and h is a homorphism from C_s to G, then the subgraph $h(C_s)$ of G has treewidth at most 2. Assume, to the

contrary, that $h(C_s)$ has treewidth more than 2 or, equivalently, $h(C_s)$ contains K_4 as a minor (see [15, Chap. 12]). Since K_4 has maximum degree 3, $h(C_s)$ contains K_4 even as a topological minor [15, Sect. 1.7]. Let M be a subgraph of $h(C_s)$ that is a subdivision of K_4. Obviously, $s \geq e(h(C_s)) \geq e(M)$. Moreover, $s \geq e(M) + 2$. Indeed, the homomorphism h determines a walk of length s via all edges of the graph $h(C_s)$. By cloning the edges traversed more than once, $h(C_s)$ can be seen as an Eulerian multigraph with s edges. Since M has four vertices of degree 3, any extension of M to such a multigraph requires adding at least 2 edges. Thus, $s \geq e(M) + 2$. Note that M is formed by six paths corresponding to the edges of K_4. Moreover, M has four cycles, each cycle consists of three paths, and each of the six paths appears in two of the cycles. It follows that $2\,e(M) \geq 4\,g(G)$. Therefore, $s \geq 2\,g(G) + 2$, yielding a contradiction.   □

It is easy to see that $K_3 \in \mathcal{R}(2)$ (this is also a formal consequence of Part 2 of Theorem 15). Using the pair G, H consisting of the 4×4-rook's graph and the Shrikhande graph, Fürer [18] proved that the complete graph K_4 is not in $\mathcal{R}(2)$. By padding G and H with new $s-4$ universal vertices, we see that $\mathcal{R}(2)$ contains K_s if and only if $s \leq 3$. Fürer's result on the non-membership of K_4 in $\mathcal{R}(2)$ admits the following generalization, whose proof is given in the full version [2].

Theorem 17. *No graph containing a unique 4-clique can be in $\mathcal{R}(2)$.*

4   Concluding Discussion

An intriguing open problem is whether Corollary 13 yields a complete description of the class $\mathcal{C}(k)$. Our Theorem 4 gives an affirmative answer in the one-dimensional case. Moreover, this theorem gives a complete description of the class $\mathcal{R}(1)$. The class $\mathcal{R}(2)$ remains a mystery. For example, it contains either finitely many matching graphs sK_2 or all of them, and we currently do not know which of these is true. In other words, is the matching number preserved by $\equiv_{2\text{-WL}}$-equivalence? Note that non-isomorphic strongly regular graphs with the same parameters cannot yield counterexamples to this. The *Brouwer-Haemers conjecture* states that every connected strongly regular graph is Hamiltonian except the Petersen graph, and Pyber [33] has shown there are at most finitely many exceptions to this conjecture. Since the Petersen graph has a perfect matching, it is quite plausible that every connected strongly regular graph has an (almost) perfect matching.

By Corollary 13, the subgraph count $\mathrm{sub}(F, G)$ is k-WL-invariant for $k = htw(F)$. Interestingly, the parameter $htw(F)$ appears in a result by Curticapean, Dell, and Marx [10] who show that $\mathrm{sub}(F, G)$ is computable in time $e(F)^{O(e(F))} \cdot v(G)^{htw(F)+1}$. An interesting area is to explore connections between k-WL-invariance and algorithmics, which are hinted by this apparent coincidence.

Which *induced* subgraphs and their counts are k-WL-invariant for different k deserves study. We note that the induced subgraph counts have been studied in the context of finite model theory by Kreutzer and Schweikardt [25].

References

1. Anderson, M., Dawar, A., Holm, B.: Solving linear programs without breaking abstractions. J. ACM **62**(6), 48:1–48:26 (2015)
2. Arvind, V., Fuhlbrück, F., Köbler, J., Verbitsky, O.: On Weisfeiler-Leman invariance: subgraph counts and related graph properties. Technical report (2018). https://arxiv.org/abs/1811.04801
3. Atserias, A., Bulatov, A.A., Dawar, A.: Affine systems of equations and counting infinitary logic. Theor. Comput. Sci. **410**(18), 1666–1683 (2009)
4. Babai, L.: Graph isomorphism in quasipolynomial time. In: Proceedings of the 48th Annual ACM Symposium on Theory of Computing (STOC 2016), pp. 684–697 (2016). https://doi.org/10.1145/2897518.2897542
5. Babai, L., Erdős, P., Selkow, S.M.: Random graph isomorphism. SIAM J. Comput. **9**(3), 628–635 (1980)
6. Beezer, R.A., Farrell, E.J.: The matching polynomial of a distance-regular graph. Int. J. Math. Math. Sci. **23**(2), 89–97 (2000)
7. Bollobás, B.: Random Graphs. Cambridge Studies in Advanced Mathematics, vol. 73, 2nd edn. Cambridge University Press, Cambridge (2001)
8. Bose, R.C., Mesner, D.M.: On linear associative algebras corresponding to association schemes of partially balanced designs. Ann. Math. Statist. **30**, 21–38 (1959)
9. Cai, J., Fürer, M., Immerman, N.: An optimal lower bound on the number of variables for graph identifications. Combinatorica **12**(4), 389–410 (1992). https://doi.org/10.1007/BF01305232
10. Curticapean, R., Dell, H., Marx, D.: Homomorphisms are a good basis for counting small subgraphs. In: Proceedings of the 49th Annual ACM SIGACT Symposium on Theory of Computing (STOC 2017), pp. 210–223. ACM (2017)
11. Dawar, A.: A restricted second order logic for finite structures. Inf. Comput. **143**(2), 154–174 (1998). https://doi.org/10.1006/inco.1998.2703
12. Dawar, A.: The nature and power of fixed-point logic with counting. SIGLOG News **2**(1), 8–21 (2015)
13. Dawar, A., Severini, S., Zapata, O.: Pebble games and cospectral graphs. Electron. Notes Discrete Math. **61**, 323–329 (2017)
14. Dell, H., Grohe, M., Rattan, G.: Lovász meets Weisfeiler and Leman. In: 45th International Colloquium on Automata, Languages, and Programming (ICALP 2018). LIPIcs, vol. 107, pp. 40:1–40:14. Schloss Dagstuhl - Leibniz-Zentrum für Informatik (2018)
15. Diestel, R.: Graph Theory. Springer, New York (2000)
16. Farrell, E.J., Guo, J.M., Constantine, G.M.: On matching coefficients. Discrete Math. **89**(2), 203–210 (1991)
17. Fürer, M.: On the power of combinatorial and spectral invariants. Linear Algebra Appl. **432**(9), 2373–2380 (2010)
18. Fürer, M.: On the combinatorial power of the Weisfeiler-Lehman algorithm. In: Fotakis, D., Pagourtzis, A., Paschos, V.T. (eds.) CIAC 2017. LNCS, vol. 10236, pp. 260–271. Springer, Cham (2017). https://doi.org/10.1007/978-3-319-57586-5_22
19. Gao, C., Lafferty, J.: Testing for global network structure using smallsubgraph statistics. Technical report (2017). http://arxiv.org/abs/1710.00862
20. Grochow, J.A., Kellis, M.: Network motif discovery using subgraph enumeration and symmetry-breaking. In: Speed, T., Huang, H. (eds.) RECOMB 2007. LNCS, vol. 4453, pp. 92–106. Springer, Heidelberg (2007). https://doi.org/10.1007/978-3-540-71681-5_7

21. Grohe, M.: Fixed-point definability and polynomial time on graphs with excluded minors. J. ACM **59**(5), 27:1–27:64 (2012). https://doi.org/10.1145/2371656. 2371662
22. Hasan, M.A., Dave, V.S.: Triangle counting in large networks: a review. Wiley Interdiscip. Rev. Data Min. Knowl. Discov. **8**(2), e1226 (2018)
23. Higman, D.: Finite permutation groups of rank 3. Math. Z. **86**, 145–156 (1964)
24. Immerman, N.: Descriptive Complexity. Graduate Texts in Computer Science. Springer, New York (1999). https://doi.org/10.1007/978-1-4612-0539-5
25. Kreutzer, S., Schweikardt, N.: On Hanf-equivalence and the number of embeddings of small induced subgraphs. In: Joint Meeting of the 23-rd EACSL Annual Conference on Computer Science Logic (CSL) and the 29-th Annual ACM/IEEE Symposium on Logic in Computer Science (LICS), pp. 60:1–60:10. ACM (2014)
26. Lovász, L.: Large Networks and Graph Limits, Colloquium Publications, vol. 60. American Mathematical Society (2012)
27. McKay, B.D., Piperno, A.: Practical graph isomorphism, II. J. Symb. Comput. **60**, 94–112 (2014)
28. Milo, R., Shen-Orr, S., Itzkovitz, S., Kashtan, N., Chklovskii, D., Alon, U.: Network motifs: simple building blocks of complex networks. Science **298**(5594), 824–827 (2002)
29. Morgan, H.L.: The generation of a unique machine description for chemical structures – a technique developed at chemical abstracts service. J. Chem. Doc. **5**(2), 107–113 (1965)
30. Morris, C., Kersting, K., Mutzel, P.: Glocalized Weisfeiler-Lehman graph kernels: global-local feature maps of graphs. In: 2017 IEEE International Conference on Data Mining (ICDM 2017), pp. 327–336. IEEE Computer Society (2017)
31. Morris, C., et al.: Weisfeiler and Leman go neural: Higher-order graph neural networks. Technical report (2018). http://arxiv.org/abs/1810.02244
32. Newman, M.E.J.: The structure and function of complex networks. SIAM Rev. **45**(2), 167–256 (2003)
33. Pyber, L.: Large connected strongly regular graphs are Hamiltonian. Technical report (2014). http://arxiv.org/abs/1409.3041
34. Scheinerman, E.R., Ullman, D.H.: Fractional Graph Theory. A Rational Approach to the Theory of Graphs. Wiley, Hoboken (1997)
35. Shervashidze, N., Schweitzer, P., van Leeuwen, E.J., Mehlhorn, K., Borgwardt, K.M.: Weisfeiler-Lehman graph kernels. J. Mach. Learn. Res. **12**, 2539–2561 (2011)
36. Ugander, J., Backstrom, L., Kleinberg, J.M.: Subgraph frequencies: mapping the empirical and extremal geography of large graph collections. In: 22nd International World Wide Web Conference (WWW 2013), pp. 1307–1318. ACM (2013)
37. Weisfeiler, B., Leman, A.: The reduction of a graph to canonical form and the algebra which appears therein. NTI Series, vol. 2, no. 9, pp. 12–16 (1968). English translation is available at https://www.iti.zcu.cz/wl2018/pdf/wl_paper_translation.pdf
38. Wormald, N.: Models of random regular graphs. In: Surveys in Combinatorics, pp. 239–298. Cambridge University Press, Cambridge (1999)

Deterministic Preparation of Dicke States

Andreas Bärtschi[1,2(✉)] and Stephan Eidenbenz[2]

[1] Center for Nonlinear Studies, Los Alamos National Laboratory, Los Alamos, USA
baertschi@lanl.gov
[2] National Security Education Center,
Los Alamos National Laboratory, Los Alamos, USA
eidenben@lanl.gov

Abstract. The Dicke state $|D_k^n\rangle$ is an equal-weight superposition of all n-qubit states with Hamming Weight k (i.e. all strings of length n with exactly k ones over a binary alphabet). Dicke states are an important class of entangled quantum states that among other things serve as starting states for combinatorial optimization quantum algorithms.

We present a deterministic quantum algorithm for the preparation of Dicke states. Implemented as a quantum circuit, our scheme uses $\mathcal{O}(kn)$ gates, has depth $\mathcal{O}(n)$ and needs no ancilla qubits. The inductive nature of our approach allows for linear-depth preparation of arbitrary symmetric pure states and – used in reverse – yields a quasilinear-depth circuit for efficient compression of quantum information in the form of symmetric pure states, improving on existing work requiring quadratic depth. All of these properties even hold for Linear Nearest Neighbor architectures.

1 Introduction

Within quantum computing, the seemingly mundane task of (efficient) state preparation is actually a separate research topic. Recall that a quantum state over n qubits is a superposition $\sum_{x \in \{0,1\}^n} c_x |x\rangle$ of all 2^n binary strings x of length n with complex weights c_x – the *amplitudes* – such that $\sum_{x \in \{0,1\}^n} |c_x|^2 = 1$. The problem of preparing an arbitrary quantum state can be solved with $\Theta(2^n)$ quantum gates [28], which can be improved to a polynomial number of gates for states which have a polynomial number of non-zero amplitudes c_x. The intriguing algorithmic question then becomes for what other classes of quantum states do polynomial-time state preparation algorithms exist? Very few results exist on this topic and we are still far from having a comprehensive solution, however, Dicke states form such a class: the Dicke state $|D_k^n\rangle$ has $\binom{n}{k}$ non-zero amplitudes, which in general is not polynomial in n.

Among different types of highly entangled states, the family of Dicke states [8] has garnered widespread attention for tasks in quantum networking [25], quantum game theory [35], quantum metrology [31] and as starting

Research presented in this article was supported by the Laboratory Directed Research and Development program of Los Alamos National Laboratory under project number 20190495DR. Los Alamos National Laboratory report LA-UR-19-22718.

L. A. Gąsieniec et al. (Eds.): FCT 2019, LNCS 11651, pp. 126–139, 2019.
https://doi.org/10.1007/978-3-030-25027-0_9

states for combinatorial optimization problems via adiabatic evolution [6]. Perhaps most promisingly – Dicke states can be used in the Quantum Alternating Operator Ansatz (QAOA) framework [11,14] for combinatorial optimization problems with hard constraints, as a starting state for the actual QAOA algorithm where they represent a superposition of all feasible solutions (in some problem variations).

Definition 1. *A Dicke state* $|D_k^n\rangle$[1] *is the equal superposition of all n-qubit states* $|x\rangle$ *with Hamming weight* $\mathrm{wt}(x) = k$,

$$|D_k^n\rangle = \binom{n}{k}^{-\frac{1}{2}} \sum\nolimits_{x \in \{0,1\}^n,\ \mathrm{wt}(x)=k} |x\rangle.$$

We have, e.g., $|D_2^4\rangle = \frac{(|1100\rangle + |1010\rangle + |1001\rangle + |0110\rangle + |0101\rangle + |0011\rangle)}{\sqrt{6}}$, a state that has been studied for its entanglement properties: from $|D_2^4\rangle$, we can generate 3-qubit W_3-states $|D_1^3\rangle$ and GHZ-class G_3-states $\frac{1}{\sqrt{2}}(|D_1^3\rangle - |D_2^3\rangle)$ by a (local) projective measurement of the same qubit [19], whereas these two states cannot be transformed into each other by stochastic local manipulations [10]; these types of basic transformations of states are non-trivial in quantum computing.

Result Overview. Despite successful experimental creation of Dicke states in physical systems such as trapped ions [15,17,20], atoms [27,30,34], photons [25,32] and superconducting qubits [33], efficient quantum *circuits* for the preparation of arbitrary Dicke states $|D_k^n\rangle$ have received little attention. In this paper, we present – as our main contribution – a circuit for deterministic preparation of Dicke states which, given as input the easily prepared classical state $|0\rangle^{\otimes n-k} |1\rangle^{\otimes k}$, prepares the Dicke state $|D_k^n\rangle$. Our circuit has depth $\mathcal{O}(n)$ – independent of k – and needs $\mathcal{O}(kn)$ gates in total. Circuit depth is equivalent to run time and gate count is a measure for overall resource needs. In fact, any difference between gate count and depth can be attributed to gate-level parallelism. Finding minimal-depth circuits is particularly crucial for Noisy Intermediate Scale Quantum (NISQ) devices, which do not allow for full error correction, and thus experience (unwanted) decoherence the longer a computation lasts. Minimizing overall gate count is crucial as each gate operation introduces noise, thus impacting result quality.

Leveraging our main result, we note (i) that all our bounds even hold for Linear Nearest Neighbor architectures, where each qubit is connected only to its two neighbors, which is a more realistic assumption for most NISQ devices than the standard all-to-all connectivity, and show (ii) that our circuit can be extended to prepare arbitrary symmetric pure states using linear circuit depth, where a state is symmetric if it is invariant under permutation of the qubits, and (iii) how to use our construction for compression of quantum information, which is the problem of compressing a symmetric pure n-qubit state into $\lceil \log(n + 1) \rceil$ qubits without information loss.

---

[1] Various symbols are used in the existing literature, e.g., $|D_k^n\rangle$, $|D_n^{(k)}\rangle$, $|{}^n_k\rangle$ or $|n; k\rangle$.

Previous Approaches. Previous work has prepared Dicke states probabilistically with success probability $\Omega(\frac{1}{\sqrt{n}})$ by applying a biased Hadamard transform to each qubit [6], followed by postselecting the Dicke state through addition of each of the n qubits into an ancilla register of size $\log n$ initialized to the $|0\rangle$ state [7] and a projective measurement thereof. A later contribution [5] uses a more involved preparation strategy – giving numerical evidence of a constant-factor improved probability – followed by a generalized parity measurement [16], also pointing out the potential use of amplitude amplification. Deterministic preparation circuits without the use of ancilla qubits have been known for the special case of W-states $|D_1^n\rangle$, either by an iterative construction of quadratic circuit size and depth [9] or by a linear number of large multi-controlled rotation gates [21]. An inductive approach to construct Dicke states up to error ε [23] uses $\Omega(\log k + \log \frac{1}{\varepsilon})$ ancilla qubits to count the Hamming weight of the qubits processed so far, to then use this register as a control for rotation gates on the next qubit, yielding a superlinear circuit size and depth overall. Our approach improves on all of these results in terms of circuit size and depth; additionally, it does not require ancilla qubits, is fully deterministic and in some cases more general.

Relation to Quantum Compression. There exists an interesting relationship between Dicke states and quantum compression. The latter can be understood through the quantum Schur-Weyl transform [1], which separates the angular momentum information of a state from its – for symmetric states trivial – permutation information.

The Schur-Weyl transform has been implemented experimentally for a separable symmetric 3-qubit state [26], i.e. a state $(\alpha|0\rangle + \beta|1\rangle)^{\otimes 3} = \sum_{\ell=0}^{3} \alpha^{3-\ell}\beta^\ell \binom{3}{\ell}^{1/2} |D_\ell^3\rangle$. A high-level description of a circuit for general n, using no ancilla qubits, has also been developed [24]. The major circuit part in [24] is of size and depth $\Theta(n^2)$ and maps each Dicke state $|D_\ell^n\rangle$ to the state $|0\rangle^{\otimes \ell-1} |1\rangle |0\rangle^{\otimes n-\ell}$. Its inverse circuit can therefore be used to prepare Dicke states with depth $\Theta(n^2)$, too. Our approach in reverse, on the other hand, will yield a quantum compression circuit of size $\mathcal{O}(n^2)$ and reduced quasilinear depth $\tilde{\mathcal{O}}(n)$, where $\tilde{\mathcal{O}}(\cdot)$ hides polylogarithmic factors due to the compression part of the circuit (mapping terms of the form $|0\rangle^{\otimes \ell-1} |1\rangle |0\rangle^{\otimes n-\ell}$ or $|0\rangle^{\otimes n-\ell} |1\rangle^{\otimes \ell}$, respectively, into $\lceil \log(n+1)\rceil$ qubits).

Outline. This article is organized as follow: In Sect. 2, we present an iterative construction of a circuit for deterministic preparation of arbitrary Dicke states. We analyze its gate count and circuit depth, and extend these bound to Linear Nearest Neighbor architectures in Sect. 3. In Sect. 4, we show how our construction can be used to create arbitrary symmetric pure states, written as a superposition of Dicke states, and we present an improved scheme for efficient compression of quantum information.

Detailed proofs for Linear Nearest Neighbor architectures can be found in the preprint of this paper [3].

2   Deterministic Dicke State Preparation

In order to prepare Dicke states, we design a unitary operator $U_{n,k}$ which, given as input the classical state $|0\rangle^{\otimes n-k} |1\rangle^{\otimes k}$ (which appears itself as a term in the superposition $|D_k^n\rangle$), generates the entire Dicke state $|D_k^n\rangle$. Additionally, $U_{n,k}$ also generates Dicke states $|D_\ell^n\rangle$ for *smaller* $\ell < k$, when given as input a string $|0\rangle^{\otimes n-\ell} |1\rangle^{\otimes \ell}$:

Definition 2. *Denote by $U_{n,k}$ any unitary satisfying for all $0 \leq \ell \leq k$:*
$$U_{n,k} |0\rangle^{\otimes n-\ell} |1\rangle^{\otimes \ell} = |D_\ell^n\rangle.$$

Having this property not only for $\ell = k$ but for all $\ell \leq k$ will allow us to build a unitary $U_{n,k}$ inductively, by making use of the following composition (also observed, e.g., in [20,22]):

Lemma 1. *Dicke states $|D_\ell^n\rangle$ have the inductive sum form*

$$|D_\ell^n\rangle = \sqrt{\tfrac{\ell}{n}} |D_{\ell-1}^{n-1}\rangle \otimes |1\rangle + \sqrt{\tfrac{n-\ell}{n}} |D_\ell^{n-1}\rangle \otimes |0\rangle.$$

Proof. We rewrite $|D_\ell^n\rangle := \binom{n}{\ell}^{-\frac{1}{2}} \sum_{x \in \{0,1\}^n,\, \mathrm{wt}(x)=\ell} |x\rangle$ as

$$|D_\ell^n\rangle = \sqrt{\tfrac{1}{\binom{n}{\ell}}} \sum_{\substack{x \in \{0,1\}^{n-1} \\ \mathrm{wt}(x)=\ell-1}} |x\rangle \otimes |1\rangle + \sqrt{\tfrac{1}{\binom{n}{\ell}}} \sum_{\substack{x \in \{0,1\}^{n-1} \\ \mathrm{wt}(x)=\ell}} |x\rangle \otimes |0\rangle$$

$$= \sqrt{\tfrac{\binom{n-1}{\ell-1}}{\binom{n}{\ell}}} |D_{\ell-1}^{n-1}\rangle \otimes |1\rangle + \sqrt{\tfrac{\binom{n-1}{\ell}}{\binom{n}{\ell}}} |D_\ell^{n-1}\rangle \otimes |0\rangle$$

$$= \sqrt{\tfrac{\ell}{n}} |D_{\ell-1}^{n-1}\rangle \otimes |1\rangle + \sqrt{\tfrac{n-\ell}{n}} |D_\ell^{n-1}\rangle \otimes |0\rangle. \qquad \square$$

The Dicke states $|D_{\ell-1}^{n-1}\rangle$ and $|D_\ell^{n-1}\rangle$ can both be prepared by the same unitary $U_{n-1,k}$ given the classical input states $|0\rangle^{\otimes n-\ell} |1\rangle^{\otimes \ell-1}$ and $|0\rangle^{\otimes n-1-\ell} |1\rangle^{\otimes \ell}$, respectively. The idea is therefore – in order to inductively design $U_{n,k}$ – to apply the composition given by Lemma 1 to the *input states* $|0\rangle^{\otimes n-\ell} |1\rangle^{\otimes \ell}$ for all $\ell \leq k$, before applying the smaller unitary $U_{n-1,k}$. Hence for $\ell \leq k$, we are looking for unitary transformations

$$|0\rangle^{\otimes n-\ell} |1\rangle^{\otimes \ell} \mapsto |0\rangle^{\otimes n-k-1} \otimes (\sqrt{\tfrac{\ell}{n}} |0\rangle^{\otimes k+1-\ell} |1\rangle^{\otimes \ell} + \sqrt{\tfrac{n-\ell}{n}} |0\rangle^{\otimes k-\ell} |1\rangle^{\otimes \ell} |0\rangle).$$

Note that this transformation acts trivially on the first $n - k - 1$ qubits. Intuitively, it can be described as taking (the last) $k+1$ (of n) qubits as an input, splitting the input term into a superposition of two parts (with amplitudes depending on n), and cyclicly shifting the second part by one position to the left. We call a unitary that simultaneously implements this transformation a *Split & Cyclic Shift* unitary $SCS_{n,k}$:

Definition 3. *Denote by* $SCS_{n,k}$ *any unitary satisfying for all* $1 \le \ell \le k$, *where* $k < n$:

$$SCS_{n,k} |0\rangle^{\otimes k+1} \qquad\quad = |0\rangle^{\otimes k+1},$$

$$SCS_{n,k} |0\rangle^{\otimes k+1-\ell} |1\rangle^{\otimes \ell} = \sqrt{\tfrac{\ell}{n}} |0\rangle^{\otimes k+1-\ell} |1\rangle^{\otimes \ell} + \sqrt{\tfrac{n-\ell}{n}} |0\rangle^{\otimes k-\ell} |1\rangle^{\otimes \ell} |0\rangle,$$

$$SCS_{n,k} |1\rangle^{\otimes k+1} \qquad\quad = |1\rangle^{\otimes k+1}.$$

Before we describe the inductive construction of the unitaries $U_{n,k}$ in terms of unitaries $SCS_{n,k}$ and $U_{n-1,k}$, we review $SCS_{n,k}$ (acting on the last $k+1$ qubits) and $U_{n-1,k}$ (acting on the first $n-1$ qubits) in comparison:

$$SCS_{n,k}: |0..0\rangle \mapsto |00..000\rangle \qquad\qquad U_{n-1,k}: |0..0\rangle \mapsto |D_0^{n-1}\rangle$$

$$|00..001\rangle \mapsto \sqrt{\tfrac{1}{n}} |00..001\rangle + \sqrt{\tfrac{n-1}{n}} |00..010\rangle \qquad |0..000..01\rangle \mapsto |D_1^{n-1}\rangle$$

$$|00..011\rangle \mapsto \sqrt{\tfrac{2}{n}} |00..011\rangle + \sqrt{\tfrac{n-2}{n}} |00..110\rangle \qquad |0..000..11\rangle \mapsto |D_2^{n-1}\rangle$$

$$\vdots \qquad\qquad\qquad\qquad \vdots$$

$$|01..111\rangle \mapsto \sqrt{\tfrac{k}{n}} |01..111\rangle + \sqrt{\tfrac{n-k}{n}} |11..110\rangle \qquad |0..001..11\rangle \mapsto |D_{k-1}^{n-1}\rangle$$

$$\underbrace{|11..111\rangle}_{k+1} \mapsto |11..111\rangle \qquad\qquad \underbrace{|0..011..11\rangle}_{n-1-k \quad k} \mapsto |D_k^{n-1}\rangle$$

2.1 Inductive Construction of $U_{n,k}$

An explicit construction of Split & Cyclic Shift unitaries in terms of standard gates will be given in Subsect. 2.2. For now, however, we will show how arbitrary $U_{n,k}$ unitaries can be constructed inductively from unitaries $SCS_{n,k}$ (acting on the last $k+1$ qubits) and $U_{n-1,k}$ (acting on the first $n-1$ qubits). Clearly, we must have $U_{1,1} = \mathrm{Id}$. We construct unitaries of the form $U_{k,k}$ by iteratively applying $SCS_{k,k-1}$ immediately before $U_{k-1,k-1}$, i.e. $U_{k,k} = (U_{k-1,k-1} \otimes \mathrm{Id}) \cdot SCS_{k,k-1}$. Arbitrary unitaries $U_{n,k}$ can be built by preceding $U_{n-1,k}$ with $SCS_{n,k}$, as shown in Fig. 1, giving $U_{n,k} = (U_{n-1,k} \otimes \mathrm{Id}) \cdot (\mathrm{Id}^{\otimes n-k-1} \otimes SCS_{n,k})$.[2]

Telescoping these recursions we get:

Lemma 2. *The inductive construction above leads to unitaries* $U_{n,k}$ *which are consistent with Definition 2:*

$$U_{n,k} := \prod_{\ell=2}^{k} \left(SCS_{\ell,\ell-1} \otimes \mathrm{Id}^{\otimes n-\ell} \right) \cdot \prod_{\ell=k+1}^{n} \left(\mathrm{Id}^{\otimes \ell-k-1} \otimes SCS_{\ell,k} \otimes \mathrm{Id}^{\otimes n-\ell} \right).$$

Proof. By induction over n we prove $\forall \ell \le k: U_{n,k} |0\rangle^{\otimes n-\ell} |1\rangle^{\otimes \ell} = |D_\ell^n\rangle$.
For the base $U_{2,2}$ we have by Definition 3 that $SCS_{2,1} |00\rangle = |00\rangle =: |D_0^2\rangle$, $SCS_{2,1} |01\rangle = \tfrac{1}{2}(|01\rangle + |10\rangle) =: |D_1^2\rangle$, $SCS_{2,1} |11\rangle = |11\rangle =: |D_2^2\rangle$ and thus for

---

[2] An inductive approach which "sandwiches" smaller unitaries has previously been used for W-states $|D_1^n\rangle$, albeit with depth $\mathcal{O}(n^2)$ [9].

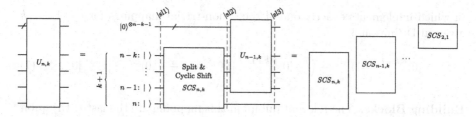

Fig. 1. Inductive construction of the unitaries $U_{n,k}$ from unitaries $SCS_{n,k}$ and $U_{n-1,k}$.

$\ell \leq 2$ that $U_{2,2} |0\rangle^{\otimes 2-\ell} |1\rangle^{\otimes \ell} := SCS_{2,1} |0\rangle^{\otimes 2-\ell} |\otimes 1\rangle^{\ell} = |D_\ell^n\rangle$. We proceed by induction,

- first considering the simpler step $U_{n-1,k} \to U_{n,k}$, where $k \leq n - 1$,
- then moving on to the step $U_{k-1,k-1} \to U_{k,k}$, which includes the case $U_{n-1,n-1} \to U_{n,n}$.

Step $U_{n-1,k} \to U_{n,k}$. We show $U_{n,k} |0\rangle^{\otimes n-\ell} |1\rangle^{\otimes \ell} = |D_\ell^n\rangle$ for all $\ell \leq k \leq n - 1$ by analyzing the three time-slices depicted in Fig. 1. As input to the circuit we have a corresponding state $|sl1\rangle = |0\rangle^{\otimes n-\ell} |1\rangle^{\otimes \ell}$ for some $\ell \in 0, \ldots, k$. Applying Lemma 1 at the end, we get

$$|sl2\rangle = (\mathrm{Id}^{\otimes n-k-1} \otimes SCS_{n,k}) |0\rangle^{\otimes n-\ell} |1\rangle^{\otimes \ell}$$

$$= |0\rangle^{\otimes n-k-1} \otimes \left(\sqrt{\tfrac{\ell}{n}} |0\rangle^{\otimes k+1-\ell} |1\rangle^{\otimes \ell} + \sqrt{\tfrac{n-\ell}{n}} |0\rangle^{\otimes k-\ell} |1\rangle^{\otimes \ell} |0\rangle \right)$$

$$= \sqrt{\tfrac{\ell}{n}} |0\rangle^{\otimes (n-1)-(\ell-1)} |1\rangle^{\otimes \ell-1} \otimes |1\rangle + \sqrt{\tfrac{n-\ell}{n}} |0\rangle^{\otimes (n-1)-\ell} |1\rangle^{\otimes \ell} \otimes |0\rangle ,$$

$$|sl3\rangle = (U_{n-1,k} \otimes \mathrm{Id}) |sl2\rangle$$

$$= \sqrt{\tfrac{\ell}{n}} |D_{\ell-1}^{n-1}\rangle \otimes |1\rangle + \sqrt{\tfrac{n-\ell}{n}} |D_\ell^{n-1}\rangle \otimes |0\rangle = |D_\ell^n\rangle .$$

Step $U_{k-1,k-1} \to U_{k,k}$. We show $U_{k,k} |0\rangle^{\otimes k-\ell} |1\rangle^{\otimes \ell} = |D_\ell^k\rangle$ for all $\ell \leq k$. Replacing k by $k-1$ and n by k in the previous analysis, we immediately get that for $\ell \leq k-1$ the state $|sl1\rangle = |0\rangle^{\otimes k-\ell} |1\rangle^{\otimes \ell}$ maps to $|sl3\rangle = |D_\ell^k\rangle$. It remains to show the same for $\ell = k$:

$$|sl3\rangle = (U_{k-1,k-1} \otimes \mathrm{Id}) SCS_{k,k-1} |sl1\rangle = (U_{k-1,k-1} \otimes \mathrm{Id}) SCS_{k,k-1} |1\rangle^{\otimes k}$$

$$= (U_{k-1,k-1} \otimes \mathrm{Id}) |1\rangle^{\otimes k} = |D_{k-1}^{k-1}\rangle \otimes |1\rangle = |1\rangle^{\otimes k} = |D_k^k\rangle . \qquad \square$$

2.2   Explicit Construction of $SCS_{n,k}$

In the following, we describe a clean construction of an arbitrary Split & Cyclic Shift unitary $SCS_{n,k}$ in terms of 1 two-qubit gate and $k-1$ three-qubit gates, each of which implements exactly one of the k non-trivial mappings for $\ell \in 1, \ldots, k$ given in Definition 3:

$$|0\rangle^{\otimes k+1-\ell} |1\rangle^{\otimes \ell} \rightarrow \sqrt{\tfrac{\ell}{n}} |0\rangle^{\otimes k+1-\ell} |1\rangle^{\otimes \ell} + \sqrt{\tfrac{n-\ell}{n}} |0\rangle^{\otimes k-\ell} |1\rangle^{\otimes \ell} |0\rangle . \qquad (1)$$

Building Blocks. The relevant qubits in this mapping are the last (nth) qubit as well as the pair of qubits in which there is a change in the binary string from 0's to 1's (the $(n-\ell)$th and $(n-\ell+1)$th qubits). Using the notation $|xy\rangle_a$ to say that qubits a and $a+1$ are in states x and y, respectively, the two- and three-qubit gates are defined and constructed by:

(i) $|00\rangle_{n-1} \quad \rightarrow |00\rangle_{n-1}$

      $|11\rangle_{n-1} \quad \rightarrow |11\rangle_{n-1}$

      $|01\rangle_{n-1} \quad \rightarrow \sqrt{\tfrac{1}{n}} |01\rangle_{n-1} + \sqrt{\tfrac{n-1}{n}} |10\rangle_{n-1}$

$(ii)_\ell$ $|00\rangle_{n-\ell} |0\rangle_n \rightarrow |00\rangle_{n-\ell} |0\rangle_n$

      $|01\rangle_{n-\ell} |0\rangle_n \rightarrow |01\rangle_{n-\ell} |0\rangle_n$

      $|00\rangle_{n-\ell} |1\rangle_n \rightarrow |00\rangle_{n-\ell} |1\rangle_n$

      $|11\rangle_{n-\ell} |1\rangle_n \rightarrow |11\rangle_{n-\ell} |1\rangle_n$

      $|01\rangle_{n-\ell} |1\rangle_n \rightarrow \sqrt{\tfrac{\ell}{n}} |01\rangle_{n-\ell} |1\rangle_n + \sqrt{\tfrac{n-\ell}{n}} |11\rangle_{n-\ell} |0\rangle_n$

The two-qubit gate (i) and the $k-1$ three-qubit gates $(ii)_\ell$ for $2 \leq \ell \leq k$ are each constructed by a (two-)controlled Y-rotation $R_y\left(2\cos^{-1}\sqrt{\tfrac{\ell}{n}}\right)$ mapping $|0\rangle \rightarrow \sqrt{\tfrac{\ell}{n}} |0\rangle + \sqrt{\tfrac{n-\ell}{n}} |1\rangle$, conjugated with a $CNOT$ on the last qubit n. Here, we use $R_y(2\theta) = \left(\begin{smallmatrix} \cos\theta & -\sin\theta \\ \sin\theta & \cos\theta \end{smallmatrix}\right)$.

Putting It All Together. Note that the states $|0\rangle^{\otimes k+1}$ and $|1\rangle^{\otimes k+1}$ remain unchanged under each of the k gates $(i), (ii)_\ell$. Furthermore, for any given $1 \leq \ell^* \leq k$, there is exactly one of the k gates $(i), (ii)_\ell$ affecting the state $|0\rangle^{\otimes k+1-\ell^*} |1\rangle^{\otimes \ell^*}$, namely the one with matching $\ell = \ell^*$. It maps $|01\rangle_{n-\ell^*} |1\rangle_n \rightarrow \sqrt{\tfrac{\ell^*}{n}} |01\rangle_{n-\ell^*} |1\rangle_n + \sqrt{\tfrac{n-\ell^*}{n}} |11\rangle_{n-\ell^*} |0\rangle_n$, implementing Eq. (1).

The resulting second term $|0\rangle^{\otimes k-\ell^*} |1\rangle^{\otimes \ell^*} |0\rangle$ remains unaffected by all gates $(ii)_\ell$ with larger $\ell > \ell^*$. Hence we can build a complete $SCS_{n,k}$ gate starting with the two-qubit gate (i) followed the $k-1$ three-qubit gates $(ii)_\ell$ order by increasing ℓ. For an illustration of $SCS_{5,3}, SCS_{4,3}, SCS_{3,2}$ and $SCS_{2,1}$ – together composing $U_{5,3}$ – see Fig. 2. The example can also be opened and verified in Quirk [12] following this link.

3   Circuit Size and Depth

We now analyze the size and depth of our circuit construction and show how to adapt the circuit to be used on Linear Nearest Neighbor (LNN) architectures, where 2-qubit gates can only be implemented between neighboring qubits:

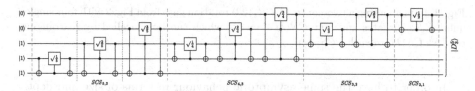

Fig. 2. Preparation of the Dicke state $|D_3^5\rangle$ with SCS gates implementing a unitary $U_{5,3}$. $\sqrt{\frac{\ell}{n}}$-gates denote Y-Rotations $R_y\left(2\cos^{-1}\sqrt{\frac{\ell}{n}}\right): |0\rangle \to \sqrt{\frac{\ell}{n}}|0\rangle + \sqrt{\frac{n-\ell}{n}}|1\rangle$.

Theorem 1. *Dicke states $|D_k^n\rangle$ can be prepared with a circuit of size $\mathcal{O}(\min(k, n-k) \cdot n)$ and depth $\mathcal{O}(n)$, even on Linear Nearest Neighbor architectures.*

Proof (Arbitrary 2-qubit gates). Note that an alternate way to prepare a Dicke state $|D_k^n\rangle$ is to prepare the Dicke state $|D_{n-k}^n\rangle$ followed by X-gates on each qubit, as $|D_k^n\rangle = X^{\otimes n}|D_{n-k}^n\rangle$. Thus we prove size $\mathcal{O}(kn)$ for Dicke states $|D_k^n\rangle$, implying size $\mathcal{O}((n-k)n)$ for $|D_k^n\rangle$, too.

We first show that the depth of our circuit construction is linear: The structure of each $SCS_{n,k}$ implementation is a stair of 2-qubit blocks interacting with its bottom qubit n. These stairs can be "pushed into each other". In particular, the 3-qubit gate $(ii)_k$ of $SCS_{n,k}$ acts on qubits $n-k, n-k+1$ and n. It can therefore be run in parallel with $k^* := \lfloor\frac{k+1}{3}\rfloor - 1$ many other 3-qubit gates, namely gate $(ii)_{k-3}$ of $SCS_{n-1,k}$ (acting on qubits $n-k+2, n-k+3, n-1$) as well as gates $(ii)_{k-6}, \ldots, (ii)_{k-3k^*}$ of $SCS_{n-2,k}, \ldots, SCS_{n-k^*,k}$, respectively. Since we can parallelize $k^* \in \mathcal{O}(k)$ stairs, the total depth is linear in the depth of gates $(i), (ii)_\ell$ (constant) and the number of gates $(i), (ii)_\ell$ ($\mathcal{O}(kn)$) divided by k^*, yielding an overall depth of $\mathcal{O}(n)$.

In light of a possible implementation, we prove the circuit size in Theorem 1 by compiling $U_{n,k}$ down to at most $5kn + \mathcal{O}(n)$ $CNOT$-gates and $4kn + \mathcal{O}(n)$ arbitrary precision R_y-gates. To build $U_{n,k}$ from SCS unitaries, we need a total of $n-1$ many 2-qubit gates (i) and $(n-k) \cdot (k-1) + \sum_{i=3}^{k}(i-2) = kn - \frac{k^2}{2} + \mathcal{O}(n)$ many 3-qubit gates $(ii)_\ell$, see Fig. 2. It remains to show that $(ii)_\ell$-gates can be implemented with 5 $CNOT$ gates and 4 R_y gates. We provide such an implementation in Fig. 3: A two-controlled $CCR_y(2\theta)$ rotation gate can be implemented with 4 $R_y(\pm\frac{\theta}{2})$ rotation gates and 4 $CNOT$s, the first one of which we can cancel by rearranging the preceeding conjugating $CNOT$ gate. $\square$

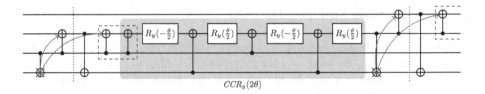

Fig. 3. Implementing a two-control $CCR_y(2\theta)$ rotation gate from Fig. 2 with four single-qubit $R_y(\pm\frac{\theta}{2})$ gates and four $CNOT$s, one of which can be cancelled by rearranging the last $CNOT$ of each SCS step.

In order to have the same asymptotic behaviour in terms of size and depth on LNN architectures, we need to slightly adapt our circuit. There are two main ideas for this: The first one is to implement the $(ii)_\ell$-gates of $SCS_{n,k}$ by "keeping the bottom qubit n close" to the two qubits $n-\ell$, $n-\ell+1$. This can be done by "sifting up" qubit n with $SWAP$ gates for the duration of $SCS_{n,k}$, then sifting it down back to its original position. The second idea is to do this in groups of $\Theta(k)$ consecutive SCS unitaries. Each group can be implemented with $\mathcal{O}(k^2)$ gates and $\mathcal{O}(k)$ depth. Overall, combining the $\mathcal{O}(\frac{n}{k})$ many groups gives circuit size $\mathcal{O}(kn)$ and depth $\mathcal{O}(n)$.

4   Symmetric Pure States and Quantum Compression

Our inductive approach yields (for $k = n$) a unitary $U_{n,n}$ which – with $\mathcal{O}(n^2)$ gates and $\mathcal{O}(n)$ depth – can be used to prepare any Dicke state $|D_\ell^n\rangle$, $1 \le \ell \le n$ for the respective input $|0\rangle^{\otimes n-\ell} |1\rangle^{\otimes \ell}$. Therefore, every superposition of these input states leads to a superposition of Dicke states. In the following, we show how this can be used to (i) prepare arbitrary symmetric pure n-qubit states in linear depth $\mathcal{O}(n)$, and to (ii) compress symmetric pure n-qubit states into $\lceil \log(n+1) \rceil$ qubits in quasilinear depth $\tilde{\mathcal{O}}(n)$ using the reverse unitary $U_{n,n}^\dagger$.

4.1   Symmetric Pure States

As the $n + 1$ different Dicke states $|D_\ell^n\rangle$ form an orthonormal basis of the fully symmetric subspace of all pure n-qubit states, every symmetric pure state can be expanded in terms of Dicke states [4], i.e. in the form $\sum_\ell e^{i\phi_\ell}\alpha_\ell |D_\ell^n\rangle$ with magnitudes $\alpha_\ell \in [0,1]$, $\alpha_0^2 + \ldots + \alpha_n^2 = 1$ and phases $\phi_\ell \in [0, 2\pi)$, $\phi_0 = 0$. We show:

Theorem 2. *Every symmetric pure n-qubit state can be prepared with a circuit of size $\mathcal{O}(n^2)$ and depth $\mathcal{O}(n)$, even on Linear Nearest Neighbor architectures.*

Proof. By Theorem 1, given as input the state $\sum_\ell e^{i\phi_\ell}\alpha_\ell |0\rangle^{\otimes n-\ell} |1\rangle^{\otimes \ell}$, the unitary $U_{n,n}$ prepares $\sum_\ell e^{i\phi_\ell}\alpha_\ell |D_\ell^n\rangle$ on LNN architectures using $\mathcal{O}(n^2)$ gates and $\mathcal{O}(n)$ depth. We prove that this input state can be constructed in linear depth and size. To this end, we define magnitudes $\beta_\ell := \alpha_\ell(1 - \alpha_0^2 - \ldots - \alpha_{\ell-1})^{-1/2}$ and angles $\psi_0 := 0$, $\psi_\ell := \phi_\ell - \phi_{\ell-1}$. The original values α, ϕ relate to the parameters β, ψ as $\phi_\ell = \sum_{k=0}^{\ell} \psi_k$ and $\alpha_\ell = \beta_\ell \cdot \prod_{k=0}^{\ell-1} \sqrt{1 - \beta_k^2}$. As already introduced in Sect. 3, we use Y-rotation gates $R_y(2\cos^{-1} \beta)$ to map $|0\rangle \to \beta |0\rangle + \sqrt{1 - \beta^2} |1\rangle$. Additionally, we use phase shift gates $R_\psi = \left(\begin{smallmatrix} 1 & 0 \\ 0 & e^{i\psi} \end{smallmatrix} \right)$ to map $|1\rangle \to e^{i\psi} |1\rangle$.

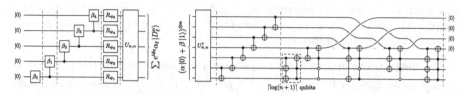

Fig. 4. (left) Arbitrary superposition of Dicke states, using Y-rotations $R_y(2\cos^{-1} \beta_\ell)$ with $\beta_\ell = \sqrt{\frac{\alpha_\ell^2}{1 - \alpha_0^2 - \ldots - \alpha_{\ell-1}^2}}$ and phase-shift gates R_{ψ_ℓ} with $\psi_\ell = \phi_\ell - \phi_{\ell-1}$, followed by unitary $U_{n,n}$. (right) Efficient compression of n identical qubits into $\lceil \log(n+1) \rceil$ qubits using unitary $U_{n,n}^\dagger$.

We start with a rotation $R_y(2\cos^{-1} \beta_0)$ on the n-th qubit. This is followed by a linear-depth stair of controlled $R_y(2\cos^{-1} \beta_\ell)$-rotations on the $(n - \ell)$th qubit, controlled by the previous qubit $n - \ell + 1$ being in state $|1\rangle$, as shown in Fig. 4 (left). At this point, we have the correct magnitudes. Finally, we add the correct phases using a layer of R_{ψ_ℓ}-phase shifts on respective qubits $n - \ell + 1$, yielding the desired symmetric state:

$$|0\rangle^{\otimes n} \xrightarrow{\beta_0} \alpha_0 |0\rangle^{\otimes n} + \sqrt{1 - \alpha_0^2} |0\rangle^{\otimes n-1} |1\rangle$$

$$\xrightarrow{\beta_1} \alpha_0 |0\rangle^{\otimes n} + \alpha_1 |0\rangle^{\otimes n-1} |1\rangle + \sqrt{1 - \alpha_0^2 - \alpha_1^2} |0\rangle^{\otimes n-2} |11\rangle$$

$$\xrightarrow{\beta_2, \ldots, \beta_{n-1}} \sum_\ell \alpha_\ell |0\rangle^{\otimes n-\ell} |1\rangle^{\otimes \ell} \xrightarrow{\psi_1, \ldots, \psi_n} \sum_\ell e^{i\phi_\ell} \alpha_\ell |0\rangle^{\otimes n-\ell} |1\rangle^{\otimes \ell}. \qquad \square$$

4.2   Quantum Compression

As symmetric pure states live in the $(n + 1)$-dimensional symmetric subspace of the full Hilbert space, they can be described with exponentially fewer dimensions than general multi-qubit states. This is the idea behind the quantum Schur-Weyl transform [1], which separates the permutation information from the angular momentum information of a state. Applied to a symmetric pure state, it will compress the angular momentum information into only $\lceil \log(n+1) \rceil$ qubits, while the rest of the qubits (the trivial permutation information) can be discarded without loss of information.

A previous approach to implement this transform for symmetric states gave a high-level description of a circuit of size and depth $\Theta(n^2)$ that needs no ancillas [24]. The major part is a transformation of Dicke states $|D_\ell^n\rangle$ to a one-hot encoding $|0\rangle^{\otimes \ell-1} |1\rangle |0\rangle^{\otimes n-\ell}$ of their Hamming weight ℓ. Substituting this part with our unitary in reverse, $U_{n,n}^\dagger$, improves the overall circuit depth to quasilinear:[3]

Theorem 3. *Every symmetric pure n-qubit state can be compressed into $\lceil \log(n+1) \rceil$ qubits with a circuit of size $\mathcal{O}(n^2)$ and depth $\tilde{\mathcal{O}}(n)$, even on Linear Nearest Neighbor architectures.*

We illustrate our approach with a particular interesting symmetric pure state, the separable state $(\alpha |0\rangle + \beta |1\rangle)^{\otimes n} = \sum \alpha^{n-\ell} \beta^\ell \binom{n}{\ell}^{1/2} |D_\ell^n\rangle$, whose compression has been implemented experimentally for $n = 3$ [26]. An implementation of our approach for $n = 7$ qubits in Quirk [12] can be found following this link.

Proof. Our compression circuit starts with the reverse unitary $U_{n,n}^\dagger$ (using size $\mathcal{O}(n^2)$ and depth $\mathcal{O}(n)$). It is followed by a mapping of states $|0\rangle^{\otimes n-\ell} |1\rangle^{\otimes \ell}$ to the one-hot encoding $|0\rangle^{\otimes n-\ell} |1\rangle |0\rangle^{\otimes \ell-1}$, which can be implemented with size and depth $\mathcal{O}(n)$ with a simple stair of $CNOT$-gates with control $n - \ell$ and target $n-\ell+1$ for increasing ℓ. Finally, the one-hot encoding $|0\rangle^{\otimes n-\ell} |1\rangle |0\rangle^{\otimes \ell-1}$ is mapped to the binary encoding $|\ell\rangle$ of ℓ (with padded leading zeroes), as illustrated in Fig. 4 (right):

$$\sum \alpha^{n-\ell} \beta^\ell \binom{n}{\ell}^{1/2} |D_\ell^n\rangle \xrightarrow{\ U_{n,n}^\dagger\ } \sum \alpha^{n-\ell} \beta^\ell \binom{n}{\ell}^{1/2} |0\rangle^{\otimes n-\ell} |1\rangle^{\otimes \ell}$$

$$\xrightarrow[\text{stair}]{CNOT} \alpha^n |0\rangle^{\otimes n} + \sum_{\ell > 0} \alpha^{n-\ell} \beta^\ell \binom{n}{\ell}^{1/2} |0\rangle^{\otimes n-\ell} |1\rangle |0\rangle^{\otimes \ell-1}$$

$$\xrightarrow[\text{change}]{\text{encoding}} \sum \alpha^{n-\ell} \beta^\ell \binom{n}{\ell}^{1/2} |\ell\rangle .$$

It remains to show that mapping each state $|0\rangle^{\otimes n-\ell} |1\rangle |0\rangle^{\otimes \ell-1}$ into the bottom $\lceil \log(n+1) \rceil$ qubits encoding $|\ell\rangle$ can be implemented with a circuit of size $\mathcal{O}(n^2)$ and depth $\tilde{\mathcal{O}}(n)$. This is done in the following way, for increasing ℓ: First, controlled on qubit $n-\ell$ we $CNOT$ into the up to $\lceil \log(n+1) \rceil$ target bottom qubits which represent the number ℓ in binary (for n and $n - 1$ these are the qubits themselves, on which we perform no operation). Then, controlled on the binary representation in the last ℓ qubits (not including padded 0s), we perform a single multi-control Toffoli on the $(n - \ell)$th qubit as target. An implementation of m-control Toffoli gates with $CNOT$ and single-qubit gates requires at least $2m$ $CNOT$ gates [29], and it is known that a $O(m)$ $CNOT$ and single-qubit gates are sufficient [2], even if no ancilla qubits are present [13]. This immediately gives us $\mathcal{O}(n \log n)$ gates and $\mathcal{O}(n \log n)$ depth. $\square$

---

[3] The referenced paper [24] provides no compilation down to standard gates and no analysis of the depth of the circuit. The latter is found together with a step-by-step comparison of our approach in the preprint of this paper [3, Appendix].

For LNN architectures, we use sifting up of processed qubits, see Fig. 4 (right). This brings the respective next qubits into direct neighborhood of the bottom $\lceil \log(n+1) \rceil$ qubits. Implementing the $CNOT$s into the bottom qubits and the multi-control Toffoli gates into the top qubits can be done with $\mathcal{O}(n \log^2 n)$ gates and depth overall.

5   Conclusions

We presented a deterministic quantum circuit for the preparation of Dicke states $|D_k^n\rangle$ with depth $\mathcal{O}(n)$ and $\mathcal{O}(kn)$ gates in total. We showed that these bounds hold for Linear Nearest Neighbor architectures, that the circuit can be extended to prepare arbitrary symmetric pure states, and that we can use it for quantum compression. For future work, the main open problem is that of characterizing the set of quantum states that can be prepared in polynomial time, of which Dicke states are one example.

Acknowledgments. We thank Yiğit Subaşı for helpful discussions.

References

1. Bacon, D., Chuang, I.L., Harrow, A.W.: Efficient quantum circuits for Schur and Clebsch-Gordan transforms. Phys. Rev. Lett. **97**(17), 170502 (2006). https://doi.org/10.1103/PhysRevLett.97.170502

2. Barenco, A., et al.: Elementary gates for quantum computation. Phys. Rev. A **52**(5), 3457–3467 (1995). https://doi.org/10.1103/PhysRevA.52.3457

3. Bärtschi, A., Eidenbenz, S.: Deterministic Preparation of Dicke States. arXiv e-prints, April 2019. https://arxiv.org/abs/1904.07358

4. Bastin, T., Thiel, C., von Zanthier, J., Lamata, L., Solano, E., Agarwal, G.S.: Operational determination of multiqubit entanglement classes via tuning of local operations. Phys. Rev. Lett. **102**(5), 053601 (2009). https://doi.org/10.1103/PhysRevLett.102.053601

5. Chakraborty, K., Choi, B.S., Maitra, A., Maitra, S.: Efficient quantum algorithms to construct arbitrary dicke states. Quantum Inf. Process. **13**(9), 2049–2069 (2014). https://doi.org/10.1007/s11128-014-0797-8

6. Childs, A.M., Farhi, E., Goldstone, J., Gutmann, S.: Finding cliques by quantum adiabatic evolution. Quantum Inf. Comput. **2**(3), 181–191 (2002). https://doi.org/10.26421/QIC2.3

7. Chuang, I.L., Modha, D.S.: Reversible arithmetic coding for quantum data compression. IEEE Trans. Inf. Theory **46**(3), 1104–1116 (2000). https://doi.org/10.1109/18.841192

8. Dicke, R.H.: Coherence in spontaneous radiation processes. Phys. Rev. **93**(1), 99–110 (1954). https://doi.org/10.1103/PhysRev.93.99

9. Diker, F.: Deterministic construction of arbitrary W states with quadratically increasing number of two-qubit gates. arXiv e-prints, June 2016. arXiv:1606.09290

10. Dür, W., Vidal, G., Cirac, J.I.: Three qubits can be entangled in two inequivalent ways. Phys. Rev. A **62**(6), 062314 (2000). https://doi.org/10.1103/PhysRevA.62.062314

11. Farhi, E., Goldstone, J., Gutmann, S.: A Quantum Approximate Optimization Algorithm. arXiv e-prints, November 2014. arXiv:1411.4028
12. Gidney, C.: Quirk: Quantum Circuit Simulator. A drag-and-drop quantum circuit simulator. https://algassert.com/quirk
13. Gidney, C.: Constructing large controlled nots/Constructing large increment gates/Using quantum gates instead of ancilla bits, June 2015. https://algassert.com/circuits/2015/06/22/Using-Quantum-Gates-instead-of-Ancilla-Bits.html
14. Hadfield, S., Wang, Z., O'Gorman, B., Rieffel, E.G., Venturelli, D., Biswas, R.: From the quantum approximate optimization algorithm to a quantum alternating operator ansatz. Algorithms 12(2), 34 (2019). https://doi.org/10.3390/a12020034
15. Hume, D.B., Chou, C.W., Rosenband, T., Wineland, D.J.: Preparation of dicke states in an ion chain. Phys. Rev. A 80(5), 052302 (2009). https://doi.org/10.1103/PhysRevA.80.052302
16. Ionicioiu, R., Popescu, A.E., Munro, W.J., Spiller, T.P.: Generalized parity measurements. Phys. Rev. A 78(5), 052326 (2008). https://doi.org/10.1103/PhysRevA.78.052326
17. Ivanov, S.S., Vitanov, N.V., Korolkova, N.V.: Creation of arbitrary dicke and NOON states of trapped-ion qubits by global addressing with composite pulses. New J. Phys. 15(2), 023039 (2013). https://doi.org/10.1088/1367-2630/15/2/023039
18. Kay, A.: Quantikz: A tikz library to typeset quantum circuit diagrams. Tutorial on the Quantikz Package. https://doi.org/10.17637/rh.7000520
19. Kiesel, N., Schmid, C., Tóth, G., Solano, E., Weinfurter, H.: Experimental observation of four-photon entangled dicke state with high fidelity. Phys. Rev. Lett. 98(6), 063604 (2007). https://doi.org/10.1103/PhysRevLett.98.063604
20. Lamata, L., López, C.E., Lanyon, B.P., Bastin, T., Retamal, J.C., Solano, E.: Deterministic generation of arbitrary symmetric states and entanglement classes. Phys. Rev. A 87(3), 032325 (2013). https://doi.org/10.1103/PhysRevA.87.032325
21. Microsoft: Quantum Katas/Superposition, March 2019. Programming exercises for learning Q# and quantum computing. https://github.com/Microsoft/QuantumKatas
22. Moreno, M.G.M., Parisio, F.: All bipartitions of arbitrary Dicke states. arXiv e-prints, January 2018. https://arxiv.org/abs/1801.00762
23. Mosca, M., Kaye, P.: Quantum networks for generating arbitrary quantum states. In: Optical Fiber Communication Conference and International Conference on Quantum Information ICQI, p. PB28, June 2001. https://doi.org/10.1364/ICQI.2001.PB28
24. Plesch, M., Bužek, V.: Efficient compression of quantum information. Phys. Rev. A 81(3), 032317 (2010). https://doi.org/10.1103/PhysRevA.81.032317
25. Prevedel, R., et al.: Experimental realization of dicke states of up to six qubits for multiparty quantum networking. Phys. Rev. Lett. 103(2), 020503 (2009). https://doi.org/10.1103/PhysRevLett.103.020503
26. Rozema, L.A., Mahler, D.H., Hayat, A., Turner, P.S., Steinberg, A.M.: Quantum data compression of a qubit ensemble. Phys. Rev. Lett. 113(16), 160504 (2014). https://doi.org/10.1103/PhysRevLett.113.160504
27. Shao, X.Q.S., Chen, L., Zhang, S., Zhao, Y.F., Yeon, K.H.: Deterministic generation of arbitrary multi-atom symmetric Dicke states by a combination of quantum Zeno dynamics and adiabatic passage. EPL (Europhys. Lett.) 90(5), 50003 (2010). https://doi.org/10.1209/0295-5075/90/50003

28. Shende, V.V., Bullock, S.S., Markov, I.L.: Synthesis of quantum-logic circuits. IEEE Trans. Comput.-Aided Des. Integr. Circ. Syst. **25**(6), 1000–1010 (2006). https://doi.org/10.1109/TCAD.2005.855930

29. Shende, V.V., Markov, I.L.: On the CNOT-cost of TOFFOLI gates. Quantum Inf. Comput. **9**(5), 461–486 (2009). https://doi.org/10.26421/QIC9.5-6

30. Stockton, J.K., van Handel, R., Mabuchi, H.: Deterministic Dicke-state preparation with continuous measurement and control. Phys. Rev. A **70**(2), 022106 (2004). https://doi.org/10.1103/PhysRevA.70.022106

31. Tóth, G.: Multipartite entanglement and high-precision metrology. Phys. Rev. A **85**(2), 022322 (2012). https://doi.org/10.1103/PhysRevA.85.022322

32. Wieczorek, W., Krischek, R., Kiesel, N., Michelberger, P., Tóth, G., Weinfurter, H.: Experimental entanglement of a six-photon symmetric dicke state. Phys. Rev. Lett. **103**(2), 020504 (2009). https://doi.org/10.1103/PhysRevLett.103.020504

33. Wu, C., Guo, C., Wang, Y., Wang, G., Feng, X.L., Chen, J.L.: Generation of Dicke states in the ultrastrong-coupling regime of circuit QED systems. Phys. Rev. A **95**(1), 013845 (2017). https://doi.org/10.1103/PhysRevA.95.013845

34. Xiao, Y.F., Zou, X.B., Guo, G.C.: Generation of atomic entangled states with selective resonant interaction in cavity quantum electrodynamics. Phys. Rev. A **75**(1), 012310 (2007). https://doi.org/10.1103/PhysRevA.75.012310

35. Özdemir, S.K., Shimamura, J., Imoto, N.: A necessary and sufficient condition to play games in quantum mechanical settings. New J. Phys. **9**(2), 43–43 (2007). https://doi.org/10.1088/1367-2630/9/2/043

Optimal Channel Utilization
with Limited Feedback

Gianluca De Marco[1(✉)], Tomasz Jurdziński[2], and Dariusz R. Kowalski[3,4]

[1] Dipartimento di Informatica, University of Salerno, Fisciano, Italy
gidemarco@unisa.it
[2] Institute of Computer Science, University of Wrocław, Wrocław, Poland
[3] School of Computer and Cyber Sciences, Augusta University, Augusta, USA
[4] SWPS University of Social Sciences and Humanities, Warsaw, Poland

Abstract. A channel with multiplicity feedback is a shared channel that in case of collision (two or more stations transmitting simultaneously) returns as a feedback the exact number of stations simultaneously transmitting. It is known that in such a model $\Theta((d\log(n/d))/\log d)$ time rounds are sufficient and necessary to identify the IDs of d transmitting stations, from an ensemble of n.

In contrast, the model with collision detection (or ternary feedback) allows only a *limited feedback* from the channel: 0 (silence), 1 (success) or 2+ (collision). In this case it is known that $\Omega(d\log(n/d))$ time rounds are necessary.

Generalizing, we can define a *feedback interval* $[x, y]$, where $0 \leq x \leq y \leq d$, such that the channel returns the exact number of transmitting stations *only if* this number is within that interval. The collision detection model corresponds to $x = 0$ and $y = 1$, while the multiplicity feedback is obtained for $x = 0$ and $y = d$.

It is natural to ask for which size of the feedback intervals we can still get the same optimal time complexity $\Theta((d\log(n/d))/(\log d))$ valid for the channel with multiplicity feedback. In this paper we show that we can still use this number of time rounds even when the interval has a substantially smaller size: namely $O(\sqrt{d\log d})$. On the other hand, we also prove that if we further reduce the size of the interval to $O\left(\frac{\sqrt{d}}{\log d}\right)$, then no protocol having time complexity $\Theta((d\log(n/d))/(\log d))$ is possible.

Keywords: Multiple-access channel · Limited feedback · Group testing · Threshold group testing · Distributed · Algorithm · Lower bound

1 Introduction

A shared channel, also called a *multiple access channel*, is one of the fundamental communication models. The formal model used in the present paper, taken as the

This work is supported by the Polish National Science Center (NCN) grants UMO-2017/25/B/ST6/02553 and UMO-2017/25/B/ST6/02010.

L. A. Gąsieniec et al. (Eds.): FCT 2019, LNCS 11651, pp. 140–152, 2019.
https://doi.org/10.1007/978-3-030-25027-0_10

basis for theoretical studies, is defined as follows (cf. the surveys by Gallager [15] and Chlebus [21]). A set of n stations are connected to the same multi-point transmission channel. Every station has an ID label that uniquely distinguish it. A subset of $d < n$ active stations have data packets and can transmit on the channel in synchronous rounds. There is no central control: stations act autonomously by means of a distributed algorithm. In this paper we consider the fundamental problem of identifying the d active stations, *i.e.* to let all the active stations to know each other's ID.

A station successfully transmits its packet, in a given time round, if and only if it is the only transmitting station in that round. Once a packet is successfully transmitted on the channel by one station, it can be heard by all other stations. When $m > 1$ stations send a packet simultaneously, *i.e.* in the same time round, these packets collide and their information content is irretrievably lost. However, in this case, all stations (also including those not active) can learn from the channel feedback that a collision occurred. The stations can recognize three possible outcomes: inactivity (no station sending), success (exactly one station sending) and collision. This makes it possible to design *adaptive* algorithms in which at any time round a station may decide whether to transmit or stay silent based on the feedback received in previous rounds. Namely, a deterministic adaptive algorithm, at each step of its execution, specifies some subset Q (often called *test* or *query*) of the n possible stations chosen as a function of the feedback obtained in previous steps. Any station that has not yet transmitted successfully, checks the chosen subset and transmits if and only if it belongs to it. In *non-adaptive* algorithms all the queries have to be designed in advance, before executing the algorithm. While adaptive algorithms are in general more powerful, non-adaptive solutions are often more desirable in that they allow all the queries to be made at once which is very useful in many practical situations.

Another strictly related important problem in this same setting is the *contention resolution*, in which the goal is to let the d active stations to transmit successfully their respective packets [5,7–10].

The identification problem considered in this paper has been studied in many variants mainly in the context of *Combinatorial search*. In this area, it is relevant to us the *Group testing problem* (see [12,13]), introduced in [11], that asks to identify d defective items in a set N of n elements. At any step, it is possible to choose a subset $Q \subseteq N$ and perform a test of the kind: "is there any defective item in Q?". The similarity with the multiple access channel setting described above is easy to observe: the d defective items correspond to the d active stations and any test Q corresponds to a time round in which a subset Q of stations attempt to transmit simultaneously. It is worth noting that while in the multiple access channel, as observed above, the outcome (channel feedback) in each time round is ternary, the outcome of a test in the Group testing problem is binary: either Q contains at least a defective item or it does not. In this sense the latter model is weaker. It is well-known that for an adaptive algorithm $\Theta(d \log(n/d))$ tests ([18]) are sufficient and necessary to identify d defectives. In contrast, any non-adaptive combinatorial group testing algorithm needs as much as $\Omega(\min\{d^2\log_d(n), n\})$ tests [2]. An efficient explicit scheme (*i.e.* constructible in polynomial time) requiring $\Theta(\min\{d^2\log(n), n\})$ has been proposed in [20]. No better upper bound for non-adaptive solutions is known even for non-explicit schemes.

Alongside the model with *ternary feedback*, Tsybakov [22] introduced the collision channel with *multiplicity feedback* (also known as the *quantitative channel*). In this model, the channel feedback makes available the "loudness" of the collision, from which it is possible to determine the actual number m of stations simultaneously transmitting. This can be implemented by means of energy detectors [16]. The problem of identifying active stations in a multiple-access channel with multiplicity feedback is also a distributed version of the counterfeit coin types of problems studied in Combinatorial search theory (see [6] for an account of this equivalence). For this reason many of the following basic results for the identification problem in multiple-access channel have been obtained in the context of Combinatorial search theory.

It is known that this problem requires $\Theta(d\log_{d+1}(n/d))$ queries in the worst case both for adaptive and non-adaptive algorithms. The lower bound can be easily inferred from the following simple argument. For any fixed parameter $d < n$ and set X, with $|X| = d$, there are $d + 1$ possible outcomes to each query Q, as $0 \le |Q \cap X| \le d$. Therefore, since there are $\binom{n}{d}$ possible subsets X of d elements, the information-theoretic lower bound tells us that any algorithm (even adaptive ones) requires at least

$$\left\lceil \log_{d+1}\binom{n}{d} \right\rceil = \Omega(d\log_{d+1}(n/d)) \tag{1}$$

queries to identify X. On the other hand, Grebinski and Kucherov [17], with a non-constructive proof, showed that there exists a non-adaptive algorithm using $O(d\log_{d+1}(n/d))$ queries in the worst case, so matching the lower bound. More recently Bshouty [1] designed a polynomial time adaptive algorithm using the same optimal asymptotic bound. A polynomial time non-adaptive algorithm, using the same optimal number of queries, is still unknown.

Throughout the paper, $[n]$ denotes the set of station's IDs, *i.e.* we let $[n] = \{1, 2, \ldots, n\}$. The family of all subsets of $[n]$ of size d will be denoted $[n]_d$.

1.1   New Results

As observed in [16], being able to detect the number of stations involved in a collision, no matter how large this number is, might be impractical as it would require an unbounded number of energy detectors. The authors of [16] proposed a model in which a limited set of energy detectors guarantees that the number of collided packets is detected only if this number is below a certain limit. Most of the literature on this subject deals with randomized adaptive protocols for the contention resolution problem.

In this paper we study the harder problem of *deterministically and non-adaptively* identifying a subset X of d active stations out of an ensemble of n

in the more general situation in which the channel returns the exact number of transmitting stations *only if* this number is within an interval $[x, y]$, called the *feedback interval*, where $0 \leq x \leq y \leq d$. The collision detection (ternary feedback) model corresponds to $x = 0$ and $y = 1$, while the multiplicity feedback is obtained for $x = 0$ and $y = d$.

This is also a generalization of the *multi-threshold group testing* introduced in [4], where we have $s > 0$ thresholds: $t_1, t_2, ..., t_s$ and the output of each test is an integer between 0 and s corresponding to which thresholds get passed by the number of defectives in the test. The first formalization for single threshold in the context of Group Testing is due to Damaschke [3]. In this case, the output of each test responds to the question: *are there at least t positive elements in the query set?* The classical version of the problem corresponds to the choice $t = 1$.

A non-adaptive algorithm with time complexity m, is viewed as a list of queries $\mathcal{Q} = Q_1, Q_2, \ldots, Q_m$, where each query Q_i is a subset of the set of IDs $[n] = \{1, 2, \ldots, n\}$. At time round i a station transmits if and only if its ID belongs to Q_i. The non-adaptivity emerges from the fact that the sequence of queries is generated beforehand the protocol's execution as a function of d and n only. In particular, the algorithm does not use the feedback: each new query is generated irrespectively of the feedback received on the previous ones. The identification process takes place at the end of the algorithm's execution and uses the feedback received on each query to uniquely determine the input set.

In this paper we prove that a feedback interval of size $O(\sqrt{d \log d})$ is sufficient to correctly identify any set X of d stations with only $O(\frac{d \log(n/d)}{\log d})$ time rounds (queries). This time complexity is *optimal* in view of the information-theoretic lower bound (1) that holds for the stronger multiplicity feedback model, *i.e.* for a feedback interval of maximum size d (nearly quadratically larger than ours), in which case the channel can correctly detect *any number* of simultaneous transmissions.

We also prove that this size of $O(\sqrt{d \log d})$ cannot be substantially reduced without negatively affect the complexity of the algorithm. Namely, we show that if the feedback interval is reduced to $O\left(\frac{\sqrt{d}}{\log d}\right)$, then any non-adaptive algorithm for the same problem would require complexity $\Omega(d \log(n/d))$.

In this extended abstract we give both upper and lower bound for the particular case of intervals centered in d/2. The complete results will be given in the full version of the paper.

2   The Upper Bound

In this section we prove that a feedback interval of size $O(\sqrt{d \log d})$ is sufficient for a non-adaptive algorithm to identify any input set $X \in [n]_d$ with the optimal number of queries: $O(\frac{d \log(n/d)}{\log d})$.

Formally, we will show that for a feedback interval $[x, y]$, with $x = d/2 - 3\sqrt{d \log d}$ and $y = d/2 + 3\sqrt{d \log d}$, there exists a sequence of queries $\mathcal{Q} =$

$(Q_1, Q_2, \ldots, Q_m)$, with $m = O(\frac{d \log(n/d)}{\log d})$, such that for any pair of sets $X, Y \in [n]_d$, $X \neq Y$, there exists a query $Q \in \mathcal{Q}$ such that

(i) $|Q \cap X|, |Q \cap Y| \in [x, y]$;

(ii) $|Q \cap X| \neq |Q \cap Y|$.

First, condition (i) guarantees that on query Q, the channel returns exactly the number of transmitting stations for both input sets X and Y. Then, condition (ii) guarantees that on query Q, X and Y cause different feedback. Therefore, for any pair of input sets X and Y there will be a time round on which the two sets have different feedback. This implies that any input set can be uniquely identified.

Let $\mathcal{Q}$ be a set of queries constructed as follows: any query $Q \in \mathcal{Q}$ is randomly and independently formed by letting $\mathbf{Pr}[x \in Q] = 1/2$ for every $x \in [n]$. Observe that for any set $X \in [n]_d$ and query $Q \in \mathcal{Q}$, $|Q \cap X|$ is a random variable taking values in the interval $[0, d]$. The following lemma holds.

Lemma 1. *For $1 \leq \ell \leq d$, let $X, Y \in [n]_\ell$, with $X \cap Y = \emptyset$. For any $Q \in \mathcal{Q}$, the probability that $|X \cap Q| = |Y \cap Q|$ is at most $\min\{1/2, 1/\sqrt{\ell}\}$.*

Proof. Fix any $Q \in \mathcal{Q}$ and $1 \leq \ell \leq d$. For any set $X \in [n]_\ell$, the probability that $|Q \cap X| = r$, for any $0 \leq r \leq \ell$, can be interpreted as the probability of having precisely r successes in a series of ℓ Bernoulli trials, where in each trial we test whether an element $x \in X$ belongs to Q. This is defined by the binomial distribution:

$$\binom{\ell}{r} p^r (1 - p)^{\ell - r},$$

where p is the probability of success, *i.e.*, the probability that any $x \in X$ belongs to Q, and $|Q \cap X|$ is the random variable equal to the total number of successes. Hence, the probability defined in the statement of the lemma writes as

$$\mathbf{Pr}[|Q \cap X| = |Q \cap Y|] = \sum_{r=0}^{\ell} \mathbf{Pr}[|Q \cap X| = r] \cdot \mathbf{Pr}[|Q \cap Y| = r]$$

$$= \sum_{r=0}^{\ell} \binom{\ell}{r}^2 p^{2r} (1 - p)^{2(\ell - r)} = \sum_{r=0}^{\ell} \binom{\ell}{r}^2 \left(\frac{1}{2}\right)^{2\ell},$$

where the last equality follows from the fact that in the random construction of Q, we set $p = 1/2$. For $\ell = 1$ this sum is $1/2$ and, since it is decreasing with the size of ℓ, it follows that $1/2$ is also an upper bound on it. In order to prove that it is at most $\min\{1/2, 1/\sqrt{\ell}\}$, it will suffice to show that it is also less than $1/\sqrt{\ell}$. In fact, since for $\ell \geq 4$, $1/\sqrt{\ell} \leq 1/2$, the lemma follows.

We have

$$\binom{\ell}{r}^2 \left(\frac{1}{2}\right)^{2\ell} = \left(\frac{1}{2}\right)^{2\ell} \binom{\ell}{\ell/2}^2 \left[\frac{(r+1)\cdots\ell/2}{(\ell/2+1)\cdots(\ell-r)}\right]^2$$

$$\leq \left(\frac{1}{2}\right)^{2\ell} \left(\frac{2^\ell}{\sqrt{\ell}}\right)^2 \left(\frac{\ell/2}{d-r}\right)^{2(\ell/2-r)} = \frac{1}{\ell}\left(1 - \frac{\ell/2-r}{\ell-r}\right)^{2(\ell/2-r)}$$

$$\leq \frac{1}{\ell} \cdot e^{-\frac{2(\ell/2-r)^2}{\ell-r}}.$$

Therefore, continuing from (2) we have

$$\mathbf{Pr}[|Q \cap X| = |Q \cap Y|] \leq \frac{1}{\ell} \sum_{r=0}^{\ell} e^{-\frac{2(\ell/2-r)^2}{\ell-r}}$$

$$\leq \frac{2}{\ell} \sum_{r=0}^{\ell/2} e^{-\frac{2(\ell/2-r)^2}{\ell-r}} = \frac{2}{\ell} \sum_{i=0}^{\sqrt{\ell}/2-1} \sum_{r=i\sqrt{\ell}+1}^{(i+1)\sqrt{\ell}} e^{-\frac{2(\ell/2-r)^2}{\ell-r}}$$

$$\leq \frac{2}{\ell} \sum_{i=0}^{\sqrt{\ell}/2-1} \sum_{r=i\sqrt{\ell}+1}^{(i+1)\sqrt{\ell}} e^{-\frac{2(\ell/2-i\sqrt{\ell})^2}{\ell-i\sqrt{\ell}}}$$

$$\leq \frac{2}{\ell} \sum_{i=0}^{\sqrt{\ell}/2-1} \sqrt{\ell} e^{-2(\sqrt{\ell}/2-i)} = \frac{2}{\sqrt{\ell}} \sum_{i=0}^{\sqrt{\ell}/2-1} e^{-2(\sqrt{\ell}/2-i)}. \quad (2)$$

Now observe that

$$\sum_{i=0}^{\sqrt{\ell}/2-1} e^{-2(\sqrt{\ell}/2-i)} = \frac{1}{e^{\sqrt{\ell}}} + \frac{1}{e^{\sqrt{\ell}-1}} + \cdots + \frac{1}{e^2} \leq \frac{1}{2}. \quad (3)$$

Finally, by substituting (3) into (2), the proof follows.   □

Considering that the expected value of the random variable $|X \cap Q|$ (Binomial distribution of parameters d and $1/2$) is $\mu = d/2$, we can use the Chernoff bound to estimate the probability that $|Q \cap X|$ falls out of the feedback interval. We will use the following formulas for the Chernoff bound (see Eq. (4.2) and (4.5) in [19]): for $0 < \delta < 1$, $\mathbf{Pr}(|Q \cap X| \geq (1+\delta)\mu) \leq e^{-\frac{\delta^2 \mu}{3}}$ and $\mathbf{Pr}(|Q \cap X| \leq (1-\delta)\mu) \leq e^{-\frac{\delta^2 \mu}{2}}$.

Letting $\mu = d/2$ and $\delta = 6\sqrt{\frac{\log d}{d}}$, we have:

$$\mathbf{Pr}(|Q \cap X| > d/2 + 3\sqrt{d \log d}) \leq e^{-\frac{36 \frac{\log d}{d} \frac{d}{2}}{3}}$$

$$= e^{-6 \log d}; \quad (4)$$

$$\mathbf{Pr}(|Q \cap X| < d/2 - 3\sqrt{d \log d}) \leq e^{-\frac{36 \frac{\log d}{d} \frac{d}{2}}{2}}$$

$$= e^{-9 \log d}. \quad (5)$$

Let $X, Y \in [n]_d$, $X \neq Y$. We say that a query $Q \in \mathcal{Q}$ *separates* X and Y if and only if X and Y get different feedback on query Q, *i.e.* when $|Q \cap X|, |Q \cap Y| \in [x, y]$ and $|Q \cap X| \neq |Q \cap Y|$. We say that a sequence of queries $\mathcal{Q}$ separates X and Y if and only if there exists a query $Q \in \mathcal{Q}$ that separates X and Y.

We are now ready to prove our upper bound.

Theorem 1. *Let the feedback interval be of size $\Theta(\sqrt{d \log d})$. For $2 < d \leq n/2$, there exists a sequence of queries $\mathcal{Q}$ of length $m = O\left(\frac{d \log(n/d)}{\log d}\right)$ that separates any pair $X, Y \in [n]_d$.*

Proof. Let $\mathcal{Q} = (Q_1, Q_2, \ldots, Q_m)$ be a sequence of m queries randomly constructed as explained in the beginning of this section, *i.e.*, for every $x \in [n]$, independently, $\mathbf{Pr}[x \in Q] = 1/2$.

In order to prove the lemma, we will show that with positive probability there exists a sequence $\mathcal{Q}$ of $m = O((d \log(n/d))/\log d)$ queries such that for any $X, Y \in [n]_d$, $\mathcal{Q}$ separates X and Y.

For a fixed set $X \subseteq [n]_q$, we define $\check{\mathcal{E}}(X; Q)$ to be the event that $|Q \cap X|$ is within the feedback interval, *i.e.* $d/2 - 3\sqrt{d \log d} \leq |Q \cap X| \leq d/2 + 3\sqrt{d \log d}$. For a fixed pair $X, Y \in [n]_d$, we let $\bar{\mathcal{E}}(X, Y; Q)$ be the event that $|X \cap Q| = |Y \cap Q|$.

Fix a set $X \in [n]_d$. We can associate to the sequence of m random queries, m independent trials where in the ith one we test whether $d/2 - 3\sqrt{d \log d} \leq |Q_i \cap X| \leq d/2 + 3\sqrt{d \log d}$, for $1 \leq i \leq m$. We consider Q_i a success if event $\check{\mathcal{E}}(X; Q)$ holds, a failure otherwise. By (4) and (5) each Q_i is a failure with probability $q \leq e^{-6 \log d} + e^{-9 \log d} < \frac{1}{d}$.

For a fixed set $X \in [n]_d$, the probability of having f failures or more in m queries is at most

$$\binom{m}{f} \left(\frac{1}{d}\right)^f.$$

Hence, for any constant $0 < \epsilon < 1$, the probability that there exists a set $X \in [n]_d$ for which there are $f = m\epsilon$ failures or more in m queries will be at most

$$\binom{n}{d}\binom{m}{f}\left(\frac{1}{d}\right)^f < 2^{d \log(ne/d) + f \log(me/f) - f \log d}$$

$$= 2^{d \log(ne/d) + m\epsilon \log(e/\epsilon) - m\epsilon \log d}. \tag{6}$$

This probability can be made less than 1 for $m = O(\frac{d \log(n/d)}{\log d})$. This means that for any constant $0 < \epsilon < 1$, there exists a constant $c > 0$ such that for

$$m \geq \frac{c \cdot d \log(n/d)}{\log d}, \tag{7}$$

with positive probability, for every set $X \in [n]_d$, there are at least $m(1 - \epsilon)$ queries of $\mathcal{Q}$ such that $|X \cap Q|$ is inside the feedback interval.

Let us now fix any pair of sets $X, Y \in [n]_d$. Let $X' = X \setminus Y$, $Y' = Y \setminus X$. Observe that for every query Q: (a) X' and Y' are disjoint and (b) $|X \cap Q| =$

$|Y \cap Q|$ if and only if $|X' \cap Q| = |Y' \cap Q|$. Hence, letting $\ell = |X'| = |Y'|$, for every query Q we have:

$$\bar{\mathcal{E}}(X,Y;Q) = \bar{\mathcal{E}}(X',Y';Q) \leq \min\left\{\frac{1}{2}, \frac{1}{\sqrt{\ell}}\right\}, \tag{8}$$

where the inequality follows from Lemma 1 applied on the disjoint sets X' and Y'. As before, we can define m independent trials corresponding to the m queries of Q. This time, in the ith trial we test whether $|Q_i \cap X'| = |Q_i \cap Y'|$ and we say that Q_i is a success if event $\bar{\mathcal{E}}(X',Y';Q_i)$ does not hold, a failure otherwise. By (8) we know that Q_i is a failure with probability $q' \leq \min\left\{1/2, 1/\sqrt{\ell}\right\}$.

For a fixed pair $X', Y' \subseteq [n]_\ell$ the probability of having g failures or more in m queries is at most

$$\binom{m}{g}\left(\min\left\{\frac{1}{2}, \frac{1}{\sqrt{\ell}}\right\}\right)^g.$$

For any constant $0 < \epsilon' < 1$, the probability that there exists a pair of sets $X', Y' \subseteq [n]_\ell$ for which there are $g = m\epsilon'$ failures or more in m queries will be at most

$$\sum_{\ell=1}^{d}\binom{n}{\ell}^2\binom{m}{g}\left(\min\left\{\frac{1}{2}, \frac{1}{\sqrt{\ell}}\right\}\right)^g$$

$$= \sum_{\ell=1}^{4}\binom{n}{\ell}^2\binom{m}{g}\left(\frac{1}{2}\right)^g + \sum_{\ell=5}^{d}\binom{n}{\ell}^2\binom{m}{g}\left(\frac{1}{\sqrt{\ell}}\right)^g$$

$$< 4 \cdot 2^{8\log(\frac{n\epsilon}{4}) + g\log(me/g) - g} + d \cdot 2^{2d\log(\frac{n\epsilon}{d}) + g\log(me/g) - \frac{g}{2}\log d}$$

$$= 2^{2 + 8\log(\frac{n\epsilon}{4}) + m\epsilon'\log(e/\epsilon') - m\epsilon'} + 2^{\log d + 2d\log(\frac{n\epsilon}{d}) + m\epsilon'\log(e/\epsilon') - \frac{m\epsilon'}{2}\log d}.$$

This probability is less than 1 for $m = O(\frac{d\log(n/d)}{\log d})$. This implies the following. For any constant $0 < \epsilon' < 1$, there exists a constant $c' > 0$ such that for

$$m \geq \frac{c' \cdot d\log(n/d)}{\log d} \tag{9}$$

with positive probability, for every pair of sets $X, Y \in [n]_d$ there are at least $m(1 - \epsilon')$ queries $Q \in \mathcal{Q}$ such that $|Q \cap X| \neq |Q \cap Y|$. In order to be able to distinguish X and Y, we need not only that $|Q \cap X| \neq |Q \cap Y|$ but also that both $|X \cap Q|$ and $|Y \cap Q|$ fall within the feedback interval.

Let us fix $\epsilon' < 1 - 2\epsilon$ and let c and c' be as in (7) and (9) respectively. Let $c'' \geq \max\{c', c''\}$. If $\mathcal{Q}$ has $m \geq \frac{c'' \cdot d\log(n/d)}{\log d}$ queries, we have that with positive probability, for every pair of sets X and Y there are

(a) at least $m(1 - \epsilon')$ queries Q such that $|Q \cap X| \neq |Q \cap Y|$;
(b) at most $2m\epsilon$ queries such that $|Q \cap X|$ or $|Q \cap Y|$ do not fall within the feedback interval.

Therefore, since ϵ and ϵ' have been chosen to satisfy $m(1-\epsilon') > 2m\epsilon$, we have proved that with positive probability for every pair X, Y there exists a query $Q \in \mathcal{Q}$ such that Q distinguishes X and Y. This concludes the proof.

$\square$

3   The Lower Bound

In this section, we prove that if the size of the interval is reduced to $O\left(\frac{\sqrt{d}}{\log d}\right)$, then we need $\Omega\left(d\log(n/d)\right)$ queries to be able to distinguish every pair of sets X and Y of size d. Formally, we will prove the following theorem.

Theorem 2. *Let the feedback interval be $\left[\frac{d}{2} - \frac{\sqrt{d}}{\log d}, \frac{d}{2} + \frac{\sqrt{d}}{\log d}\right]$. Any deterministic non-adaptive algorithm identifying any set $X \in [n]_d$ requires $\Omega(d\log(n/d))$ queries in the worst case.*

We consider a uniform distribution over $[n]_d$. We first estimate the probability for a fixed query Q that a random set $X \in [n]_d$ has intersection with Q within the feedback interval.

Lemma 2. *Let X be a random set uniformly chosen in $[n]_d$. For any fixed query Q, the probability that $|X \cap Q|$ falls within the feedback interval is $O(1/\log d)$.*

Proof. Let $q = |Q|$, $x = \frac{d}{2} - \frac{\sqrt{d}}{\log d}$ and $y = \frac{d}{2} + \frac{\sqrt{d}}{\log d}$. The probability stated in the Lemma can be evaluated as follows.

$$\mathbf{Pr}\left(x \le |Q \cap X| \le y\right) = \sum_{i=x}^{y} \frac{\binom{q}{i} \cdot \binom{n-q}{d-i}}{\binom{n}{d}} \tag{10}$$

$$\le \sum_{i=x}^{y} \frac{\binom{n/2}{i} \cdot \binom{n/2}{d-i}}{\binom{n}{d}} \tag{11}$$

To prove (11), note that for $q > n/2$:

$$\binom{q}{i} \cdot \binom{n-q}{d-i}$$

$$= \binom{n/2}{i} \cdot \binom{n/2}{d-i} \cdot \frac{(n/2+1)\cdot\ldots\cdot q}{(n/2-i+1)\cdot\ldots\cdot(q-i)} \cdot \frac{(n-q-d+i+1)\cdot\ldots\cdot(n/2-d+i)}{(n-q+1)\cdot\ldots\cdot n/2}$$

$$\le \frac{\binom{n/2}{i} \cdot \binom{n/2}{d-i}}{\binom{n}{d}}.$$

Now we use the following Stirling approximation for $n!$ (see [14], Sect. 2.9):

$$\sqrt{2\pi}n^{n+1/2}e^{-n+1/(12n+1)} < n! < \sqrt{2\pi}n^{n+1/2}e^{-n+1/(12n)}.$$

Hence,

$$\binom{n/2}{d/2}^2 < \frac{2\pi(n/2)}{2\pi(d/2) \cdot 2\pi(n-d)/2} \cdot \frac{(n/2)^n}{(d/2)^d (n-d)^{n-d}/2^{n-d}} \cdot e^{1/(3n)-2/(6d+1)-2/(6n-6d+1)}$$

$$\leq c \cdot \frac{n}{\pi d(n-d)} \cdot \frac{n^n}{d^d(n-d)^{n-d}}$$

for some constant $c \geq e$. Similarly,

$$\binom{n}{d} \geq \sqrt{\frac{2\pi n}{2\pi d \cdot 2\pi(n-d)}} \cdot \frac{n^n}{d^d(n-d)^{n-d}} \cdot e^{1/(12n+1)-1/(12d)-1/(12n-12d)}$$

$$\geq c' \cdot \sqrt{\frac{n}{2\pi d(n-d)}} \cdot \frac{n^n}{d^d(n-d)^{n-d}}$$

for some constant $c' \leq 1/e$. Hence, assuming that $d \leq a \cdot n$ for some constant $a < 1$, we get

$$\frac{\binom{n/2}{d/2} \cdot \binom{n/2}{d/2}}{\binom{n}{d}} \leq (c/c') \sqrt{\frac{2}{\pi(1-a)}} \cdot \frac{1}{\sqrt{d}} \cdot$$

We also have for any $i < d/2$:

$$\binom{n/2}{i} = \binom{n/2}{d/2} \cdot \frac{(i+1) \cdot \ldots \cdot (d/2)}{((n/2)-(d/2)+1) \cdot \ldots \cdot ((n/2)-i)}$$

and

$$\binom{n/2}{d-i} = \binom{n/2}{d/2} \cdot \frac{((n/2)-d+i+1) \cdot \ldots \cdot ((n/2)-(d/2))}{((d/2)+1) \cdot \ldots \cdot (d-i)}$$

hence

$$\binom{n/2}{i} \cdot \binom{n/2}{d-i} = \binom{n/2}{d/2}^2 \cdot \frac{(i+1) \cdot \ldots \cdot (d/2)}{((n/2)-(d/2)+1) \cdot \ldots \cdot ((n/2)-i)}$$

$$\cdot \frac{((n/2)-d+i+1) \cdot \ldots \cdot ((n/2)-(d/2))}{((d/2)+1) \cdot \ldots \cdot (d-i)}$$

$$= \binom{n/2}{d/2}^2 \cdot \Pi_{j=1}^{d/2-i} \frac{i+j}{(n/2)-(d/2)+j} \cdot \frac{(n/2)-d+i+j}{(d/2)+j}$$

$$< \binom{n/2}{d/2}^2 \cdot \tag{12}$$

The last inequality holds because for every $i \leq d/2$ and every $1 \leq j \leq d/2 - i$, the difference between the denominator and the nominator of

$$\frac{i+j}{(n/2)-(d/2)+j} \cdot \frac{(n/2)-d+i+j}{(d/2)+j}$$

in the product is

$$(n/2) \cdot ((d/2) - i) + (i + j)(d - i - 2j) - d^2/4 \,,$$

which is a decreasing function in j for all $j \geq 0$, provided $d < 3i$, which holds for sufficiently large d as $i \geq x = \frac{d}{2} - \sqrt{\frac{d}{\log n}}$. Therefore, this function is at least the value for $j = d/2 - j$, which is

$$(n/2) \cdot ((d/2) - i) + (d/2)i - d^2/4 \,,$$

for all the considered values of j. This value is however always positive for $i < d/2$, because as a function of i it is decreasing for all positive i, provided $d \leq n/2$, and its value for $i = d/2$ is 0. It follows that each factor in the product is smaller than 1, which complete the proof of Eq. (12).

The proof of the counterpart of Eq. (12) for $i > d/2$ is by symmetry argument, as

$$\binom{n/2}{i} \cdot \binom{n/2}{d-i}$$

for $i > d/2$ is equal to

$$\binom{n/2}{j} \cdot \binom{n/2}{d-i}$$

for $j = d - i < d/2$, and the latter was proved to be smaller than

$$\binom{n/2}{d/2}^2 \,.$$

Consequently, continuing from (11):

$$\sum_{i=x}^{y} \frac{\binom{n/2}{i} \cdot \binom{n/2}{d-i}}{\binom{n}{d}} < (y - x + 1) \frac{\binom{n/2}{d/2} \cdot \binom{n/2}{d/2}}{\binom{n}{d}} \tag{13}$$

$$= (2\frac{\sqrt{d}}{\log d} + 1) \cdot (c/c')\sqrt{\frac{2}{\pi(1 - a)}} \cdot \frac{1}{\sqrt{d}} = O(1/\log d) \,. \tag{14}$$

$\square$

Now, let us consider an arbitrary algorithm represented by any set of m queries $Q_1, \ldots, Q_m$. Assume that, the algorithm being correct, the sequence $Q_1, \ldots, Q_m$ separates each pair $X, Y \in [n]_d$. For the sake of analysis, we introduce the notion of a *bad query* with respect to a set Q, i.e., a query whose intersection with X has size in the feedback interval.

Definition 1. (Bad query). *A query set $Q \subseteq [n]$ is bad with respect to a set $X \in [n]_d$ when $|Q \cap X|$ falls within the feedback interval.*

As a consequence of Lemma 2, there exists a constant $c > 0$ such that, the expected number of bad queries for a random set $X \in [n]_d$ is at most $cm/\log d$.

Let $\mathcal{X}$ be the family of those sets $X \in [n]_d$ for which the actual number of bad queries is at most $2cm/\log d$. By Markov's inequality and the fact that our estimations are done for the uniform distribution over $[n]_d$ (having size $\binom{n}{d}$) the size of $\mathcal{X}$ is at least $\frac{1}{2}\binom{n}{d}$.

According to the assumption that $Q_1, \ldots, Q_m$ separate each pair $X, Y \in [n]_d$, there must be at least $\frac{1}{2}\binom{n}{d}$ different feedbacks to be able to distinguish all elements of $\mathcal{X}$. On the other hand, the feedback for any set $X \in \mathcal{X}$ can be encoded as follows. For each query Q_i, one bit encodes whether it is a bad query with respect to X. Then, if the query is bad, the feedback is encoded in

$$\log\left(2\frac{\sqrt{d}}{\log d}\right)$$

bits. Therefore, the size of the encoding of the feedback is

$$O\left(\log(d/\log d) \cdot \frac{2cm}{\log d} + 2(m - \frac{2cm}{\log d})\right) = O(m),$$

since $c > 0$ is constant and $d < n$. This in turn implies that the number of different feedbacks for elements of $\mathcal{X}$ is $2^{O(m)}$. On the other hand, the size of $\mathcal{X}$ is

$$\Omega\left(\frac{1}{2}\binom{n}{d}\right) = 2^{\Omega(d\log(n/d))}.$$

This requires $m = \Omega(d\log(n/d))$ and the proof of Theorem 2 is completed.

4   Conclusions and Open Problems

We studied the impact of limited number of thresholds on utilization of a shared communication channel, showing that around $\sqrt{d}$ thresholds (up to a logarithmic factor) are necessary and enough to achieve similar utilization as with all thresholds for d active users. Further study could concentrate on shrinking the polylogarithmic gap between lower and upper bound on the number of thresholds, as well as on general application to combinatorial testing with restricted measurements.

References

1. Bshouty, N.H.: Optimal algorithms for the coin weighing problem with a spring scale. In: COLT (2009)
2. Chaudhuri, S., Radhakrishnan, J.: Deterministic restrictions in circuit complexity. In: Proceedings of the Twenty-eighth Annual ACM Symposium on Theory of Computing, STOC 1996, pp. 30–36, ACM. New York (1996)
3. Damaschke, P.: Threshold group testing. Electron. Notes Discrete Math. **21**, 265–271 (2005)

4. De Marco, G., Jurdziński, T., Różański, M., Stachowiak, G.: Subquadratic non-adaptive threshold group testing. In: Klasing, R., Zeitoun, M. (eds.) FCT 2017. LNCS, vol. 10472, pp. 177–189. Springer, Heidelberg (2017). https://doi.org/10.1007/978-3-662-55751-8_15

5. De Marco, G., Kowalski, D.R.: Towards power-sensitive communication on a multiple-access channel. In: 2010 International Conference on Distributed Computing Systems, ICDCS 2010, Genova, Italy, 21–25 June 2010, pp. 728–735 (2010)

6. De Marco, G., Kowalski, D.R.: Searching for a subset of counterfeit coins: randomization vs determinism and adaptiveness vs non-adaptiveness. Random Struct. Algorithms 42(1), 97–109 (2013)

7. De Marco, G., Kowalski, D.R.: Fast nonadaptive deterministic algorithm for conflict resolution in a dynamic multiple-access channel. SIAM J. Comput. 44(3), 868–888 (2015)

8. De Marco, G., Kowalski, D.R.: Contention resolution in a non-synchronized multiple access channel. Theor. Comput. Sci. 689, 1–13 (2017)

9. De Marco, G., Kowalski, D.R., Stachowiak, G.: Brief announcement: deterministic contention resolution on a shared channel. In 32nd International Symposium on Distributed Computing, DISC 2018, New Orleans, LA, USA, 15–19 October 2018, pp. 44:1–44:3 (2018)

10. De Marco, G., Stachowiak, G.: Asynchronous shared channel. In: Proceedings of the ACM Symposium on Principles of Distributed Computing, PODC 2017, Washington, DC, USA, 25–27 July 2017, pp. 391–400 (2017)

11. Dorfman, R.: The detection of defective members of large populations. Ann. Math. Stat. 14(4), 436–440 (1943)

12. Du, D.-Z., Hwang, F.K.: Combinatorial Group Testing and Its Applications, vol. 12, January 2000

13. Du, D-Z., Hwang, F.K.: Pooling Designs and Nonadaptive Group Testing, Jaunary 2006

14. Feller, W.: An Introduction to Probability Theory and Its Applications, vol. 1. Wiley, Hoboken (1968)

15. Gallager, R.: A perspective on multiaccess channels. IEEE Trans. Inf. Theor. 31(2), 124–142 (2006)

16. Georgiadis, L., Papantoni-Kazakos, P.: A collision resolution protocol for random access channels with energy detectors. IEEE Trans. Commun. 30(11), 2413–2420 (1982)

17. Grebinski, V., Kucherov, G.: Optimal reconstruction of graphs under the additive model. Algorithmica 28(1), 104–124 (2000)

18. Hwang, F.K.: A method for detecting all defective members in a population by group testing. J. Am. Stat. Assoc. 67(339), 605–608 (1972)

19. Mitzenmacher, M., Upfal, E.: Probability and Computing: Randomized Algorithms and Probabilistic Analysis. Cambridge University Press, New York (2005)

20. Porat, E., Rothschild, A.: Explicit nonadaptive combinatorial group testing schemes. IEEE Trans. Inf. Theory 57(12), 7982–7989 (2011)

21. Chlebus, B.S.: Randomized communication in radio networks, pp. 401–456, January 2001

22. Tsybakov, B.S.: Resolution of a conflict of known multiplicity. Prob. Inf. Transm. 16, 01 (1980)

RETRACTED CHAPTER: Complete Disjoint CoNP-Pairs but No Complete Total Polynomial Search Problems Relative to an Oracle

Titus Dose[✉]

Institute of Computer Science, University of Würzburg, Würzburg, Germany
titus.dose@uni-wuerzburg.de

Abstract. Consider the following conjectures:
- TFNP: the set TFNP of all total polynomial search problems has no complete problems with respect to polynomial reductions.
- DisjCoNP: there exists no many-one complete disjoint coNP-pair.

We construct an oracle relative to which TFNP holds and DisjCoNP does not hold. This partially answers a question by Pudlák [12], who lists several conjectures and asks for oracles that show corresponding relativized conjectures to be different. As there exists a relativizable proof for the implication DisjCoNP ⇒ TFNP [12], relative to our oracle the conjecture TFNP is strictly stronger than DisjCoNP.

Keywords: Total polynomial search problem · Disjoint coNP-pair · Oracle

1 Introduction

The main motivation for the present paper is an article by Pudlák [12] that is "motivated by the problem of finding finite versions of classical incompleteness theorems", investigates major conjectures in the field of proof complexity, discusses their relations, and in particular draws new connections between the conjectures. Among others Pudlák conjectures the non-existence

- of P-optimal proof systems for any coNP-complete (resp., NP-complete) sets, denoted by CON (resp., SAT),
- of complete disjoint NP-pairs (resp., coNP-pairs) with respect to polynomial many-one reductions, denoted by DisjNP (resp., DisjCoNP),
- of complete total polynomial search problems with respect to polynomial reductions, denoted by TFNP,
- and of polynomial many-one complete problems for UP (resp., NP ∩ coNP), denoted by UP (resp., NP ∩ coNP).

The main conjectures of these are CON and TFNP. We give some background on these main conjectures and on the notion of disjoint pairs. The first main conjecture CON has an interesting connection to some finite version of an incompleteness statement. Consider CON^N, the nonuniform version of CON, i.e., the

The original version of this chapter was retracted: The retraction note to this chapter is available at https://doi.org/10.1007/978-3-030-25027-0_25

conjecture that no coNP-complete set has optimal proof systems. Denote by $\mathrm{Con}_T(n)$ the finite consistency of a theory T, i.e., $\mathrm{Con}_T(n)$ is the statement that T has no proofs of contradiction of length $\leq n$. Krajíček and Pudlák [7] raise the conjectures CON and $\mathrm{CON^N}$ and show that the latter is equivalent to the statement that there is no finitely axiomatized theory S which proves the finite consistency $\mathrm{Con}_T(n)$ for every finitely axiomatized theory T by a proof of polynomial length in n. In other words, $\neg\mathrm{CON^N}$ expresses that a weak version of Hilbert's program (to prove the consistency of all mathematical theories) is possible [11]. Correspondingly, $\neg$CON is equivalent to the existence of a theory S such that, for any fixed theory T, proofs of $\mathrm{Con}_T(n)$ in S can be constructed in polynomial time in n [7].

The conjecture TFNP, together with the class TFNP, was introduced by Megiddo and Papadimitriou and is known (i) to be implied by the non-existence of disjoint coNP-pairs and known (ii) to imply that no NP-complete set has P-optimal proof systems.

The notion of disjoint NP-pairs, i.e., pairs (A, B) with $A \cap B = \emptyset$ and $A, B \in$ NP, was introduced by Even, Selman, and Yacobi [3,4]. Razborov [13] connects it with the concept of propositional proof systems (pps), i.e., proof systems for the set of propositional tautologies TAUT, defining for each pps f a disjoint NP-pair, the so-called canonical pair of f, and showing that the canonical pair of an optimal pps f is complete. Hence, putting it contrapositively, DisjNP $\Rightarrow$ $\mathrm{CON^N}$.

For a graphical overview over the implications between the above conjectures we refer to Fig. 1. In contrast to the many implications only very few oracles were known separating two of the relativized conjectures [12], which is why Pudlák asks for further oracles showing relativized conjectures to be different.

Khaniki [6] partially answers this question: besides showing two of the conjectures to be equivalent he presents two oracles showing that SAT and CON as well as TFNP and CON are independent in relativized worlds. To be more precise, relative to the one oracle, there exist P-optimal propositional proof systems but no many-one complete disjoint coNP-pairs, where the latter implies TFNP [12] and SAT. Relative to the other oracle, there exist no P-optimal propositional proof systems and each total polynomial search problem has a polynomial-time solution, where the latter implies $\neg$SAT. Hence, this oracle shows that there is no relativizable proof for the implication CON $\Rightarrow$ SAT. In another paper [1], the author extends this by showing that there is even no relativizable proof for the weaker implication DisjNP $\Rightarrow$ SAT (recall that DisjNP implies CON in a relativizable way), which —together with the first oracle by Khaniki— shows that even DisjNP is independent of DisjCoNP, TFNP, and CON in relativized worlds.

Dose and Glaßer [2] construct an oracle O that also separates some of the above relativized conjectures. Relative to O there exist no many-one complete disjoint NP-pairs, UP, the class of problems accepted by NP-machines with at most one accepting path for any given input, has many-one complete problems, and NP $\cap$ coNP has no many-one complete problems. In particular, relative to O, there do not exist any P-optimal propositional proof systems. Thus, among others, O shows that the conjectures CON and UP as well as NP $\cap$ coNP and UP cannot be proven equivalent with relativizable proofs.

The present paper adds one more oracle to this list proving that there is no relativizable proof for the implication TFNP $\Rightarrow$ DisjCoNP, i.e., relative to the oracle, TFNP has no complete problems with respect to polynomial reductions, but there exists a many-one complete disjoint coNP-pair. As Pudlák [12] proves the converse implication to hold relative to all oracles, the statement TFNP is strictly stronger than DisjCoNP relative to our oracle.

2 Preliminaries

Throughout this paper let Σ be the alphabet $\{0, 1\}$. We denote the length of a word $w \in \Sigma^*$ by $|w|$. Let $\Sigma^{\leq n} = \{w \in \Sigma^* \mid |w| \leq n\}$. The empty word is denoted by ε and the i-th letter of a word w for $0 \leq i < |w|$ is denoted by $w(i)$, i.e., $w = w(0)w(1)\cdots w(|w| - 1)$. If v is a prefix of w, i.e., $|v| \leq |w|$ and $v(i) = w(i)$ for all $0 \leq i < |v|$, then we write $v \sqsubseteq w$. For any finite set $Y \subseteq \Sigma^*$, let $\ell(Y) \stackrel{df}{=} \sum_{w \in Y} |w|$.

$\mathbb{N}$ (resp., $\mathbb{N}^+$) denotes the set of natural numbers (resp., positive natural numbers). The set of primes is denoted by $\mathbb{P} = \{2, 3, 5, \ldots\}$.

We identify Σ^* with $\mathbb{N}$ via the polynomial-time computable, polynomial-time invertible bijection $w \mapsto \sum_{i < |w|} (1 + w(i))2^i$, which is a variant of the dyadic encoding. Hence, notations, relations, and operations for Σ^* are transferred to $\mathbb{N}$ and vice versa. In particular, $|n|$ denotes the length of $n \in \mathbb{N}$. We eliminate the ambiguity of the expressions 0^i and 1^i by always interpreting them over Σ^*.

Let $\langle \cdot \rangle : \bigcup_{i > 0} \mathbb{N}^i \to \mathbb{N}$ be an injective, polynomial-time computable, polynomial-time invertible pairing function such that $|\langle u_1, \ldots, u_n \rangle| = 2(|u_1| + \cdots + |u_n| + n)$.

Given two sets A and B, $A - B$ denotes the set difference between A and B. The complement of a set A relative to the universe U is denoted by $\overline{A} = U - A$. The universe will always be apparent from the context.

FP, P, and NP denote standard complexity classes [10]. Define $\text{co}\mathcal{C} = \{A \subseteq \Sigma^* \mid \overline{A} \in \mathcal{C}\}$ for a class $\mathcal{C}$. If $A, B \in$ NP (resp., $A, B \in$ coNP) and $A \cap B = \emptyset$, then we call (A, B) a disjoint NP-pair (resp., a disjoint coNP-pair). The set of all disjoint NP-pairs (resp., coNP-pairs) is denoted by DisjNP (resp., DisjCoNP).

We also consider all these complexity classes in the presence of an oracle O and denote the corresponding classes by FP^O, P^O, NP^O, and so on.

Let M be a Turing machine. $M^D(x)$ denotes the computation of M on input x with D as an oracle. For an arbitrary oracle D we let $L(M^D) = \{x \mid M^D(x)$ accepts$\}$, where —as usual— we say that a nondeterministic machine accepts some input x if and only if it accepts x on some path.

For a deterministic polynomial-time Turing transducer F (i.e., a Turing machine computing a function), depending on the context, $F^D(x)$ either denotes the computation of F on input x with D as an oracle or the output of this computation.

Definition 1 ([2]). *A sequence (M_i) is called* standard enumeration *of nondeterministic, polynomial-time oracle Turing machines, if it has the following properties:*

1. *All M_i are nondeterministic, polynomial-time oracle Turing machines.*
2. *For all oracles D and all inputs x the computation $M_i^D(x)$ stops within $|x|^i + i$ steps.*
3. *For every nondeterministic, polynomial-time oracle Turing machine M there exist infinitely many $i \in \mathbb{N}$ such that for all oracles D it holds that $L(M^D) = L(M_i^D)$.*
4. *There exists a nondeterministic, polynomial-time oracle Turing machine M such that for all oracles D and all inputs x it holds that $M^D(\langle i, x, 0^{|x|^i + i}\rangle)$ nondeterministically simulates the computation $M_i^D(x)$.*

Analogously we define standard enumerations of deterministic, polynomial-time oracle Turing machines and deterministic, polynomial-time oracle Turing transducers.

Throughout this paper, we fix some standard enumerations. Let $M_1, M_2, \ldots$ be a standard enumeration of nondeterministic polynomial-time oracle Turing machines and note that for every oracle D, the sequence (M_i^D) represents an enumeration of the languages in NP^D. Let $F_1, F_2, \ldots$ be a standard enumeration of polynomial-time oracle Turing transducers. Moreover, we let $P_1, P_2, \ldots$ be a standard enumeration of deterministic polynomial-time oracle Turing machines.

Let Z be an oracle and $A, B, C, D \in \Sigma^*$ such that $A \cap B = C \cap D = \emptyset$. In this paper we always use the following reducibility for disjoint pairs [13]. $(A, B) \leq_m^{\text{pp}, Z} (C, D)$, i.e., (A, B) is polynomially many-one reducible to (C, D), if there exists $f \in \text{FP}^Z$ with $f(A) \subseteq C$ and $f(B) \subseteq D$. We say that (C, D) is $\leq_m^{\text{pp}, Z}$-hard ($\leq_m^{\text{pp}, Z}$-complete) for DisjCoNP^Z if $(A, B) \leq_m^{\text{pp}, Z} (C, D)$ for all $(A, B) \in \text{DisjCoNP}^Z$ (and $(C, D) \in \text{DisjCoNP}^Z$).

Let us define total polynomial search problems [8] relative to some oracle D (for $D = \emptyset$ we obtain the unrelativized version and D can be simply omitted for all notations). Major parts of the following definitions are copied from [12].

Definition 2. *A total polynomial search problem relative to D is given by a pair (p, R), where p is a polynomial and $R \in \text{P}^D$ such that for all $x \in \mathbb{N}$ there exists $y \in \mathbb{N}$ with $|y| \leq p(|x|)$ and $\langle x, y\rangle \in R$. The computation task is: with access to D, for a given x, find y with $|y| \leq p(|x|) \wedge \langle x, y\rangle \in R$. The class of all total polynomial search problems relative to D will be denoted by TFNP^D.*

Definition 3. *Let R and S be total polynomial search problems relative to D. We say that R is polynomially reducible to S relative to D if and only if R can be solved in polynomial time using two oracles, namely D and one oracle that gives solutions to S. We call R complete for TFNP^D if all $S \in \text{TFNP}^D$ are polynomially reducible to R relative to D.*

We say that R is polynomially many-one reducible to S relative to D if and only if there are functions $f, g \in \text{FP}^D$ such that for all x and z it holds $S(\langle f^D(x), z\rangle) \Rightarrow R(\langle x, \langle g^D(x, z)\rangle\rangle)$. We call R many-one complete for TFNP^D if all $S \in \text{TFNP}^D$ are polynomially many-one reducible to R relative to D.

The following proposition is due to Jeřábek. A proof can be found in [12].

Proposition 1 (Jeřábek). TFNP *has complete problems if and only if it has many-one complete problems.*

Corollary 1. *Relative to any oracle D, the following assertions are equivalent.*

1. TFNP^D *has complete problems.*
2. TFNP^D *has many-one complete problems.*
3. $(n \mapsto n^i + i, L(P_i^D))$ *for some $i > 0$ is many-one complete for TFNP^D.*

Proof. 1 and 2 are equivalent as the proof of Proposition 1 in [12] is relativizable. 3 trivially implies 2. We show $2 \Rightarrow 3$. Let (p, R) be many-one complete for TFNP^D. Choose i such that (i) $\forall_{n \in \mathbb{N}} p(n) \leq n^i + i$ and (ii) $L(P_i^D) = R$ (such i exists by Definition 1). Then by the choice of i, the pair $T = (n \mapsto n^i + i, L(P_i^D))$ is a total polynomial search problem relative to D and (p, R) can be polynomially many-one reduced to T via the identity.

Since we focus on the question of whether there exist complete total polynomial search problems, Corollary 1 shows that we only need to consider problems of the form $(n \mapsto n^i + i, L(P_i^D))$. We occasionally use $L(P_i^D)$ as an abbreviation for this pair. Hence, $L(P_i^D)$ denotes a total polynomial search problem if and only if for each x there exists y with $|y| \leq |x|^i + i$ and $\langle x, y \rangle \in L(P_i^D)$.

Let us introduce some quite specific notations that are designed for the construction of oracles [2]. The domain and range of a function t are denoted by $\text{dom}(t)$ and $\text{ran}(t)$, respectively. The support $\text{supp}(t)$ of a real-valued function t is the subset of the domain that consists of all values that t does not map to 0. We say that a partial function t is injective on its support if $t(i, j) = t(i', j')$ for $(i, j), (i', j') \in \text{supp}(t)$ implies $(i, j) = (i', j')$. If a partial function t is not defined at point x, then $t \cup \{x \mapsto y\}$ denotes the extension of t that at x has value y.

If A is a set, then $A(x)$ denotes the characteristic function at point x, i.e., $A(x)$ is 1 if $x \in A$, and 0 otherwise. An oracle $D \subseteq \mathbb{N}$ is identified with its characteristic sequence $D(0)D(1)\cdots$, which is an ω-word. In this way, $D(i)$ denotes both, the characteristic function at point i and the i-th letter of the characteristic sequence, which are the same. A finite word w describes an oracle that is partially defined, i.e., only defined for natural numbers $x < |w|$. We can use w instead of the set $\{i \mid w(i) = 1\}$ and write for example $A = w \cup B$, where A and B are sets. For nondeterministic oracle Turing machines M we use the following phrases: a computation $M^w(x)$ *definitely accepts*, if at least one path of the computation $M^w(x)$ accepts and all queries on this path are $< |w|$. A computation $M^w(x)$ *definitely rejects*, if all paths of this computation reject and all queries on these paths are $< |w|$. For deterministic oracle Turing machines P we say: A computation $P^w(x)$ *definitely accepts* (resp., *definitely rejects*), if it accepts (resp., rejects) and the queries are $< |w|$.

For a deterministic or nondeterministic Turing machine M we say that the computation $M^w(x)$ *is defined*, if it definitely accepts or definitely rejects. For a transducer F, the computation $F^w(x)$ *is defined*, if all queries are $< |w|$.

3   Oracle Construction

The following theorem guarantees that there is no relativizable proof for the implication TFNP $\Rightarrow$ DisjCoNP, which is interesting as the converse implication has a relativizable proof [12]. Thus, TFNP is strictly stronger than DisjCoNP relative to the oracle that we construct below.

The basic idea for the encoding of a coNP-complete pair originates from an article by Glaßer et al. [5].

Theorem 1. *There exists an oracle O relative to which the following holds:*

- *DisjCoNP^O has a $\leq_m^{pp,O}$-complete pair.*
- *TFNP^O has no complete problem.*

Proof. Let D be a (possibly partial) oracle and p be some prime. We define:

$$A^D = \{\langle 0^m, 0^t, x\rangle \mid \forall_{y,|y|=|\langle 0^m,0^t,x\rangle|} y0\langle 0^m, 0^t, x\rangle \in D\}$$
$$B^D = \{\langle 0^m, 0^t, x\rangle \mid \forall_{y,|y|=|\langle 0^m,0^t,x\rangle|} y1\langle 0^m, 0^t, x\rangle \in D\}$$
$$R_p^D = \{\langle 0^n, y\rangle \mid \exists_{k>0} n = p^k, |y| = 2n, y \in D\} \cup \{\langle x, y\rangle \mid \forall_{k>0} x \neq 0^{p^k}\}$$

Note the following:

- (A^D, B^D) is a disjoint coNP-pair, if and only if for each $m, t, x \in \mathbb{N}$ there exist $b \in \{0,1\}$ and y of length $|\langle 0^m, 0^t, x\rangle|$ with $yb\langle 0^m, 0^t, x\rangle \notin D$.
- $(n \mapsto 2n, R_p^D)$ is a total polynomial search problem if and only if for each $k \in \mathbb{N}^+$ it holds $D \cap \{y \mid |y| = 2p^k\} \neq \emptyset$.

Preview of Construction. On the one hand, for all $i \neq j$ the construction tries to achieve that $\overline{L(M_i)} \cap \overline{L(M_j)} \neq \emptyset$. If this is not possible, then $(\overline{L(M_i)}, \overline{L(M_j)})$ inherently is a disjoint coNP-pair. Once we know this, we start to encode this pair into the pair (A, B). Thus, finally (A, B) will be a $\leq_m^{pp,O}$-complete disjoint coNP-pair.

On the other hand, for all i the construction intends to ensure that $L(P_i)$ is not a total polynomial search problem, i.e., there exists x such that for no y of length $\leq |x| + i$ it holds that $P_i(\langle x, y\rangle)$ accepts. If this is not possible, then $L(P_i)$ inherently is a total polynomial search problem and in that case, we choose a prime p and diagonalize against all pairs of FP-functions f and g making sure that R_p is not polynomially many-one reducible to $L(P_i)$ via f and g.

During the construction we maintain a growing collection of requirements. These are represented in a partial function belonging to $\mathcal{T} := \{t : \mathbb{N}^+ \times \mathbb{N}^+ \to \mathbb{N} \cup \{-p \mid p \in \mathbb{P}\} \mid \text{dom}(t) \text{ is finite and } t \text{ is injective on its support}\}$. If an oracle satisfies the properties defined by some $t \in \mathcal{T}$, then we will call it t-valid.

For i, j with $i \neq j$ and $t(i,j) > 0$ we define $c(i,j,x) = \langle 0^{t(i,j)}, 0^{|x|^{i+j}+i+j}, x\rangle$. A partial oracle w is called *t-valid* for $t \in \mathcal{T}$ if it satisfies the following properties.

V1 For all $i, j \in \mathbb{N}^+$ with $i \neq j$, if $t(i,j) = 0$, then there exists x such that $M_i^w(x)$ and $M_j^w(x)$ definitely reject.

(meaning: $(\overline{L(M_i)}, \overline{L(M_j)})$ is not a disjoint coNP-pair.)

V2 $A^w \cap B^w = \emptyset$.

V3 For all $i, j \in \mathbb{N}^+$ and $x \in \mathbb{N}$ with $i \neq j$ and $|x| \geq t(i, j) > 0$,

1. if $M_i^w(x)$ rejects, then w contains all words $y0c(i, j, x) < |w|$ with $|y| = |c(i, j, x)|$.

2. if $M_j^w(x)$ rejects, then w contains all words $y1c(i, j, x) < |w|$ with $|y| = |c(i, j, x)|$.

3. if $M_i^w(x)$ accepts, then w does not contain all words $y0c(i, j, x)$ with $|y| = |c(i, j, x)|$, i.e., for one such word α, $w(\alpha) = 0$ or $w(\alpha)$ is undefined).

4. if $M_j^w(x)$ accepts, then w does not contain all words $y1c(i, j, x)$ with $|y| = |c(i, j, x)|$, i.e., for one such word α, $w(\alpha) = 0$ or $w(\alpha)$ is undefined).

   (meaning: if $t(i, j) > 0$, then the pair $(\overline{L(M_i)}, \overline{L(M_j)})$ is encoded into the pair (A, B) from stage $t(i, j)$ on.)

V4 For all $i \in \mathbb{N}^+$ with $t(i, i) = 0$, there exists x such that for all y of length $\leq |x|^i + i$, the computation $P_i^w(\langle x, y \rangle)$ definitely rejects.

   (meaning: $L(P_i)$ is not a total polynomial search problem.)

V5 For all $i \in \mathbb{N}^+$ with $t(i, i) = -p$ for some prime and for each 0^{p^k} for $k \in \mathbb{N}^+$, if w is defined for all words of length p^k, then there exists $y \in w$ with $|y| = 2p^k$.

   (meaning: R_p will finally be a total polynomial search problem.)

The two subsequent claims follow directly from the definition of t-valid and $c(i, j, x)$. The first claim refers to V3 and implies that once we have made an encoding for a computation $M_i^w(x)$ or $M_j^w(x)$, the computation cannot change anymore. We will apply this result several times without mentioning it explicitly.

Claim 1. *In V3, the computations $M_i^w(x)$ and $M_j^w(x)$ are defined if w is defined for all words of length $\leq |c(i, j, x)|$.*

Claim 2. *Let $t, t' \in \mathcal{T}$ such that t' is an extension of t. For oracles $w \in \Sigma^*$, if w is t'-valid, then w is t-valid.*

Oracle construction: Let T be an enumeration of $(\mathbb{N}^+ \times \mathbb{N}^+) \cup \{(i, r, r') \mid i, r, r' \in \mathbb{N}^+\}$ having the property that (i, i) appears earlier than (i, r, r') for all i, r, r'. Each element of T stands for a task. We treat the tasks in the order specified by T and after treating a task we remove it and possibly other tasks from T. We start with the nowhere defined function t_0 and the t_0-valid oracle $w_0 = \varepsilon$. Then we define functions $t_1, t_2, \ldots$ in $\mathcal{T}$ such that t_{i+1} is an extension of t_i and partial oracles $w_0 \subsetneq w_1 \subsetneq w_2 \subsetneq \ldots$ such that each w_i is t_i-valid. Finally, we choose $O = \bigcup_{i=0}^{\infty} w_i$ (note that O is totally defined since in each step we strictly extend the oracle) and $t = \lim_{i \to \infty} t_i$. We describe step $s > 0$, which starts with a t_{s-1}-valid oracle w_{s-1} and extends it to a t_s-valid $w_s \supsetneq w_{s-1}$. Each task is immediately deleted from the task list T after it is treated. We will argue later that the construction is possible.

– task (i, j) with $i \neq j$: Let $t' = t_{s-1} \cup \{(i,j) \mapsto 0\}$. If there exists a t'-valid $v \sqsupsetneq w_{s-1}$, then let $t_s = t'$ and let w_s be the minimal t'-valid $v \sqsupsetneq w_{s-1}$. Otherwise, let $z = |w_{s-1}|$, let $t_s = t_{s-1} \cup \{(i,j) \mapsto |z| + 1\}$, and choose $w_s = w_{s-1}b$ for $b \in \{0,1\}$ such that w_s is t_s-valid.

– task (i, i): Let $t' = t_{s-1} \cup \{(i,i) \mapsto 0\}$. If there exists a t'-valid $v \sqsupsetneq w_{s-1}$, then let $t_s = t'$, $w_s = v$, and delete all $(i, \cdot, \cdot)$ from the task list T. Otherwise, let $z = |w_{s-1}|$, let $t_s = t_{s-1} \cup \{(i,i) \mapsto -p\}$ for some prime p greater than $|z|$ and all primes p' with $-p' \in \mathrm{ran}(t_{s-1})$, and choose $w_s = w_{s-1}b$ for $b \in \{0,1\}$ such that w_s is t_s-valid.

– task (i, r, r'): It holds $t_{s-1}(i,i) = -p$ for a prime p, since otherwise, this task would have been deleted in the treatment of task (i,i). Define $t_s = t_{s-1}$ and choose a t_s-valid $w_s \sqsupsetneq w_{s-1}$ such that for some $n, z \in \mathbb{N}$ the following holds:
   • $F_r^{w_s}(0^n)$ and $F_{r'}^{w_s}(\langle 0^n, z\rangle)$ are defined.
   • $|z| \leq |F_r^{w_s}(0^n)|^i + i$ and $P_i^{w_s}(\langle F_r^{w_s}(0^n), z\rangle)$ definitely accepts.
   • $\langle 0^n, F_{r'}^{w_s}(\langle 0^n, z\rangle)\rangle \notin R_p^v$ for all $v \sqsupseteq w_s$.
   (meaning: R_p is not polynomially many-one reducible to $L(P_i)$ via $(F_r, F_{r'})$)

For proving this construction to be possible we need the following two claims.

Claim 3. *Let $s > 0$ and $w \sqsupseteq w_s$ be t_s-valid. For all x and all positive and distinct i, j with $t_s(i,j) > 0$, if $M_i^w(x)$ and $M_j^w(x)$ are defined, then $M_i^w(x)$ or $M_j^w(x)$ accepts.*

Proof. Assume that for some i, j, x both computations definitely reject. Let $s' \leq s$ be the unique step, where the task (i, j) is treated. By Claim 2, the oracle w is $t_{s'-1}$-valid and as $M_i^w(x)$ and $M_j^w(x)$ definitely reject, it is even t'-valid for $t' = t_{s'-1} \cup \{(i,j) \mapsto 0\}$. But then the construction would have defined $t_{s'}(i,j) = 0$, a contradiction to $t_{s'}(i,i) = t_s(i,i) > 0$.

Claim 4. *Let $s > 0$ and $w \sqsupseteq w_s$ be t_s-valid. Then for $z = |w|$ it holds:*

1. *if $z = y0c(i,j,x)$, $i, j \in \mathbb{N}^+$, $|y| = |c(i,j,x)|$, and $|x| \geq t(i,j) > 0$:*
   (a) *if $M_i^w(x)$ rejects, then $w1$ is t_s-valid.*
   (b) *if $M_i^w(x)$ accepts and $y \neq 1^{|c(i,j,x)|}$, then $w1$ is t_s-valid.*
   (c) *if $M_i^w(x)$ accepts, $y = 1^{|c(i,j,x)|}$, and there exists $y' < y$ with $|y'| = |y|$ and $y'0c(i,j,x) \notin w$, then $w1$ is t_s-valid.*
   (d) *if $M_i^w(x)$ accepts, then $w0$ is t_s-valid.*
2. *if $z = y1c(i,j,x)$, $i, j \in \mathbb{N}^+$, $|y| = |c(i,j,x)|$, and $|x| \geq t(i,j) > 0$:*
   (a) *if $M_j^w(x)$ rejects, then $w1$ is t_s-valid.*
   (b) *if $M_j^w(x)$ accepts and $y \neq 1^{|c(i,j,x)|}$, then $w1$ is t_s-valid.*
   (c) *if $M_j^w(x)$ accepts, $y = 1^{|c(i,j,x)|}$, and there exists $y' < y$ with $|y'| = |y|$ and $y'1c(i,j,x) \notin w$, then $w1$ is t_s-valid.*
   (d) *if $M_j^w(x)$ accepts, then $w0$ is t_s-valid.*
3. *if $|z|$ is odd and not of the form $yb\langle 0^m, 0^t, x\rangle$ for $b \in \{0,1\}$ and $m, t, x, y \in \mathbb{N}$ with $|y| = |\langle 0^m, 0^t, x\rangle|$, then $w0$ and $w1$ are t_s-valid.*

4. if $|z| = 2p^k$ for a prime p with $-p \in \mathrm{ran}(t)$ and $k \in \mathbb{N}^+$:
   (a) $w1$ is t_s-valid.
   (b) if $z \neq 1^{2p^k}$, then $w0$ is t_s-valid.
   (c) if $z = 1^{2p^k}$ and there exists $z' \in w$ of length $2p^k$, then $w0$ is t_s-valid.
5. if $|z| = 2n$ for some $n \in \mathbb{N}$ and $n \neq p^k$ for all primes p with $-p \in \mathrm{ran}(t)$ and all $k \geq 1$, then $w0$ and $w1$ are t_s-valid.
6. in all other cases $w0$ is t_s-valid.

Proof. Observe that V1 and V4 are not affected by extending the oracle.

1. In all subcases, V5 is still satisfied as we extend the oracle only for a word of odd length. Moreover, V2 still holds as wb for $b \in \{0,1\}$ does not contain any word of the form $\tilde{y}1c(i,j,x)$ with $|\tilde{y}| = c(i,j,x)$. In the subcases (a) and (d), the extended oracle still satisfies V3. In case (b), $w1$ satisfies V3 as $w1$ is not defined for $1^{|c(i,j,x)|}0c(i,j,x)$ yet. In case (c), $y'0c(i,j,x) \in w1$ and thus V3 is satisfied by $w1$.
2. In all subcases, V5 is still satisfied as we extend the oracle only for a word of odd length. In the subcases (a) and (d), V3 is satisfied by the extended oracle. In (b), V3 is satisfied as $w1$ is not defined for $1^{|c(i,j,x)|}1c(i,j,x)$ yet. In (c) it holds $y'1c(i,j,x) \notin w1$ and therefore, $w1$ satisfies V3. It remains to argue for V2.
   By Claim 1, the computations $M_i^w(x)$ and $M_j^w(x)$ are defined and by Claim 3, at least one of them accepts. As we have already proven that in the present case all the extended oracles satisfy V3 we obtain that none of these contains all words of the form $\tilde{y}bc(i,j,x)$ for $b \in \{0,1\}$ and $|\tilde{y}| = |c(i,j,x)|$. Thus, the extended oracles satisfy V2.
3. This follows immediately from the definition of t-valid.
4. Here V2 and V3 are not affected as we extend the oracle only for words of even length. Moreover, in case (a) V5 trivially holds. In (b), V5 holds as $w0$ is not defined for 1^{2p^k} yet. In (c), V5 holds since $z' \in w \sqsubseteq w0$.
5. This follows immediately from the definition of t-valid.
6. Here $z = ybc\langle 0^m, 0^t, x \rangle$ for $b \in \{0,1\}$ and $m,t,x,y \in \mathbb{N}$ with $|y| = |\langle 0^m, 0^t, x \rangle|$, but $\langle 0^m, 0^t, x \rangle \neq c(i,j,x)$ for all distinct $i,j \in \mathbb{N}^+$ with $|x| \geq t(i,j) > 0$, which implies that $w0$ satisfies V3. Clearly $w0$ satisfies V2. As $|z|$ is odd, V5 is satisfied by $w0$.

We now argue that the construction described above is possible, i.e., in each step s, the choices of t_s and w_s with the required properties are possible. Assume this is not true and let s be minimal such that the construction fails in step s.

Assume step s treats a task $(i,j) \in \mathbb{N}^2$. Then $t_{s-1}(i,j)$ is undefined as the unique treatment of the task (i,j) takes place in step s. Hence, the definition of t_s is possible. If the construction defines $t_s = t'$, then it is clearly possible. Otherwise, by the choice of $t_s(i,j)$, the t_{s-1}-valid oracle w_{s-1} is even t_s-valid. Then by Claim 4, we can extend w_{s-1} by one bit and obtain a t_s-valid $w_s \sqsupsetneq w_{s-1}$. This contradicts the assumption that the construction fails in step s.

Now assume step s treats a task (i,r,r'). Then $t_s = t_{s-1}$ and $t_s(i,i) = -p$ for some prime p. Choose $n = p^k$ such that w_{s-1} is undefined for all words of length $\geq 2n$ and $2^n > 4 \cdot q(n)$, where q is defined by

$$\alpha \mapsto [\alpha^r + r] + \left[\left(2 \cdot (\alpha + (\alpha^r + r)^i + i + 1)\right)^{r'} + r'\right] + \left[\left(2 \cdot (\alpha^r + r + (\alpha^r + r)^i + i + 1)\right)^i + i\right].$$

Note that by the choice of q, for each oracle D and for all z of length $\leq |F_r^D(0^n)|^i + i$, it holds that $q(n)$ is not less than the sum of the running times of the three computations $F_r^D(0^n)$, $F_{r'}^D(\langle 0^n, z \rangle)$, and $P_i^D(\langle F_r^D(0^n), z \rangle)$. In particular, for each oracle D, $q(n)$ is not less than $\ell(Y)$, where Y is the set of all oracle queries of the three mentioned computations.

By Claim 4, there exists an oracle $u \supsetneq w_{s-1}$ that is t_s-valid, defined for all words of length $< 2n$ and undefined for all other words. By Claim 2, u is valid for the function $t' \in \mathcal{T}$ undefined for (i,i) and equal to t_s on all other input.

Now let $u' \supsetneq u$ be the minimal t'-valid oracle defined for all words of length $q(n)$ (such an oracle exists according to Claim 4). Hence, the computations $F_r^{u'}(0^n)$, $F_{r'}^{u'}(\langle 0^n, z \rangle)$, and $P_i^{u'}(\langle F_r^{u'}(0^n), z \rangle)$ for all z of length $\leq (n^r + r)^i + i$ are defined. Note that u' does not contain any word of length $\geq$ (cf. Claim 4.5).

We show that there exists a word z of length $\leq |F_r^{u'}(0^n)|^i + i$ such that $P_i^{u'}(\langle F_r^{u'}(0^n), z \rangle)$ accepts: for a contradiction, assume that $P_i^{u'}(\langle F_r^{u'}(0^n), z \rangle)$ rejects for all z of length $\leq |F_r^{u'}(0^n)|^i + i$ (by the choice of u' it even definitely rejects). Let s' be the minimal step for which $t_{s'}(i,i)$ is defined. Due to Claim 2 the oracle u' is $t_{s'-1}$-valid and by our assumption, it even is t''-valid for $t'' = t_{s'-1} \cup \{(i,i) \mapsto 0\}$. But then the construction would have defined $t_{s'} = t''$, a contradiction to $t_{s'}(i,i) = t_s(i,i) = -1 \neq 0$.

Hence, we can fix some word μ of length $\leq |F_r^{u'}(0^n)|^i + i$ such that $P_i^{u'}(\langle F_r^{u'}(0^n), \mu \rangle)$ definitely accepts. Let U' be the set of all oracle questions asked by the computations $F_r^{u'}(0^n)$, $F_{r'}^{u'}(\langle 0^n, \mu \rangle)$, and $P_i^{u'}(\langle F_r^{u'}(0^n), \mu \rangle)$. Thus, $\ell(U') \leq q(n)$. Let $U = (U' \cup \{F_{r'}^{u'}(\langle 0^n, \mu \rangle)\}) \cap \Sigma^{\geq 2n}$. Then $\ell(U) \leq 2q(n)$. Define $Q_0(U) = U$ and for $j \in \mathbb{N}$, define $Q_{j+1}(U)$ as the set

$$\bigcup_{\substack{y0c(i',j',x') \in Q_j(U), \\ i' \neq j', t_s(i',j') > 0, \\ |y| = |c(i',j',x')|}} \{q \mid |q| \geq 2n, \text{ the least accepting path of } M_{i'}^{u'}(x') \text{ queries } q\} \cup$$

$$\bigcup_{\substack{y1c(i',j',x') \in Q_j(U), \\ i' \neq j', t_s(i',j') > 0, \\ |y| = |c(i',j',x')|}} \{q \mid |q| \geq 2n, \text{ the least accepting path of } M_{j'}^{u'}(x') \text{ queries } q\}.$$

Note that some computation $M_{i'}^{u'}(x')$ might reject or it might not ask any question on its least accepting path. Therefore, $Q_{j+1}(U)$ is not necessarily non-empty.

Furthermore, define $Q(U) = \bigcup_{j=0}^{\infty} Q_j(U)$.

Claim 5 $\ell(Q(U)) \leq 2\ell(U)$.

Proof. We show that for all $j \in \mathbb{N}$ it holds $\ell(Q_{j+1}(U)) \leq 1/2 \cdot \ell(Q_j(U))$. Then $\sum_{j=0}^{m} 1/2^j \leq 2$ for all $m \in \mathbb{N}$ implies $\ell(Q(U)) \leq 2 \cdot \ell(U)$.

Consider an element α of $Q_j(U)$. Assume $\alpha = ybc(i', j', x')$ for $b \in \{0, 1\}$, distinct $i', j' > 0$, and $x', y \in \mathbb{N}$ with $|y| = |c(i', j', x')|$ (otherwise, α generates no elements in $Q_{j+1}(U)$). By symmetry, it suffices to consider the case $b = 0$. If $M_{i'}^{u'}(x')$ rejects, then α generates no elements in $Q_{j+1}(U)$. Assume that $M_{i'}^{u'}(x')$ accepts. Then $Q_{j+1}(U)$ consists of all queries of length $\geq 2n$ that are asked on the least accepting path of $M_{i'}^{u'}(x')$. Recall $c(i', j', x') = \langle 0^{t_s(i',j')}, 0^{|x'|^{i'+j'}+i'+j'}, x' \rangle$. $M_{i'}^{u'}(x')$ runs for at most $|x'|^{i'} + i'$ steps and it holds $|x'|^{i'} + i' \leq |x'|^{i'+j'} + i' + j' \leq |c(i', j', x')| \leq 1/2 \cdot |\alpha|$. Hence, the sum of the lengths of all queries on the least accepting path of $M_{i'}^{u'}(x')$ is $\leq |\alpha|/2$. Thus,

$$\ell(Q_{j+1}(U)) \leq \sum_{\substack{y0c(i',j',x') \in Q_j(U), \\ i' \neq j', t_s(i',j') > 0, \\ |y| = |c(i',j',x')|}} |y0c(i',j',x')|/2 + \sum_{\substack{y1c(i',j',x') \in Q_j(U), \\ i' \neq j', t_s(i',j') > 0, \\ |y| = |c(i',j',x')|}} |y1c(i',j',x')|/2 \leq 1/2 \cdot \ell(Q_j(U)).$$

Claim 6. *There exists a t_s-valid $v \sqsupsetneq u$ defined for all words of length $q(n)$ that satisfies $v(q) = u'(q)$ for all $q \in Q(U)$.*

Proof. Due to Claim 5 and $\ell(U) \leq 2q(n)$, it holds $\ell(Q(U)) \leq 4q(n)$. As by the choice of n it holds $2^n > 4q(n) \geq \ell(Q(U)) \geq |Q(U)|$, there exists a word $z' \in \Sigma^{2n}$ that is not in $Q(U)$. Let v' be the minimal oracle defined for all words of length $2n$ and containing z', i.e., interpreting u and v' as sets we have $v' = u \cup \{z'\}$. As u is t_s-valid, Claim 4.4 yields that v' is t_s-valid. Furthermore, $v'(q) = u'(q)$ for all $q \in Q(U)$ with $q < |v'|$, since u' contains no words of length $2n$, $v' \cap \Sigma^{2n} = \{z'\}$, and $z' \notin Q(U)$.

For technical reasons we introduce the following notion. We say that an oracle w respects blocks if the following holds for the greatest word z that w is defined for (i.e., $z = |w| - 1$):

- if $|z|$ is even, then $z = 1^{|z|}$.
- if $z = yb\langle 0^m, 0^t, x \rangle$ for $b \in \{0, 1\}$ and $m, t, x, y \in \mathbb{N}$ with $|y| = |\langle 0^m, 0^t, x \rangle|$, then $y = 1^{|y|}$ and $b = 1$.

That means, if w respects blocks, then for each block, w is either defined for all words of the block or for no word of the block, where a block is either a set of the form Σ^{2r} for $r \in \mathbb{N}$ or a set of the form $\{yb\langle 0^m, 0^t, x \rangle \mid |y| = |\langle 0^m, 0^t, x \rangle|, b \in \{0, 1\}\}$ for fixed $m, t, x \in \mathbb{N}$. We now start with v' and successively extend the current oracle blockwise (resp., bitwise if the next word is not contained in a block). At this point we refer to the definition of the bijection over which we identify words and number. This bijection defines an order on the set of words such that e.g. for fixed words m, t, x, sets of the form $\{yb\langle 0^m, 0^t, x \rangle \mid |y| = |\langle 0^m, 0^t, x \rangle|, b \in \{0, 1\}\}$ are intervals. Note that u' respects blocks. It suffices to prove the following assertion.

For each t_s-valid $w \sqsupseteq v'$ with $|w| < |u'|$ that respects blocks and satisfies $w(q) = u'(q)$ for all $q \in Q(U)$ with $q < |w|$, there exists a t_s-valid $w' \sqsupsetneq w$ with $|w'| \leq |u'|$ that respects blocks and satisfies $w'(q) = u'(q)$ for all $q \in Q(U)$ with $q < |w'|$. (1)

In each step, we will extend the current oracle by no more than one block. Hence, we will finally receive an oracle of length $|u'|$, i.e., the final oracle is defined for all words of length $q(n)$. Let w be an oracle according to (1) and z be the least word that w is not defined for (i.e., $z = |w|$). Then the following cases are possible.

Case 1: $|z|$ is odd, but not of the form $yb\langle 0^m, 0^t, x\rangle$ for $b \in \{0,1\}$ and $m, t, x, y \in \mathbb{N}$ with $|y| = |\langle 0^m, 0^t, x\rangle|$. In this case, by Claim 4.3, both $w0$ and $w1$ are t_s-valid. Consequently, if $z \in u'$, then we choose $w' = w1$, otherwise, we choose $w' = w0$. Then w' respects blocks (note that z is not contained in a block) and satisfies $w'(q) = u'(q)$ for all $q \in Q(U)$ with $q < |w'|$.

Case 2: $z = 0^{|\langle 0^m, 0^t, x\rangle|} 0 \langle 0^m, 0^t, x\rangle$ for $m, t, x \in \mathbb{N}$, but it does not hold: $\langle 0^m, 0^t, x\rangle = c(i', j', x)$ for distinct and positive i' and j' with $|x| \geq t_s(i', j') > 0$. Let $Y_0 = \{y0\langle 0^m, 0^t, x\rangle \mid |y| = |\langle 0^m, 0^t, x\rangle|\}$, $Y_1 = \{y1\langle 0^m, 0^t, x\rangle \mid |y| = |\langle 0^m, 0^t, x\rangle|\}$, and $Y = Y_0 \cup Y_1$. Recall that u' is t'-valid. Hence, by V2, $(u' \cap Y_0) \neq Y_0$ or $(u' \cap Y_1) \neq Y_1$. By symmetry, it suffices to consider the case $(u' \cap Y_0) \neq Y_0$. Let $w' \sqsupseteq w$ be the minimal oracle that is defined for all words in Y and contains all words in $u' \cap Q(U) \cap Y$, i.e., when interpreting the oracle as sets it holds $w' = w \cup (u' \cap Q(U) \cap Y)$. Clearly w' respects blocks and $w'(q) = u'(q)$ for all $q \in Q(U)$ with $q < |w'|$. We show that w' is t_s-valid. As $(u' \cap Y_0) \neq Y_0$ and $w' \cap Y \subseteq u'$, it also holds $(w' \cap Y_0) \neq Y_0$. Hence, $\langle 0^m, 0^t, x\rangle \notin A^{w'}$ and thus w' satisfies V2. V1 and V4 are not affected by extending w to w'. Furthermore, w' satisfies V5, since we only extended the oracle for words of odd length. Finally, V3 also holds, since by assumption, it does not hold that $\langle 0^m, 0^t, x\rangle = c(i', j', x)$ for distinct and positive i' and j' with $|x| \geq t_s(i', j') > 0$.

Case 3: $z = 0^{|c(i', j', x)|} 0 c(i', j', x)$ for positive and distinct i' and j' and $x \in \mathbb{N}$ with $|x| \geq t_s(i', j') > 0$. We define $Y_0 = \{y0c(i', j', x) \mid |y| = |c(i', j', x)|\}$, $Y_1 = \{y1c(i', j', x) \mid |y| = |c(i', j', x)|\}$, and $Y = Y_0 \cup Y_1$. The computations $M_{i'}^w(x)$ and $M_{j'}^w(x)$ are defined by Claim 1. By Claim 3, one of the computations accepts. By symmetry, it is sufficient to consider the case that $M_{i'}^w(x)$ definitely accepts. We now consider two subcases.

Case 3a: $M_{j'}^w(x)$ definitely rejects. Choose $w' \sqsupseteq w$ to be the minimal oracle that is defined for all words in Y and contains all words in $u' \cap Q(U) \cap Y$ and all words in Y_1, i.e., interpreting oracles as sets it holds $w' = w \cup Y_1 \cup (u' \cap Q(U) \cap Y)$. As $|c(i', j', x)| \geq n$ and thus $|Y_0| \geq 2^n > 4q(n) \geq |Q(U)|$, there exists a word in $Y_0 - w$. Hence, by the statements 1 and 2 of Claim 4, w' is t_s-valid. Clearly w' respects blocks. It remains to show that $w'(q) = u'(q)$ for all $q \in Q(U)$ with $q < |w'|$. For all $q < |w|$ this holds by assumption. By the choice of w', the assertion also holds for all $q \in Y_0$. If $q \in Q(U) \cap Y_1$, then $q \in w'$, and we have to show that $q \in u'$. For a contradiction, assume $q \notin u'$. Then, as u' is t'-valid and $|x| \geq t'(i', j') = t_s(i', j') > 0$, $M_{j'}^{u'}(x)$ accepts (cf. V3). By the choice of $Q(U)$, all queries q' answered on the least accepting path of $M_{j'}^{u'}(x)$ are in $Q(U)$. Moreover, it holds $|q'| < |c(i', j', x)|$ for all these queries as $M_{j'}^{u'}(x)$ runs for $\leq |x|^{j'} + j'$ steps. But then for all these queries q', w is defined for q' and by assumption, $w(q') = u'(q')$. It follows that $M_{j'}^w(x)$ accepts on the same path that $M_{j'}^{u'}(x)$ accepts on, a contradiction. Hence, $q \in u'$ and therefore, w' and u' agree on all queries in $Q(U)$ that are less than $|w'|$.

Case 3b: $M_{j'}^w(x)$ definitely accepts. Choose $w' \supsetneq w$ to be the minimal oracle that is defined for all words in Y and contains all words in $u' \cap Q(U) \cap Y$, i.e., interpreting oracles as sets it holds $w' = w \cup (u' \cap Q(U) \cap Y)$. Clearly, $w'(q) = u'(q)$ for all words $q < |w'|$ and w' respects blocks.

Moreover, $|c(i', j', x)| \geq n$, $|Y_0| = |Y_1| \geq 2^n > 4q(n) \geq |Q(U)|$, and thus there exists a word in $Y_0 - w'$ and a word in $Y_1 - w'$. Hence, by the statements 1 and 2 of Claim 4, the oracle w' is t_s-valid.

Case 4: $|z|$ is even, i.e., $|z| = 2r$ for some $r \geq n$. If $r \neq p^k$ for all primes p with $-p \in \mathrm{ran}(t_s)$ and all $k \geq 1$, then choose $w' \supsetneq w$ to be the minimal oracle that is defined for all words of length $\leq 2r$ and contains all words in $u' \cap Q(U) \cap \Sigma^{2r}$, i.e., $w' = w \cup (u' \cap Q(U) \cap \Sigma^{2r})$ when the oracles are interpreted as sets. Then w' respects blocks, $w'(q) = u'(q)$ for all $q \in Q(U)$ with $q \leq |w'|$, and by Claim 4.5, the oracle w' is t_s-valid.

Now consider the case that $r = p^k$ for some prime p with $-p \in \mathrm{ran}(t_s)$ and some $k \geq 1$. As $r \geq n$ and $2^n > 4q(n) \geq |Q(U)|$, there exists a word $y \in \Sigma^{2r} - Q(U)$. Choose $w' \supsetneq w$ to be the minimal oracle that is defined for all words in Σ^{2r} and contains y and all words in $u' \cap Q(U) \cap \Sigma^{2r}$, i.e., interpreting oracles as sets it holds $w' = w \cup \{y\} \cup (u' \cap Q(U) \cap \Sigma^{2r})$. Then w' respects blocks, $w'(q) = u'(q)$ for all $q \in Q(U)$ with $q < |w'|$ (since $y \notin Q(U)$), and by Claim 4.4, the oracle w' is t_s-valid.

This shows (1) and consequently finishes the proof of Claim 6.

As all queries of length $\geq 2n$ of the computations $F_r^{u'}(0^n)$, $F_{r'}^{u'}(\langle 0^n, \mu \rangle)$, and $P_i^{u'}(\langle F_r^{u'}(0^n), \mu \rangle)$ are in $Q(U)$, u and v agree on these queries by Claim 6, and $u' \cap \Sigma^{<2n} = v \cap \Sigma^{<2n} = u$, it holds $F_r^{u'}(0^n) = F_r^v(0^n)$, $F_{r'}^{u'}(\langle 0^n, \mu \rangle) = F_{r'}^v(\langle 0^n, \mu \rangle)$, and $P_i^{u'}(\langle F_r^{u'}(0^n), \mu \rangle) = P_i^v(\langle F_r^v(0^n), \mu \rangle)$. Because v is defined for all words of length $\leq q(n)$, all these computations are defined. In particular,

$$F_r^v(0^n) \text{ and } F_{r'}^v(\langle 0^n, \mu \rangle) \text{ are defined,} \tag{2}$$

$$|\mu| \leq |F_r^v(0^n)| + i, \text{ and } P_i^v(\langle F_r^v(0^n), \mu \rangle) \text{ definitely accepts.} \tag{3}$$

We show

$$\langle 0^n, F_{r'}^v(\langle 0^n, \mu \rangle) \rangle \notin R_p^{v'} \text{ for all } v' \supseteq v: \tag{4}$$

Fix $v' \supseteq v$ and write $z = F_{r'}^v(\langle 0^n, \mu \rangle) = F_{r'}^{u'}(\langle 0^n, \mu \rangle)$. If $|z| = 2n$, then $z \in Q(U)$ and as v is defined for z, it holds $v'(z) = v(z) = u'(z) = 0$ (cf. Claim 6), since u does not contain any word of length $2n$. Hence, $\langle 0^n, z \rangle \notin R_p^{v'}$. Otherwise, $\langle 0^n, z \rangle \notin R_p^{v'}$ as $R_p^{v'}$ only contains words of the form $\langle 0^\kappa, y \rangle$ for y of length 2κ, $\kappa \in \mathbb{N}$. This shows (4).

The fact that v is t_s-valid by Claim 6 and the assertions (2), (3), and (4) show that the task (i, r, r') can be treated as described in the construction, a contradiction to the assumption that the construction fails in step s. This shows that the oracle construction described above is possible.

The $\leq_m^{\mathrm{pp}, O}$-completeness of (A^O, B^O) for DisjCoNP^O can be shown by a straightforward argument. Hence, the subsequent claim finishes the proof of Theorem 1.

Claim 7. TFNP^O *has no complete problem.*

Proof. By Corollary 1, it suffices to show that no total polynomial search problem of the form $(n \mapsto n^i + i, L(P_i^O))$ is many-one complete. For a contradiction, assume there exists i such that $(n \mapsto n^i + i, L(P_i^O))$ is many-one complete for TFNP^O. Hence, $t(i, i) = -p$ for some prime p and by V5, $(n \mapsto 2n, R_p^O)$ is a total polynomial search problem that is polynomially many-one reducible to $L(P_i^O)$ via F_r^O and $F_{r'}^O$ for some r and r', i.e., for all x and z it holds

$$P_i^O(\langle F_r^O(x), z\rangle) \text{ accepts } \Rightarrow \langle x, F_{r'}^O(\langle x, z\rangle)\rangle \in R_p^O. \tag{5}$$

Let s be the step that treats the task (i, r, r'). Then by construction there exist $n, z \in \mathbb{N}$ such that the following holds:

- $F_r^{w_s}(0^n)$ and $F_{r'}^{w_s}(\langle 0^n, z\rangle)$ are defined.
- $P_i^{w_s}(\langle F_r^{w_s}(0^n), z\rangle)$ definitely accepts.
- $\langle 0^n, F_{r'}^{w_s}(\langle 0^n, z\rangle)\rangle \notin R_p^v$ for all $v \sqsupseteq w_s$.

As O is an extension of w_s, this is a contradiction to (5), which finishes the proof of Claim 7 and hence the proof of Theorem 1. ∎

4   Conclusion

The following figure illustrates the current state of the art regarding the conjectures presented by Pudlák [12].

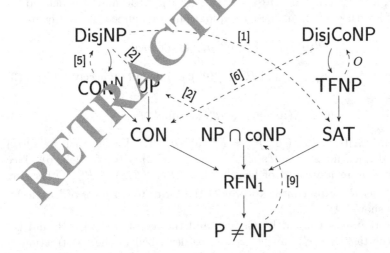

Fig. 1. Consider the conjectures as the relativized versions of the original statements. For the definition of RFN$_1$ we refer to [12]. Note, however, that RFN$_1$ is equivalent to CON $\vee$ SAT [6]. Solid arrows mean implications. A dashed arrow from some conjecture A to another conjecture B means that there is an oracle X against the implication A $\Rightarrow$ B, i.e., relative to X, it holds A $\wedge \neg$B.

References

1. Dose, T.: P-optimal proof systems for each set in NP but no complete disjoint NP-pairs relative to an oracle. arXiv e-prints. arXiv:1904.06175, April 2019
2. Dose, T., Glaßer, C.: NP-completeness, proof systems, and disjoint NP-pairs. Technical Report 19–050, Electronic Colloquium on Computational Complexity (ECCC) (2019)
3. Even, S., Selman, A.L., Yacobi, J.: The complexity of promise problems with applications to public-key cryptography. Inf. Control **61**, 159–173 (1984)
4. Even, S., Yacobi, Y.: Cryptocomplexity and NP-completeness. In: de Bakker, J., van Leeuwen, J. (eds.) ICALP 1980. LNCS, vol. 85, pp. 195–207. Springer, Heidelberg (1980). https://doi.org/10.1007/3-540-10003-2_71
5. Glaßer, C., Selman, A.L., Sengupta, S., Zhang, L.: Disjoint NP-pairs. SIAM J. Comput. **33**(6), 1369–1416 (2004)
6. Khaniki, E.: New relations and separations of conjectures about incompleteness in the finite domain. arXiv e-prints. arXiv:1904.01362, April 2019
7. Krajíček, J., Pudlák, P.: Propositional proof systems, the consistency of first order theories and the complexity of computations. J. Symbolic Logic **54**, 1063–1079 (1989)
8. Megiddo, N., Papadimitriou, C.H.: On total functions, existence theorems and computational complexity. Theor. Comput. Sci. **81**(2), 317–324 (1991)
9. Ogiwara, M., Hemachandra, L.: A complexity theory of feasible closure properties. J. Comput. Syst. Sci. **46**, 295–325 (1993)
10. Papadimitriou, C.M.: Computational complexity. Addison-Wesley, Reading (1994)
11. Pudlák, P.: On the lengths of proofs of consistency. In: Pudlák, P. (ed.) Collegium Logicum Collegium Logicum (Annals of the Kurt-Gödel-Society), vol. 2, pp. 65–86. Springer, Vienna (1996). https://doi.org/10.1007/978-3-7091-9461-4_5
12. Pudlák, P.: Incompleteness in the finite domain. Bull. Symbolic Logic **23**(4), 405–441 (2017)
13. Razborov, A.A.: On provably disjoint NP-pairs. Electron. Colloq. Comput. Complex. (ECCC), **1**(6), 2 p. (1994)

Algorithms

An Efficient Algorithm for the Fast Delivery Problem

Iago A. Carvalho[1] ⓘ, Thomas Erlebach[2(✉)] ⓘ, and Kleitos Papadopoulos[2] ⓘ

[1] Department of Computer Science, Universidade Federal de Minas Gerais,
Belo Horizonte, Brazil
iagoac@dcc.ufmg.br
[2] Department of Informatics, University of Leicester, Leicester, UK
te17@leicester.ac.uk, kleitospa@gmail.com
https://iagoac.github.io

Abstract. We study a problem where k autonomous mobile agents are initially located on distinct nodes of a weighted graph (with n nodes and m edges). Each autonomous mobile agent has a predefined velocity and is only allowed to move along the edges of the graph. We are interested in delivering a package, initially positioned in a source node s, to a destination node y. The delivery is achieved by the collective effort of the autonomous mobile agents, which can carry and exchange the package among them. The objective is to compute a delivery schedule that minimizes the delivery time of the package. In this paper, we propose an $\mathcal{O}(kn \log(kn) + km)$ time algorithm for this problem. This improves the previous state-of-the-art $\mathcal{O}(k^2m + kn^2 + \text{APSP})$ time algorithm for this problem, where APSP stands for the running-time of an algorithm for the All-Pairs Shortest Paths problem.

Keywords: Mobile agents · Dijkstra's algorithm ·
Polynomial-time algorithm · Time-dependent shortest paths

1 Introduction

Enterprises, such as DHL, UPS, Swiss Post, and Amazon, are now delivering goods and packages to their clients using *autonomous drones* [1,17]. Those drones depart from a base (which can be static, such as a warehouse [15], or mobile, such as a truck or a van [16]) and deliver the package into their clients' houses or in the street. However, packages are not delivered to a client that is too far from the drone's base due to the energy limitations of such autonomous aerial vehicles.

In the literature, we find some proposals for delivering packages over a longer distance. One of them, proposed by Hong, Kuby, and Murray [15], is to install recharging bases in several spots, which allows a drone to stop, recharge, and continue its path. However, this strategy may result in a delayed delivery, because drones may stop several times to recharge during a single delivery.

© Springer Nature Switzerland AG 2019
L. A. Gąsieniec et al. (Eds.): FCT 2019, LNCS 11651, pp. 171–184, 2019.
https://doi.org/10.1007/978-3-030-25027-0_12

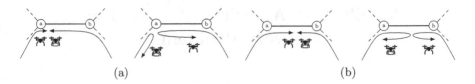

Fig. 1. (a) Package exchange on a node; (b) package exchange on an edge.

A manner to overcome this limitation is to use a *swarm* of drones. The idea of this technique is to position drones in recharging bases all over the delivery area. Therefore, a package can be delivered from one place to another through the collective effort of such drones, which can exchange packets among them to achieve a faster delivery. One may note that, when not carrying a package, a drone is stationed in its recharging base, waiting for the next package arrival. The problem of computing a package delivery schedule with minimum delivery time for a single package is called the FASTDELIVERY problem [4].

We can model the input to the FASTDELIVERY problem as a graph $G = (V, E)$ with $|V| = n$ and $|E| = m$, with a positive length l_e associated with each edge $e \in E$, and a set of k *autonomous mobile agents* (e.g., autonomous drones) located initially on distinct nodes $p_1, p_2, \ldots, p_k$ of G. Each agent i has a predefined velocity $v_i > 0$. Mobile agent i can traverse an edge e of the graph in l_e/v_i time. The package handover between agents can be done on the nodes of the graph or in any point of the graph's edges, as exemplified in Fig. 1. The objective of FASTDELIVERY is to deliver a single package, initially located in a source node $s \in V$, to a target node $y \in V$ while minimizing the delivery time $\mathcal{T}$.

Bärtschi et al. [4] also consider the case where each agent i is additionally associated with a weight $\omega_i > 0$ and consumes energy $\omega_i \cdot l_e$ when traversing edge e. For this model, the total energy consumption $\mathcal{E}$ of a solution becomes relevant as well, and one can consider the objective of minimizing $\mathcal{E}$ among all solutions that have the minimum delivery time $\mathcal{T}$ (or vice versa), or of minimizing a convex combination $\varepsilon \cdot \mathcal{T} + (1 - \varepsilon) \cdot \mathcal{E}$ for a given $\varepsilon \in (0, 1)$.

1.1   Related Work

The problem of delivering packages through a swarm of autonomous drones has been studied in the literature. The work of Bärtschi et al. [3] considers the problem of delivering packages while minimizing the total energy consumption of the drones. In their work, all drones have the same velocity but may have different weights, and the package's exchanges between drones are restricted to take place on the graph's nodes. They show that this problem is NP-hard when an arbitrary number of packages need to be delivered, but can be solved in polynomial time for a single package, with complexity $\mathcal{O}(k + n^3)$.

When minimizing only the delivery time $\mathcal{T}$, one can solve the problem of delivering a single package with autonomous mobile agents with different velocities in polynomial-time: Bärtschi et al. [4] gave an $\mathcal{O}(k^2m + kn^2 + \text{APSP})$ algorithm for this problem, where APSP stands for the time complexity of the All-Pairs Shortest Paths problem.

Some work in the literature considered the minimization of both the total delivery time and the energy consumption. It was shown that the problem of delivering a single package with autonomous agents of different velocities and weights is solvable in polynomial-time when lexicographically minimizing the tuple $(\mathcal{E}, \mathcal{T})$ [5]. On the other hand, it is NP-hard to lexicographically minimize the tuple $(\mathcal{T}, \mathcal{E})$ or a convex combination of both parameters [4].

A closely related problem is the Budgeted Delivery Problem (BDP) [2,8,9], in which a package needs to be delivered by a set of energy-constrained autonomous mobile agents. In BDP, the objective is to compute a route to deliver a single package while respecting the energy constraints of the autonomous mobile agents. This problem is weakly NP-hard in line graphs [9] and strongly NP-hard in general graphs [8]. A variant of this problem is the Returning Budgeted Delivery Problem (RBDP) [2], which has the additional constraint that the energy-constrained autonomous agents must return to their original bases after carrying the package. Surprisingly, this new restriction makes RBDP solvable in polynomial time in trees. However, it is still strongly NP-hard even for planar graphs.

A variant of the classical search problem (also known as the cow-path problem) where an agent aims to reach the location of a target as quickly as possible and the search space contains additional *expulsion points* has recently been studied by Gasieniec et al. [13]. Visiting an expulsion point updates the speed of the agent to the maximum of its current speed and the expulsion speed associated with that expulsion point. They present online and offline algorithms for one- and two-dimensional search.

1.2 Our Contribution

This paper deals with the FASTDELIVERY problem. We focus on the first objective, *i.e.*, computing delivery schedules with the minimum delivery time. We provide an $\mathcal{O}(kn \log(kn) + km)$ time algorithm for FASTDELIVERY, which is more efficient than the previous $\mathcal{O}(k^2 m + kn^2 + \text{APSP})$ time algorithm for this problem [4].

Preliminaries are presented in Sect. 2. We then describe our algorithm to solve FASTDELIVERY in Sect. 3. The algorithm uses as a subroutine, called once for each edge of G, an algorithm for a problem that we refer to as FASTLINEDELIVERY, which is presented in Sect. 4.

2 Preliminaries

As mentioned earlier, in the FASTDELIVERY problem we are given an undirected graph $G = (V, E)$ with $n = |V|$ nodes and $m = |E|$ edges. Each edge $e \in E$ has a positive length l_e. We assume that a path can start on a node or in some point in the interior of an edge. Analogously, it can end on another node or in some point in the interior of an edge. The length of a path is equal to the sum of the lengths of its edges. If a path starts or ends at a point in the interior of an edge, only the portion of its length that is traversed by the path is counted. For example, a path that is entirely contained in an edge $e = \{u, v\}$ of length $l_e = 10$ and starts at distance 2 from u and ends at distance 5 from u has length 3.

We are also given k mobile agents, which are initially located at nodes $p_1, p_2, \ldots, p_k \in V$. Each agent i has a positive velocity (or speed) v_i, $1 \leq i \leq k$. A single package is located initially (at time 0) on a given source node $s \in V$ and needs to be delivered to a given target node $y \in V$. An agent can pick up the package in one location and drop it off (or hand it to another agent) in another one. An agent with velocity v_i takes time d/v_i to carry a package over a path of length d. The objective of FASTDELIVERY is to determine a schedule for the agents to deliver the package to node y as quickly as possible, *i.e.*, to minimize the time when the package reaches y.

We assume that there is at most one agent on each node. This assumption can be justified by the fact that, if there were several agents on the same node, we would use only the fastest one among them. Therefore, as already observed in [4], after a preprocessing step running in time $\mathcal{O}(k + |V|)$, we may assume that $k \leq n$.

The following lemma from [4] establishes some useful properties of an optimal delivery schedule for the mobile agents.

Lemma 1 (Bärtschi et al., 2018). *For every instance of* FASTDELIVERY, *there is an optimum solution in which (i) the velocities of the involved agents are strictly increasing, and (ii) no involved agent arrives on its pick-up location earlier than the package (carried by the preceding agent).*

3   Algorithm for the Fast Delivery Problem

Bärtschi et al. [4] present a dynamic programming algorithm that computes an optimum solution for FASTDELIVERY in time $\mathcal{O}(k^2 m + kn^2 + \text{APSP}) \subseteq \mathcal{O}(k^2 n^2 + n^3)$, where APSP denotes the time complexity of an algorithm for solving the all-pairs shortest path problem. In this paper we design an improved algorithm with running time $\mathcal{O}(km + nk \log(nk)) \subseteq \mathcal{O}(n^3)$ by showing that the problem can be solved by adapting the approach of Dijkstra's algorithm for edges with time-dependent transit times [10,11].

For any edge $\{u, v\}$, we denote by $a_t(u, v)$ the earliest time for the package to arrive at v if the package is at node u at time t and needs to be carried over the edge $\{u, v\}$. We refer to the problem of computing $a_t(u, v)$, for a given value of t that represents the earliest time when the package can reach u, as FASTLINEDELIVERY. In Sect. 4, we will show that FASTLINEDELIVERY can be solved in $\mathcal{O}(k)$ time after a preprocessing step that spends $\mathcal{O}(k \log k)$ time per node. Our preprocessing calls PREPROCESSRECEIVER(v) once for each node $v \in V \setminus \{s\}$ at the start of the algorithm. Then, it calls PREPROCESSSENDER(u, t) once for each node $u \in V$, where t is the earliest time when the package can reach u. Both preprocessing steps run in $\mathcal{O}(k \log k)$ time per node. Once both preprocessing steps have been carried out, a call to FASTLINEDELIVERY(u, v, t) computes $a_t(u, v)$ in $\mathcal{O}(k)$ time.

Algorithm 1. Algorithm for FASTDELIVERY

Data: graph $G = (V, E)$ with positive edge lengths l_e and source node $s \in V$,
target node $y \in V$; k agents with velocity v_i and initial location p_i for
$1 \le i \le k$

Result: earliest arrival time $dist(y)$ for package at destination

```
 1  begin
 2  |   compute d(p_i, v) for 1 ≤ i ≤ k and all v ∈ V;
 3  |   construct list A(v) of agents in order of increasing arrival times and
    |     velocities for each v ∈ V;
 4  |   PREPROCESSRECEIVER(v) for all v ∈ V \ {s};
 5  |   dist(s) ← t_s;                    /* time when first agent reaches s */
 6  |   dist(v) ← ∞ for all v ∈ V \ {s};
 7  |   final(v) ← false for all v ∈ V;
 8  |   insert s into priority queue Q with priority dist(s);
 9  |   while Q not empty do
10  |   |   u ← node with minimum dist value in Q;
11  |   |   delete u from Q;
12  |   |   final(u) ← true;
13  |   |   if u = y then
14  |   |   |   break;
15  |   |   end
16  |   |   t ← dist(u);                  /* time when package reaches u */
17  |   |   PREPROCESSSENDER(u, t);
18  |   |   forall the neighbors v of u with final(v) = false do
19  |   |   |   a_t(u, v) ← FASTLINEDELIVERY(u, v, t);
20  |   |   |   if a_t(u, v) < dist(v) then
21  |   |   |   |   dist(v) ← a_t(u, v);
22  |   |   |   |   if v ∈ Q then
23  |   |   |   |   |   decrease priority of v to dist(v);
24  |   |   |   |   else
25  |   |   |   |   |   insert v into Q with priority dist(v);
26  |   |   |   |   end
27  |   |   |   end
28  |   |   end
29  |   end
30  |   return dist(y);
31  end
```

Algorithm 1 shows the pseudo-code for our solution for FASTDELIVERY. Initially, we run Dijkstra's algorithm to solve the single-source shortest paths problem for each node where an agent is located initially (line 2). This takes time $\mathcal{O}(k(n \log n + m))$ if we use the implementation of Dijkstra's algorithm with

Fibonacci heaps as priority queue [12] and yields the distance $d(p_i, v)$ (with respect to edge lengths l_e) between any node p_i where an agent is located and any node $v \in V$. From this we compute, for every node v, the earliest time when each mobile agent can arrive at that node: The earliest possible arrival time of agent i at node v is $a_i(v) = d(p_i, v)/v_i$. Then, we create a list of the arrival times of the k agents on each node (line 3). For each node, we sort the list of the k agents by ascending arrival time in $\mathcal{O}(k \log k)$ time, or $\mathcal{O}(nk \log k)$ in total for all nodes. We then discard from the list of each node all agents that arrive at the same time or after an agent that is strictly faster. If several agents with the same velocity arrive at the same time, we keep one of them arbitrarily. Let $A(v)$ denote the resulting list for node v. Those lists will be used in the solution of the FASTLINEDELIVERY problem described in Sect. 4.

For each node v, we maintain a value $dist(v)$ that represents the current upper bound on the earliest time when the package can reach v (lines 5 and 6). The algorithm maintains a priority queue containing nodes that have a finite $dist$ value, with the $dist$ value as the priority (line 8). In each step, a node u with minimum $dist$ value is removed from the priority queue (lines 10 and 11), and the node becomes *final* (line 12). Nodes that are not final are called *non-final*. The $dist$ value of a final node will not change any more and represents the earliest time when the package can reach the node (line 16). After u has been removed from the priority queue, we compute for each non-final neighbor v of u the time $a_t(u, v)$, where $t = dist(u)$, by solving the FASTLINEDELIVERY problem (line 19). If v is already in Q, we compare $a_t(u, v)$ with $dist(v)$ and, if $a_t(u, v) < dist(v)$, update $dist(v)$ to $dist(v) = a_t(u, v)$ and adjust the priority of v in Q accordingly (line 23). On the other hand, if v is not yet in Q, we set $dist(v) = a_t(u, v)$ and insert v into Q (line 25).

Let t_s be the earliest time when an agent reaches s (or 0, if an agent is located at s initially). Let i' be that agent. As the package must stay at s from time 0 to time t_s, we can assume that i' brings the package to s at time t_s. Therefore, we initially set $dist(s) = t_s$ and insert s into the priority queue Q with priority t_s. The algorithm terminates when y becomes final (line 14) and returns the value $dist(y)$, *i.e.*, the earliest time when the package can reach y. The schedule that delivers the package to y by time $dist(y)$ can be constructed in the standard way, by storing for each node v the predecessor node u such that $dist(v) = a_{dist(u)}(u, v)$ and the schedule of the solution to FASTLINEDE-LIVERY$(u, v, dist(u))$.

Theorem 1. *Algorithm 1 computes an optimal solution to the* FASTDELIVERY *problem in* $\mathcal{O}(nk \log(nk) + mk)$ *time.*

Proof. It is easy to see that $a_t(u, v) \le a_{t'}(u, v)$ holds for $t' \ge t$ in our setting, because the agents that transport the package from u to v starting at time t would also be available to transport the package from u to v at time $t' \ge t$ following the same schedule, shifted by $t' - t$ time units. Thus, the network has the FIFO property (or non-overtaking property), and it is known that the modified Dijkstra algorithm is correct for such networks [11].

Furthermore, we can observe that concatenating the solutions of FAST-LINEDELIVERY (which are computed by Algorithm 4 in Sect. 4 and which are correct by Theorem 2 in Sect. 4) over the edges of the shortest path from s to y determined by Algorithm 1 indeed gives a feasible solution to FASTDELIVERY: Assume that the package reaches u at time t while being carried by agent i and is then transported from u to v over edge $\{u, v\}$, reaching v at time $a_t(u, v)$. The only agents involved in transporting the package from u to v in the solution returned by FASTLINEDELIVERY(u, v, t) will have velocity at least v_i because agent i arrives at u before time t, i.e., $a_i(u) \leq t$, and hence no slower agent would be used to transport the package from u to v. These agents have not been involved in transporting the package from s to u by property (i) of Lemma 1, except for agent i who is indeed available at node u from time t.

The running time of the algorithm consists of the following components: Computing standard shortest paths with respect to the edge lengths l_e from the locations of the agents to all other nodes takes $\mathcal{O}(k(n \log n + m))$ time. The time complexity of the Dijkstra algorithm with time-dependent transit times for a graph with n nodes and m edges is $\mathcal{O}(n \log n + m)$. The only extra work performed by our algorithm consists of $\mathcal{O}(k \log k)$ pre-processing time for each node and $\mathcal{O}(k)$ time per edge for solving the FASTLINEDELIVERY problem, a total of $\mathcal{O}(nk \log k + mk)$ time. $\qquad \square$

4   An Algorithm for Fast Line Delivery

In this section we present the solution to FASTLINEDELIVERY that was used as a subroutine in the previous section. We consider the setting of a single edge $e = \{u, v\}$ with end nodes u and v. The objective is to deliver the package from node u to node v over edge e as quickly as possible. In our illustrations, we use the convention that v is drawn on the left and u is drawn on the right. We assume that the package reaches u at time t (where t is the earliest possible time when the package can reach u) while being carried by agent i_0.

As discussed in the previous section, let $A(v) = (a_1, a_2, \ldots, a_\ell)$ be the list of agents arriving at node v in order of increasing velocities and increasing arrival times. For $1 \leq i \leq \ell$, denote by t_i the time when a_i reaches v, and by v_i the velocity of agent a_i. We have $t_i < t_{i+1}$ and $v_i < v_{i+1}$ for $1 \leq i < \ell$.

Let $B(u) = (b_1, b_2, \ldots, b_r)$ be the list of agents with increasing velocities and increasing arrival times arriving at node u, starting with the agent i_0 whose arrival time is set to t. The list $B(u)$ can be computed from $A(u)$ in $\mathcal{O}(k)$ time by discarding all agents slower than i_0 and setting the arrival time of i_0 to t. For $1 \leq i \leq r$, let t'_i denote the time when b_i reaches w, and let v'_i denote the velocity of b_i. We have $t'_i < t'_{i+1}$ and $v'_i < v'_{i+1}$ for $1 \leq i < r$.

As k is the total number of agents, we have $\ell \leq k$ and $r \leq k$. In the following, we first introduce a geometric representation of the agents and their potential movements in transporting the package from u to v (Sect. 3) and then present the algorithm for FASTLINEDELIVERY (Sect. 4.2).

4.1   Geometric Representation and Preprocessing

Figure 2 shows a geometric representation of how agents $a_1, \ldots, a_\ell$ move towards u if they start to move from v to u immediately after they arrive at v. The vertical axis represents time, and the horizontal axis represents the distance from v (in the direction towards u or, more generally, any neighbor of v). The movement of each agent a_i can be represented by a line with the line equation $y = t_i + x/v_i$ (i.e., the y value is the time when agent a_i reaches the point at distance x from v). After an agent is overtaken by a faster agent, the slower agent is no longer useful for picking up the package and returning it to v, so we can discard the part of the line of the slower agent that lies to the right of such an intersection point with the line of a faster agent. After doing this for all agents (only the fastest agent a_ℓ does not get overtaken and will not have part of its line discarded), we obtain a representation that we call the *relevant arrangement* Ψ of the agents $a_1, \ldots, a_\ell$. In the relevant arrangement, each agent a_i is represented by a line segment that starts at $(0, t_i)$, lies on the line $y = t_i + x/v_i$, and ends at the first intersection point between the line for a_i and the line of a faster agent a_j, $j > i$. For the fastest agent a_ℓ, there is no faster agent, and so the agent is represented by a half-line. One can view the relevant arrangement as representing the set of all points where an agent from $A(v)$ travelling towards u could receive the package from a slower agent travelling towards v.

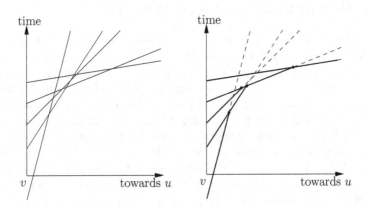

Fig. 2. Geometric representation of agents moving from u towards v (left), and their relevant arrangement with removed half-lines shown dashed (right)

The relevant arrangement has size $\mathcal{O}(k)$ because each intersection point can be charged to the slower of the two agents that create the intersection. It can be computed in $\mathcal{O}(k \log k)$ time using a sweep-line algorithm very similar to the algorithm by Bentley and Ottmann [6] for line segment intersection. The relevant arrangement is created by a call to PREPROCESSRECEIVER(v) (see Algorithm 2).

For the agents in the list $B(u) = (b_1, \ldots, b_r)$ that move from u towards v, we use a similar representation. However, in this case we only need to determine the

---

Algorithm 2. Algorithm PREPROCESSRECEIVER(v)

Data: Node v (and list $A(v)$ of agents arriving at v)
Result: Relevant arrangement Ψ
1 Create a line $y = t_i + x/v_i$ for each agent a_i in $A(v)$;
2 Use a sweep-line algorithm (starting at $x = 0$, moving towards larger x values) to construct the relevant arrangement Ψ;

Algorithm 3. Algorithm PREPROCESSSENDER(u, t)

Data: Node u (and list $A(u)$ of agents arriving at u), time t when package arrives at u (carried by agent i_0)
Result: Lower envelope L of agents carrying package away from u
1 $B(u) \leftarrow A(u)$ with agents slower than i_0 removed and arrival time of i_0 set to t;
2 Create a line $y = t_i' - x/v_i'$ for each agent b_i in $B(v)$;
3 Use a sweep-line algorithm (starting at $x = 0$, moving towards smaller x values) to construct the lower envelope L;

---

lower envelope of the lines representing the agents. See Fig. 3 for an example. The lower envelope L can be computed in $\mathcal{O}(k \log k)$ time (e.g., using a sweep-line algorithm, or via computing the convex hull of the points that are dual to the lines [7, Sect. 11.4]). The call PREPROCESSSENDER(u, t) (see Algorithm 3) determines the list $B(u)$ from $A(u)$ and t in $\mathcal{O}(k)$ time and then computes the lower envelope of the agents in $B(u)$ in time $\mathcal{O}(k \log k)$. When we consider a particular edge $e = \{u, v\}$, we place the lower envelope L in such a way that the position on the x-axis that represents u is at $x = l_e$. We say in this case that the lower envelope is *anchored* at $x = l_e$. Algorithm 3 creates the lower envelope anchored at $x = 0$, and the lower envelope anchored at $x = l_e$ can be obtained by shifting it right by l_e.

4.2   Main Algorithm

Assume we have computed the relevant arrangement Ψ of the agents in the list $A(v) = (a_1, \ldots, a_\ell)$ and the lower envelope L of the lines representing the agents in the list $B(u) = (b_1, b_2, \ldots, b_r)$.

The lower envelope L of the agents in $B(u)$ represents the fastest way for the package to be transported from u to v if only agents in $B(u)$ contribute to the transport and these agents move from u towards v as quickly as possible. At each time point during the transport, the package is at the closest point to v that it can reach if only agents in $B(u)$ travelling from u to v contribute to its transport. We say that such a schedule where the package is as close to v as possible at all times is *fastest and foremost* (with respect to a given set of agents).

The agents in $A(v)$ can potentially speed up the delivery of the package to v by travelling towards u, picking up the package from a slower agent that is currently carrying it, and then turning around and moving back towards v

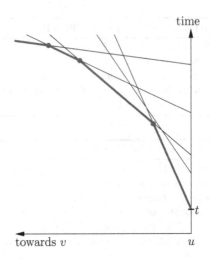

Fig. 3. Geometric representation of agents moving from u towards v (lower envelope highlighted)

as quickly as possible. By considering intersections between L and the relevant arrangement Ψ of $A(v)$, we can find all such potential handover points. More precisely, we trace L from u (*i.e.*, $x = d(u,v)$) towards v (*i.e.*, $x = 0$). Assume that q is the first point where a handover is possible. If a faster agent j from $A(v)$ can receive the package from a slower agent i at point q of L, we update L by computing the lower envelope of L and the half-line representing the agent j travelling from point q towards v. If the intersection point is with an agent j from $A(v)$ that is not faster than the agent i that is currently carrying the package, we ignore the intersection point. We then continue to trace L towards v and process the next intersection point in the same way. We repeat this step until we reach v (*i.e.*, $x = 0$). The final L represents an optimum solution to the FASTLINEDELIVERY problem, and the y-value of L at $x = 0$ represents the arrival time of the package at v. See Algorithm 4 for pseudo-code of the resulting algorithm.

An illustration of step 7 of Algorithm 4, which updates L by incorporating a faster agent from $A(v)$, is shown in Fig. 4. Note that the time for executing this step is $\mathcal{O}(g)$, where g is the number of segments removed from L in the operation. As a line segment corresponding to an agent can only be removed once, the total time spent in executing step 7 (over all executions of step 7 while running Algorithm 4) is $\mathcal{O}(k)$.

Finally, we need to analyze how much time is spent in finding intersection points with line segments of the relevant arrangement Ψ while following the lower envelope L from u to v. See Fig. 5 for an illustration. We store the relevant arrangement using standard data structures for planar arrangements [14], so that we can follow the edges of each face in clockwise or counter-clockwise direction efficiently (*i.e.*, we can go from one edge to the next in constant time) and move

---

Algorithm 4. Algorithm FASTLINEDELIVERY(u, v, t)

Data: Edge $e = \{u, v\}$, earliest arrival time t of package at u, lists $A(u)$
    and $A(v)$
Result: Earliest time when package reaches v over edge $\{u, v\}$
    /* Assume PREPROCESSRECEIVER(v) and PREPROCESSSENDER(u,t) have
    already been called.   */
1 $L \leftarrow$ lower envelope of agents $B(u)$ anchored at $x = l_e$;
2 $\Psi \leftarrow$ relevant arrangement of $A(v)$;
3 start tracing L from u (*i.e.*, $x = l_e$) towards v (*i.e.*, $x = 0$);
4 **while** v *(i.e., $x = 0$) is not yet reached* **do**
5 $\quad$ $q \leftarrow$ next intersection point of L and Ψ;
   $\quad$ /* assume q is intersection of agent i from L and agent j from Ψ
   $\quad$ */
6 $\quad$ **if** $v_j > v_i$ **then**
7 $\quad\quad$ replace L by the lower envelope of L and the line for agent j moving left
   $\quad\quad$ from point q;
8 $\quad$ **else**
9 $\quad\quad$ ignore q
10 $\quad$ **end**
11 **end**
12 **return** *y-value of L at $x = 0$*

---

from an edge of a face to the instance of the same edge in the adjacent face in constant time. This representation also allows us to to trace the lower envelope of Ψ in time $\mathcal{O}(k)$.

First, we remove from Ψ all line segments corresponding to agents that are not faster than i_0 (recall that i_0 is the agent that brings the package to node u at time t). Then, in order to find the first intersection point q_1 between L and Ψ, we can trace L and the lower envelope of Ψ from u towards v in parallel until they meet. One may observe that L cannot be above the lower envelope of Ψ at u because otherwise an agent faster than i_0 reaches u before time t, and that agent could pick up the package from i_0 before time t and deliver it to u before time t, a contradiction to t being the earliest arrival time for the package at u. This takes $\mathcal{O}(k)$ time. After computing one intersection point q_i (and possibly updated L as shown in Fig. 4), we find the next intersection point by following the edges on the inside of the next face in counter-clockwise direction until we hit L again at q_{i+1}. This process is illustrated by the dashed arrow in Fig. 5, showing how q_2 is found starting from q_1. Hence, the total time spent in finding intersection points is bounded by the initial size of L and the number of edges of all the faces of the relevant arrangement, which is $\mathcal{O}(k)$.

Theorem 2. *Algorithm 4 solves* FASTLINEDELIVERY(u, v, t) *in $\mathcal{O}(k)$ time, assuming that* PREPROCESSRECEIVER(v) *and* PREPROCESSSENDER(u, t), *which take time $\mathcal{O}(k \log k)$ each, have already been executed.*

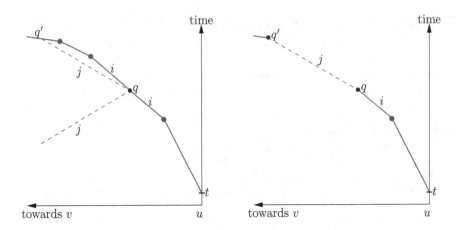

Fig. 4. Agent i meets a faster agent j at intersection point q (left). The part of L from q to q' has been replaced by a line segment representing agent j carrying the package towards v (right).

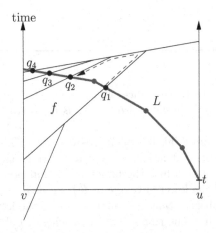

Fig. 5. Intersection points q_1, q_2, q_3, q_4 between the lower envelope L (highlighted in bold) and the relevant arrangement Ψ. Point q_2 is found from q_1 by simultaneously tracing L and the edges of the face f of Ψ in counter-clockwise direction.

Proof. The claimed running time follows from the discussion above. Correctness follows by observing that the following invariant holds: If the algorithm has traced L up to position (x_0, y_0), then the current L represents the fastest and foremost solution for transporting the package from u to v using only agents in $B(u)$ and agents from $A(v)$ that can reach the package by time y_0.   □

Acknowledgments. Iago A. Carvalho was financed in part by the Coordenação de Aperfeiçoamento de Pessoal de Nível Superior - Brasil (CAPES) - Finance Code 001.

References

1. Bamburry, D.: Drones: designed for product delivery. Des. Manag. Rev. **26**(1), 40–48 (2015)
2. Bärtschi, A., et al.: Collaborative delivery with energy-constrained mobile robots. Theor. Comput. Sci. (2017). https://doi.org/10.1016/j.tcs.2017.04.018
3. Bärtschi, A., et al.: Energy-efficient delivery by heterogeneous mobile agents. In: 34th Symposium on Theoretical Aspects of Computer Science (STACS 2017). LIPIcs, vol. 66, p. 10, Schloss Dagstuhl, Leibniz-Zentrum für Informatik (2017). https://doi.org/10.4230/LIPIcs.STACS.2017.10
4. Bärtschi, A., Graf, D., Mihalák, M.: Collective fast delivery by energy-efficient agents. In: Potapov, I., Spirakis, P., Worrell, J. (eds.) 43rd International Symposium on Mathematical Foundations of Computer Science (MFCS 2018). LIPIcs, vol. 117, pp. 56:1–56:16, Schloss Dagstuhl-Leibniz-Zentrum für Informatik (2018). https://doi.org/10.4230/LIPIcs.MFCS.2018.56
5. Bärtschi, A., Tschager, T.: Energy-efficient fast delivery by mobile agents. In: Klasing, R., Zeitoun, M. (eds.) FCT 2017. LNCS, vol. 10472, pp. 82–95. Springer, Heidelberg (2017). https://doi.org/10.1007/978-3-662-55751-8_8
6. Bentley, J.L., Ottmann, T.: Algorithms for reporting and counting geometric intersections. IEEE Trans. Comput. **28**(9), 643–647 (1979). https://doi.org/10.1109/TC.1979.1675432
7. de Berg, M., Cheong, O., van Kreveld, M.J., Overmars, M.H.: Computational Geometry: Algorithms and Applications, 3rd edn. Springer, Heidelberg (2008)
8. Chalopin, J., Das, S., Mihal'ák, M., Penna, P., Widmayer, P.: Data delivery by energy-constrained mobile agents. In: Flocchini, P., Gao, J., Kranakis, E., Meyer auf der Heide, F. (eds.) ALGOSENSORS 2013. LNCS, vol. 8243, pp. 111–122. Springer, Heidelberg (2014). https://doi.org/10.1007/978-3-642-45346-5_9
9. Chalopin, J., Jacob, R., Mihalák, M., Widmayer, P.: Data delivery by energy-constrained mobile agents on a line. In: Esparza, J., Fraigniaud, P., Husfeldt, T., Koutsoupias, E. (eds.) ICALP 2014. LNCS, vol. 8573, pp. 423–434. Springer, Heidelberg (2014). https://doi.org/10.1007/978-3-662-43951-7_36
10. Cooke, K., Halsey, E.: The shortest route through a network with time-dependent internodal transit times. J. Math. Anal. Appl. **14**(3), 493–498 (1966). https://doi.org/10.1016/0022-247X(66)90009-6
11. Delling, D., Wagner, D.: Time-dependent route planning. In: Ahuja, R.K., Möhring, R.H., Zaroliagis, C.D. (eds.) Robust and Online Large-Scale Optimization. LNCS, vol. 5868, pp. 207–230. Springer, Heidelberg (2009). https://doi.org/10.1007/978-3-642-05465-5_8
12. Fredman, M.L., Tarjan, R.E.: Fibonacci heaps and their uses in improved network optimization algorithms. J. ACM **34**(3), 596–615 (1987). https://doi.org/10.1145/28869.28874
13. Gąsieniec, L., Kijima, S., Min, J.: Searching with increasing speeds. In: Izumi, T., Kuznetsov, P. (eds.) SSS 2018. LNCS, vol. 11201, pp. 126–138. Springer, Cham (2018). https://doi.org/10.1007/978-3-030-03232-6_9
14. Goodrich, M., Ramaiyer, K.: Geometric data structures. In: Sack, J.R., Urrutia, J. (eds.) Handbook of Computational Geometry, pp. 463–489. Elsevier Science (2000)
15. Hong, I., Kuby, M., Murray, A.: A deviation flow refueling location model for continuous space: a commercial drone delivery system for urban areas. In: Griffith, D.A., Chun, Y., Dean, D.J. (eds.) Advances in Geocomputation. AGIS, pp. 125–132. Springer, Cham (2017). https://doi.org/10.1007/978-3-319-22786-3_12

16. Murray, C.C., Chu, A.G.: The flying sidekick traveling salesman problem: optimization of drone-assisted parcel delivery. Transp. Res. Part C: Emerg. Technol. **54**, 86–109 (2015). https://doi.org/10.1016/j.trc.2015.03.005
17. Regev, A.: Drone deliveries are no longer pie in the sky, April 2018. https://www.forbes.com/sites/startupnationcentral/2018/04/10/drone-deliveries-are-no-longer-pie-in-the-sky/

Extension of Some Edge Graph Problems: Standard and Parameterized Complexity

Katrin Casel[1,2], Henning Fernau[2], Mehdi Khosravian Ghadikolaei[3], Jérôme Monnot[3], and Florian Sikora[3(✉)]

[1] Hasso Plattner Institute, University of Potsdam, 14482 Potsdam, Germany
[2] Universität Trier, Fachbereich 4, Informatikwissenschaften, 54286 Trier, Germany
{casel,fernau}@informatik.uni-trier.de
[3] Université Paris-Dauphine, PSL University, CNRS, LAMSADE,
75016 Paris, France
{mehdi.khosravian-ghadikolaei,jerome.monnot,
florian.sikora}@lamsade.dauphine.fr

Abstract. We consider *extension* variants of some edge optimization problems in graphs containing the classical EDGE COVER, MATCHING, and EDGE DOMINATING SET problems. Given a graph $G = (V, E)$ and an edge set $U \subseteq E$, it is asked whether there exists an inclusion-wise *minimal* (resp., *maximal*) feasible solution E' which satisfies a given property, for instance, being an edge dominating set (resp., a matching) and containing the *forced* edge set U (resp., avoiding any edges from the *forbidden* edge set $E \setminus U$). We present hardness results for these problems, for restricted instances such as bipartite or planar graphs. We counterbalance these negative results with parameterized complexity results. We also consider the *price of extension*, a natural optimization problem variant of extension problems, leading to some approximation results.

Keywords: Extension problems · Edge cover · Matching · Edge domination · NP-completeness · Parameterized complexity · Approximation

1 Introduction

We consider *extension problems* related to several classical edge optimization problems in graphs, namely EDGE COVER, MAXIMUM MATCHING and EDGE DOMINATING SET. Informally, in an extension version of an edge optimization problem, one is given a graph $G = (V, E)$ as well as a subset of edges $U \subseteq E$, and the goal is to *extend* U to a minimal (or maximal) solution (if possible).

Such variants of problems are interesting for efficient enumeration algorithms or branching algorithms (see more examples of applications in [11]).

Related Work. Extension versions have been studied for classical optimization problems, for example, the minimal extension of 3-HITTING SET [9], minimal

© Springer Nature Switzerland AG 2019
L. A. Gąsieniec et al. (Eds.): FCT 2019, LNCS 11651, pp. 185–200, 2019.
https://doi.org/10.1007/978-3-030-25027-0_13

DOMINATING SET [2,8] or VERTEX COVER [1]. Extensions show up quite naturally in quite a number of situations. For instance, when running a search tree algorithm, usually parts of the constructed solution are fixed. It is highly desirable to be able to prune branches of the search tree as early as possible. Hence, it would be very nice to tell efficiently if such a solution part can be extended to a valid (minimal) solution. When trying to enumerate all minimal solutions, the same type of problem arises and has known applications in so-called flashlight algorithms [24]. Another type of area where extension problems show up is linked to Latin squares [13] (and similar combinatorial questions), or also coloring extensions in graphs [7]. In a recent paper, we investigated the complexity of extension versions of VERTEX COVER and INDEPENDENT SET, i.e., classical *vertex* graph problems [12], and we give a first *systematic* study of this type of problems in [11], providing quite a number of different examples of extension problems. For extension variants of automata-related problems, see [17].

Organization of the Paper. After giving some definitions in Sect. 2, we prove that generalization of these problems remain NP-complete, even in bipartite graphs of bounded degree and with some constraints on the forced set of edges. Having a planar embedding does not help much either, as we show in Sect. 4 that these problems remain hard on subcubic bipartite planar graphs. Motivated by these negative results, we study the parameterized complexity of these problems in Sect. 5 and the approximability of a natural optimization version in Sect. 6. Due to lack of space the proofs of statements marked with (∗) are deferred to the full version of the paper.

2   Definitions

Graph Definitions. We consider simple undirected graphs only, to which we refer to as *graphs*. Let $G = (V, E)$ be a graph and $S \subseteq V$; $N_G(S) = \{v \in V : \exists u \in S, vu \in E\}$ denotes the *neighborhood* of S in G and $N_G[S] = S \cup N_G(S)$ denotes the *closed neighborhood* of S. For singleton sets $S = \{s\}$, we simply write $N_G(s)$ or $N_G[s]$, even omitting G if clear from context. The cardinality of $N_G(s)$ is called *degree* of s, denoted $d_G(s)$. If 3 upper-bounds the degree of all vertices, we speak of *subcubic graphs*. For a subset of edges S, $V(S)$ denotes the vertices incident to S. A vertex set S induces the graph $G[S]$ with vertex set S and $e \in E$ being an edge in $G[S]$ iff both endpoints of e are in S. If $S \subseteq E$ is an edge set, then $\overline{S} = E \setminus S$, edge set S induces the graph $G[V(S)]$, while $G_S = (V, S)$ denotes the *partial graph induced by* S; in particular, $G_{\overline{S}} = (V, E \setminus S)$.

A vertex set S is *independent* if S is a set of pairwise non-adjacent vertices. An edge set S is called an *edge cover* if the partial graph G_S is spanning and it is a *matching* if S is a set of pairwise non-adjacent edges. An edge set S is *minimal* (resp., *maximal*) with respect to a graph property if S satisfies the graph property and any proper subset $S' \subset S$ of S (resp., any proper superset $S' \supset S$ of S) does not satisfy the graph property. A graph $G = (L \cup R, E)$ is called *bipartite* if its vertex set decomposes into two independent sets L and R. The *line graph* $L(G) = (V', E')$ of a graph $G = (V, E)$ is a simple graph where each vertex of $L(G)$ represents an edge of G and two vertices of $L(G)$ are adjacent if and only if their corresponding edges share a common vertex in G. Hence, it is exactly the intersection graph of the edges of G. It is well known the class of line graphs is a subclass of *claw-free* graphs (i.e., without $K_{1,3}$ as induced subgraph).

Problem Definitions. Let $G = (V, E)$ be a graph where the minimum degree is at least $r \geq 1$. We assume r is a fixed constant (but all results given here hold even if r depends on the graph). An *r-degree constrained partial subgraph* is defined as an edge subset $S \subseteq E$ such that none of the vertices in V is incident to more than r edges in S. The problem of finding such a set S of size at least k is termed r-DCPS. An *r-degree edge-cover* is defined as a subset of edges such that each vertex of G is incident to at least $r \geq 1$ distinct edges $e \in S$, leading to the decision problem r-EC, determining if such a set of size at most k exists. For the particular cases of $r = 1$, 1-DCPS corresponds to the famous MATCHING problem and 1-EC is also known as the EDGE COVER problem.

The optimization problem associated to r-DCPS, denoted MAX r-DCPS, consists of finding an edge subset E' of maximum cardinality that is a solution to r-DCPS. MAX r-DCPS is known to be solvable in polynomial time even for the edge weighted version (here, we want to maximize the weight of E') [19]. When additionally the constraint r is not uniform and depends on each vertex (i.e., at most $b(v) = r_v$ edges incident to vertex v), MAX r-DCPS is usually known as SIMPLE b-MATCHING and remains solvable in polynomial time even for the edge-weighted version (Theorem 33.4, Chap. 33 of Volume A in [27]).

A well-studied optimization version of a generalization of r-EC, known as the MIN LOWER-UPPER-COVER PROBLEM (MINLUCP), is the following. Given a graph $G = (V, E)$ and two functions $a, b: V \rightarrow \mathbb{N}$ such that for all $v \in V$, $0 \leq a(v) \leq b(v) \leq d_G(v)$, find a subset $M \subseteq E$ such that the partial graph $G_M = (V, M)$ induced by M satisfies $a(v) \leq d_{G_M}(v) \leq b(v)$ (such a solution will be called a *lower-upper-cover*), minimizing its cardinality $|M|$ among all such solutions (if any). Hence, an r-EC solution corresponds to a lower-upper-cover with $a(v) = r$ and $b(v) = d_G(v)$ for every $v \in V$. MINLUCP is known to be solvable in polynomial time even for edge-weighted graphs (Theorem 35.2 in Chap. 35 of Volume A in [27]).

We are considering the following *extension problems* associated to r-DCPS and r-EC.

EXT r-DCPS
Input: A graph $G = (V, E)$ and $U \subseteq E$.
Question: Does there exist an edge set $S \subseteq E$ with $S \subseteq U$ such that the partial graph G_S has maximum degree at most r and is maximal in G?

EXT r-EC
Input: A graph $G = (V, E)$ and $U \subseteq E$.
Question: Does there exist an edge set $S \subseteq E$ with $S \supseteq U$ such that the partial graph G_S has minimum degree at least r and is minimal in G?

An r-*edge dominating set* $S \subseteq E$ of a simple graph $G = (V, E)$ is a set S of edges such that for any edge $e \in E$ of G, at least r edges of S are incident to e (by definition, an edge dominates itself one time). The MINIMUM r-EDGE DOMINATING SET problem (MIN r-EDS for short) consists in finding an r-edge dominating set of minimum size. Notice that there is a feasible solution if and only if $r \leq \min_{xy \in E}(d_G(x) + d_G(y) - 1)$. Obviously, 1-EDS is the classical EDGE DOMINATING SET problem (EDS), which is NP-hard in general graphs (problem [GT2] in [20]). The generalization to r-EDS has been studied in [3,4] (under the name b-EDS) from an approximation point of view. However, to the best of our knowledge, r-EDS for every $r \geq 2$ was not proved NP-hard so far. As associated extension problem, we formally study the following problem.

EXT r-EDS
Input: Given a simple graph $G = (V, E)$ and $U \subseteq E$.
Question: Is there a subset $S \subseteq E$ such that $U \subseteq S$ and S is a minimal r-edge dominating set?

For an edge extension problem π, $ext_\pi(G, U)$ denotes the set of *extremal extensions* of U (i.e., minimal or maximal depending on the context). For a minimal version, U corresponds to a subset of *forced* edges (i.e., each minimal solution has to contain U) while for a maximal version, $E \setminus U$ corresponds to a subset of *forbidden* edges (i.e., each maximal solution has to contain no edges from $E \setminus U$). Sometimes, the set $ext_\pi(G, U)$ is empty, which makes the question of the existence of such extensions interesting. Hence, for $\pi \in \{$EXT r-DCPS, EXT r-EC, EXT r-EDS$\}$, the extension problems ask if $ext_\pi(G, U) \neq \emptyset$. We call $|U|$ the *standard parameter* when considering these problems as parameterized. We may drop the subscript π if clear from context.

3   Complexity Results

The results given in this section are based on a reduction from 2-BALANCED 3-SAT, $(3, B2)$-SAT for short. An instance $(\mathcal{C}, \mathcal{X})$ of $(3, B2)$-SAT is a set $\mathcal{C}$ of CNF clauses defined over a set $\mathcal{X}$ of Boolean variables such that each clause has exactly 3 literals and each variable appears exactly twice as a negative and twice as a positive literal in $\mathcal{C}$. The bipartite graph associated to $(\mathcal{C}, \mathcal{X})$ is $BP = (C \cup X, E(BP))$, with $C = \{c_1, \dots, c_m\}$, $X = \{x_1, \dots, x_n\}$ and $E(BP) = \{c_j x_i : x_i$ or $\neg x_i$ is a literal of $c_j\}$. $(3, B2)$-SAT is NP-hard by [5, Theorem 1].

Theorem 1. (∗) *For every fixed $r \geq 1$, EXT r-DCPS is NP-complete in bipartite graphs with maximum degree $\max\{3, r + 1\}$, even if $\overline{U}$ is an induced matching for $r \geq 2$ or an induced collection of paths of length at most 2 for $r = 1$.*

Proof. Let $r = 1$. For the technical details for the case $r > 1$, we refer to the long version of this paper. Consider an instance of $(3, B2)$-SAT with clauses $\mathcal{C} = \{c_1, \dots, c_m\}$ and variables $\mathcal{X} = \{x_1, \dots, x_n\}$. We build a bipartite graph $G = (V, E)$ of maximum degree 3 as follows:

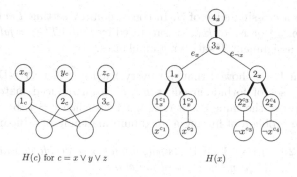

$H(c)$ for $c = x \vee y \vee z$ $H(x)$

Fig. 1. The Gadgets $H(c)$ and $H(x)$. Edges not in U are drawn as thicker lines.

- For each clause $c = x \vee y \vee z$, where x, y, z are literals, introduce a subgraph $H(c) = (V_c, E_c)$ with 8 vertices and 9 edges. V_c contains three specified vertices x_c, y_c and z_c corresponding to literals of the clause c. Moreover, $\overline{U}_c = \{x_c 1_c, y_c 2_c, z_c 3_c\}$ is a set of three forbidden edges included in $H(c)$. The gadget $H(c)$ is illustrated in the left part of Fig. 1.
- For each variable x, introduce 12 new vertices. They induce the subgraph $H(x) = (V_x, E_x)$ illustrated in Fig. 1. The vertex set V_x contains four special vertices x^{c_1}, x^{c_2}, $\neg x^{c_3}$ and $\neg x^{c_4}$, where it is implicitly assumed that variable x appears as a positive literal in clauses c_1, c_2 and as a negative literal in clauses c_3, c_4. Finally, there are two sets of *free* edges (non-forbidden edges): $F_x = \{e_x\} \cup \{2_x^{c_3} \neg x^{c_3}, 2_x^{c_4} \neg x^{c_4}\}$ and $F_{\neg x} = \{e_{\neg x}\} \cup \{1_x^{c_1} x^{c_1}, 1_x^{c_2} x^{c_2}\}$. Hence, the forbidden edges U_x in $H(x)$ are given by $\overline{U}_x = E_x \setminus (F_x \cup F_{\neg x})$.
- We interconnect $H(x)$ and $H(c)$, where x is a literal of clause c, by adding edge $x_c x^c$ if x appears as a positive literal and edge $x_c \neg x^c$ if x appears as a negative literal. We call these edges *crossing edges*.

We set $U = E \setminus ((\bigcup_{c \in C} \overline{U}_c) \cup (\bigcup_{x \in X} \overline{U}_x))$. This construction is computable within polynomial time and G is a bipartite graph of maximum degree 3. We claim that there is a truth assignment of I which satisfies all clauses iff there is a maximal matching $S \subseteq U$ of G.

If T is a truth assignment of I which satisfies all clauses, then we add the set of edges $x_c x^c$ and F_x if $T(x) = true$; otherwise, we add the edge $x_c \neg x^c$ and all edges in $F_{\neg x}$. For each clause c, we choose one literal l_c which satisfies the clause; then, we add 2 edges saturating vertices 1_c, 2_c and 3_c and which are not incident to the edge of $\overline{U}_c$ saturating l_c. For instance, assume it is y; then, we add two edges saturating vertices 1_c and 3_c and the white vertices in $H(c)$. The resulting matching S is maximal with $S \cap \overline{U} = \emptyset$.

Conversely, assume the existence of a maximal matching S with $S \subseteq U$. Hence, for each variable $x \in X$ exactly one edge between e_x and $e_{\neg x}$ is in S (in order to block edge $3_x 4_x$). If it is $e_x \in S$ (resp., $e_{\neg x} \in S$), then $F_x \subset S$ (resp., $F_{\neg x} \subset S$). Hence, S does not contain any crossing edges saturating $\neg x^c$ (resp., x^c). Now for each clause $c = x \vee y \vee z$, at least one vertex among x_c, y_c, z_c must be adjacent to a crossing edge of S. In conclusion, by setting $T(x) = true$ if at least one vertex x^{c_1} or x^{c_2} of $H(x)$ is saturated by S and $T(x) = false$ otherwise, we get a valid assignment T satisfying all clauses.   □

In Theorem 1, we showed that, for every fixed $r \geq 2$, EXT r-DCPS is hard even when the set of forbidden edges $E \setminus U$ is an induced matching. In the following, we prove the same result does not hold when $r = 1$, by reducing this problem to the problem of finding a maximum matching in a bipartite graph.

Proposition 2. (∗) EXT 1-DCPS *is polynomial-time decidable when the forbidden edges* $\overline{U} = E \setminus U$ *form an induced matching.*

Remark 3. Proposition 2 can be extended to the case where $\overline{U}$ is a matching and $G_{\overline{U}}$ does not contain an alternating path of length at least 5. The complexity of EXT 1-DCPS when $\overline{U}$ is a matching remains unsettled.

In [12], several results are proposed for the extension of the independent set problem (EXT IS for short) in bipartite graphs, planar graphs, chordal graphs, etc. Here, we deduce a new result for a subclass of claw-free graphs.

Corollary 4. EXT IS *is NP-complete restricted to line graphs of bipartite graphs of maximum degree 3.*

Proof. Let $G = (V, E)$ be a bipartite graph of maximum degree 3 and $L(G) = (V', E')$ its line graph. It is well known that any matching S of G corresponds to an independent set $S' = L(S)$ of G' and vice versa. In particular, S is a maximal matching of G iff $L(S)$ is a maximal independent set. Hence, (G, U) is a yes-instance of EXT 1-DCPS iff $(L(G), L(U))$ is a yes-instance of EXT IS. Theorem 1 with $r = 1$ concludes the proof.   □

A reduction from $(3, B2)$-SAT can also be used to show the following.

Theorem 5. (∗) *For every fixed* $r \geq 1$, EXT r-EC *is NP-complete in bipartite graphs with maximum degree* $r + 2$, *even if the forced edge set* U *is a matching.*

4   Planar Graphs

All reductions given in this section are from 4-BOUNDED PLANAR 3-CONNECTED SAT (4P3C3SAT for short), the restriction of EXACT 3-SATISFIABILITY[1] to clauses in $\mathcal{C}$ over variables in $\mathcal{X}$, where each variable occurs in at most four clauses (at least one time but at most two times negated) and the associated bipartite graph BP (explained in Sect. 3) is planar of maximum degree 4. This restriction is also NP-complete [23]; in the following, we always assume that the planar graph comes with an embedding in the plane. This gives us a planar *variable-clause-graph* G, corresponding to the original SAT instance I. The additional technical

---

[1] Addressing the problem to decide whether there is a truth assignment setting exactly one literal in each clause to true.

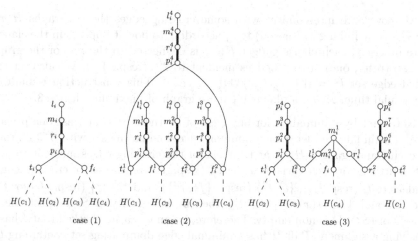

Fig. 2. Variable gadgets $H(x_i)$ of Theorem 6. Cases (1), (2), (3) are corresponding to $H(x_i)$, depending on how x_i appears (as a negative or positive literal) in the four clauses (here, case 3 is rotated). Bold edges denote elements of U_{x_i}. Crossing edges are marked by dashed lines.

difficulties come with the embeddings that need to be preserved. Suppose that a variable x_i appears in at most four clauses c_1, c_2, c_3, c_4 of I such that in the induced (embedded) subgraph $G_i = G[\{x_i, c_1, c_2, c_3, c_4\}]$, $c_1 x_i, c_2 x_i, c_3 x_i, c_4 x_i$ is an anticlockwise ordering of edges around x_i. By looking at G_i and considering how variable x_i appears as a negative or positive literal in the four clauses c_1, c_2, c_3, c_4 in I, the construction should handle the three following cases: (1): $x_i \in c_1, c_2$ and $\neg x_i \in c_3, c_4$; (2): $x_i \in c_1, c_3$ and $\neg x_i \in c_2, c_4$; (3): $x_i \in c_1, c_2, c_3$ and $\neg x_i \in c_4$. All other cases are included in these cases by rotations and/or interchanging x_i with $\neg x_i$.

Theorem 6. *For any $r \geq 1$,* EXT r-EDS *is* NP-*complete for planar bipartite graphs of maximum degree $r + 2$.*

Proof. Consider first $r = 1$, corresponding to EXT EDS. Given an instance I of 4P3C3SAT with clause set $\mathcal{C} = \{c_1, \ldots, c_m\}$ and variable set $\mathcal{X} = \{x_1, \ldots, x_n\}$, we build a planar bipartite graph $H = (V_H, E_H)$ with maximum degree 3 together with a set $U \subseteq E_H$ of forced edges as an instance of EXT EDS.

For each variable x_i we introduce a corresponding gadget $H(x_i)$ as depicted in Fig. 2, the forced edge set U_{x_i} contains $\{m_i r_i, r_i p_i\}$ for case (1), $\{p_i^j r_i^j, r_i^j m_i^j : 1 \leq j \leq 4\}$ for case (2) and $\{p_i^1 p_i^2, p_i^2 p_i^3, p_i^5 p_i^6, p_i^6 p_i^7, m_i^2 f_i\}$ for case (3).

For each clause $c_j \in \mathcal{C}$, we construct a clause gadget $H(c_j)$ as depicted on the right, and a forced edge set U_{c_j}, each clause gadget $H(c_j)$ contains 8 vertices and 7 edges where $|U_{c_j}| = 2$. Edges in U are drawn in bold.

$H(c)$ for clause $c = \ell_1 \vee \ell_2 \vee \ell_3$

Moreover, we interconnect with some crossing edges the subgraphs $H(x_i)$ and $H(c_j)$ by linking x_i (or $\neg x_i$) to c_j according to how it appears in the clause. More precisely, each clause gadget $H(c_j)$ is connected to the rest of the graph via two (resp., one) crossing edges incident to $2'_{c_j}$ (resp., $1'_{c_j}$). We also set the forced edge set $U = (\bigcup_{x_i \in \mathcal{X}} U_{x_i}) \cup (\bigcup_{c_j \in \mathcal{C}} U_{c_j})$. This construction is built in polynomial time, giving a planar bipartite graph of maximum degree 3.

Note that by minimality, for any edge of U, there exist at least one private edge to dominate. So, let S be a minimal edge dominating set with $S \supseteq U$, then for each clause gadget $H(c)$, at least one of the crossing edges incident to it is in S. Further, for each variable x, let c_t^x (resp., c_f^x) be the set of crossing edges incident to t_i (resp., f_i), $\{t_i^1, t_i^2\}$ (resp., $\{f_i^1, f_i^2\}$), and $\{t_i^1, t_i^2\}$ (resp., f_i) for the case 1, 2 and 3 of $H(x)$ respectively, then by minimality of S, at most one of $(S \cap c_t^x)$ or $(S \cap c_f^x)$ is non-empty. Therefore, it can be easily checked that I has a satisfying assignment T iff H has a minimal edge dominating set containing U. For $r \geq 2$, we start with the instance $I = (H, U)$ given in the above construction for $r = 1$. Recall $H = (V_H, E_H)$ is a bipartite graph with bipartition $V_H = L \cup R$, while $U \subseteq E_H$ is a subset of forced edges. Now, for each vertex v of the left part L, we add the gadget $B_r(v)$ depicted to the right. Denote by H' the resulting bipartite graph and consider $I' = (H', U)$ as an instance of EXT r-EDS.

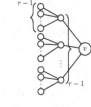

$B_r(v)$

Let $B = \bigcup_{v \in L} B_r(v)$ be the added edges from H to H'. Note that any r-EDS S' of H' must contain B. Moreover, S' is a minimal r-EDS of H' iff $S' \setminus B$ is a minimal EDS of H.   □

Remark 7. Reconsidering the previous construction that reduces the case when $r > 1$ to the case when $r = 1$, and using the NP-hardness of EDS in bipartite graphs [6,31], we deduce NP-hardness of r-EDS for all $r \geq 1$.

In [22], several results are proposed for the enumeration of minimal dominating sets in line-graphs. Here, we strengthen these results by showing that extending a given vertex set to a minimal dominating set (a problem we call EXT DS) in line graphs of a planar bipartite subcubic graphs is already a hard problem.

Corollary 8. EXT DS *is* NP-*complete, even when restricted to line graphs of planar bipartite subcubic graphs.*

Proof. Let $G = (V, E)$ be a bipartite graph of maximum degree 3 and $L(G) = (V', E')$ its line graph. It is well known that any edge dominating set S of G corresponds to a dominating set $S' = L(S)$ of G' and vice versa. In particular, S is a minimal edge dominating set of G iff $L(S)$ is a minimal dominating set. Hence, (G, U) is a yes-instance of EXT EDS iff $(L(G), L(U))$ is a yes-instance of EXT DS. Theorem 6 with $r = 1$ concludes the proof.   □

The two next statements appear to be only strengthening Theorems 1 and 5 in the particular case of $r = 1$, but the details behind can be different indeed.

Theorem 9. (∗) EXT 1-EC *is* NP-*complete for planar bipartite subcubic graphs.*

Theorem 10. (∗) EXT 1-DCPS *is* NP-*complete even for planar bipartite subcubic graphs.*

5   Parameterized Perspective

The next result is quite simple and characterizes the yes-instances of EXT r-EC.

Lemma 11. (∗) $ext(G, U) \neq \emptyset$ *iff there is an* r-*EC solution* $G' = (V, E')$ *where* $E' \supseteq U$ *such that* $S_{G'} = \{v \in V(U) : d_{G'}(v) > r\}$ *is an independent set of* G_U.

This structural property can be used to design an FPT-algorithm for EXT r-EC. More precisely, our proposed algorithm lists all $3^{|U|}$ many independent sets of $G[U]$ included in $V(U)$ from an instance $I = (G, U)$ of EXT r-EC. In each case, we produce an equivalent instance of MINLUCP that can be solved in polynomial time which gives the following result.

Theorem 12. (∗) EXT r-EC, *with standard parameter, is in* FPT.

For EXT r-DCPS, we can also exploit structural properties of yes-instances and use the polynomial solvability of SIMPLE b-MATCHING to show the following.

Theorem 13. (∗) EXT r-DCPS, *parameterized by the number of forbidden edges* $\overline{U}$, *is in* FPT.

When bounding the degree of the graphs, we can consider an even smaller parameter and obtain feasibility results.

Proposition 14. (∗) *For graphs with maximum degree* $r + 1$, EXT r-DCPS *is polynomial-time decidable when* $r = 1$ *and is in* FPT *with respect to the number of isolated edges in* $\overline{U}$ *for* $r \geq 2$.

Remark 15. For graphs with maximum degree $r + 1$, EXT r-DCPS with $r \geq 2$ is parameterized equivalent to SAT with respect to the number of isolated edges in $E \setminus U$ and variables, respectively.

Theorem 16. *For any* $r \geq 1$, EXT r-EDS *(with standard parameter) is* $W[1]$-*hard, even when restricted to bipartite graphs.*

Proof. We only consider $r = 1$. For $r \geq 2$, we can use the gadget $B_r(v)$ as in Theorem 6. The hardness result comes from a reduction from EXT VC on bipartite graphs, the extension version of VERTEX COVER; see [12]. Let $I = (G, U)$ be an instance of EXT VC, where $G = (V, E)$ is a bipartite graph with partition (V_1, V_2) of V and $U \subseteq V$, the question of VERTEX COVER is to decide if G has a minimal vertex cover S with $U \subseteq S$. We build an instance $I' = (G', U')$

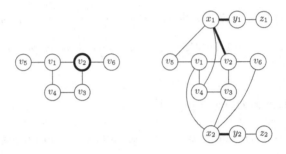

Fig. 3. (G, U) as an instance of EXT VC is shown on the left, with $V_1 = \{v_2, v_4, v_5\}$ and $V_2 = \{v_1, v_3, v_6\}$ and $U = \{v_2\}$. The constructed instance (G', U') of EXT EDS is shown on the right. The vertices and edges of U and U' are in marked with bold lines.

of EXT EDS as follows. Let us first construct a new graph $G' = (V', E')$ with $V' = V \cup \{x_i, y_i, z_i \colon i = 1, 2\}$ and

$$E' = E \cup \bigcup_{i=1,2} \left(\{x_i y_i, y_i z_i\} \cup \{v x_i \colon v \in V_i\}\right).$$

G' is bipartite with partition into $V_1' = V_1 \cup \{x_2, y_1, z_2\}$ and $V_2' = V_2 \cup \{x_1, y_2, z_1\}$. Set $U' = \{u x_1 \colon u \in U \cap V_1\} \cup \{u x_2 \colon u \in U \cap V_2\} \cup \{x_1 y_1, x_2 y_2\}$ so, $|U'| = |U| + 2$. This construction is illustrated in Fig. 3. We claim that (G', U') is a yes-instance of EXT EDS if and only if (G, U) is a yes-instance of EXT VC.

Suppose (G, U) is a yes-instance for EXT VC; so there exists a minimal vertex cover S for G with $U \subseteq S$. The set $S' = \{v x_1 \colon v \in V_1 \cap S\} \cup \{v x_2 \colon v \in V_2 \cap S\} \cup \{x_1 y_1, x_2 y_2\}$ is an edge dominating set of G' which includes U' because S contains U. Since S is minimal, S' is minimal, too; observe that private edges of a vertex $v \in S \cap V_1$ (i.e. an edge vu with $u \notin S \cap V_1$) translate to private edges of $v x_1 \in S'$, analogously for $x \in S \cap V_2$. By construction, $y_i z_i$ is a private edge for $x_i y_i$, $i = 1, 2$.

Conversely, suppose S' is a minimal edge dominating set of G' containing U'. Since S' is minimal, then for each $e \in S'$ there is a private edge set $S_e \subseteq E'$, $S_e \neq \emptyset$, which is dominated only by e. Moreover, we have, for $i \in \{1, 2\}$:

$$\forall v \in V_i \left((v x_i \in S') \iff (\forall u \in V_{3-i}(vu \notin S' \cap E))\right)$$

since S' is minimal and $\{x_1 y_1, x_2 y_2\} \subseteq U'$. We now show how to safely modify S' such that $S' \cap E = \emptyset$. If it is not already the case, there is some edge, w.l.o.g., $e = uv \in S' \cap E$ with $u \in V_1$ and $v \in V_2$. In particular from the above observations, we deduce $u \notin U$, $v \notin U$ and $S_e \subseteq E$. Modify S' by the following procedure.

- If the private solution set $S_e \setminus \{e\}$ contains some edges incident to u and some edges incident to v, then $e \in S'$ will be replaced by $u x_1$ and $v x_2$;
- if every edge in the private solution S_e is adjacent to u, replace e in S' by $u x_1$, otherwise if every edge in the private solution S_e is adjacent to v, replace e in S' by $v x_2$.

The case distinction is necessary to guarantee that S' stays a minimal edge dominating set after each modification step. We repeat this procedure until $S' \cap E = \emptyset$. At the end of the process, every vertex $v \in V$ covers the same set of edges as vx_1 or vx_2 dominates. Hence, by setting $S = \{v \in V : vx_1 \in S' \text{ or } vx_2 \in S'\}$, we build a minimal vertex cover of G containing U.   □

Remark 17. Note that the procedure of local modifications given in Theorem 16 does not preserve optimality, but only inclusion-wise minimality.

6   Price of Extension

Considering the possibility that some set U might not be extensible to any minimal solution, one might ask how far U is from an extensible set. This concept, introduced in [11], is called *Price of Extension (PoE)*. A similar approach has already been studied in the past called *the Price of Connectivity* in [10] in the context of connectivity. This notion has been introduced in [10] for MIN VC which is defined as the maximum ratio between the connected vertex cover number and the vertex cover number. Here, the goal of studying PoE is to measure how far efficiently computable extensible subsets of the given presolution U are to U or to the largest possible extensible subsets of U. To formalize this, we define optimization problems corresponding to EXT r-EC and EXT r-EDS. Actually, since we mainly propose negative results, we only focus on $r = 1$ considering the problems:

MAX EXT EC
Input: A connected graph $G = (V, E)$ and a set of edges $U \subseteq E$.
Solution: Minimal edge cover S of G.
Output: Maximize $|S \cap U|$.

MAX EXT EDS
Input: A graph $G = (V, E)$ and a set of edges $U \subseteq E$.
Solution: Minimal edge dominating set S of G.
Output: Maximize $|S \cap U|$.

For $\Pi = $ MAX EXT EC or $\Pi = $ MAX EXT EDS, we denote the value of an optimal solution by $opt_\Pi(G, U)$. Since for all of them $opt_\Pi(G, U) \leq |U|$ with equality iff (G, U) is a yes-instance of the extension variant, we deduce from our previous results that MAX EXT EC and MAX EXT EDS are NP-hard. In the particular case $U = E$, MAX EXT EDS is exactly the problem called UPPER EDS where the goal is to find the largest minimal edge dominating set; UPPER EDS can be also viewed as UPPER DS in line graphs. In [25], it is shown that UPPER EDS is NP-hard in bipartite graphs. Very recently, an NP-hardness proof for planar graphs of bounded degree, an **APX**-completeness for graphs of max degree 6 and a tight $\Omega\left(n^{\varepsilon - 1/2}\right)$-inapproximation for general graphs and for any constant $\varepsilon \in (0, \frac{1}{2})$, are given in [18].

The price of extension PoE is defined exactly as the ratio of approximation, i.e., $\frac{apx}{opt}$. We say that Π admits a polynomial ρ-PoE if for every instance (G, U), we can compute a solution S of G in polynomial time which satisfies $\text{PoE}(S) \geq \rho$.

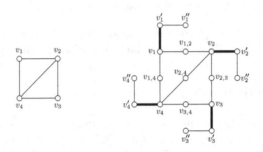

Fig. 4. On the left side, an instance of MAX IS and on the right side, the corresponding instance of MAX EXT EC. Bold edges of H are the set of forced edges U.

Theorem 18. *For any constant $\varepsilon \in (0, \frac{1}{2})$ and any $\rho \in \Omega\left(\Delta^{\varepsilon-1}\right)$ and $\rho \in \Omega\left(n^{\varepsilon-\frac{1}{2}}\right)$, MAX EXT EC does not admit a polynomial ρ-PoE for general graphs of n vertices and maximum degree Δ, unless $\mathsf{P} = \mathsf{NP}$.*

Proof. The proof is based on a reduction from the maximum independent set problem (MAX IS for short). Given a graph $G = (V, E)$ with n vertices and m edges where $V = \{v_1, \ldots, v_n\}$, as an instance of MAX IS, we build a connected bipartite graph $H = (V_H, E_H)$ as follows: for each $v_i \in V$, add a P_3 with edge set $\{v_i v_i', v_i' v_i''\}$, and for each edge $e = v_i v_j \in E$ with $i < j$, add a middle vertex $v_{i,j}$ and connect v_i to v_j via $v_{i,j}$. Consider $I = (H, U)$ as instance of MAX EXT EC, where the forced edge subset is given by $U = \{v_i v_i' : 1 \leq i \leq n\}$. Clearly, H is a bipartite graph with $|V_H| = 3n + m$ vertices, $|E_H| = 2(m + n)$ edges and $\Delta(H) = \Delta(G) + 1$. An example of this construction is illustrated in Fig. 4.

We claim that there is a solution of size k for MAX EXT EC on (H, U) iff G has an independent set of size k. Suppose that S is a maximal independent set of G of size k. For each $e \in E$, let $v^e \in V \setminus S$ be a vertex which covers e. Clearly, $S' = \{v_{i,j} v^e : e = v_i v_j \in E\} \cup \{v_i' v_i'' : v_i \in V\} \cup \{v_i v_i' : v_i \in S\}$ is a minimal edge cover of H with $|S' \cap U| = k$. Conversely, suppose S' is a minimal edge cover of H such that $|S' \cap U| = k$. $\{v_i' v_i'' : v_i \in V\}$ is a part of every edge cover since, v_i'' for $v_i \in V$ are leaves of H. Moreover, for each $e = v_i v_j \in E$ with $i < j$, at least one edge between $v_i v_{i,j}$ or $v_j v_{i,j}$ belongs to any edge cover of H. Furthermore, if $v_i v_{i,j} \in S$, by minimality we deduce that $v_i v_i' \notin S'$. Hence, for each $v_i v_j \in E$, at most one of $v_i v_i', v_j v_j'$ can be in S'. Hence, $S = \{v_i : v_i v_i' \in S'\}$ is an independent set of G with size k.

Using the strong inapproximability results for MAX IS given in [28,32], observing $\Delta(H) = \Delta(G) + 1$ and $|V_H| \leq 2|V|^2$, we obtain the claimed result. $\square$

Using result given in [18], an $\Omega\left(n^{\varepsilon-1/2}\right)$-inapproximation can be immediately deduced for MAX EXT EDS. The next result is obtained by a simple approximation preserving reduction from MAX EXT VC to MAX EXT EDS.

Theorem 19. $(*)$ *For any constant* $\varepsilon \in (0,1)$ *and any* $\rho \in \Omega\left(n^{\varepsilon-1}\right)$, MAX EXT EDS *does not admit a polynomial* ρ-PoE *for general graphs of* n *vertices, unless* P = NP.

In contrast to the last hardness result, we give a simple approximation depending on the maximum degree $\Delta(G)$.

Theorem 20. MAX EXT EDS *is* $\frac{1}{\Delta(G_U)+1}$-*approximable for instance* (G, U) *of maximum degree* Δ.

Proof. Let $(G = (V, E), U)$ be an instance of MAX EXT EDS, where the maximum degree of partial subgraph G_U induced by U is bounded by Δ. Compute a maximum matching M of G_U and transform it into a maximal matching M' of G containing M. It is well known that any maximal matching is an edge dominating set. Obviously, $(\Delta(G_U) + 1)|M| \geq |U| \geq opt_{\text{MAX EXT EDS}}(G, U)$ since G_U is $(\Delta(G_U) + 1)$-edge colorable. $\square$

Considering EXT 1-DCPS, we need to adapt the notion of the price of extension because we have to consider subset of forbidden edges (i.e., $\overline{U}$); more precisely, we want to increase $|U|$ as few as possible. Hence, the optimization problem called MIN EXT 1-DCPS is defined as follows:

MIN EXT 1-DCPS
Input: A graph $G = (V, E)$ and a set of edges $U \subseteq E$.
Solution: Maximal matching S of G.
Output: Minimize $|U \cup S|$.

Recall that PoE is meant to measure how far efficiently computable extensible subsets are from the given presolution U or to the largest possible extensible subsets of U. We say that MIN EXT 1-DCPS admits a polynomial ρ-PoE if for every instance (G, U), we can compute a solution S of G in polynomial time which satisfies $\text{PoE}(S) = \frac{apx}{opt} \leq \rho$. In the particular case $U = \emptyset$, MIN EXT 1-DCPS is exactly the well known problem MINIMUM MAXIMAL MATCHING where the goal is to find the smallest maximal matching. In [14,15], it is shown that MINIMUM MAXIMAL MATCHING is hard to approximate with a factor better than 2 and 1.18, assuming Unique Games Conjecture (UGC) and P $\neq$ NP, respectively. We complement this bound by showing the following.

Theorem 21. $(*)$ *A 2-approximation for* MIN EXT 1-DCPS *can be computed in polynomial time.*

7 Conclusions

We have undertaken some study on several complexity aspects of extension variants of edge graph problems. Our results should be useful in particular to the (input-sensitive) enumeration algorithms community that has so far not put that much attention on edge graph problems; we are only aware of [21] in this direction. Conversely, output-sensitive enumeration algorithms, e.g., for matchings have been around for more than twenty years [29]. Some thoughts on edge cover enumeration can be found in [30]. Our research might also inspire to revisit exact and/or parameterized algorithms on EDGE DOMINATION; previous papers like [16] or [26] did not focus on special graph classes, where we see some potentials for future research.

References

1. Bazgan, C., Brankovic, L., Casel, K., Fernau, H.: On the complexity landscape of the domination chain. In: Govindarajan, S., Maheshwari, A. (eds.) CALDAM 2016. LNCS, vol. 9602, pp. 61–72. Springer, Cham (2016). https://doi.org/10.1007/978-3-319-29221-2_6
2. Bazgan, C., et al.: The many facets of upper domination. Theor. Comput. Sci. **717**, 2–25 (2018)
3. Berger, A., Fukunaga, T., Nagamochi, H., Parekh, O.: Approximability of the capacitated b-edge dominating set problem. Theor. Comput. Sci. **385**(1–3), 202–213 (2007)
4. Berger, A., Parekh, O.: Linear time algorithms for generalized edge dominating set problems. Algorithmica **50**(2), 244–254 (2008)
5. Berman, P., Karpinski, M., Scott, A.D.: Approximation hardness of short symmetric instances of MAX-3SAT. ECCC (049) (2003)
6. Bertossi, A.A.: Dominating sets for split and bipartite graphs. Inf. Process. Lett. **19**(1), 37–40 (1984)
7. Biró, M., Hujter, M., Tuza, Z.: Precoloring extension. I. Interval graphs. Disc. Math. **100**(1–3), 267–279 (1992)
8. Bonamy, M., Defrain, O., Heinrich, M., Raymond, J.-F.: Enumerating minimal dominating sets in triangle-free graphs. In: Niedermeier, R., Paul, C. (eds.) STACS, Dagstuhl, Germany. Leibniz International Proceedings in Informatics (LIPIcs), vol. 126, pp. 16:1–16:12. Schloss Dagstuhl-Leibniz-Zentrum fuer Informatik (2019)
9. Boros, E., Gurvich, V., Hammer, P.L.: Dual subimplicants of positive Boolean functions. Optim. Meth. Softw. **10**(2), 147–156 (1998)
10. Cardinal, J., Levy, E.: Connected vertex covers in dense graphs. Theor. Comput. Sci. **411**(26–28), 2581–2590 (2010)
11. Casel, K., Fernau, H., Khosravian Ghadikoei, M., Monnot, J., Sikora, F.: On the complexity of solution extension of optimization problems. CoRR, abs/1810.04553 (2018)
12. Casel, K., Fernau, H., Ghadikoalei, M.K., Monnot, J., Sikora, F.: Extension of vertex cover and independent set in some classes of graphs. In: Heggernes, P. (ed.) CIAC 2019. LNCS, vol. 11485, pp. 124–136. Springer, Cham (2019). https://doi.org/10.1007/978-3-030-17402-6_11

13. Colbourn, C.J.: The complexity of completing partial Latin squares. Disc. Appl. Math. **8**(1), 25–30 (1984)
14. Dudycz, S., Lewandowski, M., Marcinkowski, J.: Tight approximation ratio for minimum maximal matching. In: Lodi, A., Nagarajan, V. (eds.) IPCO 2019. LNCS, vol. 11480, pp. 181–193. Springer, Cham (2019). https://doi.org/10.1007/978-3-030-17953-3_14
15. Escoffier, B., Monnot, J., Paschos, V.T., Xiao, M.: New results on polynomial inapproximabilityand fixed parameter approximability of edge dominating set. Theory Comput. Syst. **56**(2), 330–346 (2015)
16. Fernau, H.: EDGE DOMINATING SET: efficient enumeration-based exact algorithms. In: Bodlaender, H.L., Langston, M.A. (eds.) IWPEC 2006. LNCS, vol. 4169, pp. 142–153. Springer, Heidelberg (2006). https://doi.org/10.1007/11847250_13
17. Fernau, H., Hoffmann, S.: Extensions to minimal synchronizing words. J. Autom. Lang. Comb. **24** (2019)
18. Fernau, H., Manlove, D.F., Monnot, J.: Algorithmic study of upper edge dominating set (2019, manuscript)
19. Gabow, H.N.: An efficient reduction technique for degree-constrained subgraph and bidirected network flow problems. In: Johnson, D.S., et al. (eds.) STOC, pp. 448–456. ACM (1983)
20. Garey, M.R., Johnson, D.S.: Computers and Intractability: A Guide to the Theory of NP-Completeness. W. H. Freeman & Co. (1979)
21. Golovach, P.A., Heggernes, P., Kratsch, D., Vilnger, Y.: An incremental polynomial time algorithm to enumerate all minimal edge dominating sets. Algorithmica **72**(3), 836–859 (2015)
22. Kanté, M.M., Limouzy, V., Mary, A., Nourine, L.: On the neighbourhood helly of some graph classes and applications to the enumeration of minimal dominating sets. In: Chao, K.-M., Hsu, T., Lee, D.-T. (eds.) ISAAC 2012. LNCS, vol. 7676, pp. 289–298. Springer, Heidelberg (2012). https://doi.org/10.1007/978-3-642-35261-4_32
23. Kratochvíl, J.: A special planar satisfiability problem and a consequence of its NP-completeness. Discrete Appl. Math. **52**, 233–252 (1994)
24. Lawler, E.L., Lenstra, J.K., Rinnooy Kan, A.H.G.: Generating all maximal independent sets: NP-hardness and polynomial-time algorithms. SIAM J. Comput. **9**, 558–565 (1980)
25. McRae, A.A.: Generalizing NP-completeness proofs for bipartite and chordal graphs. Ph.D. thesis, Clemson University, Department of Computer Science, South Carolina (1994)
26. van Rooij, J.M.M., Bodlaender, H.L.: Exact algorithms for edge domination. Algorithmica **64**(4), 535–563 (2012)
27. Schrijver, A.: Combinatorial Optimization: Polyhedra and Efficiency. Springer, Heidelberg (2003)
28. Trevisan, L.: Non-approximability results for optimization problems on bounded degree instances. In: Vitter, J.S., Spirakis, P.G., Yannakakis, M. (eds.) STOC, pp. 453–461. ACM (2001)
29. Uno, T.: Algorithms for enumerating all perfect, maximum and maximal matchings in bipartite graphs. In: Leong, H.W., Imai, H., Jain, S. (eds.) ISAAC 1997. LNCS, vol. 1350, pp. 92–101. Springer, Heidelberg (1997). https://doi.org/10.1007/3-540-63890-3_11

30. Wang, J., Chen, B., Feng, Q., Chen, J.: An efficient fixed-parameter enumeration algorithm for weighted edge dominating set. In: Deng, X., Hopcroft, J.E., Xue, J. (eds.) FAW 2009. LNCS, vol. 5598, pp. 237–250. Springer, Heidelberg (2009). https://doi.org/10.1007/978-3-642-02270-8_25
31. Yannakakis, M., Gavril, F.: Edge dominating sets in graphs. SIAM J. Appl. Math. **38**(3), 364–372 (1980)
32. Zuckerman, D.: Linear degree extractors and the inapproximability of max clique and chromatic number. Theory Comput. **3**(1), 103–128 (2007)

Space Efficient Algorithms
for Breadth-Depth Search

Sankardeep Chakraborty[1], Anish Mukherjee[2], and Srinivasa Rao Satti[3(✉)]

[1] RIKEN Center for Advanced Intelligence Project,
1-4-1 Nihonbashi, Chuo-ku, Tokyo, Japan
sankar.chakraborty@riken.jp
[2] Chennai Mathematical Institute,
H1 SIPCOT IT Park, Siruseri, Chennai, India
anish@cmi.ac.in
[3] Seoul National University,
1 Gwanak-ro, Gwanak-gu, Seoul, South Korea
ssrao@cse.snu.ac.kr

Abstract. Continuing the recent trend, in this article we design several space-efficient algorithms for two well-known graph search methods. Both these search methods share the same name *breadth-depth search* (henceforth BDS), although they work entirely in different fashion. The classical implementation for these graph search methods takes $O(m+n)$ time and $O(n \lg n)$ bits of space in the standard word RAM model (with word size being $\Theta(\lg n)$ bits), where m and n denotes the number of edges and vertices of the input graph respectively. Our goal here is to beat the space bound of the classical implementations, and design $o(n \lg n)$ space algorithms for these search methods by paying little to no penalty in the running time. Note that our space bounds (i.e., with $o(n \lg n)$ bits of space) do not even allow us to explicitly store the required information to implement the classical algorithms, yet our algorithms visits and reports all the vertices of the input graph in correct order.

1 Introduction

Graph searching is an efficient and widely used bookkeeping method for exploring the vertices and edges of a graph. Given a graph G, a typical graph search method starts with an arbitrary vertex v in G, marks v as visited, and systematically explores other unvisited vertices of G by iteratively traversing the edges incident with a previously visited vertex. The ordering in which the next vertex is chosen from an already visited vertex yields different vertex orderings of the graph. Two of the most popular and widely used graph search methods are *depth-first search* (DFS) and *breadth-first search* (BFS). BFS tries to explore an untraversed edge incident with the least recently visited vertex, whereas DFS tries to explore an untraversed edge with the most recently visited vertex. Both of

This work was partially supported by JST CREST Grant Number JPMJCR1402.

L. A. Gąsieniec et al. (Eds.): FCT 2019, LNCS 11651, pp. 201–212, 2019.
https://doi.org/10.1007/978-3-030-25027-0_14

these search methods have been successfully employed as backbones for designing other powerful and efficient graph algorithms. Researchers have also devised other graph search methods [9], and explored their properties to design efficient graph algorithms. For example, Tarjan and Yannakakis [17] introduced a graph search method, called *maximum cardinality search* (MCS) and used it to design a linear time algorithm for chordal graph recognition and other related problems.

Our focus here is to study another graph search method, namely *breadth-depth search* (BDS) from the point of view of making it space efficient. We note that, two very different graph search strategies exist in the literature, but surprisingly, under the same name. Historically, Horowitz and Sahni [13], in 1984, defined BDS and demonstrated its applications to branch-and-bound strategies. Henceforth we will refer to this version of BDS as BDS_{hs} after Horowitz and Sahni. Greenlaw, in his 1993 paper [12], proved that BDS_{hs} is inherently sequential by showing it is P-complete. Almost a decade later, Jiang [14], in 1993, defined another graph search method, under same name BDS, while designing an I/O- and CPU-optimal algorithm for decomposing a directed graph into its strongly connected components (SCC). In particular, he devised and used BDS (note that, this is different from BDS_{hs} [13] as we will see shortly) to give an alternate algorithm for SCC recognition. We will refer to this version of BDS as BDS_j after Jiang. Implementing either of these algorithms takes $O(m+n)$ time and $O(n \lg n)$ bits of space in the standard word RAM model, where m and n denotes the number of edges and vertices of the input graph respectively. Our goal in this paper is to improve the space bound of the classical implementations without sacrificing too much in the running time.

1.1   Motivation and Related Work

Recently, designing space efficient algorithms has become enormously important due to their applications in the presence of fast growth of "big data" and the escalation of specialized handheld mobile devices and embedded systems that have a limited supply of memory i.e., devices like Rasberry Pi which has a huge use case in IoT related applications. Even if these mobile devices and embedded systems are designed with large supply of memory, it might be useful to restrict the number of write operations. For example, on flash memory, writing is a costly operation in terms of speed, and it also reduces the reliability and longevity of the memory. Keeping all these constraints in mind, it makes sense to consider algorithms that do not modify the input and use only a limited amount of work space. One computational model that has been proposed in algorithmic literature to study space efficient algorithms, is the read-only memory (ROM) model. Here we focus on space efficient implementations of BDS in such settings.

Starting with the paper of Asano et al. [1] who showed how one can implement DFS using $O(n)$ bits in ROM, improving on the naive $O(n \lg n)$-bit implementation, the recent series of papers [2,4,6–8,11] presented such space-efficient algorithms for a variety of other basic and fundamental graph problems: namely BFS, maximum cardinality search, topological sort, connected components, minimum spanning tree, shortest path, dynamic DFS, recognition of outerplanar graph and chordal graphs among others. We add to this small yet rapidly growing body of space-efficient algorithm design literature by providing such algorithms for both the BDS algorithms, BDS_{hs} and BDS_j. In this process, we also want to draw attention to the fact that, even though these two search methods have same name, they work essentially in different manner. To the best of our knowledge, surprisingly this fact does not seem to be mentioned anywhere in the literature.

We conclude this section by briefly mentioning some very recent works on designing space efficient algorithms for various other algorithmic problems: Longest increasing subsequence [15], geometric computation [3] among many others.

1.2   Model of Computation and Input Representation

Like all the recent research that focused on designing space-efficient graph algorithms (as in [1,2,6–8,15,16]), here also we assume the standard word RAM model for the working memory with words size $w = \Theta(\lg n)$ bits where constant time operations can be supported on $\Theta(\lg n)$-bit words, and the input graph G is given in a read-only memory with a limited read-write working memory, and write-only output. We count space in terms of the number of bits in workspace used by the algorithms. Throughout this paper, let $G = (V, E)$ denote a graph on $n = |V|$ vertices and $m = |E|$ edges where $V = \{v_1, v_2, \cdots, v_n\}$. We also assume that G is given in an adjacency array representation, i.e., an array of length $|V|$ where the i-th entry stores a pointer to an array that stores all the neighbors of the i-th vertex. For the directed graphs, we assume that the input representation has both in/out adjacency array for all the vertices.

1.3   Our Main Results and Organization of the Paper

We start off by introducing BDS_j and BDS_{hs} in Sects. 2 and 3 respectively as defined in [13] and [14] along with presenting their various space efficient implementations before concluding in Sect. 4 with some concluding remarks and future directions. Our main results can be compactly summarized as follows.

Theorem 1. *Given a graph G with n vertices and m edges, the BDS_j and BDS_{hs} traversals of G can be performed in randomized $O(m \lg^* n)$ time[1] using $O(n)$ bits with high probability $((1-1/n^c)$, for some fixed constant $c)$, or $O(m+n)$ time using $O(n \lg(m/n))$ bits, respectively.*

1.4   Preliminaries

We use the following theorem repeatedly in our algorithms.

Theorem 2 [10]. *Given a universe U of size u, there exists a dynamic dictionary data structure storing a subset $S \subseteq U$ of cardinality at most n using space*

---

[1] We use lg to denote logarithm to the base 2.

$n \lg(u/n) + nr$ *bits where* $r \in O(\lg n)$ *denotes the size of the satellite data attached with elements of* U. *This data structure can support membership, retrieval (of the satellite data), insertion, and deletion of any element along with its satellite data in* $O(1)$ *time with probabality* $(1 - 1/n^c)$, *for some fixed constant* c.

2   Breadth-Depth Search of Jiang

A BDS_j traversal of a graph G walks through the vertices of G and processes each vertex one by one according to the following rule. Suppose the most recently traversed edge is (u, w). If w still has an unvisited edge, then select this edge to traverse. Otherwise choose an unvisited edge incident on the node most recently visited that still has unvisited edges. At this point (see line 7 of the pseudocode for BDS_j provided below) we also say that the node w is being expanded. Note that a vertex v might be visited many times via different edges and here we are only interested in the last visit to the vertex (in contrast to the BFS and DFS where only the first visit to the vertex is considered) when v is expanded.

To implement this, more specifically to capture the fact of last visit, Jiang used an *adaptive stack* (abbreviated as *adp-stack* in the pseudocode below). A stack is called adaptive if pushing a node into the stack removes the older version of the node, if it was present in the stack earlier. We refer to the PUSH operation in an adaptive stack as ADPPUSH in the pseudocode. One way to implement an adaptive stack is via using a doubly linked list L i.e., the algorithm stores the vertices in L along with an array P of n pointers, one for each vertex, pointing to it's location in L. Now to push adaptively a vertex v_i, we first insert v_i into L. Assuming it already belongs to L, go to $P[v_i]$ to update it so that it now points to the new location in L, and delete the older entry from L. Otherwise, $P[v_i]$ is empty, and is now updated to the newly inserted location of v_i. All of these can be done in $O(1)$ time. Popping a vertex v_i is straightforward as we have to delete the node from L and update $P[v_i]$ to NULL. We also maintain in a bitmap of size n, call it *visited*, information regarding whether a vertex v is visited or not. Then using all this auxiliary structure, it is easy to see that BDS_j can be implemented in $O(m + n)$ time using $O(n)$ words or equivalently $O(n \lg n)$ bits of space (becuase of storing the list L and the array P). This concludes a brief description of BDS_j as well as its implementation. Jiang also showed, using BDS_j, how one can perform topological sort and strongly connected component decomposition. For detailed description, readers are referred to his paper [14]. Our focus here is to implement BDS_j space efficiently.

In what follows, we illustrate a bit more on the inner working details of BDS_j with the help of an example. Following the convention, as in the recent papers [1,2], here also in BDS_j we output the vertices as and when they are expanded (note that, if reporting in any other order is required, it can be done so with straighforward modification in our algorithms). Hence the root will be output at the very first step, followed by its rightmost child and so on. Towards designing space efficient algorithms for BDS_j, we first note its similarities with DFS traversal method. Taking the graph G of Fig. 1(a) as running example where

Algorithm 1. $BDS_j(v)$

```
 1: EMPTY(visited); EMPTY(adp-stack);
 2: ADPPUSH(v, adp-stack);
 3: while ISNOEMPTY(adp-stack) do
 4:     w := TOP(adp-stack);
 5:     if w ∉ visited then
 6:         visited := visited ∪{w}
 7:         for all u in adj[w] do
 8:             if u ∉ visited then
 9:                 ADPPUSH(u, adp-stack);
10:             end if
11:         end for
12:     else POP(adp-stack);
13:     end if
14: end while
```

(say) the root is s, and assuming that the adjacency list of every vertex is lexicographically sorted in the order of their labels, DFS would have put s first in the stack, followed by pushing a then d and so on. As a result, these three vertices would come first in the output of DFS and so on. BDS_j works in a slightly different manner. More specifically, BDS_j pushes a, b and c into the stack (with a at the bottom and c at the top), and then expands c (see the pseudocode for BDS_j). The node b will again be discovered while expanding c, and due to the adaptivity of the stack, the older entry of b which was inserted into the stack due to the expansion of s, will be removed (with a new entry of b added to the stack). This phenomenon will be repeated again while expanding g. Eventually b will be discovered from e and expanded. See the final BDS_j tree in Fig. 1(c). To enable expanding a vertex during the last visit (instead of the first visit which is the case for BFS and DFS), Jiang used the adaptive stack. As analyzed previously, the bottleneck factor in the space consumption of BDS_j is the adaptive stack. Our main observation is that we can get rid of the adaptive stack and still perform BDS_j traversal of the graph G correctly. More specifically, in what follows we describe how to implement BDS_j space efficiently using a standard stack (without the adaptive push operation), along with some bookkeeping, yet producing the same vertex ordering as Jiang's BDS_j.

2.1   Using $O(n)$ Bits and $O(m \lg N)$ Time

Note that, a vertex v could be in one of the three states during the entire execution of the algorithm, (i) unvisited, (ii) expanded but not all of its children are expanded yet, and (iii) completed i.e., it is expanded as well as all of its children, if any. In our space efficient implementation of BDS_j, we denote them by color white, grey and black respectively, and store this information using an array (say) C of size $O(n)$ bits. Along with this, we also store the last $O(n/\lg n)$ vertices that are grey in a (normal i.e., not adaptive) stack S. We divide the stack S into blocks of size $O(n/\lg n)$ vertices where the first block refers to the

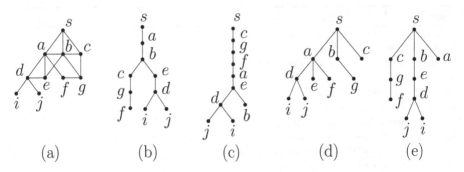

Fig. 1. (a) Graph G. We output the vertices when they are visited (for DFS and BFS) or expanded (for BDS) for the first time in any graph search method. The adjacency lists are assumed to be ordered in the sorted order of their labels. (b) DFS tree of G and the resulting output for this DFS traversal is $s, a, b, c, g, f, e, d, i, j$. (c) BDS_j tree of G and the resulting output for this BDS_j traversal is $s, c, g, f, a, e, d, j, i, b$. (d) BFS tree of G and the resulting output for this BFS traversal is $s, a, b, c, d, e, f, g, i, j$. (e) BDS_{hs} tree of G and the resulting output for this BDS_{hs} traversal is $s, c, g, f, b, e, d, j, i, a$.

first set of $O(n/\lg n)$ vertex labels that are pushed into S, the second block refers to the second bunch of $O(n/\lg n)$ vertex labels pushed into S during BDS_j and so on. Thus, there are $O(\lg n)$ blocks in total, and we always store the last block. Moreover, for every block we store the first and last element that are pushed in S in a separate smaller sized stack T. Thus, we need overall $O(n)$ bits.

Now armed with these data structures, we start by marking the root s as grey and pushing in S. Note that, as and when a vertex v gets expanded, i.e., turns grey, we can also output v (i.e., report v as the next vertex in BDS_j order). At the next step, instead of inserting all of s's white neighbors as in Jiang's BDS_j implementation, we insert only the rightmost white neighbor c into the stack, and change its color (from white) to grey (see Fig. 1(c)). Observe crucially that by delaying the insertion of other white neighbors at once, we are essentially removing the need of adaptivity from the stack as now elements are pushed only when they are expanded, not multiple times as in BDS_j. Thus, we scan c's adjacency list from right to left and insert the first white neighbor into the stack, mark it as grey in the C array, and continue. We call this phase of the algorithm as *forward step* i.e., the phase in which we discover new vertices of the graph and insert them in S. At some point during the execution of the algorithm, when we arrive at a vertex v such that none of v's neighbors are white, then we color the vertex v as black, and we pop it from the stack. If the stack is still non-empty, then the parent of v (in the BDS_j tree) would be at the top of the stack, and we continue the BDS_j from this vertex. On the other hand, if the stack becomes empty after removing v, we need to reconstruct it to the state such that it holds the last $O(n/\lg n)$ grey vertices after all the pops done so far. We refer to the following phase of the algorithm as *reconstruction step*. For this, we basically repeat the same algorithm but with one twist which also enables us now to skip some of the vertices during this reconstruction phase. In detail, we again start with an empty stack, insert the root s first and scan its adjacency list from the rightmost entry to skip all the black vertices and insert into the stack the rightmost grey vertex. Then the repeat the same for this newly inserted vertex into the stack until we reconstruct the last $O(n/\lg n)$ grey vertices. As we have stored the first and last vertices of each of the blocks in T, we know when to stop this reconstruction procedure. Another equivalent way to achieve the same effect is to recolor all the grey vertices back to white, while retaining the colors of all the other (black and white) vertices, and repeat the same algorithm. It is not hard to see that this procedure correctly reconstructs the latest set of grey vertices in the stack S. We continue this process until all the vertices become black. Obviously this procedure takes $O(n)$ bits of space. To bound the running time, note that, whenever this procedure tries to reconstruct, $O(n/\lg n)$ vertices have changed their colors to black, and they are not going to be inserted again into the stack. As this can happen only for $O(\lg n)$ rounds, and since in each round we might spend $O(m)$ time to scan the adjacency list and insert correct vertices into the stack, overall this procedure takes $O(m \lg n)$ time. We conclude this section by mentioning that a similar kind of idea was used in [1] to provide space efficient DFS implementation, but we emphasize that ours algorithm is markedly different than [1] from the point of view of introducing *delayed insertion* of vertices into the stack, and thus removing the adaptivity from the stack, both the features not present in DFS. In what follows, we describe an improved algorithm generalizing the ideas developed in this section.

2.2   Using $O(n)$ Bits and $O(m \lg^* N)$ Time

In this section we first describe an algorithm that uses $O(n \lg \lg \lg n)$ bits to performs BDS_j in $O(m + n)$ time with high probability, and modify it later to get an even improved algorithm. To obtain this, we first divide the stack S into $O(\lg n/\lg \lg \lg n)$ blocks of size $n \lg \lg \lg n/\lg n$ vertices each. We group $(\lg n/\lg \lg n)$ blocks into a super-block; thus there are $O(\lg \lg n/\lg \lg \lg n)$ super-blocks, each having $O(n \lg \lg \lg n/\lg \lg n)$ vertices. For each vertex v, we store its (a) color, (b) super-block ID (SID), if it is in S, (and -1 if it is not added to S yet, i.e., if it is white), and (c) the number of groups of m/n vertices that have been explored with v as the current vertex. We also keep track of the first and the last element of each block, as well as super-block, and these takes up negligible (poly-logarithmic) space. We describe the algorithm below in detail.

The algorithm is similar to the BDS_j algorithm of Sect. 2.1 with the following changes. The *forward step* remains mostly the same except updating the Items (b) and (c) above after every insertion of a vertex into the stack S. More specifically, whenever a vertex is inserted into the stack, we store its SID in an array (Item (b) above), and also update the information regarding Item (c) above (also stored in a separate array). In addition, we store the nodes in the topmost two blocks of the top super-block of the stack. We also maintain the block IDs (BIDs) of all the vertices belonging to the topmost two super-blocks using the dictionary structure of Theorem 2.

The *reconstruction step* changes significantly as we cannot really afford $O(m)$ time for the reconstruction of each super-block (like in Sect. 2.1); rather we would ideally like to spend time proportional to the size of the super-block, hence resulting in an optimal linear time algorithm. In order to achieve this, we do the following. As we have stored the first element of all the super-blocks, we can start by pushing that element (say v) into a temporary stack. We obtain the next vertex by determining (by consulting Item (c) above) the first grey vertex, say u, that belongs to this super-block (as we can check from its SID) from the right endpoint of v's adjacency array, and that is not already inserted in the current reconstruction procedure (can be checked from the dictionary structure of Theorem 2). Now we repeat the same in u's list until we reconstruct the whole super-block. Note that, simultaneously we are also inserting the BIDs for every vertex in the structure of Theorem 2. We should mention one point at this time, the necessity of dynamic dictionary comes from the fact that we need to quickly find the BID information associated with the vertices in order to decide whether to insert any particular vertex in the stack or not. For performing this task very efficiently both time and space wise, having a simple array is not enough and thus, the requirement of more powerful dynamic dictionary structure. Due to the space limitations, we may need to discard all other blocks inside a super-block except the topmost two. Once we reconstruct the required blocks, the algorithm can proceed normally. Now all that is left is to determine the time and the space complexity of this procedure. Space requirement of our algorithm is $O(n \lg \lg \lg n)$ bits which is dominated by the SID, topmost two blocks inside the top super-block and the dictionary structure.

To bound the number of reconstructions, note that, each time we reconstruct a super-block, the previous super-block's $O(n \lg \lg \lg n / \lg \lg n)$ vertices change their color to black and get popped from the stack, hence they will never be pushed again. Thus, the number of restorations (denoted by q) is bounded by $O(\lg \lg n / \lg \lg \lg n)$. Now if the degree of a vertex v is v_d, then we spend $O(min\{v_d, m/n\})$ time on v searching for its correct neighbor in our algorithm due to the information stored in Item (c) above. To bound the running time of the algorithm, note that over q reconstructions and over all vertices of degree at most m/nq, we spend $O(qn(m/nq)) = O(m)$ time, and for vertices having degree larger than m/nq, over q such reconstructions, we spend $O(q(n/q)(m/n)) = O(m)$ time. Observe that, this running time is randomized linear because of the use of dynamic dictionary[2] of Theorem 2. This concludes the description of the BDS_j algorithm taking randomized $O(m + n)$ time and using $O(n \lg \lg \lg n)$ bits with high probabality $(1 - 1/n^c)$ for some constant c.

Before generalizing this algorithm, let us define some notations that are going to be used in what follows. The function $\lg^{(k)} n$ is defined as applying

---

[2] Our algorithm performs atmost $O(m + n)$ insertion/deletion/retrieval during its entire execution using the dictionary of Theorem 2 which takes $O(1)$ time with a probability of $(1 - 1/n^c)$ (where $c \geq 3$) for each insertion/deletion/retrieval. Thus, the probability that our algorithm takes more than $O(m + n)$ time is $(1/n^{c-2})$ by union bound rule.

the logarithm function on n repeatedly k times i.e., $\lg\lg\ldots$ *(k times)* $\ldots\lg n$. Similarly $\lg^* n$ (also known as *iterated logarithm*) is the number of times the logarithm function is iteratively applied till the result is less than or equal to 2. It's easy to see that $\lg^{(\lg^* n)} n$ is always a constant for any n. Like the previous algorithm, this algorithm also uses the data structures of Item (a), (b), and (c) along with a hierarchy of levels (instead of just two levels like the previous algorithm). For some k (which we will fix later), we set the size of k-th level blocks as $O(n/(\lg^{(k)} n)^2)$, and we divide the k-th level blocks into $(k+1)$-th level blocks. Thus, the number of k-th level blocks inside a $(k+1)$-th level block is $O(\{(\lg^{(k)} n)/(\lg^{(k+1)} n)\}^2)$, where $k = 1$ means the smallest level blocks. We store the dynamic dictionary for k-th level at the $(k+1)$-th level for every k, and the space required for storing the dictionary at level k is given by $O((n/(\lg^{(k+1)} n)^2)(\lg\{(\lg^{(k)} n)/(\lg^{(k+1)} n)\}^2)) = o(n)$ bits. Through the entire execution of the algorithm, we always maintain the topmost two smallest level blocks along with other data structures. The *forward step* as well as the *reconstruction step* of the algorithm remains exactly the same other than modifying/storing informations at each level of the data structures suitably. As the work involved at each such level is simply one of the four operations from {insertion/deletion/membership/retrieval} (all takes $O(1)$ time with high probability) at the dynamic dictionaries of the corresponding levels, by similar analysis as before, the final running time of the algorithm simply becomes m times the overall number of levels of data structure that we maintain during the execution of the algorithm, and this can be bounded by $O(mk)$. Also, we can bound the overall space requirement as $O(n\lg^{(k+1)} n) + o(n)$ bits. Now choosing $k + 1 = \lg^* n$, our algorithm takes $O(n)$ bits of space and $O(m\lg^* n)$ running time, and this concludes the description of the algorithm.

In what follows, we specially focus on designing space efficient algorithms for BDS_j when the input graph is sparse (i.e., $m = O(n)$). Studying such graphs is very important not only from theoretical perspective but also from practical point of view. These graphs appear very frequently in most of the realistic network scenario, like Road networks and the Internet, in real world applications.

2.3   Using $O(n\lg(m/n))$ Bits and $O(m+n)$ Time

In this section, we show how one can obtain linear bits and linear time algorithm for BDS_j for sparse graphs. For this we use the following lemma from [2].

Lemma 1 ([2]). *Given the adjacency array representation of a graph G, using $O(m)$ time, one can construct an auxiliary structure of size $O(n\lg(m/n))$ bits that can store a "pointer" into an arbitrary position within the adjacency array of each vertex. Also, updating any of these pointers takes $O(1)$ time.*

The idea is to store *parent* pointers into the adjacency array of each vertex using the representation of Lemma 1. More specifically, for an undirected graph, whenever the BDS_j expands a vertex u to reach v following the edge (u,v), u becomes the parent of v in the BDS_j tree, and at that time, we scan the adjacency array of v to find u and store a pointer to that position (within the adjacency array of v). For every vertex v in G, we can also store another pointer marking how far in v's list BDS_j has already processed. This pointer will start from the very end of every list, gradually moves towards the left, and at the end of the algorithm, will point to the first vertex of list. We also maintain color information in a bitmap of size $O(n)$ bits. Given this pointer representation, it is easy to see how to implement BDS_j in $O(m+n)$ time. The main advantage of this algorithm of ours is, note that, we don't even need to maintain any explicit stack to implement this process. We can extend similar idea for doing BDS_j in directed graphs by setting up parent pointers (which are used during backtracking) in the in-adjacency list of every vertex and use the other pointer to mark progress in the out-adjacency list. With this, we complete the proof of Theorem 1 for BDS_j.

3   Breadth-Depth Search of Horowitz and Sahni

This version of BDS works as follows. The algorithm starts by pushing the root (i.e., the starting vertex) into the stack S initially. At every subsequent step, the algorithm pops the topmost vertex v of S, and pushes all its unvisited neighbors into S. See Fig. 1(e) for an example. Note crucially that, due to the popping of the parent while pushing the children in S, during backtracking the next vertex to be expanded is always at the top of the stack S. This stack S could grow to contain $O(n)$ vertices, thus the classical implementation of this procedure takes $O(m+n)$ time and $O(n \lg n)$ bits of space. See [13] for a detailed description. In what follows, we show how to implement this BDS_{hs} space efficiently.

3.1   Using $O(n)$ Bits and $O(m \lg^* N)$ Time

To implement BDS_{hs} using $O(n)$ bits, we crucially change the way we handle the stack during the execution of the algorithm. More specifically, we will not pop immediately the vertex v which is going to be expanded at the very next step (as done in [13]), rather keep it in the stack S instead for later use. We refer to this technique as the *delayed removal* of the vertices. Even though this is different than the *delayed insertion* technique (which was crucially used for BDS_j's implementation), it is worth emphasizing that by introducing delayed removal of the vertices, the behaviour/operation of the stack in BDS_{hs} becomes pretty similar to the one in BDS_j (as it will be clear from the next paragraph), thus we can reuse previously developed ideas for BDS_j to obtain space efficient implementation of BDS_{hs}. In addition to this change, we use three colors as we did in the previous BDS_j implementation with the exact same meaning attached to them, and store this information in an array C. Also, we always store the last block of $O(n/\lg n)$ grey vertices of S.

   In detail, we start by marking the root, say s, as visited, coloring it grey and inserting it into S. This is followed by inserting all of s's unvisited white neighbors into S, change them to grey in C. Now s's rightmost child (say v) is at the top of the stack and we insert in S all of v's white neighbors without popping v, also simultaneously marking them visited, and coloring v as grey. This process is repeated until we arrive at the vertex u all of whose neighbors are visited; at this point we make u to be black and pop it from the stack. The vertex which is below u in the stack (say p) is either its parent (if u is the first child of its parent) or its previous sibling. We actually don't know which case it is, but it does not matter – we simply continue the search from p. The case when p is the previous sibling of u is handled the same way by the original algorithm as well as ours. In the case when p is the parent of u, all the other children of p are colored black (since u is the first child of p), and hence our algorithm colors p as black and pops it from S. Reconstructions are also handled in a similar fashion as in Sect. 2.1. I.e., we recolor the grey vertices back to white, and start executing the same algorithm from root but we don't insert the black vertices again. This ensures that, if a vertex has become black already, its subtree will not be explored again, and once we restore the latest block of $O(n/\lg n)$ vertices, we start executing the normal algorithm. Clearly, we are using $O(n)$ bits of space. Since the reconstruction happens only $O(\lg n)$ times, and each time we spend $O(m)$ time, overall this procedure takes $O(m \lg n)$ time. Generalizing this strategy by creating hierarchy of levels and then using dynamic dictionary at each levels like we did for BDS_j in Sect. 2.2, we can similarly obtain an implementation of BDS_{hs} taking $O(n)$ bits and $O(m \lg^* n)$ time. This completes the description of the algorithms taking $O(n)$ bits.

3.2   Using $O(n \lg(m/n))$ Bits and $O(m+n)$ Time

We can use Lemma 1 to store parent pointers in the adjacency array of every vertex, and another pointer to mark the progress of BDS_{hs} so far in a similar way as we did for BDS_j in Sect. 2.3. It is easy to see that with these structures, and additional color array, using $O(m+n)$ time and $n \lg(m/n)$ bits, we can implement BDS_{hs}. One can also extend this to the directed graphs as mentioned in Sect. 2.3. With this, we complete the proof of the Theorem 1 for BDS_{hs}.

4   Conclusions

We obtained space-efficient as well as time-efficient implementations for two graph search methods, both are known under the same name, breadth-depth search even though they perform entirely differently. The main idea behind our algorithm is the introduction of the *delayed insertion* and the *delayed removal* techniques for better managing the elements of the stack, and finally we use the classical blocking idea carefully to obtain the space-time efficient implementations. We think that these ideas might be of independent interest while designing similar space-time efficient algorithms for other existing graph search methods in the literature. We believe this is an important research direction as these search methods form basis of many important graph/AI algorithms.

We leave with two concrete open problems, is it possible to design a) $o(n)$ space and polynomial time algorithms, and b) $O(n)$ bits and $O(m+n)$ time algorithms (deterministic or randomized) for both the BDS implementations? Another interesting direction would be to study these graph search methods in the recently introduced in-place [6] model where changing the input is also allowed in a restricted manner unlike the ROM model which is what we have focused in this paper.

References

1. Asano, T., et al.: Depth-first search using $O(n)$ bits. In: Ahn, H.-K., Shin, C.-S. (eds.) ISAAC 2014. LNCS, vol. 8889, pp. 553–564. Springer, Cham (2014). https://doi.org/10.1007/978-3-319-13075-0_44
2. Banerjee, N., Chakraborty, S., Raman, V., Satti, S.R.: Space efficient linear time algorithms for BFS, DFS and applications. Theory Comput. Syst. (2018)
3. Banyassady, B., et al.: Improved time-space trade-offs for computing voronoi diagrams. In: 34th STACS, pp. 9:1–9:14 (2017)
4. Chakraborty, S.: Space efficient graph algorithms. Ph.D. thesis. The Institute of Mathematical Sciences, HBNI, India (2018)
5. Chakraborty, S., Jo, S., Satti, S.R.: Improved space-efficient linear time algorithms for some classical graph problems. CoRR, abs/1712.03349 (2017)
6. Chakraborty, S., Mukherjee, A., Raman, V., Satti, S.R.: A framework for in-place graph algorithms. In: 26th ESA, pp. 13:1–13:16 (2018)
7. Chakraborty, S., Raman, V., Satti, S.R.: Biconnectivity, st-numbering and other applications of DFS using O(n) bits. J. Comput. Syst. Sci. **90**, 63–79 (2017)
8. Chakraborty, S., Satti, S.R.: Space-efficient algorithms for maximum cardinality search, its applications, and variants of BFS. J. Comb. Optim. **37**(2), 465–481 (2018)
9. Corneil, D.G., Krueger, R.: A unified view of graph searching. SIAM J. Discrete Math. **22**(4), 1259–1276 (2008)
10. Demaine, E.D., der Heide, F.M., Pagh, R., Pătraşcu, M.: De dictionariis dynamicis pauco spatio utentibus. In: Correa, J.R., Hevia, A., Kiwi, M. (eds.) LATIN 2006. LNCS, vol. 3887, pp. 349–361. Springer, Heidelberg (2006). https://doi.org/10.1007/11682462_34
11. Elmasry, A., Hagerup, T., Kammer, F.: Space-efficient basic graph algorithms. In: 32nd STACS, pp. 288–301 (2015)
12. Greenlaw, R.: Breadth-depth search is P-complete. Parallel Process. Lett. **3**, 209–222 (1993)
13. Horowitz, E., Sahni, S.: Fundamentals of Computer Algorithms. Computer Science Press (1978)
14. Jiang, B.: I/O-and CPU-optimal recognition of strongly connected components. Inf. Process. Lett. **45**(3), 111–115 (1993)
15. Kiyomi, M., Ono, H., Otachi, Y., Schweitzer, P., Tarui, J.: Space-efficient algorithms for longest increasing subsequence. In: 35th STACS, pp. 44:1–44:15 (2018)
16. Lincoln, A., Williams, V.V., Wang, J.R., Williams, R.R.: Deterministic time-space trade-offs for k-sum. In: 43rd ICALP, pp. 58:1–58:14 (2016)
17. Tarjan, R.E., Yannakakis, M.: Simple linear-time algorithms to test chordality of graphs, test acyclicity of hypergraphs, and selectively reduce acyclic hypergraphs. SIAM J. Comput. **13**(3), 566–579 (1984)

Circular Pattern Matching
with k Mismatches

Panagiotis Charalampopoulos[1] , Tomasz Kociumaka[2,3] , Solon P. Pissis[4] ,
Jakub Radoszewski[3] , Wojciech Rytter[3] , Juliusz Straszyński[3] ,
Tomasz Waleń[3] , and Wiktor Zuba[3(✉)]

[1] Department of Informatics, King's College London, London, UK
panagiotis.charalampopoulos@kcl.ac.uk
[2] Department of Computer Science, Bar-Ilan University, Ramat Gan, Israel
[3] Institute of Informatics, University of Warsaw, Warsaw, Poland
{kociumaka,jrad,rytter,jks,walen,w.zuba}@mimuw.edu.pl
[4] CWI, Amsterdam, The Netherlands
solon.pissis@cwi.nl

Abstract. The k-mismatch problem consists in computing the Hamming distance between a pattern P of length m and every length-m substring of a text T of length n, if this distance is no more than k. In many real-world applications, any cyclic shift of P is a relevant pattern, and thus one is interested in computing the minimal distance of every length-m substring of T and any cyclic shift of P. This is the circular pattern matching with k mismatches (k-CPM) problem. A multitude of papers have been devoted to solving this problem but, to the best of our knowledge, only average-case upper bounds are known. In this paper, we present the first non-trivial worst-case upper bounds for the k-CPM problem. Specifically, we show an $\mathcal{O}(nk)$-time algorithm and an $\mathcal{O}(n + \frac{n}{m} k^5)$-time algorithm. The latter algorithm applies in an extended way a technique that was very recently developed for the k-mismatch problem [Bringmann et al., SODA 2019].

1 Introduction

Pattern matching is a fundamental problem in computer science [15]. It consists in finding all substrings of a text T of length n that match a pattern P of length

P. Charalampopoulos—Supported by a Studentship from the Faculty of Natural and Mathematical Sciences at King's College London and an A. G. Leventis Foundation Educational Grant.

T. Kociumaka—Supported by ISF grants no. 824/17 and 1278/16 and by an ERC grant MPM under the EU's Horizon 2020 Research and Innovation Programme (grant no. 683064).

J. Radoszewski and J. Straszyński—Supported by the "Algorithms for text processing with errors and uncertainties" project carried out within the HOMING program of the Foundation for Polish Science co-financed by the European Union under the European Regional Development Fund.

© Springer Nature Switzerland AG 2019
L. A. Gąsieniec et al. (Eds.): FCT 2019, LNCS 11651, pp. 213–228, 2019.
https://doi.org/10.1007/978-3-030-25027-0_15

m. In many real-world applications, a measure of similarity is usually introduced allowing for *approximate* matches between the given pattern and substrings of the text. The most widely-used similarity measure is the Hamming distance between the pattern and all length-m substrings of the text.

Computing the Hamming distance between P and all length-m substrings of T has been investigated for the past 30 years. The first efficient solution requiring $\mathcal{O}(n\sqrt{m}\log m)$ time was independently developed by Abrahamson [1] and Kosaraju [30] in 1987. The k-mismatch version of the problem asks for finding only the substrings of T that are close to P, specifically, at Hamming distance at most k. The first efficient solution to this problem running in $\mathcal{O}(nk)$ time was developed in 1986 by Landau and Vishkin [31]. It took almost 15 years for a breakthrough result by Amir et al. improving this to $\mathcal{O}(n\sqrt{k\log k})$ [2]. More recently, there has been a resurgence of interest in the k-mismatch problem. Clifford et al. gave an $\mathcal{O}((n/m)(k^2\log k)+n\text{polylog}n)$-time algorithm [13], which was subsequently improved further by Gawrychowski and Uznański to $\mathcal{O}((n/m)(m+k\sqrt{m})\text{polylog}n)$ [21]. In [21], the authors have also provided evidence that any further progress in this problem is rather unlikely.

The k-mismatch problem has also been considered on compressed representations of the text [10,11,19,37], in the parallel model [18], and in the streaming model [13,14,35]. Furthermore, it has been considered in non-standard stringology models, such as the parameterized model [23] and the order-preserving model [20].

In many real-world applications, such as in bioinformatics [4,7,22,25] or in image processing [3,32–34], any cyclic shift (rotation) of P is a relevant pattern, and thus one is interested in computing the minimal distance of every length-m substring of T and any cyclic shift of P, if this distance is no more than k. This is the circular pattern matching with k mismatches (k-CPM) problem. A multitude of papers [5,6,8,9,17,24] have thus been devoted to solving the k-CPM problem but, to the best of our knowledge, only average-case upper bounds are known; i.e. in these works the assumption is that text T is uniformly random. The main result states that, after preprocessing pattern P, the average-case optimal search time of $\mathcal{O}(n\frac{k+\log m}{m})$ [12] can be achieved for certain values of the error ratio k/m (see [9,17] for more details on the preprocessing costs).

In this paper, we draw our motivation from (i) the importance of the k-CPM problem in real-world applications and (ii) the fact that no (non-trivial) worst-case upper bounds are known. Trivial here refers to running the fastest-known algorithm for the k-mismatch problem [21] separately for each of the m rotations of P. This yields an $\mathcal{O}(n(m+k\sqrt{m})\text{polylog}n)$-time algorithm for the k-CPM problem. This is clearly unsatisfactory: it is a simple exercise to design an $\mathcal{O}(nm)$-time or an $\mathcal{O}(nk^2)$-time algorithm. In an effort to tackle this unpleasant situation, we present two much more efficient algorithms: a simple $\mathcal{O}(nk)$-time algorithm and an $\mathcal{O}(n+\frac{n}{m}k^5)$-time algorithm. Our second algorithm applies in an extended way a technique that was developed very recently for k-mismatch pattern matching in grammar compressed strings by Bringmann et al. [11].

Our Approach. We first consider a simple version of the problem (called ANCHOR-MATCH) in which we are given a position in T (an *anchor*) which belongs to potential k-mismatch circular occurrences of P. A simple $\mathcal{O}(k)$ time algorithm is given (after linear-time preprocessing) to compute all relevant occurrences. By considering separately each position in T as an anchor we obtain an $\mathcal{O}(nk)$-time algorithm. The concept of an anchor is extended to the so called *matching-pairs*: when we know a pair of positions, one in P and the other in T, that are aligned. Then comes the idea of a *sample* P', which is a fragment of P of length $\Theta(m/k)$ which supposedly exactly matches a corresponding fragment in T. We choose $\mathcal{O}(k)$ samples and work for each of them and for windows of T of size $2m$. As it is typical in many versions of pattern matching, our solution is split into the periodic and non-periodic cases. If P' is non-periodic the sample occurs only $\mathcal{O}(k)$ times in a window and each occurrence gives a matching-pair (and consequently two possible anchors). Then we perform ANCHOR-MATCH for each such anchor. The hard part is the case when P' is periodic. Here we compute all exact occurrences of P' and obtain $\mathcal{O}(k)$ groups of occurrences, each one being an arithmetic progression. Now each group is processed using the approach "few matches or almost periodicity" of Bringmann et al. [11]. In the latter case periodicity is approximate, allowing up to k mismatches.

2 Preliminaries

Let $S = S[0]S[1] \cdots S[n-1]$ be a *string* of length $|S| = n$ over an integer alphabet Σ. The elements of Σ are called *letters*. For two positions i and j on S, we denote by $S[i..j] = S[i] \cdots S[j]$ the *fragment* of S that starts at position i and ends at position j (it equals the empty string ε if $j < i$). A *prefix* of S is a fragment that starts at position 0, i.e. of the form $S[0..j]$, and a *suffix* is a fragment that ends at position $n-1$, i.e. of the form $S[i..n-1]$. For an integer k, we define the kth *power* of S, denoted by S^k, as the string obtained from concatenating k copies of S. S^∞ denotes the string obtained by concatenating infinitely many copies. If S and S' are two strings of the same length, then by $S =_k S'$ we denote the fact that S and S' have at most k mismatches, that is, that the Hamming distance between S and S' does not exceed k.

We say that a string S has period q if $S[i] = S[i+q]$ for all $i = 0, \ldots, |S|-q-1$. String S is periodic if it has a period q such that $2q \leq |S|$. We denote the smallest period of S by $\mathsf{per}(S)$.

For a string S, by $\mathsf{rot}_x(S)$ for $0 \leq x < |S|$, we denote the string that is obtained from S by moving the prefix of S of length x to its suffix. We call the string $\mathsf{rot}_x(S)$ (or its representation x) a *rotation* of S. More formally, we have

$$\mathsf{rot}_x(S) = VU, \text{ where } S = UV \text{ and } |U| = x.$$

2.1 Anatomy of Circular Occurrences

In what follows, we denote by m the length of the pattern P and by n the length of the text T. We say that P has a k-*mismatch circular occurrence* (in short k-*occurrence*) in T at position p if $T[p..p+m-1] =_k \mathsf{rot}_x(P)$ for some rotation x.

In this case, the position x in the pattern is called the *split point* of the pattern and $p + (m - x) \bmod m$ [1] is called the *anchor* in the text (see Fig. 1).

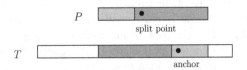

Fig. 1. The anchor and the split point for a k-occurrence of P in T.

In other words, if $P = UV$ and its rotation VU occurs in T, then the first position of V in P is the split point of this occurrence, and the first position of U in T is the anchor of this occurrence.

For an integer z, let us denote $\mathbf{W}_z = [z .. z + m - 1]$ (*window* of size m). For a k-occurrence at position p with rotation x, we introduce a set of pairs of positions in the fragment of the text and the corresponding positions from the original (unrotated) pattern:

$$M(p,x) = \{(i, (i - p + x) \bmod m) : i \in \mathbf{W}_p\}.$$

The pairs $(i, j) \in M(p, x)$ are called *matching pairs* of an occurrence p with rotation x. In particular, $(p + ((m - x) \bmod m), 0) \in M(p, x)$. An example is provided in Fig. 2.

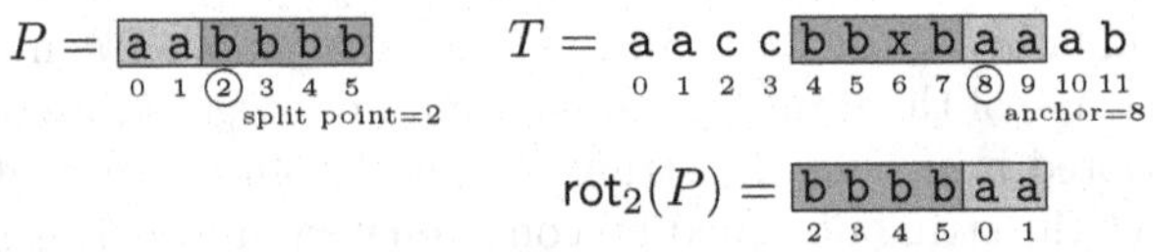

$$P = \boxed{\text{a a b b b b}} \qquad T = \text{a a c c} \boxed{\text{b b x b a a}} \text{a b}$$

$$\begin{array}{ccccccc} \text{a} & \text{a} & \text{b} & \text{b} & \text{b} & \text{b} \\ 0 & 1 & ② & 3 & 4 & 5 \end{array}$$
split point=2

$T = $ a a c c b b x b a a a b
0 1 2 3 4 5 6 7 ⑧ 9 10 11
anchor=8

$$\text{rot}_2(P) = \boxed{\text{b b b b a a}}$$
2 3 4 5 0 1

Fig. 2. A 1-occurrence of $P = $ aabbbb in text $T = $ aaccbbxbaaab at position $p = 4$ with rotation $x = 2$; $M(4, 2) = \{(4, 2), (5, 3), (6, 4), (7, 5), (8, 0), (9, 1)\}$.

2.2   Internal Queries in a Text

Let T be a string of length n called text. The length of the longest common prefix (suffix) of strings U and V is denoted by $\mathsf{lcp}(U, V)$ ($\mathsf{lcs}(U, V)$). There is a well-known efficient data structure answering such queries over suffixes (prefixes) of a given text in $\mathcal{O}(1)$ time after $\mathcal{O}(n)$-time preprocessing. It consists of the suffix array and a data structure for range minimum queries; see [15]. Using the kangaroo method [18,31], longest common prefix (suffix) queries can handle mismatches; after an $\mathcal{O}(n)$-time preprocessing of the text, longest common prefix (suffix) queries with up to k mismatches can be answered in $\mathcal{O}(k)$ time.

---

[1] The modulo operation is used to handle the trivial rotation with $x = 0$.

An Internal Pattern Matching (IPM) query, for two given fragments F and G of the text, such that $|G| \leq 2|F|$, computes the set of all occurrences of F in G. If there are more than two occurrences, they form an arithmetic sequence with difference $\mathsf{per}(F)$. For a text of length n, a data structure for IPM queries can be constructed in $\mathcal{O}(n)$ time and answers queries in $\mathcal{O}(1)$ time (see [29] and [26, Theorem 1.1.4]). It can be used to compute all occurrences of a given fragment F of length p in T, expressed as a union of $\mathcal{O}(n/p)$ pairwise disjoint arithmetic sequences with difference $\mathsf{per}(F)$, in $\mathcal{O}(n/p)$ time.

3   An $\mathcal{O}(nk)$-time Algorithm

We first introduce an auxiliary problem in which one wants to compute all k-occurrences of P in T with a given anchor a.

ANCHOR-MATCH PROBLEM

Input: Text T of length n, pattern P of length m, positive integer k, and position a.

Output: Find all k-occurrences p of P in T with anchor a.

Lemma 1. *After $\mathcal{O}(n)$-time preprocessing, the answer to* ANCHOR-MATCH *problem, represented as a union of $\mathcal{O}(k)$ intervals, can be computed in $\mathcal{O}(k)$ time.*

Proof. In the preprocessing we prepare a data structure for lcp and lcs queries in $P\#T$, for a special symbol $\#$ that does not occur in P and T.

The processing of each query is split into $k + 1$ phases. In the jth phase, we compute the interval $[l_j \mathbin{..} r_j]$ such that for every $p \in [l_j \mathbin{..} r_j]$ there exists a k-occurrence p in T that has an anchor at a and the number of mismatches between $T[p \mathbin{..} a - 1]$ and the suffix of P of equal length is exactly j.

Let us consider the conditions for interval $[l_j \mathbin{..} r_j]$ (see also Fig. 3):

C1 $[l_j \mathbin{..} r_j] \subseteq [a - m + 1 \mathbin{..} a)$ since occurrences must contain anchor a,

C2 $[l_j \mathbin{..} r_j] \subseteq [a - 1 - s_j \mathbin{..} a - 1 - s_{j-1})$, where s_i is the length of the longest common suffix of $T[0 \mathbin{..} a - 1]$ and P with exactly i mismatches, since we need exactly j mismatches in $T[p \mathbin{..} a - 1]$,

C3 $[l_j \mathbin{..} r_j] \subseteq [a - m \mathbin{..} a + p_{k-j} - m)$, where p_{k-j} is the length of the longest common prefix of $T[a \mathbin{..} n - 1]$ and P with at most $k - j$ mismatches, since we cannot exceed k mismatches in total.

Using the kangaroo method [18, 31], the values s_j, p_j for all $0 \leq j \leq k$ can be computed in $\mathcal{O}(k)$ time in total. Then the interval $[l_j \mathbin{..} r_j]$ is a simple intersection of the above conditions, which can be computed in $\mathcal{O}(1)$ time. $\square$

Proposition 2. *k-CPM can be solved in $\mathcal{O}(nk)$ time and $\mathcal{O}(n)$ space.*

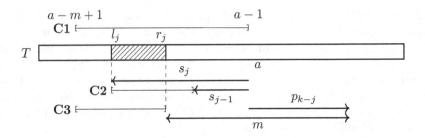

Fig. 3. An illustration of the setting in Lemma 1.

Proof. We invoke the algorithm of Lemma 1 for all $a \in [0\mathinner{.\,.}n-1]$ and obtain $\mathcal{O}(nk)$ intervals of k-occurrences of P in T. Instead of storing all the intervals, we count how many intervals start and end at each position of the text. We can then compute the union of the intervals by processing these counts left to right. □

4   An $\mathcal{O}(n + \frac{n}{m}k^5)$-time Algorithm

In this section, we assume that $m \le n \le 2m$ and aim at an $\mathcal{O}(n + k^5)$-time algorithm.

A *(deterministic) sample* is a short segment P' of the pattern P. An occurrence in the text without any mismatch is called *exact*. We introduce a problem of SAMPLE-MATCHING that consists in finding all k-occurrences of P in T such that P' matches exactly a fragment of length $|P'|$ in T.

We split the pattern P into $k+2$ fragments of length $\left\lfloor \frac{m}{k+2} \right\rfloor$ or $\left\lceil \frac{m}{k+2} \right\rceil$ each. One of those fragments will occur exactly in the text (up to k fragments may occur with a mismatch and at most one fragment will contain the split point). Let us henceforth fix a sample P' as one of these fragments, let p' be its starting position in P, and let $m' = |P'|$.

We assume that the split point x in P is to the right of P', i.e., that $x \ge p' + m'$. The opposite case—that $x < p'$—can be handled analogously.

4.1   Matching Non-periodic Samples

Let us assume that P' is non-periodic. We introduce a problem in which, intuitively, we compute all k-occurrences of P in T which align $T[i]$ with $P[j]$.

PAIR-MATCH PROBLEM

Input: Text T of length n, pattern P of length m, positive integer k, and two integers $i \in [0\mathinner{.\,.}n-1]$ and $j \in [0\mathinner{.\,.}m-1]$.

Output: The set $A(i,j)$ of all positions in T where we have a k-mismatch occurrence of $\mathrm{rot}_x(P)$ for some x such that (i,j) is a matching pair.

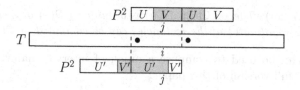

Fig. 4. The two possible anchors for the matching pair of positions (i, j) are shown as bullet points. A possible k-occurrence of P in T corresponding to the left (resp. right) anchor is shown below T (above T, resp.).

Lemma 3. *After $\mathcal{O}(n)$-time preprocessing, the* PAIR-MATCH *problem can be solved in $\mathcal{O}(k)$ time, where the output is represented as a union of $\mathcal{O}(k)$ intervals.*

Proof. The PAIR-MATCH problem can be essentially reduced to the ANCHOR-MATCH problem, since for a given matching pair of characters in P and T, there are at most two ways of choosing the anchor depending on the relation between j and a split point: these are $i-j$ (if $i-j \geq 0$) and $i+|P|-j$ (if $i+|P|-j < |T|$); see Fig. 4. We then have to take the intersection of the answer with $[i-m+1 \mathinner{.\,.} i]$ to ensure that the k-occurrence contains position i. □

Lemma 4. *After $\mathcal{O}(n)$-time preprocessing, the* SAMPLE-MATCHING *problem for a non-periodic sample can be solved in $\mathcal{O}(k^2)$ time and outputs a union of $\mathcal{O}(k^2)$ intervals of occurrences.*

Proof. If P' is non-periodic, then it has $\mathcal{O}(k)$ occurrences in T, which can be computed in $\mathcal{O}(k)$ time after an $\mathcal{O}(n)$-time preprocessing using IPM queries [26, 29] in $P\#T$. Let j be the starting position of P' in P and i be a starting position of an occurrence of P' in T. For each of the $\mathcal{O}(k)$ such pairs (i, j), the computation reduces to the PAIR-MATCH problem for i and j. The statement follows by Lemma 3. □

4.2 Simple Geometry of Arithmetic Sequences of Intervals

Before we proceed with showing how to efficiently handle periodic samples, we present algorithms that will be used in the proofs for handling regular sets of intervals. For an interval I and integer r, let $I \oplus r = \{i+r : i \in I\}$. We define

$$\mathsf{Chain}_q(I, a) = I \cup (I \oplus q) \cup (I \oplus 2q) \cup \cdots \cup (I \oplus aq).$$

This set is further called an *interval chain*. Note that it can be represented in $\mathcal{O}(1)$ space (with four integers: a, q, and the endpoints of I).

For a given value of q, let us fit the integers from $[1 \mathinner{.\,.} n]$ into the cells of a grid of width q so that the first row consists of numbers 1 through q, the second of numbers $q + 1$ to $2q$, etc. Let us call this grid $\mathcal{G}_q$. A chain Chain_q can be conveniently represented in the grid $\mathcal{G}_q$ using the following lemma; it was stated in [28] and its proof can be found in the full version of that paper [27].

Lemma 5 ([27,28]). *The set* $\mathsf{Chain}_q(I, a)$ *is a union of* $\mathcal{O}(1)$ *orthogonal rectangles in* $\mathcal{G}_q$. *The coordinates of the rectangles can be computed in* $\mathcal{O}(1)$ *time.*

Lemma 6 can be used to compute a union of interval chains; its proof is deferred to the full version of this paper.

Lemma 6. *Assume that we are given* m *interval chains whose elements are subsets of* $[0..n]$. *The union of these chains, expressed as a subset of* $[0..n]$, *can be computed in* $\mathcal{O}(n + m)$ *time.*

We will also use the following auxiliary lemma.

Lemma 7. *Let* X *and* Z *be intervals and* q *be a positive integer. Then the set* $Z' := \{z \in Z : \exists_{x \in X}\, z \equiv x \pmod{q}\}$, *represented as a disjoint sum of at most three interval chains with difference* q, *can be computed in* $\mathcal{O}(1)$ *time.*

Proof. If $|X| \geq q$, then $Z' = Z$ is an interval and thus an interval chain. If $|X| < q$, then Z' can be divided into disjoint intervals of length smaller than or equal to $|X|$. The intervals from the second until the penultimate one (if any such exist), have length $|X|$. Hence, they can be represented as a single chain, as the first element of each such interval is equal mod q to the first element of X. The two remaining intervals can be treated as chains as well. $\square$

4.3 Matching Periodic Samples

Let us assume that P' is periodic, i.e., it has a period q with $2q \leq |P'|$. A fragment of a string S containing an inclusion-maximal arithmetic sequence of occurrences of P' in a string S with difference q is called here a P'-run. If P' matches a fragment in the text, then the match belongs to a P'-run. For example, the underlined substring of $S = \mathsf{bb\underline{abababa}aa}$ is a P'-run for $P' = \mathsf{abab}$.

Lemma 8. *If a string* P' *is periodic, the number of* P'-runs *in the text is* $\mathcal{O}(k)$ *and they can all be computed in* $\mathcal{O}(k)$ *time after* $\mathcal{O}(n)$-*time preprocessing.*

Proof. We construct the data structure for IPM queries on $P\#T$. This allows us to compute the set of all occurrences of P' in T as a collection of $\mathcal{O}(k)$ arithmetic sequences with difference $\mathsf{per}(P')$. We then check for every two consecutive sequences if they can be joined together. This takes $\mathcal{O}(k)$ time and results in $\mathcal{O}(k)$ P'-runs. $\square$

For two equal-length strings S and S', we denote the set of their *mismatches* by

$$\mathsf{Mis}(S, S') = \{i = 0, \ldots, |S| - 1 : S[i] \neq S'[i]\}.$$

Let $Q = S[i..j]$. We say that position a in S is a *misperiod* with respect to the fragment $S[i..j]$ if $S[a] \neq S[b]$ where b is the unique position such that $b \in [i..j]$ and $|Q| \mid b - a$. We define the set $\mathsf{LeftMisper}_k(S, i, j)$ as the set of k maximal misperiods that are smaller than i and $\mathsf{RightMisper}_k(S, i, j)$ as the set

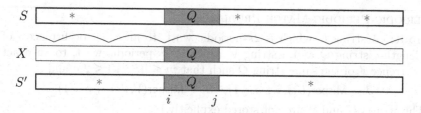

Fig. 5. Let S, S', and X be equal-length strings such that X is a factor of Q^∞ and $S[i..j] = S'[i..j] = X[i..j] = Q$. The asterisks in S denote the positions in $\mathsf{Mis}(S, X)$, or equivalently, the misperiods with respect to $S[i..j]$. Similarly for S'. One can observe that $\mathsf{Mis}(S, X) \cap \mathsf{Mis}(S', X) = \emptyset$ and that $\mathsf{Mis}(S, X) \cup \mathsf{Mis}(S', X) = \mathsf{Mis}(S, S')$.

of k minimal misperiods that are greater than j. Each of the sets can have less than k elements if the corresponding misperiods do not exist. We further define

$$\mathsf{Misper}_k(S, i, j) = \mathsf{LeftMisper}_k(S, i, j) \cup \mathsf{RightMisper}_k(S, i, j)$$

and $\mathsf{Misper}(S, i, j) = \bigcup_{k=0}^{\infty} \mathsf{Misper}_k(S, i, j)$.

The following lemma captures the main combinatorial property behind the new technique of Bringmann et al. [11]. Its proof is deferred to the full version of this paper; the intuition is shown in Fig. 5.

Lemma 9. *Assume that $S =_k S'$ and that $S[i..j] = S'[i..j]$. Let*

$$I = \mathsf{Misper}_{k+1}(S, i, j) \text{ and } I' = \mathsf{Misper}_{k+1}(S', i, j).$$

If $I \cap I' = \emptyset$, then $\mathsf{Mis}(S, S') = I \cup I'$, $I = \mathsf{Misper}(S, i, j)$, and $I' = \mathsf{Misper}(S', i, j)$.

A string S is *k-periodic w.r.t. an occurrence i of Q* if $|\mathsf{Misper}(S, i, i+|Q|-1)| \leq k$. In particular, in the conclusion of the above lemma S and S' are $|I|$-periodic and $|I'|$-periodic, respectively, w.r.t. $Q = S[i..j] = S'[i..j]$. This notion forms the basis of the following auxiliary problem in which we search for k-occurrences in which the rotation of the pattern and the fragment of the text are k-periodic for the same period Q.

Let U and V be two strings and J and J' be sets containing positions in U and V, respectively. We say that length-m fragments $U[p..p+m-1]$ and $V[x..x+m-1]$ are *(J, J')-disjoint* if the sets $(\mathbf{W}_p \cap J) \oplus (-p)$ and $(\mathbf{W}_x \cap J') \oplus (-x)$ are disjoint. For example, if $J = \{2, 4, 11, 15, 16, 17\}$, $J' = \{5, 6, 15, 18, 19\}$, and $m = 12$, then $U[3..14]$ and $V[6..17]$ are (J, J')-disjoint for:

$$U = \quad \text{ab•} \boxed{\text{a•b abc ab• abc}} \text{•••}$$
$$V = \text{abc ab•} \boxed{\text{•bc abc abc •bc}} \text{••c}$$

PERIODIC-PERIODIC-MATCH PROBLEM

Input: A string U which is $2k$-periodic w.r.t. to an exact occurrence i of a length-q string Q and a string V which is $2k$-periodic w.r.t. to an exact occurrence i' of the same string Q such that $m \le |U|, |V| \le 2m$ and

$$J = \mathsf{Misper}(U, i, i + q - 1), \quad J' = \mathsf{Misper}(V, i', i' + q - 1).$$

(The strings U and V are not stored explicitly.)

Output: The set of positions p in U for which there exists a (J, J')-disjoint k-occurrence $U[p..p + m - 1]$ of $V[x..x + m - 1]$ for x such that

$$i - p \equiv i' - x \pmod{q}.$$

Intuitively, the condition on the output of the problem corresponds to the fact that the k-mismatch periodicity is aligned. We defer the solution to this problem to Lemma 12. Let us now show how it can be used to solve SAMPLE-MATCHING for a periodic sample.

Let us define

$$\text{PAIRS-MATCH}(T, I, P, J) = \bigcup_{i \in I, j \in J} \text{PAIR-MATCH}(T, i, P, j).$$

Let A be a set of positions in a string S and m be a positive integer. We then denote $A \bmod m = \{a \bmod m : a \in A\}$ and by $\mathsf{frag}_A(S)$ we denote the fragment $S[\min A .. \max A]$. We provide a pseudocode of an algorithm that computes all k-occurrences of P such that P' matches a fragment of a given P'-run below.

Data: A periodic fragment P' of pattern P, a P'-run R in text T, $q = \mathsf{per}(P')$, and k.

Result: A compact representation of k-occurrences of P in T including all k-occurrences where P' in P matches a fragment of R in T.

Let $R = T[s..s + |R| - 1]$;

$J := \mathsf{Misper}_{k+1}(T, s, s + q - 1)$; { $\mathcal{O}(k)$ time }

$J' := \mathsf{Misper}_{k+1}(P^2, m + p', m + p' + q - 1)$; { $\mathcal{O}(k)$ time }

$U := \mathsf{frag}_J(T)$; $V := \mathsf{frag}_{J'}(P^2)$;

$Y := \text{PERIODIC-PERIODIC-MATCH}(U, V)$; { $\mathcal{O}(k^2)$ time }

$Y := Y \oplus \min(J)$;

$J' := J' \bmod m$;

$X := \text{PAIRS-MATCH}(T, J, P, J')$; { $\mathcal{O}(k^3)$ time }

return $X \cup Y$;

Algorithm 1. Run-Sample-Matching

Lemma 10. *After $\mathcal{O}(n)$-time preprocessing, algorithm Run-Sample-Matching works in $\mathcal{O}(k^3)$ time and returns a compact representation that consists of $\mathcal{O}(k^3)$ interval chains.*

Proof. See the pseudocode. The sets J and J' can be computed in $\mathcal{O}(k)$ time:

Claim. If S is a string of length n, then the sets $\mathsf{RightMisper}_k(S, i, j)$ and $\mathsf{LeftMisper}_k(S, i, j)$ can be computed in $\mathcal{O}(k)$ time after $\mathcal{O}(n)$-time preprocessing.

Proof. For $\mathsf{RightMisper}_k(S, i, j)$, we use the kangaroo method [18,31] to compute the longest common prefix with at most k mismatches of $S[j + 1 .. n - 1]$ and U^∞ for $U = S[i .. j]$. The value $\mathsf{lcp}(X^\infty, Y)$ for a substring X and a suffix Y of a string S, occurring at positions a and b, respectively, can be computed in constant time as follows. If $\mathsf{lcp}(S[a .. n - 1], S[b .. n - 1]) < |X|$ then we are done. Otherwise the answer is given by $|X| + \mathsf{lcp}(S[b .. n - 1], S[b + |X| .. n - 1])$. The computations for $\mathsf{LeftMisper}_k(S, i, j)$ are symmetric. □

The $\mathcal{O}(k^3)$ and $\mathcal{O}(k^2)$ time complexities of computing X and Y follow from Lemmas 3 and 12, respectively (after $\mathcal{O}(n)$-time preprocessing). The sets X and Y consist of $\mathcal{O}(k^3)$ intervals and $\mathcal{O}(k^2)$ interval chains. The claim follows. □

The correctness of the algorithm follows from Lemma 9. A detailed proof of the following lemma is deferred to the full version of this paper.

Lemma 11. *Assume $n \le 2m$. Let P' be a periodic sample in P with smallest period q and R be a P'-run in T. Let X and Y be defined as in the pseudocode of Run-Sample-Matching. Then $X \cup Y$ is a set of k-occurrences of P in T which is a superset of the solution to SAMPLE-MATCH for P' in R.*

4.4   Solution to Periodic-Periodic-Match Problem

Lemma 12. *We can compute in $\mathcal{O}(k^2)$ time a set of k-occurrences of P in T represented as $\mathcal{O}(k^2)$ interval chains that is a superset of the solution to PERIODIC-PERIODIC-MATCH problem.*

Proof. We reduce our problem to the following abstract problem (see also Fig. 6).

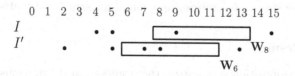

Fig. 6. An instance of the ABSTRACT PROBLEM with $m = 6$, $k = 3$, $q = 3$, $\delta = 2$, $I = \{4, 5, 9, 15\}$ and $I' = \{2, 5, 7, 8, 13\}$. $8 \in A$, since for 6, we have that $|\mathbf{W_8} \cap I| + |\mathbf{W_6} \cap I'| \le 3$, $8 \equiv 2 + 6 \pmod 3$, $\mathbf{W_8} \subseteq (4, 15)$ and $\mathbf{W_6} \subseteq (2, 13)$.

Abstract Problem

Input: Positive integers m, k, q, δ and two sets I and I' such that $2 \leq |I|, |I'| \leq 2k + 4$.

Output: The set A of integers z for which there exists z' such that:

1. $|\mathbf{W}_z \cap I| + |\mathbf{W}_{z'} \cap I'| \leq k$
2. $z \equiv \delta + z' \pmod{q}$
3. $\mathbf{W}_z \subseteq (\min I, \max I)$, $\mathbf{W}_{z'} \subseteq (\min I', \max I')$.

Claim. Periodic-Periodic-Match can be reduced in $\mathcal{O}(k)$ time to the Abstract Problem so that if z belongs to the solution to the Abstract Problem then $p = z$ is a solution to Periodic-Periodic-Match, which potentially may not satisfy the third condition of the problem.

Proof. Let the parameters m, k and q remain unchanged. We set $I = J \cup \{-1, |U|\}$, $I' = J' \cup \{-1, |V|\}$, and $\delta = i - i'$. □

Claim. Abstract Problem can be solved in $\mathcal{O}(k^2)$ time with the output represented as a collection of $\mathcal{O}(k^2)$ interval chains.

Proof. Let us denote $Z = (\min I, \max I - m + 1)$, $Z' = (\min I', \max I' - m + 1)$. We partition the set Z into intervals such that for all z in an interval, the set $\mathbf{W}_z \cap I$ is the same. For this, we use a sliding window approach. We generate events corresponding to x and $x - m + 1$ for all $x \in I$ and sort them. When z crosses an event, the set $\mathbf{W}_z \cap I$ changes. Thus we obtain a partition of Z into intervals $Z_1, \ldots, Z_{n_1}$ for $n_1 \leq 4k$. We obtain a similar partition of Z' into intervals $Z'_1, \ldots, Z'_{n_2}$ for $n_2 \leq 4k$.

Let us now fix Z_j and $Z'_{j'}$ (see also Fig. 7). First we check if condition 1 is satisfied for $z \in Z_j$ and $z' \in Z'_{j'}$. If so, we compute the set $X = \{(\delta + z') \bmod q : z' \in Z'_{j'}\}$. It is a single circular interval and can be computed in constant time.

The sought result is $\{z \in Z_j : z \bmod q \in X\}$. By Lemma 7, this set can be represented as a union of three chains and, as such, can be computed in $\mathcal{O}(1)$ time. The conclusion follows. □

This completes the proof of the lemma. □

In the solution we do not check if the sets $(\mathbf{W}_p \cap J) \oplus (-p)$ and $(\mathbf{W}_x \cap J') \oplus (-x)$ are disjoint. However, a k'-occurrence is found for some $k' < k$ otherwise.

4.5   Main Result

The following proposition summarizes the results from the previous subsections.

Proposition 13. *If $m \leq n \leq 2m$, k-CPM can be solved in $\mathcal{O}(n + k^5)$ time.*

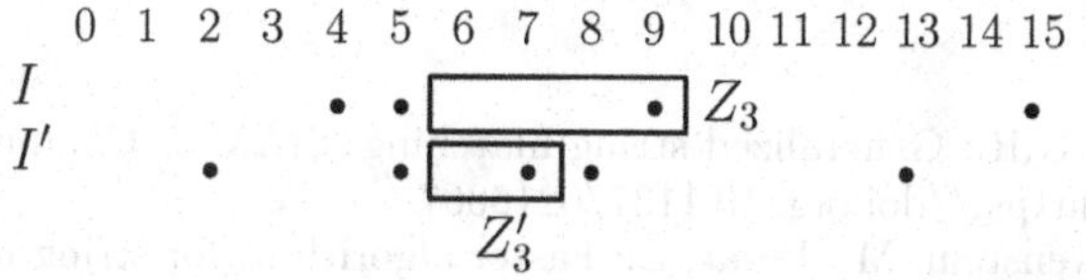

0 1 2 3 4 5 6 7 8 9 10 11 12 13 14 15

Fig. 7. The same instance of the ABSTRACT PROBLEM as in Fig. 6. For $Z_3 = \{6, 7, 8, 9\}$ and $Z_3' = \{6, 7\}$ we get $X = \{0, 2\}$ and hence the sought result is $\{6, 8, 9\}$.

Proof. There are $k + 2$ ways to choose a sample P' in the pattern.

If the sample P' is not periodic, we use the algorithm of Lemma 4 for SAMPLE MATCHING in $\mathcal{O}(k^2)$ time (after $\mathcal{O}(n)$-time preprocessing). It returns a representation of k-occurrences as a union of $\mathcal{O}(k^2)$ intervals.

If the sample P' is periodic, we need to find all P'-runs in T. By Lemma 8, there are $\mathcal{O}(k)$ of them and they can all be computed in $\mathcal{O}(k)$ time (after $\mathcal{O}(n)$-time preprocessing). For every such P'-run R, we apply the Run-Sample-Matching algorithm. Its correctness follows from Lemma 11. By Lemma 10, it takes $\mathcal{O}(k^3)$ time and returns $\mathcal{O}(k^3)$ interval chains of k-occurrences of P in T (after $\mathcal{O}(n)$-time preprocessing). Over all P'-runs, this takes $\mathcal{O}(k^4)$ time after the preprocessing.

In total, SAMPLE MATCHING takes $\mathcal{O}(k^4)$ time for a given sample (after preprocessing), $\mathcal{O}(n + k^5)$ time in total, and returns $\mathcal{O}(k^5)$ intervals and interval chains of k-occurrences. Let us note that an interval is a special case of an interval chain. Hence, in the end, we apply Lemma 6 to compute the union of all chains of occurrences in $\mathcal{O}(n + k^5)$ time.   □

We use the standard trick: splitting the text into $\mathcal{O}(n/m)$ fragments, each of length $2m$ (perhaps apart from the last one), starting at positions equal to 0 mod m. We need to ensure that the data structures for answering lcp, lcs, and other internal queries over each such fragment of the text can be constructed in $\mathcal{O}(m)$ time in the case when our input alphabet Σ is large. As a preprocessing step we hash the letters of the pattern using perfect hashing. For each key, we assign a rank value from $\{1, \ldots, m\}$. This takes $\mathcal{O}(m)$ (expected) time and space [16]. When reading a fragment F of length (at most) $2m$ of the text we look up its letters using the hash table. If a letter is in the hash table we replace it in F by its rank value; otherwise we replace it by rank $m + 1$. We can now construct the data structures in $\mathcal{O}(m)$ time and the whole algorithm is implemented in $\mathcal{O}(m)$ space. If $\Sigma = \{1, \ldots, n^{\mathcal{O}(1)}\}$, the same bounds can be achieved deterministically using [36]. We combine Propositions 2 and 13 to get our final result.

Theorem 14. *Circular Pattern Matching with k Mismatches can be solved in $\mathcal{O}(\min(nk, \, n + \frac{n}{m} k^5))$ time and $\mathcal{O}(m)$ space.*

Our algorithms output all positions in the text where some rotation of the pattern occurs with k mismatches. It is not difficult to extend the algorithms to output, for each of these positions, a corresponding rotation of the pattern.

References

1. Abrahamson, K.R.: Generalized string matching. SIAM J. Comput. **16**(6), 1039–1051 (1987). https://doi.org/10.1137/0216067
2. Amir, A., Lewenstein, M., Porat, E.: Faster algorithms for string matching with k mismatches. J. Algorithms **50**(2), 257–275 (2004). https://doi.org/10.1016/S0196-6774(03)00097-X
3. Ayad, L.A.K., Barton, C., Pissis, S.P.: A faster and more accurate heuristic for cyclic edit distance computation. Pattern Recognit. Lett. **88**, 81–87 (2017). https://doi.org/10.1016/j.patrec.2017.01.018
4. Ayad, L.A.K., Pissis, S.P.: MARS: improving multiple circular sequence alignment using refined sequences. BMC Genomics **18**(1), 86 (2017). https://doi.org/10.1186/s12864-016-3477-5
5. Azim, M.A.R., Iliopoulos, C.S., Rahman, M.S., Samiruzzaman, M.: A filter-based approach for approximate circular pattern matching. In: Harrison, R., Li, Y., Măndoiu, I. (eds.) ISBRA 2015. LNCS, vol. 9096, pp. 24–35. Springer, Cham (2015). https://doi.org/10.1007/978-3-319-19048-8_3
6. Azim, M.A.R., Iliopoulos, C.S., Rahman, M.S., Samiruzzaman, M.: A fast and lightweight filter-based algorithm for circular pattern matching. In: Baldi, P., Wang, W. (eds.) 5th ACM Conference on Bioinformatics, Computational Biology, and Health Informatics, BCB 2014, pp. 621–622. ACM (2014). https://doi.org/10.1145/2649387.2660804
7. Barton, C., Iliopoulos, C.S., Kundu, R., Pissis, S.P., Retha, A., Vayani, F.: Accurate and efficient methods to improve multiple circular sequence alignment. In: Bampis, E. (ed.) SEA 2015. LNCS, vol. 9125, pp. 247–258. Springer, Cham (2015). https://doi.org/10.1007/978-3-319-20086-6_19
8. Barton, C., Iliopoulos, C.S., Pissis, S.P.: Fast algorithms for approximate circular string matching. Algorithms Mol. Biol. **9**, 9 (2014). https://doi.org/10.1186/1748-7188-9-9
9. Barton, C., Iliopoulos, C.S., Pissis, S.P.: Average-case optimal approximate circular string matching. In: Dediu, A.-H., Formenti, E., Martín-Vide, C., Truthe, B. (eds.) LATA 2015. LNCS, vol. 8977, pp. 85–96. Springer, Cham (2015). https://doi.org/10.1007/978-3-319-15579-1_6
10. Bille, P., Fagerberg, R., Gørtz, I.L.: Improved approximate string matching and regular expression matching on Ziv-Lempel compressed texts. ACM Trans. Algorithms **6**(1), 3:1–3:14 (2009). https://doi.org/10.1145/1644015.1644018
11. Bringmann, K., Wellnitz, P., Künnemann, M.: Few matches or almost periodicity: faster pattern matching with mismatches in compressed texts. In: Chan, T.M. (ed.) 30th Annual ACM-SIAM Symposium on Discrete Algorithms, SODA 2019, pp. 1126–1145. SIAM (2019). https://doi.org/10.1137/1.9781611975482.69
12. Chang, W.I., Marr, T.G.: Approximate string matching and local similarity. In: Crochemore, M., Gusfield, D. (eds.) CPM 1994. LNCS, vol. 807, pp. 259–273. Springer, Heidelberg (1994). https://doi.org/10.1007/3-540-58094-8_23
13. Clifford, R., Fontaine, A., Porat, E., Sach, B., Starikovskaya, T.: The k-mismatch problem revisited. In: Krauthgamer, R. (ed.) 27th Annual ACM-SIAM Symposium on Discrete Algorithms, SODA 2016, pp. 2039–2052. SIAM (2016). https://doi.org/10.1137/1.9781611974331.ch142
14. Clifford, R., Kociumaka, T., Porat, E.: The streaming k-mismatch problem. In: Chan, T.M. (ed.) 30th Annual ACM-SIAM Symposium on Discrete Algorithms, SODA 2019, pp. 1106–1125. SIAM (2019). https://doi.org/10.1137/1.9781611975482.68

15. Crochemore, M., Hancart, C., Lecroq, T.: Algorithms on Strings. Cambridge University Press, Cambridge (2007). https://doi.org/10.1017/cbo9780511546853
16. Fredman, M.L., Komlós, J., Szemerédi, E.: Storing a sparse table with $\mathcal{O}(1)$ worst case access time. J. ACM **31**(3), 538–544 (1984). https://doi.org/10.1145/828.1884
17. Fredriksson, K., Navarro, G.: Average-optimal single and multiple approximate string matching. ACM J. Exp. Algorithmics **9**(1.4), 1–47 (2004). https://doi.org/10.1145/1005813.1041513
18. Galil, Z., Giancarlo, R.: Parallel string matching with k mismatches. Theor. Comput. Sci. **51**, 341–348 (1987). https://doi.org/10.1016/0304-3975(87)90042-9
19. Gawrychowski, P., Straszak, D.: Beating $\mathcal{O}(nm)$ in approximate LZW-compressed pattern matching. In: Cai, L., Cheng, S.-W., Lam, T.-W. (eds.) ISAAC 2013. LNCS, vol. 8283, pp. 78–88. Springer, Heidelberg (2013). https://doi.org/10.1007/978-3-642-45030-3_8
20. Gawrychowski, P., Uznański, P.: Order-preserving pattern matching with k mismatches. Theor. Comput. Sci. **638**, 136–144 (2016). https://doi.org/10.1016/j.tcs.2015.08.022
21. Gawrychowski, P., Uznański, P.: Towards unified approximate pattern matching for Hamming and L_1 distance. In: Chatzigiannakis, I., Kaklamanis, C., Marx, D., Sannella, D. (eds.) Automata, Languages, and Programming, ICALP 2018. LIPIcs, vol. 107, pp. 62:1–62:13. Schloss Dagstuhl-Leibniz-Zentrum für Informatik (2018). https://doi.org/10.4230/LIPIcs.ICALP.2018.62
22. Grossi, R., Iliopoulos, C.S., Mercas, R., Pisanti, N., Pissis, S.P., Retha, A., Vayani, F.: Circular sequence comparison: algorithms and applications. Algorithms Mol. Biol. **11**, 12 (2016). https://doi.org/10.1186/s13015-016-0076-6
23. Hazay, C., Lewenstein, M., Sokol, D.: Approximate parameterized matching. ACM Trans. Algorithms **3**(3), 29 (2007). https://doi.org/10.1145/1273340.1273345
24. Hirvola, T., Tarhio, J.: Bit-parallel approximate matching of circular strings with k mismatches. ACM J. Exp. Algorithmics **22**, 1–5 (2017). https://doi.org/10.1145/3129536
25. Iliopoulos, C.S., Pissis, S.P., Rahman, M.S.: Searching and indexing circular patterns. In: Elloumi, M. (ed.) Algorithms for Next-Generation Sequencing Data, pp. 77–90. Springer, Cham (2017). https://doi.org/10.1007/978-3-319-59826-0_3
26. Kociumaka, T.: Efficient data structures for internal queries in texts. Ph.D. thesis, University of Warsaw, October 2018. https://www.mimuw.edu.pl/~kociumaka/files/phd.pdf
27. Kociumaka, T., Radoszewski, J., Rytter, W., Straszyński, J., Waleń, T., Zuba, W.: Efficient representation and counting of antipower factors in words (2018). http://arxiv.org/abs/1812.08101
28. Kociumaka, T., Radoszewski, J., Rytter, W., Straszyński, J., Waleń, T., Zuba, W.: Efficient representation and counting of antipower factors in words. In: Martín-Vide, C., Okhotin, A., Shapira, D. (eds.) LATA 2019. LNCS, vol. 11417, pp. 421–433. Springer, Cham (2019). https://doi.org/10.1007/978-3-030-13435-8_31
29. Kociumaka, T., Radoszewski, J., Rytter, W., Waleń, T.: Internal pattern matching queries in a text and applications. In: Indyk, P. (ed.) 26th Annual ACM-SIAM Symposium on Discrete Algorithms, SODA 2015, pp. 532–551. SIAM (2015). https://doi.org/10.1137/1.9781611973730.36
30. Kosaraju, S.: Efficient string matching (1987, manuscript)
31. Landau, G.M., Vishkin, U.: Efficient string matching with k mismatches. Theor. Comput. Sci. **43**, 239–249 (1986). https://doi.org/10.1016/0304-3975(86)90178-7

32. Palazón-González, V., Marzal, A.: On the dynamic time warping of cyclic sequences for shape retrieval. Image Vision Comput. **30**(12), 978–990 (2012). https://doi.org/10.1016/j.imavis.2012.08.012

33. Palazón-González, V., Marzal, A.: Speeding up the cyclic edit distance using LAESA with early abandon. Pattern Recognit. Lett. **62**, 1–7 (2015). https://doi.org/10.1016/j.patrec.2015.04.013

34. Palazón-González, V., Marzal, A., Vilar, J.M.: On hidden Markov models and cyclic strings for shape recognition. Pattern Recognit. **47**(7), 2490–2504 (2014). https://doi.org/10.1016/j.patcog.2014.01.018

35. Porat, B., Porat, E.: Exact and approximate pattern matching in the streaming model. In: 50th Annual IEEE Symposium on Foundations of Computer Science, FOCS 2009, pp. 315–323. IEEE Computer Society (2009). https://doi.org/10.1109/FOCS.2009.11

36. Ružić, M.: Constructing efficient dictionaries in close to sorting time. In: Aceto, L., Damgård, I., Goldberg, L.A., Halldórsson, M.M., Ingólfsdóttir, A., Walukiewicz, I. (eds.) ICALP 2008. LNCS, vol. 5125, pp. 84–95. Springer, Heidelberg (2008). https://doi.org/10.1007/978-3-540-70575-8_8

37. Tiskin, A.: Threshold approximate matching in grammar-compressed strings. In: Holub, J., Zdárek, J. (eds.) Prague Stringology Conference 2014, PSC 2014, pp. 124–138. Department of Theoretical Computer Science, Faculty of Information Technology, Czech Technical University in Prague (2014). http://www.stringology.org/event/2014/p12.html

Succinct Representations of Finite Groups

Bireswar Das, Shivdutt Sharma[(✉)], and P. R. Vaidyanathan

IIT Gandhinagar, Gandhinagar, India
{bireswar,shiv.sharma,pr.vaidyanathan}@iitgn.ac.in

Abstract. The Cayley table representation of a group uses $O(n^2)$ words for a group of order n and answers multiplication queries in time $O(1)$ in word RAM model. It is interesting to ask if there is a $o(n^2)$ space representation of groups that still has $O(1)$ query-time. We show that for any δ, $\frac{1}{\log n} \leq \delta \leq 1$, there is an $\mathcal{O}(\frac{n^{1+\delta}}{\delta})$ space representation for groups of order n with $\mathcal{O}(\frac{1}{\delta})$ query-time.

We also show that for Dedekind groups, simple groups and several group classes defined in terms of semidirect product, there are linear space representation to answer multiplication queries in logarithmic time.

Farzan and Munro (ISSAC'06) defined a model for group representation and gave a succinct data structure for abelian groups with constant query-time. They asked if their result can be extended to categorically larger group classes. We show we can construct data structures in their model to represent Hamiltonian groups and extensions of abelian groups by cyclic groups to answer multiplication queries in constant time.

1 Introduction

Groups are algebraic objects which arise in mathematics and in computer science. Group theory has many important applications in physics, chemistry, and materials science. Group theory has been used elegantly in proving various important results in computer science, such as Barrington's theorem [3], results on the graph isomorphism problem [1,15] etc.

Algorithms for computational group theoretic problems are essential building blocks for many of the computer algebra systems such as GAP, SageMath, Singular etc. Some of the fundamental group theoretic algorithms were designed based on the ideas of Sims and Schreier (see [19]). Various computational group theoretic problems such as the group isomorphism problem, set stabilizer problem for permutation groups are also interesting from a purely complexity theoretic point of view for their connection with the graph isomorphism problem [8].

Two of the most commonly used ways of representing groups are via generators in the permutation group setting and via Cayley tables. Several interesting problems such as the group isomorphism problem, various property testing problems, the group factoring problem etc., have been studied for groups represented by their Cayley tables [5,10–12,18,21,22].

While a multiplication query for a group of order n can be answered in constant time in the Cayley table representation, the space required to store the

© Springer Nature Switzerland AG 2019
L. A. Gąsieniec et al. (Eds.): FCT 2019, LNCS 11651, pp. 229–242, 2019.
https://doi.org/10.1007/978-3-030-25027-0_16

table is $\mathcal{O}(n^2 \log n)$ bits or $\mathcal{O}(n^2)$ words in word-RAM model, which is prohibitively large. It is interesting to know if there are data structures to store a group using $o(n^2)$ words but still supporting constant query-time for multiplication. We construct a data structure that has constant query-time but uses just $\mathcal{O}(n^{1.05})$ words to represent the group. In fact, our result is more general and offers several other interesting space versus query-time trade-offs.

We note that there are succinct representations of groups such as the generator-relator representation (see [16]), polycyclic representation [20] etc., that store groups succinctly. However, answering multiplication queries generally takes too much time. For example with a polycyclic representation of a cyclic group it takes linear time to answer a multiplication query.

An easy information theoretic lower bound [9] states that to represent a group of order n, at least $n \log n$ bits (or $\Omega(n)$ words in word RAM model) are needed. We do not know if in general it is possible to use only $\mathcal{O}(n)$ words to store a group while supporting fast query-time. We show that for restricted classes of groups such as Dedekind groups, simple groups it is possible to construct data structures that use only $\mathcal{O}(n)$ space and answer multiplication query in $\mathcal{O}(\log n)$ time.

In the past succinct representation of groups has been studied for restricted classes of groups [9,13]. Farzan and Munro [9] defined a model of computation in which a compression algorithm is applied to the group to get a succinct canonical form for the group. The query processing unit in their model assumes that the group elements to be multiplied are in given by their labels. They also assume that the query processing architecture supports an extra operation called bit-reversal. In their model they show that for abelian groups, the query processing unit needs to store only constant number of words in order to answer multiplication queries in constant time. Farzan and Munro ask if their results can be extended to categorically larger classes of groups. We show that we can design succinct data structures with same space bounds and query-time for Hamiltonian groups and Z-groups. Hamiltonian groups are nonabelian groups all of whose subgroups are normal. Z-groups are groups all of whose Sylow subgroups are cyclic. There are many interesting nonabelian groups in the class of Z-groups. We also show that in their model constant query-time can be achieved for larger classes of groups defined in terms of semidirect products provided the query processing unit is allowed to use linear space.

2   Preliminary

In this section, we describe some of the group-theoretic definitions and background used in the paper. For more details see [4–7].

For a group G, the number of elements in G or the *order* of G is denoted by $|G|$. Let $x \in G$ be an element of group G, then $\mathrm{ord}_G(x)$ denotes the order of the element x in G, which is the smallest power i of x such that $x^i = e$, where e is the identity element of the group G. For a subset $S \subseteq G$, $\langle S \rangle$ denotes the subgroup generated by the set S.

A group *homomorphism* from $(G, \cdot)$ to $(H, \times)$ is a function $\varphi : G \longrightarrow H$ such that $\forall g_1, g_2 \in G, \varphi(g_1 \cdot g_2) = \varphi(g_1) \times \varphi(g_2)$. A bijective homomorphism is called an *isomorphism*. Let $\mathrm{Aut}(H)$ denote *automorphism* group of H, $\mathrm{Aut}(H) = \{\sigma \mid \sigma : A \longrightarrow A$ is an isomorphism $\}$. The set of all automorphism from a H to H under function composition forms a group. Two elements a and b with the conditions, $a = 1, a^2 = (ab)^2 = b^2$ generates a nonabelian group of order 8 known as *quaternian group*. A group is said be a *simple* if every non-trivial subgroup of it is not a normal subgroup. Let G be a finite group and A, B be subgroups of G. Then G is a *direct product* of A and B, denoted $G = A \times B$, if 1) $A \trianglelefteq G$ and $B \trianglelefteq G$, 2) $|G| = |A||B|$, 3) $A \cap B = \{e\}$.

Let A and B be two groups and let $\varphi : B \longrightarrow \mathrm{Aut}(A)$ be a homomorphism. The semidirect product of A and B with respect to φ, denoted $A \rtimes_\varphi B$, is a group whose underlying set is $A \times B$ and the group multiplication is define as follows: Let $(a_1, b_1), (a_2, b_2) \in A \times B$. The multiplication of (a_1, b_1) and (a_2, b_2) is defined as $(a_1(\varphi(b_1)(a_2)), b_1 b_2)$. It is routine to check that the resulting structure is indeed a group. A group G is called the semidirect product of two of its subgroups A and B if there exists $\varphi : B \longrightarrow \mathrm{Aut}(A)$ such that $G \cong A \rtimes_\varphi B$.

A group G is said to be *abelian* if $ab = ba, \forall a, b \in G$. The fundamental theorem for finitely generated abelian groups implies that a finite abelian group G can be decomposed as a direct product $G = G_1 \times G_2 \times \ldots \times G_t$, where each G_i is a cyclic group of order p^j for some prime p and integer $j \geq 1$. If a_i generates the cyclic group G_i for $i = 1, 2, 3, \ldots, t$ then the elements $a_1, a_2, \ldots, a_t$ are called a *basis* of G.

A group H is *Hamiltonian* if every subgroup of H is normal. It is a well known fact that [4] a group is Hamiltonian if and only if

- G is the quaternion group Q_8; or,
- G is the direct product of Q_8 and B, of Q_8 and A, or of Q_8 and B and A,

where A is an abelian group of odd order k and B is an elementary abelian 2-group.[1] A group is *Dedekind* if it is either abelian or Hamiltonian.

Let p^k is the highest power of a prime p dividing the order of a finite group G, a subgroup of G of order p^k is called a *Sylow p-subgroup* of G.

Z-groups are groups all of whose Sylow subgroups are cyclic.

Let G be a group with n elements. A sequence $(g_1, \ldots, g_k)$ of k group elements is said to be *cube generating set* of G if

$$G = \{g_1^{\epsilon_1} g_2^{\epsilon_2} \cdots g_k^{\epsilon_k} \mid \epsilon_i \in \{0, 1\}\} \qquad (1)$$

Let $G = \langle S \rangle$. The *Cayley graph* of the group G on generating set S is the directed graph $X = (V, E)$ where $V = G$ and $E = \{(g, gs) \mid g \in G, s \in S\}$. Additionally, every edge $(g, gs) \, \forall g \in G, \, \forall s \in S$ is labeled with s. We denote diameter(G, S) as the graph diameter of the Cayley graph of group G on generating set S.

---

[1] An elementary abelian 2-group is an abelian group in which every nontrivial element has order 2.

We use the % symbol as the modulo operator such that $a \% b$ denotes the remainder of a when divided by b.

2.1 Model of Computation

The model of computation we follow is the word RAM model, where random access can be done in constant time. Each register and memory unit can store $\mathcal{O}(\log n)$ bits where n is the input size (in our case n is the order of the given group). These memory units are called words. The arithmetic, logic and comparison operations on $\mathcal{O}(\log n)$ bits words take constant time. Unless stated otherwise we assume that the elements of the group are encoded as $1, 2, \ldots, n$.

The group G and its Cayley table are already known and we are allowed to preprocess the input in finite time in order to generate the required data structures for the multiplication operation. The time and space required in the preprocessing phase is not counted towards the space complexity and query time of the data structure. The space complexity is measured in terms of the number of words required for the storage of the data structure. The *multiplication query* for a group G takes two elements x and y, and it has to return $z = xy$.

We note that in this model the inverses of each element can be trivially stored in $\mathcal{O}(n)$ space. Thus, we primarily focus on the problem of answering multiplication queries using data structures that uses less space.

3 Our Results

In Sects. 4 and 5 we present succinct data structures for various group classes in standard word RAM model. Our results for various group classes are summarized in the table below. We start with a representation for general groups and then move towards more restricted group classes such as Dedekind groups, groups which are *almost* abelian and then finally consider simple groups (Table 1).

Table 1. Table of results

Group class	Space (in words)	Query time
General groups	$\mathcal{O}\left(\frac{n^{1+\delta}}{\delta}\right)$, $\frac{1}{\log n} \leq \delta \leq 1$	$\mathcal{O}\left(\frac{1}{\delta}\right)$
Dedekind groups	$\mathcal{O}(n)$	$\mathcal{O}(\log n)$
Semidirect product of two abelian groups	$\mathcal{O}(n)$	$\mathcal{O}(\log n)$
Simple groups	$\mathcal{O}(n)$	$\mathcal{O}(\log n)$
Groups whose every proper subgroup is abelian	$\mathcal{O}(n)$	$\mathcal{O}(n)$

In Sect. 6 we study succinct representation of groups in the model defined by Farzan and Munro [9]. Our results in this model are listed below.

Theorem 1. *There is a representation of Hamiltonian groups such that multiplication operation can be performed using $\mathcal{O}(1)$ space in $\mathcal{O}(1)$ time.*

Theorem 2. *There is a representation of Z-groups such that multiplication operation can be performed using $\mathcal{O}(1)$ space in $\mathcal{O}(1)$ time.*

Theorem 3. *There is a representation of groups $G = A \rtimes C_m$ such that multiplication operation can be performed in using $\mathcal{O}(|A|)$ space in $\mathcal{O}(1)$ time.*

4   Succinct Representation of Finite Groups

In this section we construct a succinct representation of finite groups that can quickly answer multiplication queries. More precisely,

Theorem 4. *Let G be a group of order n. Then for any δ such that $\frac{1}{\log n} \leq \delta \leq 1$, there is a representation of G that uses $\mathcal{O}(\frac{n^{1+\delta}}{\delta})$ space and answers multiplication queries in time $\mathcal{O}(\frac{1}{\delta})$.*

Proof. Let $\{g_1 g_2 \ldots g_k\}$ be a cube generating set for the group G. A cube generating set could be found by brute-force. For each $g \in G$ and $i \in [k]$ we fix $\epsilon_i(g)$ such that $g = \prod_{i=1}^{k} g_i^{\epsilon_i(g)}$.

Let $h \in G$. To compute the product hg, we first compute $x_1 = hg_1^{\epsilon_1(g)}$. Inductively, we compute $x_i = x_{i-1}g_i^{\epsilon_i(g)}$. Here $x_k = hg$. Note that $g_i^{\epsilon_i(g)}$ is either g_i or identity. In the later case there is actually no multiplication. With suitable data structures this method has query time $\mathcal{O}(k)$. However, to obtain a general result that gives interesting space versus query-time trade-offs we take the following route.

First, divide the k-length sequence $g_1^{\epsilon_1(g)} g_2^{\epsilon_2(g)} \cdots g_k^{\epsilon_k(g)}$ into l sized blocks as shown below.

$$g = \boxed{g_1^{\epsilon_1(g)} \cdots g_l^{\epsilon_l(g)}} \quad \boxed{g_{l+1}^{\epsilon_{l+1}(g)} \cdots g_{2l}^{\epsilon_{2l}(g)}} \quad \cdots \quad \boxed{g_r(g)^{\epsilon_r(g)} \cdots g_k^{\epsilon_k(g)}}$$
$$\underleftarrow{\quad l \quad}\overrightarrow{} \qquad \underleftarrow{\quad l \quad}\overrightarrow{} \qquad\qquad \underleftarrow{\quad l \quad}\overrightarrow{}$$

There are 2^l possible products in each block and each such product will be an element of the group G. We will store the result of the multiplication of every element $g \in G$ with each possible l-length combination from each block. Each element g can be seen as a sequence of m words $w_1(g), \ldots, w_m(g)$, where $m = \lceil \frac{k}{l} \rceil$ and

$$w_i(g) = \prod_{j=(i-1)\ell+1}^{i\ell} g_j^{\epsilon_j(g)} \tag{2}$$

Let $s_i(g)$ be the number whose binary representations is $\epsilon_{(i-1)l+1}(g) \ldots \epsilon_{il}(g)$. The number $s_i(g)$ can be viewed as a representation of the word $w_i(g)$.

Data Structures: In order to perform the multiplication, we will use the following data structures which are constructed during the preprocessing phase.

1. Word Arrays: For each $g \in G$ an array $\mathcal{W}_g$ of length m. The ith element $\mathcal{W}_g[i]$ in the array is set to $s_i(g)$.
2. Multiplication Arrays: For each $g \in G$ and $i \in [m]$ an array $\mathcal{A}_g^{(i)}$ of length 2^l. The jth element of $\mathcal{A}_g^{(i)}$ is computed as follows. First we compute the binary representation $\epsilon_1 \epsilon_2 \dots \epsilon_l$ of $j-1$ (possibly padding 0's in the left to make it an l-bit binary number). We set $\mathcal{A}_g^{(i)}[j] = g \, g_{(i-1)l+1}^{\epsilon_1} g_{(i-1)l+2}^{\epsilon_2} \cdots g_{il}^{\epsilon_l}$.

Query Time: Given h and g, we want to compute hg. First we obtain the sequence $s_1(g), \dots, s_m(g)$ from the word array $\mathcal{W}_g$. By design, this sequence corresponds to $w_1(g), \dots, w_m(g)$ and $g = w_1(g) \dots w_m(g)$. Now access array $\mathcal{A}_h^{(1)}[s_1(g)]$ to get the multiplication of the element h with word $w_1(g)$ to obtain x_1. Next access $\mathcal{A}_{x_1}^{(2)}[s_2(g)]$ to obtain $x_2 = x_1 w_2(g)$. Now repeat this process until we get the final result. The runtime is $\mathcal{O}(m)$ as we need to access the word arrays and the multiplication arrays $\mathcal{O}(m)$ times.

Space Complexity: The space use by the word arrays $\mathcal{W}$ is $\mathcal{O}(nm)$. The space used by the multiplication arrays $\mathcal{A}_g^{(i)}$, $i \in [m]$, $g \in G$ is $\mathcal{O}(2^l mn)$ as each array has length 2^l. The overall space is $\mathcal{O}(nm + 2^l mn)$ which is $\mathcal{O}(2^l mn)$.

Erdős and Renyi [7] showed that for any group of G of order n, there are cube generating sets of length $\mathcal{O}(\log n)$. If we set $l = \delta \log n$, then space used by our representation will be $\mathcal{O}(\frac{n^{1+\delta}}{\delta})$ words and the query time will be $\mathcal{O}(\frac{1}{\delta})$. Notice that as $l \geq 1$, we need $\delta \geq 1/\log n$. □

Corollary 1. *There is a representation of groups such that multiplication query can be answered in $\mathcal{O}(\frac{\log n}{\log \log n})$ time using $\mathcal{O}(\frac{n(\log n)^2}{\log \log n})$ space.*

Proof. Set $\delta = \mathcal{O}(\frac{\log \log n}{\log n})$ in Theorem 4.

Corollary 2. *There is a representation of groups such that multiplication query can be answered in $\mathcal{O}(\log n)$ time using $\mathcal{O}(n \log n)$ space.*

Proof. Set $\delta = \mathcal{O}(\frac{1}{\log n})$ in Theorem 4.

Corollary 3. *There is a representation of groups such that multiplication query can be answered in $\mathcal{O}(1)$ time using $\mathcal{O}(n^{1.05})$ space.*

Proof. Set $\delta = \frac{1}{20}$ in Theorem 4.

5   Succinct Representation for Restricted Group Classes

In many of the results in this paper, group elements are treated as tuples. For example, if $\{g_1, \dots, g_k\}$ is a cube generating set for a group G, $(\epsilon_1, \dots, \epsilon_k)$ is a representation of the group element $g_1^{\epsilon_1} \cdots g_k^{\epsilon_k}$. For many of the data structures we design, we want a way of encoding these tuples which can be stored efficiently. We also want to retrieve the group element from its encoding efficiently.

Forward and Backward Map: Let G be a group, $c_1, \ldots, c_k$ be k integers each greater than 1 with $\prod_i c_i = \mathcal{O}(n)$ and $F : G \longrightarrow [c_1] \times \cdots \times [c_k]$ be an injective map. Suppose $F(g) = (\alpha_1, \ldots, \alpha_k)$. Let $\bar{b}_i$ be the $\lceil \log c_i \rceil$-bit binary encoding of α_i (possibly some 0's padded on the left to make it a $\lceil \log c_i \rceil$-bit string). The concatenation $\mathbf{b} = \bar{b}_1 \ldots \bar{b}_k$ of the $\bar{b}_i$'s encodes $(\alpha_1, \ldots, \alpha_k)$. The encoding $\mathbf{b}$ can be stored in constant number of words as $\sum_i \lceil \log c_i \rceil = \mathcal{O}(\log n)$. Thus, F can be stored in an array $\mathcal{F}$, indexed by the group elements using $\mathcal{O}(n)$ words by setting $\mathcal{F}[g] = \mathbf{b}$. We call this *the forward map*. We also store an array $\mathcal{B}$, called *the backward map*, of dimension $c_1 \times \cdots \times c_k$ such that $\mathcal{B}[\alpha_1] \cdots [\alpha_k] = g$ if $F(g) = (\alpha_1, \cdots, \alpha_k)$.[2] Finally we also store each c_i in separate words, which could be used to extract $(\alpha_1, \cdots, \alpha_k)$ from $\mathcal{F}[g]$ in $\mathcal{O}(\log n)$ time. Notice, that while the access to $\mathcal{F}$ is constant time, the access time for $\mathcal{B}$ is $\mathcal{O}(k)$ which is $\mathcal{O}(\log n)$.

Theorem 5. *There is a representation of abelian groups using $\mathcal{O}(n)$ space such that a multiplication query can be answered in $\mathcal{O}(\log n)$ time.*

Proof. Let G be an abelian group. By structure theorem of finite abelian groups, G can be decomposed as a direct product $\langle g_1 \rangle \times \ldots \times \langle g_t \rangle$ of cyclic subgroups g_i's, where $g_i \in G$ is a generator for $\langle g_i \rangle$ of order c_i. This gives an injective mapping $F : G \longrightarrow [c_1] \times \cdots \times [c_k]$ with $\prod_i c_i = \mathcal{O}(n)$. This allows us to use the forward and backward maps discussed above. Multiplication of two group elements g and h with $F(g) = (\alpha_1, \ldots, \alpha_k)$ and $F(h) = (\beta_1, \ldots, \beta_k)$ can be done by just computing $(\alpha_i + \beta_i)\%c_i$ for all i and then consulting the backward map. $\qquad \square$

Theorem 6. *There is a data structure for representing Hamiltonian groups using $\mathcal{O}(n)$ space with $\mathcal{O}(\log n)$ multiplication query time.*

Proof. Let G be a Hamiltonian group. We know that G can be decomposed as $G = Q_8 \times A \times B$, where Q_8 is a quaternion group, B is an elementary abelian 2-group and A is an abelian group of odd order [4]. This decomposition can be obtained in polynomial time [11]. This decomposition gives us a bijection f from G to $Q_8 \times A \times B$. Both the bijection f and its inverse f^{-1} can be stored in linear space. To multiply two group elements x and y, we first find $f(x)$ and $f(y)$ which will of the form (q_1, a_1, b_1) and (q_2, a_2, b_2) respectively where $q_1, q_2 \in Q_8, a_1, a_2 \in A$ and $b_1, b_2 \in B$. Now to compute $xy = (q_1 q_2, a_1 a_2, b_1 b_2)$, we can use the result of Theorem 5 to take care of the abelian components in $\mathcal{O}(\log n)$ time. As the order of Q_8 is $\mathcal{O}(1)$, the multiplication in the first component can be found in $\mathcal{O}(1)$ time. To obtain the final result we use the reverse map f^{-1}. $\qquad \square$

We now proceed to study the nonabelian groups in which every proper subgroup is abelian. This requirement is similar to that of Dedekind groups where

---

[2] If $(\alpha_1, \cdots, \alpha_k) \notin \mathrm{Image}(F)$, then the value could be arbitrary.

every subgroup is required to be normal. These groups were introduced and studied by Miller and Moreno [17]. They also provided a classification for such groups.

Theorem 7. *There is a representation of groups whose every proper subgroup is abelian, using $\mathcal{O}(n)$ space such that multiplication query can be answered in $\mathcal{O}(n)$ time.*

Proof. Let G be a group whose every proper subgroup is abelian. Assume that group G is nonabelian. First we prove that such groups can be generated by *two elements*. Let M be any proper maximal subgroup of G. By definition of the group class M is abelian. Let $z \notin M$. Then by maximality of M, $\langle M \cup \{z\} \rangle = G$. Let $x \in M$ such that $xz \neq zx$. Such an element exists, as otherwise G will be abelian. As x and z do not commute they must generate the whole group as any proper subgroup is abelian. It is easy to see that verifying and therefore finding if two given elements generates a group can be done in polynomial time.

The succinct data structure for G will be the Cayley graph of G with the generating set $\{x, z\}$. To compute gh, first we compute a representation of h in terms of x and z. This can be done by traversing a path in the Cayley graph from the identity element to h (which will be of length at most n) and noting down the edge labels. Next we traverse the graph starting from g and following the exact same sequence of edge labels obtained in the first traversal.

The space taken to store the Cayley graph in the form of it's adjacency list is $\mathcal{O}(n)$ as there are only $2n$ edges. Answering a multiplication query, which involves two graph traversals, can be done in $\mathcal{O}(n)$ time. $\square$

Let $\mathcal{A}$ and $\mathcal{B}$ be two group classes. Let $\mathcal{G}_{\mathcal{A},\mathcal{B}} = \{G \mid G = A \rtimes_\varphi B, A \in \mathcal{A}, B \in \mathcal{B}, \text{and } \varphi \text{ is a homomorphism from } B \text{ to } \text{Aut}(A)\}$.

Theorem 8. *Let $\mathcal{A}, \mathcal{B}$ be two group classes. Suppose we are given data structures $D_{\mathcal{A}}$ and $D_{\mathcal{B}}$ for group classes $\mathcal{A}$ and $\mathcal{B}$ respectively. Let $S(D_{\mathcal{A}}, m_1)$, $S(D_{\mathcal{B}}, m_2)$ denote the space required by the data structures $D_{\mathcal{A}}, D_{\mathcal{B}}$ to represent groups of order m_1, m_2 from $\mathcal{A}, \mathcal{B}$ respectively. Let $Q(D_{\mathcal{A}}, m_1), Q(D_{\mathcal{B}}, m_2)$ denote the time required by the data structures $D_{\mathcal{A}}, D_{\mathcal{B}}$ to answer multiplication queries for groups of order m_1, m_2 from $\mathcal{A}, \mathcal{B}$ respectively. Then there is a representation of groups in $\mathcal{G}_{\mathcal{A},\mathcal{B}}$ such that multiplication query can be answered in $\mathcal{O}(\log n + Q(D_{\mathcal{A}}, |A|) + Q(D_{\mathcal{B}}, |B|))$ time and $\mathcal{O}(n + S(D_{\mathcal{A}}, |A|) + S(D_{\mathcal{B}}, |B|))$ space.*

Proof. First we describe the preprocessing phase. Given group G, finding two groups A, B and a homomorphism φ such that $G = A \rtimes_\varphi B$ can be done in finite time where $\varphi : B \longrightarrow \text{Aut}(A)$. Without loss of generality, one can assume that elements of the group A are numbered from 1 to $|A|$. For each element $b \in B$, we store its image $\varphi(b) \in \text{Aut}(A)$ in the array T_b indexed by elements of group A. Let $T = \{T_b \mid b \in B\}$ be the set of $|B|$ arrays.

Now we move on to the querying phase. Let g_1 and g_2 be the two elements to be multiplied. Let $g_1 = (a_1, b_1)$ and $g_2 = (a_2, b_2)$ such that $a_1, a_2 \in A$ and $b_1, b_2 \in B$ which can be obtained using the forward map array. The result of the multiplication query g_1g_2 is $(a_1(\varphi(b_1)(a_2)), b_1b_2)$. The only *non-trivial* computation involved here is computing $\varphi(b_1)(a_2)$, which can be obtained using array T_{b_1}. Let $T_{b_1}(a_2) = a_3$, then the result of the multiplication query g_1g_2 is (a_1a_3, b_1b_2) the components of which can be computed using data structures for A and B respectively to obtain (a_4, b_4) where $a_4 = a_1a_3$ and $b_4 = b_1b_2$. Finally using the backward map we can obtain the resultant element.

The data structures we use are – forward map array, $|B|$ many arrays T (each of size $|A|$), data structures for B and A and the backward map. Thus the overall space required is $\mathcal{O}(n + |B||A| + S(D_A, |A|) + S(D_B, |B|) + n)$ which is $\mathcal{O}(n + S(D_A, |A|) + S(D_B, |B|))$.

The query time constitutes of the time required to get a representation of g_1 and g_2 as (a_1, b_1) and (a_2, b_2) respectively using the forward array, time required to compute $(a_1(\varphi(b_1)(a_2)), b_1b_2)$ using T_{b_1}, time required to compute a_1a_3 and b_1b_2 and time required to access the backward map array to obtain the resultant element. Thus the overall time required is $\mathcal{O}(1 + 1 + Q(D_A, |A|) + Q(D_B, |B|) + \log n)$ which is $\mathcal{O}(\log n + Q(D_A, |A|) + Q(D_B, |B|))$. $\qquad\square$

We now present a corollary which directly follows from the above theorem.

Corollary 4. *For groups which can be decomposed as a semidirect product of two abelian subgroups, there is a succinct data structure that uses $\mathcal{O}(n)$ space and answers queries in time $\mathcal{O}(\log n)$.*

The group class mentioned in the above corollary includes extension of abelian groups by cyclic group and has been studied in the context of the group isomorphism problem [14].

Simple groups serve as the building blocks for classifying finite groups. We next present a succinct representation of simple groups.

Theorem 9. *There is a representation of simple groups using $\mathcal{O}(n)$ space such that multiplication query can be answered in $\mathcal{O}(\log n)$ time.*

Proof. The case of abelian groups is already discussed in Theorem 5. We assume that group is nonabelian. Babai, Kantor and Lubotsky [2] proved that there is a constant c such that every nonabelian finite simple group has a set S of size at most 14 generators such that the diameter of the Cayley graph of G with respect to S has diameter at most $c\log n$. Such a generating set can be found by iterating over all possible subsets of size 14. Let $G = \langle S \rangle = \langle s_1, \cdots, s_{14} \rangle$. Each $g \in G$ can be represented by the edge labels in one of fixed shortest paths from the identity to g in the Cayley graph. By the result of Babai, Kantor and Lubotsky [2] the length of the path is $\mathcal{O}(\log n)$. The edge labels are from $\{1, 2, \ldots, 14\}$ indicating the generators associated with the edge. This representation of each element by the sequence of edge labels can be stored using a forward map. We also store a multiplication table M of dimension $|G| \times [14]$. We set $M[g][i] = gg_i$. To multiply two elements g and h we consult the forward map for the representation of h and then use M to compute the gh in $\mathcal{O}(\log n)$ time. $\qquad\square$

6   Representation in the Model of Farzan and Munro

In this section we use the model of computation defined by Farzan and Munro [9] for the succinct representation of abelian groups. We describe the model briefly here. For further details about this model, refer to [9]. Farzan and Munro use a Random Access Model (RAM) where a word is large enough to hold the value of the order of the input group n. The model also assumes the availability of bit-reversal as one of the native operations which can be performed in $\mathcal{O}(1)$ time.

Given a group G, *the compression algorithm* is defined as the process that takes G as an input and outputs a succinct representation (called *compressed form*) of G. The *labeling* of elements of group G (based on the compression) is a representation of the elements. Let A be an abelian group and t be the number of cyclic factors in the structural decomposition of A. We denote by $\mathcal{L}_A : A \longrightarrow \mathbb{N}^t$ the labeling of elements as per Farzan and Munro's labeling [9].

We denote by *outside user*, the entity responsible for the preprocessing operations such as compression, labeling etc. We denote by *query processing unit*, the entity responsible for performing the actual multiplication. The query processing unit is responsible for storing the compressed form of the group G. The outside user is responsible for supplying to the query processing unit the labels of the group elements to be multiplied. The query processing unit returns the label of the result of the multiplication query. The space and time required in the compression and labeling phase is not counted towards the algorithm's space complexity and query time. Thus, in the following sections, we only consider the space and time consumed by the query processing unit.

Theorem 10 ([9]). *There is a representation of finite abelian group of order n that uses constant number of words and answers multiplication queries in constant time.*

Answering the question posed by [9], we design data structures similar to the ones used in Theorem 10, for Hamiltonian groups and Z-groups. We also come up with a representation for groups which can be expressed as a semidirect product of an abelian group with a cyclic group.

6.1   Hamiltonian Groups

Theorem 1. *There is a representation of Hamiltonian groups such that multiplication query can be answered in $\mathcal{O}(1)$ time using $\mathcal{O}(1)$ space.*

Proof. Let G be a Hamiltonian group. We know that G can be decomposed as $G = Q_8 \times C$, where Q_8 is a quaternion group and C is abelian (See Sect. 2). The compressed form of group G is same as the compressed form of the abelian group C. Every element $g \in G$ has a representation of the form (q, d) where $q \in Q_8$ and $d \in C$. The elements $q \in Q_8$ are assigned labels from the set $\{1, \ldots, 8\}$. Since C is abelian, we use $\mathcal{L}_C(d)$ as the label for element $d \in C$. Since the order of Q_8 is constant, storing its entire Cayley table in some arbitrary but fixed representation requires $\mathcal{O}(1)$ space.

Given two elements r, s of G such that $r = (q_1, d_1)$ and $s = (q_2, d_2)$ where $q_1, q_2 \in Q_8$ and $d_1, d_2 \in C$. The result of rs is $(q_1 q_2, d_1 d_2)$. The multiplication of q_1 and q_2 can be computed in $\mathcal{O}(1)$ time using the stored Cayley table. After obtaining the labels $\mathcal{L}_C(d_1)$ and $\mathcal{L}_C(d_2)$ of elements d_1 and d_2 respectively, we can perform Farzan's multiplication algorithm to obtain the result of the multiplication of d_1 and d_2 in $\mathcal{O}(1)$ time. The overall space required is $\mathcal{O}(1)$ words. □

6.2 Z-groups

We now consider Z-groups which are semidirect product of two cyclic groups. This group class contains the groups studied by Le Gall [14]. We exploit the fact that every automorphism of a cyclic group is a cyclic permutation.

Theorem 2. *There is a representation of Z-groups using $\mathcal{O}(1)$ space such that multiplication query can be answered in $\mathcal{O}(1)$ time.*

Proof. Let $G = C_m \rtimes_\varphi C_d$ be a Z-group where $\varphi : C_d \longrightarrow \mathrm{Aut}(C_m)$ is a homomorphism and $C_m = \langle g \rangle$ and $C_d = \langle h \rangle$. Without loss of generality, we assume that the elements of C_m are numbered from the set $[m]$ in the natural cyclic order starting from g. Let $\sigma_j := \varphi(h^j)$. Let $\sigma_j(c)$ denote the image of the element $c \in C_m$ under the automorphism σ_j. In the compression process we first obtain a decomposition of G as $C_m \rtimes_\varphi C_d$. The compressed form of group G comprises of the two integers m and d and the compressed form of φ which is $\sigma_1(g)$. In the labeling phase, we label each element $t \in G$, such that $t = (g^i, h^j)$ as $(i, (\sigma_j(g), j))$ where $i \in [m]$ and $j \in [d]$. Note that, with this labeling (computed by the outside user), representing any element from G takes $\mathcal{O}(1)$ words.

In the querying phase, given two elements $r, s \in G$ such that $r = (g^{i_1}, h^{j_1})$ and $s = (g^{i_2}, h^{j_2})$ where $i_1, i_2 \in [m]$ and $j_1, j_2 \in [d]$. The result of rs which is $(g^{i_1} \varphi(h^{j_1})(g^{i_2}), h^{j_1} h^{j_2}) = (g^{i_1} \sigma_{j_1}(g^{i_2}), h^{j_1 + j_2})$. To compute $\sigma_{j_1}(g^{i_2})$, first obtain $\sigma_{j_1}(g)$ from the label of $r = (g^{i_1}, h^{j_1})$. Now to compute $\sigma_{j_1}(g^{i_2})$ we need to perform one integer multiplication operation $(\sigma_{j_1}(g) \times i_2) \% m$. Now the problem of multiplication reduces component-wise to the cyclic case. The multiplication query can thus be answered in $\mathcal{O}(1)$ time using $\mathcal{O}(1)$ space to store the orders of C_m and C_d. □

6.3 Semidirect Product Classes

A natural way to construct nonabelian groups is by the extension of abelian groups. The groups which can be formed by semidirect product extension of abelian groups by cyclic groups has been studied by Le Gall [14]. We denote $\mathcal{G}$ to be the class of groups which can be written as $G = A \rtimes C_m$, where A is an abelian group and C_m is a cyclic group. It is easy to see that the group class $\mathcal{G}$ is *categorically larger* than abelian groups as it contains all abelian groups as well as some nonabelian groups. Without loss of generality, assume that the elements of the group A are numbered from 1 to $|A|$.

Fact 1. Any permutation can be decomposed as a composition of disjoint cycles [6].

Lemma 1. *Given an abelian group A, a permutation π on the set $\{1,\ldots,|A|\}$ and an element $g \in A$, there exists a representation of π such that $\pi^d(g)$ for any element $g \in A$ and $d \in [m]$ can be computed in $\mathcal{O}(1)$ time using $\mathcal{O}(|A|)$ words of space.*

Proof. Let $\pi = \pi_1 \circ \pi_2 \circ \cdots \circ \pi_l$ be the decomposition of π into disjoint cycles $\pi_i, i \in [l]$. Such a decomposition of the input permutation π can be computed in polynomial time. Let $\mathcal{C}_1, \ldots \mathcal{C}_l$ be arrays corresponding to the cycles $\pi_1, \ldots, \pi_l$ respectively. For any cycle π_i store the elements of the cycle in $\mathcal{C}_i$ in the same order as they appear in π_i, starting with the least element of π_i. Construct an array $\mathcal{B}$ indexed by the elements of group A, storing for each $g \in A$, $\mathcal{B}[g] = (j, r)$, where $j \in [l]$ and $r \in \{0, \ldots, |\mathcal{C}_j| - 1\}$ such that $\mathcal{C}_j[r] = g$.

Now, in order to compute $\pi^d(g)$, first we obtain j and r from $\mathcal{B}$ such that the g appears in the cycle π_j at the rth index. Then we compute $r' := (r + d) \% |\mathcal{C}_j|$ and finally return $\mathcal{C}_j[r']$. This requires $\mathcal{O}(1)$ time as the involved operations are one word operations. Note that we require overall $\mathcal{O}(|A|)$ space to store the arrays $\mathcal{C}_1, \ldots, \mathcal{C}_l$ as $\sum_i |\mathcal{C}_i| = |A|$, and the space required for $\mathcal{B}$ is also $\mathcal{O}(|A|)$. ∎

Theorem 3. *There is a representation of groups $G \in \mathcal{G}$ such that multiplication query can be answered in $\mathcal{O}(1)$ time using $\mathcal{O}(|A|)$ space.*

Proof. Let $G \in \mathcal{G}$ be such that $G = A \rtimes_\varphi C_m$ where $\varphi : C_m \longrightarrow \text{Aut}(A)$ is a homomorphism. In the compression process, we first obtain the decomposition of group G as $A \rtimes_\varphi C_m$. The compressed form of group G comprises of the compressed form of group A, the integer m and the succinct representation of the homomorphism φ (described below).

Let $C_m = \langle g \rangle$ and $\pi := \varphi(g)$. Note that π is a bijection on the set $\{1, \ldots, |A|\}$. Using Lemma 1 we can store π in $\mathcal{O}(|A|)$ words, such that $\pi^d(a)$ for $a \in A$ can be computed in $\mathcal{O}(1)$ time. This forms the succinct representation(compressed form) of φ. Since the data structure used above is a part of the compressed form of group G, the query processing unit is responsible for storing it. We label each element $h \in G$ such that $h = (a, g^i)$ as $((\mathcal{L}_A(a), a), i)$ where $a \in A$, $c \in C_m$. This labeling requires $\mathcal{O}(1)$ words of space for each element $h \in G$.

In the querying phase, given two elements $r = (a_1, c_1)$ and $s = (a_2, c_2)$ of G such that $a_1, a_2 \in A$ and $c_1, c_2 \in C_m$, the result of rs is $(a_1(\varphi(c_1)(a_2)), c_1 c_2)$. Let $c_1 = g^{k_1}$ and $c_2 = g^{k_2}$. Now $\varphi(c_1)(a_2)$ which is $\pi^{k_1}(a_2)$ can be computed using Lemma 1 in $\mathcal{O}(1)$ time. Let $a_3 := \pi^{k_1}(a_2)$, then $rs = (a_1 a_3, c_1 c_2)$. Since the query processing unit is storing the labels of all the elements of A, it can obtain the label for the element a_3. After obtaining the labels $\mathcal{L}_A(a_1)$ and $\mathcal{L}_A(a_3)$ of elements a_1 and a_3 respectively, we can perform Farzan's multiplication algorithm to obtain the result of the multiplication of a_1 and a_3 in $\mathcal{O}(1)$ time. Let $c_3 = g^{k_3}$ be result of multiplication of c_1 and c_2. Then $k_3 = (k_1 + k_2) \% m$ can be computed in $\mathcal{O}(1)$ time.

In the query processing unit, we are storing the elements of the group A along with their labels, which takes $\mathcal{O}(|A|)$ words of space. The query processing unit also needs $\mathcal{O}(|A|)$ space to store the data structures from Lemma 1. Thus the total space required is $\mathcal{O}(|A|)$. $\qquad\square$

References

1. Babai, L.: Graph isomorphism in quasipolynomial time. In: Proceedings of the Forty-Eighth Annual ACM Symposium on Theory of Computing, pp. 684–697. ACM (2016). https://doi.org/10.1145/2897518.2897542
2. Babai, L., Kantor, W.M., Lubotsky, A.: Small-diameter cayley graphs for finite simple groups. Eur. J. Comb. **10**, 507–522 (1989). https://doi.org/10.1016/S0195-6698(89)80067-8
3. Barrington, D.A.: Bounded-width polynomial-size branching programs recognize exactly those languages in NC^1. J. Comput. Syst. Sci. **38**, 150–164 (1989). https://doi.org/10.1016/0022-0000(89)90037-8
4. Carmichael, R.D.: Introduction to the Theory of Groups of Finite Order. GINN and Company (1937)
5. Chen, L., Fu, B.: Linear and sublinear time algorithms for the basis of abelian groups. Theor. Comput. Sci. **412**, 4110–4122 (2011). https://doi.org/10.1016/j.tcs.2010.06.011
6. Dummit, D.S., Foote, R.M.: Abstract Algebra, vol. 3. Wiley, Hoboken (2004)
7. Erdös, P., Rényi, A.: Probabilistic methods in group theory. J. d'Analyse Math. **14**, 127–138 (1965). https://doi.org/10.1007/BF02806383
8. Eugene, M.: Permutation groups and polynomial-time computation. In: Groups and Computation: Workshop on Groups and Computation, 7–10 October 1991, vol. 11, p. 139. American Mathematical Society (1993)
9. Farzan, A., Munro, J.I.: Succinct representation of finite abelian groups. In: Proceedings of the Symbolic and Algebraic Computation, International Symposium, ISSAC 2006, Genoa, Italy, 9–12 July 2006, pp. 87–92. ACM (2006). https://doi.org/10.1145/1145768.1145788
10. Kavitha, T.: Linear time algorithms for abelian group isomorphism and related problems. J. Comput. Syst. Sci. **73**, 986–996 (2007). https://doi.org/10.1016/j.jcss.2007.03.013
11. Kayal, N., Nezhmetdinov, T.: Factoring groups efficiently. In: Albers, S., Marchetti-Spaccamela, A., Matias, Y., Nikoletseas, S., Thomas, W. (eds.) ICALP 2009. LNCS, vol. 5555, pp. 585–596. Springer, Heidelberg (2009). https://doi.org/10.1007/978-3-642-02927-1_49
12. Kumar, S.R., Rubinfeld, R.: Property testing of abelian group operations (1998)
13. Leedham-Green, C.R., Soicher, L.H.: Collection from the left and other strategies. J. Symb. Comput. **9**, 665–675 (1990). https://doi.org/10.1016/S0747-7171(08)80081-8
14. Le Gall, F.: An efficient quantum algorithm for some instances of the group isomorphism problem. In: 27th International Symposium on Theoretical Aspects of Computer Science, vol. 5, pp. 549–560 (2010). https://doi.org/10.4230/LIPIcs.STACS.2010.2484
15. Luks, E.M.: Isomorphism of graphs of bounded valence can be tested in polynomial time. J. Comput. Syst. Sci. **25**, 42–65 (1982). https://doi.org/10.1016/0022-0000(82)90009-5

16. Magnus, W., Karrass, A., Solitar, D.: Combinatorial group theory: presentations of groups in terms of generators and relations. Courier Corporation (2004)

17. Miller, G.A., Moreno, H.C.: Non-abelian groups in which every subgroup is abelian. Trans. Am. Math. Soc. **4**, 398–404 (1903). https://doi.org/10.2307/1986409

18. Qiao, Y., Sarma M.N., J., Tang, B.: On isomorphism testing of groups with normal hall subgroups. In: 28th International Symposium on Theoretical Aspects of Computer Science (STACS 2011), pp. 567–578 (2011). https://doi.org/10.4230/LIPIcs.STACS.2011.567

19. Seress, Á.: Permutation Group Algorithms, vol. 152. Cambridge University Press, Cambridge (2003)

20. Sims, C.C.: Computation with Finitely Presented Groups, vol. 48. Cambridge University Press, Cambridge (1994)

21. Vikas, N.: An $\mathcal{O}(n)$ algorithm for abelian p-group isomorphism and an $\mathcal{O}(n \log n)$ algorithm for abelian group isomorphism. J. Comput. Syst. Sci. **53**, 1–9 (1996). https://doi.org/10.1006/jcss.1996.0045

22. Wilson, J.B.: Existence, algorithms, and asymptotics of direct product decompositions, I. Groups Complex. Cryptol. **4**, 33–72 (2012). https://doi.org/10.1515/gcc-2012-0007

On the Tractability of Covering a Graph with 2-Clubs

Riccardo Dondi[1(✉)] and Manuel Lafond[2]

[1] Università degli Studi di Bergamo, Bergamo, Italy
riccardo.dondi@unibg.it
[2] Université de Sherbrooke, Québec, Canada
manuel.lafond@USherbrooke.ca

Abstract. Covering a graph with cohesive subgraphs is a classical problem in theoretical computer science. In this paper, we prove new complexity results on the Min 2-Club Cover problem, a variant recently introduced in the literature which asks to cover the vertices of a graph with a minimum number of 2-clubs (which are induced subgraphs of diameter at most 2). First, we answer an open question on the decision version of Min 2-Club Cover that asks if it is possible to cover a graph with at most two 2-clubs, and we prove that it is W[1]-hard when parameterized by the distance to a 2-club. Then, we consider the complexity of Min 2-Club Cover on some graph classes. We prove that Min 2-Club Cover remains NP-hard on subcubic planar graphs, W[2]-hard on bipartite graphs when parameterized by the number of 2-clubs in a solution and fixed-parameter tractable on graphs having bounded treewidth.

1 Introduction

Covering a graph with cohesive subgraphs, in particular cliques, is a relevant problem in theoretical computer science with many practical applications. Two classical problems in this direction are the Minimum Clique Cover problem [13], and the Minimum Clique Partition problem [13], which are known to be NP-hard [16] even in restricted cases [5,6,10,24]. These two problems are based on the clique model and ask for cliques that cover the input graph. Other definitions of cohesive graphs have been considered in the literature, some of them called *relaxed cliques* [18], and rather ask for subgraphs that are "close" to a clique. For example, while each pair of distinct vertices in a clique are at distance exactly one, an *s-club*, where s is an integer greater than or equal to one, relaxes this constraint and is defined as a subgraph whose vertices are at distance at most s from each other. A 1-club is a clique, so a natural step towards generalizing cliques using distances is to study the $s = 2$ case, especially given that 2-clubs have relevant practical applications in social network analysis and bioinformatics [1,3,9,20–22]. Hence, in this paper, we focus on 2-clubs.

Finding 2-clubs and, more generally s-clubs, of maximum size, a problem known as Maximum s-Club, has been extensively studied in the literature. Maximum s-Club is NP-hard, for each $s \geqslant 1$ [4]. Furthermore, the decision version of

© Springer Nature Switzerland AG 2019
L. A. Gąsieniec et al. (Eds.): FCT 2019, LNCS 11651, pp. 243–257, 2019.
https://doi.org/10.1007/978-3-030-25027-0_17

the problem that asks whether there exists an s-club larger than a given size in a graph of diameter $s + 1$ is NP-complete, for each $s \geqslant 1$ [3]. Maximum s-Club has also been studied in the parameterzied complexity framework. The problem is fixed-parameter tractable when parameterized by the size of an s-club [7,19,25]. Moreover the problem has been studied for structural parameters and in chordal graphs and weakly chordal graphs [14,15]. The approximation complexity of the problem has also been considered. Maximum s-Club on an input graph $G = (V, E)$ is approximable within factor $|V|^{1/2}$, for every $s \geqslant 2$ [2] and not approximable within factor $|V|^{1/2-\varepsilon}$, for each $\varepsilon > 0$ and $s \geqslant 2$ [2].

Recently, the relaxation approach of s-clubs has been applied to the Minimum Clique Partition problem in order to cover a graph with s-clubs instead of cliques. More precisely, the Min s-Club Cover problem asks for a minimum collection $\{C_1, \ldots, C_h\}$ of subsets of vertices (possibly not disjoint) whose union contains every vertex, and such that every C_i, $1 \leqslant i \leqslant h$, is an s-club. This problem has been considered in [8], in particular for $s = 2, 3$. The decision version of the problem is NP-complete when it asks whether it is possible to cover a graph with two 3-clubs, and whether is possible to cover a graph with three 2-clubs [8]. Min 3-Club Cover on an input graph $G = (V, E)$ has been shown to be not approximable within factor $|V|^{1-\varepsilon}$, for each $\varepsilon > 0$, while Min 2-Club Cover on an input graph $G = (V, E)$ is approximable within factor $O(|V|^{1/2} \log^{3/2} |V|)$ and not approximable within factor $|V|^{1/2-\varepsilon}$ [8].

In this paper, we present results on the complexity of Min 2-Club Cover. First, in Sect. 3 we answer an open question on the decision version of Min 2-Club Cover that asks if it is possible to cover a graph with at most two 2-clubs, and we prove that it is not only NP-hard, but W[1]-hard even when parameterized by the parameter "distance to 2-club". Notice that in contrast, the problem that asks if it possible to cover a graph with two cliques is in P [12]. Our hardness is obtained through an intermediate problem, called the Steiner-2-Club, which asks whether a given subset of k vertices belongs to *some* 2-club. We show that this latter problem is W[1]-hard when paramterized by k. Then, we consider the complexity of Min 2-Club Cover on some graph classes. In Sect. 4 we prove that Min 2-Club Cover is NP-hard on subcubic planar graphs. In Sect. 5 we prove that Min 2-Club Cover on a bipartite graph $G = (V, E)$ is W[2]-hard when parameterized by the number of 2-clubs in a solution and not approximable within factor $\Omega(\log(|V|))$. Finally, we prove in Sect. 6 that Min 2-Club Cover is fixed-parameter tractable on graphs having bounded treewidth. We start in Sect. 2 by giving some definitions and by defining formally the Min 2-Club Cover problem. Some of the proofs are omitted due to space constraints.

2   Preliminaries

Given a graph $G = (V, E)$ and a subset $W \subseteq V$, we denote by $G[W]$ the subgraph of G induced by V'. Given two vertices $u, v \in V$, the distance between u and v in G, denoted by $d_G(u, v)$, is the number of edges on a shortest path from u to v. The diameter of a graph $G = (V, E)$ is the maximum distance between two vertices of V. Given a graph $G = (V, E)$ and a vertex $v \in V$, we denote by $N_G(v)$ the set of neighbors of v, that is $N_G(v) = \{u : \{v, u\} \in E\}$. We denote by $N_G[v]$ the closed neighborhood of V, that is $N_G[v] = N_G(v) \cup \{v\}$. Given a set $V' \subseteq V$, we denote by $N(V') = \{u : \{v, u\} \in E, v \in V'\} \setminus V'$.

Definition 1. *Given a graph $G = (V, E)$, a subset $V' \subseteq V$, such that $G[V']$ has diameter at most 2, is a 2-club.*

Notice that a 2-club must be connected, and that $d_{G[V']}(u, v)$ might differ from $d_G(u, v)$. Now we present the definition of the problem we are interested in, called Minimum 2-Club Cover.

Problem 1. Minimum 2-Club Cover (Min 2-Club Cover)
Input: A graph $G = (V, E)$.
Output: A minimum cardinality collection $\mathcal{C} = \{V_1, \ldots, V_h\}$ such that, for each i with $1 \leqslant i \leqslant h$, $V_i \subseteq V$, V_i is a 2-club, and, for each vertex $v \in V$, there exists a set V_j, with $1 \leqslant j \leqslant h$, such that $v \in V_j$.

Notice that the 2-clubs in $\mathcal{C} = \{V_1, \ldots, V_h\}$ do not have to be disjoint. We denote by 2-Club Cover(h), with $1 \leqslant h \leqslant |V|$, the decision version of Min 2-Club Cover that asks whether there exists a cover of G consisting of at most h 2-clubs.

3   W[1]-Hardness of 2-Club Cover(2) for Parameter Distance to 2-club

In this section, we show that 2-Club Cover(2), i.e. deciding if a graph can be covered by two 2-clubs, is W[1]-hard for the parameter "distance to 2-club", which is the number of vertices to be removed from the input graph $G = (V, E)$ such that the resulting graph is a 2-club. Note that Max s-Club is FPT for this parameter [25]. In order to prove this result, we first prove the W[1]-hardness of an intermediate decision problem, called the Steiner-2-Club (whose W[1]-hardness may be of independent interest), even in a restricted case. Then we present a reduction from this restriction of Steiner-2-Club to 2-Club Cover(2).

Problem 2. Steiner-2-Club
Input: A graph $G_s = (V_s, E_s)$, and a set $X_s \subseteq V_s$.
Output: Does there exist a 2-club in G_s that contains every vertex of X_s?

We call X_s the set of *terminal vertices*. We show that Steiner-2-Club is W[1]-hard for parameter $|X_s|$, by a parameter-preserving reduction from Multicolored Clique.

Problem 3. Multicolored Clique
Input: A graph $G_c = (V_c, E_c)$, where V_c is partitioned into k independent sets $V_{c,1}, \ldots, V_{c,k}$ (hereafter called the *color classes*).
Output: Does there exist a clique $V'_c \subseteq V_c$ such that $|V'_c| = k$?

Note that V_c' has one vertex per color class, that is, for each $1 \leqslant i \leqslant k$, $|V_c' \cap V_{c,i}| = 1$. It is well-known that Multicolored Clique is W[1]-hard for parameter k [11].

Our proof holds on a restriction of Steiner-2-Club called Restricted Steiner-2-Club, where the set X_s is an independent set, $|X_s| > 4$, and each vertex in $V_s \setminus X_s$ has at most 2 neighbors in X_s. We start by giving an hardness result for Restricted Steiner-2-Club.

Theorem 2. *The* Restricted Steiner-2-Club *problem is W[1]-hard with respect to the number of terminal vertices* $|X_S|$.

We can now prove the hardness of 2-Club Cover(2).

Theorem 3. *The* 2-Club Cover(2) *problem is W[1]-hard for the parameter distance to 2-club.*

Proof. Let $(G_s = (V_s, E_s), X_s)$ be an instance of Restricted Steiner-2-Club, where $k = |X_s|$ and $V_s = \{v_1, \ldots, v_n\}$. It follows from Theorem 2 that Restricted Steiner-2-Club is W[1]-hard when parameterized by k. Recall that in Restricted Steiner-2-Club $|X_s| = k > 4$. Finally, denote $V_s' = V_s \setminus X_s$.

Starting from $(G_s = (V_s, E_s), X_s)$, we construct an instance $G = (V, E)$ of 2-Club Cover(2), where $V = H \uplus W \uplus Y \uplus Z$ (here $\uplus$ means disjoint union). First, we define the sets H, W, Y, Z and the edges of the subgraphs $G[H]$, $G[W]$, $G[Y]$ and $G[Z]$, then the remaining edges of G. The subgraph $G[H] = (H, E_H)$, is defined as follows:

$$H = \{h_i : v_i \in V_s\} \quad E_H = \{\{h_i, h_j\} : \{v_i, v_j\} \in E_s\}$$

hence H is a copy of G_s. Moreover, define $H_X \subseteq H$ as follows

$$H_X = \{h_i \in H : v_i \in X_s\}$$

Notice that H_X is an independent set in G.

The subgraph $G[W] = (W, E_W)$ is a complete graph containing a vertex for each two vertices v_i, v_j in V_s', with $1 \leqslant i < j \leqslant n - k$, defined as follows:

$$W = \{w_{i,j} : v_i, v_j \in V_s'\} \quad E_W = \{\{w_{i,j}, w_{h,l}\} : w_{i,j}, w_{h,l} \in W\}.$$

The subgraph $G[Y] = (Y, E_Y)$ is also complete and has a vertex for each $v_i \in V_s'$. It is defined as follows:

$$Y = \{y_i : v_i \in V_s'\} \quad E_Y = \{\{y_i, y_j\} : y_i, y_j \in Y\}.$$

The subgraph $G[Z] = (Z, E_Z)$ is the same as $G[Y]$, and is defined as follows:

$$Z = \{z_i : v_i \in V_s'\} \quad E_Z = \{\{z_i, z_j\} : z_i, z_j \in Z\}.$$

Finally, we define the edges in E between two vertices that belong to different sets in H, W, Y and Z. Informally, every edge between W and Y is present, and every edge between Y and Z is present (point 1 and 2 below). Each vertex $w_{i,j}$ of W is connected with vertices h_i and h_j of H (point 3 below). Similarly, each vertex of Y is connected with the corresponding vertex of H (point 4 below).

1. $\{w_{i,j}, y_l\} \in E$, for each $w_{i,j} \in W$ and each $y_l \in Y$
2. $\{y_i, z_j\} \in E$, for each $y_i \in Y$ and each $z_j \in Z$
3. For each $\{v_i, v_j\}$ of V'_s, with $i < j$, $\{h_i, w_{i,j}\} \in E$ and $\{h_j, w_{i,j}\} \in E$, where $h_i, h_j \in H \setminus H_X$ and $w_{i,j} \in W$
4. For each $v_i \in V'_s$, $\{h_i, y_i\} \in E$, with $h_i \in H \setminus H_X$ and $y_i \in Y$.

Notice that, by construction, $W \cup Y$ and $Y \cap Z$ are cliques.

We first prove that $G = (V, E)$ has a distance to 2-club of exactly k. First note that a vertex of H_X and vertex of Z are at distance three in G, since they don't share any common neighbour in G. It follows that to obtain a 2-club from G, either all the vertices of H_X or all the vertices of Z have to be removed from G. This implies a distance of at least k from a 2-club, since $|H_X| = k$.

Next we prove in the following claim that $V \setminus H_X$ is a 2-club.
Claim (1). $V \setminus H_X$ is a 2-club of G.
Thus we have shown that $V \setminus H_X$ is a 2-club in G and that G has distance at most $|H_X| = k$ from a 2-club.

In order to complete the prove, we have to show that there exists a solution of Restricted Steiner-2-Club on instance (G_s, X_s) if and only G can be covered by two 2-clubs.

First assume that Restricted Steiner-2-Club on instance (G_s, X_s) admits a 2-club C_s containing X_s. Then, we claim that $V \setminus H_X$ and $C = \{h_i \in H : v_i \in C_s\}$ are a solution of 2-Club Cover(2) on instance G, that is they are two 2-clubs of G and cover every vertex of V. First notice that, since $X_s \subseteq C_s$, then $H_X \subseteq C$ and thus $V = C \cup (V \setminus H_X)$ as desired. It remains to show that C and $V \setminus H_X$ are 2-clubs of G. By Claim 1, we already know that $V \setminus H_X$ is a 2-club of G. Moreover, since $G[H]$ is isomorphic to G_s and C_s is a 2-club of G_s, C is also 2-club of G.

Conversely, suppose that $G = (V, E)$ can be covered by two 2-clubs C_1 and C_2. First, notice that vertices of H_X and vertices of Z are at distance 3 from each other. It follows that one of these 2-clubs, say C_1, satisfies $H_X \subseteq C_1$, while the other, in our case C_2, satisfies $Z \subseteq C_2$. We claim that $(W \cup Y) \cap C_1 = \emptyset$. Assume that there exists a vertex $w_{i,j} \in W \cap C_1$, where $v_i, v_j \in V'_s$ are the vertices of G_s corresponding to $w_{i,j}$. Consider a common neighbor r of $w_{i,j}$ and any $h_l \in H_X$ in C_1. By construction, then $r \in H \setminus H_X$, since h_l has only neighbors in $H \setminus H_X$. It follows that $r = h_i$ or $r = h_j$, as the only vertices of $H \setminus H_X$ adjacent to $w_{i,j}$ are h_i or h_j. This holds for each $h_l \in H_X$, thus $H_X \subseteq N(h_i) \cup N(h_j)$. By construction, since $G[H]$ and G_s are isomorphic, then $v_i, v_j \in X_s$ have at most two neighbors in $V_s \setminus X_s$ and $h_i, h_j \in H \setminus H_X$ have at most two neighbors in H_X. Since $N(w_{i,j}) \cap (H \setminus H_X) = \{h_i, h_j\}$, it follows that $H_X \subseteq N(h_1) \cup N(h_2)$, thus $|H_X| \leqslant 4$, while $|H_X| > 4$, a contradiction, thus there is no vertex $w_{i,j} \in C_1$.

Assume that there exists a vertex $y_i \in Y \cap C_1$, where $v_i \in V'_s$ is the vertex of G_s corresponding to y_i. By construction, the common neighbor of of each $h_j \in H_X$ and vertex $y_i \in Y$ is $h_i \in H \setminus H_X$. This implies that $H_X \subseteq N(h_i)$, again reaching a contradiction since h_i has at most 2 neighbors in H_X, while $|H_X| > 4$. We can conclude that there is no vertex $y_i \in C_1$.

We have proved that $(W \cup Y) \cap C_1 = \emptyset$ and thus $C_1 \subseteq H$. Define a 2-club $C_s \subseteq V_s$ of G_s as follows: $C_s = \{v_i : h_i \in C_1\}$. Since C_1 is a 2-club of G, and $G[H]$ is isomorphic to G_s, it follows that C_s is a 2-club of G_s. Moreover, $H_X \subseteq C_1$, implying that $X_s \subseteq C_s$. Thus C_s is a solution of Restricted Steiner-2-Club, implying that 2-Club Cover(2) is W[1]-hard when parameterized by distance to a 2-club.   □

4   Hardness of Min 2-Club Cover in Subcubic Planar Graphs

In this section we prove that Min 2-Club Cover is NP-hard even if the input graph is connected, has maximum degree 3 (i.e. a subcubic graph) and it is planar. We present a reduction from the Minimum Clique Partition problem on planar subcubic graphs (we denote this restriction by Min Subcubic Planar Clique Partition).

Problem 4. (Min Subcubic Planar Clique Partition)
Input: A planar subcubic graph $G = (V_P, E_P)$.
Output: A partition of V_P into a minimum number of cliques of G_P.

Min Subcubic Planar Clique Partition is known to be NP-hard [5].
We first prove that subcubic graphs have a specific type of matching [1], which will be useful for our reduction. A triangle in a graph is a clique of size 3.

Lemma 4. *Let $G_P = (V_P, E_P)$ be a connected subcubic graph that is not isomorphic to K_4. Then there is a matching $F_P \subseteq E_P$ in G_P that can be computed in polynomial time, with the following properties:*

- *every triangle of G_P contains exactly one edge of F_P;*
- *every edge of F_P is contained in some triangle of G_P.*

We are now ready to describe our reduction. Informally, an instance G of Min 2-Club Cover, is constructed starting from $G_P = (V_P, E_P)$ by subdividing every edge of $E_P \setminus F_P$, and, for every vertex obtained by the subdivision of an edge, by connecting it to a new dangling path of length two.

Next, we define the graph G formally. Given a instance $G_P = (V_P, E_P)$ of Min Subcubic Planar Clique Partition, where $V_P = \{u_1, \ldots, u_n\}$, we first compute a matching F_P of G_P that satisfies the requirements of Lemma 4. Then, define $G = (V, E)$, an instance of Min 2-Club Cover, where $V = V' \cup V_1 \cup V_B$ as follows. First, define V' as $V' = \{v_i : u_i \in V_P\}$.
For each edge $\{u_i, u_j\} \in E_P \setminus F_P$, with $1 \leqslant i < j \leqslant n$, define:

$$V_1 = \{v_{i,j,1} : \{u_i, u_j\} \in E_P \setminus F_P\} \qquad V_B = \{v_{i,j,2}, v_{i,j,3} : \{u_i, u_j\} \in E_P \setminus F_P\}$$

---

[1] Recall that a matching is a set of edges that share no endpoint.

Next, we define the edge set E of G

$$
\begin{aligned}
E = & \{\{v_i, v_j\} : v_i, v_j \in V', \{u_i, u_j\} \in F_P\} \quad \cup \\
& \{\{v_i, v_{i,j,1}\}, \{v_j, v_{i,j,1}\} : v_i, v_j \in V', v_{i,j,1} \in V_1, \{u_i, u_j\} \in E_P \setminus F_P\} \quad \cup \\
& \{\{v_{i,j,t}, v_{i,j,t+1}\} : v_{i,j,t}, v_{i,j,t+1} \in V, t \in \{1, 2\}\}
\end{aligned}
$$

Notice that G has maximum degree three, since G_P has maximum degree three. Indeed, the vertices in V' have the same degree as the corresponding vertices in G_P, those in V_1 have degree exactly three and those in V_B degree at most 2. Moreover, since G_P is planar, then also G is planar. Indeed, the vertices of V_B cannot belong to a subdivision of a K_5 or a $K_{3,3}$, since they don't belong to a cycle of G. Hence, it is sufficient to consider the subgraph $G[V' \cup V_1]$. The vertices in V_1 cannot belong to a K_5 or a $K_{3,3}$, since they have degree two in $G[V' \cup V_1]$. But then, if $G[V' \cup V_1]$ is a subdivision of a K_5 or a $K_{3,3}$, the same property holds for G, since the vertices of V_1 are obtained by subdiving edges of G_P, a contradiction to the planarity of G_P.

For the remainder of this section, set $q = |E_P| - |F_P|$, that is q is the number of edges of G_P that were subdivided in the construction of G.

Lemma 5. *Given a planar cubic graph G_P instance of* Min Subcubic Planar Clique Partition, *consider the corresponding instance G of* Min 2-Club Cover. *If there exists a clique partition $\mathcal{C} = \{C_{P,1}, \ldots, C_{P,k}\}$ of G_P with k cliques, then there exists a solution of* Min 2-Club Cover *on instance G consisting of $q + k$ 2-clubs.*

Proof. Recall that G_P is a subcubic graph. Note that if $\mathcal{C} = \{C_{P,1}, \ldots, C_{P,k}\}$ is a clique partition of G_P, then each $C_{P,i}$, with $1 \leqslant i \leqslant k$, is either a triangle, two adjacent vertices or a singleton vertex of G_P, since we have assumed that G_P is not a K_4. For each $C_{P,i} \in \mathcal{C}$, with $1 \leqslant i \leqslant k$, we define a corresponding 2-club C_i in G, If $C_{P,i} = \{u_j\}$, with $1 \leqslant j \leqslant n$, that is it is a singleton, then define $C_i = \{v_j\}$, with $v_j \in V'$. Consider the case that $C_{P,i} = \{u_j, u_l\}$, with $1 \leqslant j, l \leqslant n$, i.e. $C_{P,i}$ is an edge of G_P. If $\{u_j, u_l\} \in F_P$, then $C_i = \{v_j, v_l\}$. If $\{u_j, u_l\} \in E_P \setminus F_P$, then $C_i = \{v_j, v_l, v_{i,l,1}\}$.

If $C_{P,i} = \{u_j, u_l, u_z\}$, then $C_{P,i}$ is a triangle in G_P. By construction of the matching F_P, there exists one edge of G connecting two vertices of v_j, v_l, v_z. Thus, in G there exists a cycle D of length 5 that contains v_j, v_l, v_z. Then D is a 2-club of G and we define $C_i = D$. Since each vertex of G_P belongs to a clique of $\{C_{P,1}, \ldots, C_{P,k}\}$, the 2-clubs $C_1 \ldots, C_k$ cover every vertex in V'. The vertices of $V_1 \cup V_B$ are covered with q 2-clubs as follows. For each vertex of V_1, define a 2-club $\{v_{i,j,1}, v_{i,j,2}, v_{i,j,3}\}$. It follows that G admits a cover with at most $q + k$ 2-clubs. $\square$

Lemma 6. *Given a graph G_P instance of* Min Subcubic Planar Clique Partition, *consider the corresponding graph G instance of* Min 2-Club Cover. *Then, any 2-club covering of G contains strictly more than q 2-clubs. Moreover, if there exists a solution $\mathcal{C} = \{C_1, \ldots, C_{q+k}\}$ of* Min 2-Club Cover *on instance G, for some $k \geqslant 1$, there exists a clique partition of G_P with at most k cliques.*

From Lemmas 5, 6 and from the NP-hardness of Min Subcubic Planar Clique Partition [5], we can conclude that Min 2-Club Cover is NP-hard on planar subcubic graphs.

Theorem 7. Min 2-Club Cover *is NP-hard on planar subcubic graphs.*

5 Hardness of **Min 2-Club Cover** on Bipartite Graphs

In this section, we show that Min 2-Club Cover is W[2]-hard when parameterized by h (the number of 2-clubs in a solution of Min 2-Club Cover) and not approximable within factor $\Omega(\log |V|)$, by giving a reduction from Minimum Set Cover to Min 2-Club Cover on bipartite graphs.

Problem 5. Minimum Set Cover (Minimum Set Cover)
Input: A set $U = \{u_1, \ldots u_n\}$ of n elements and a collection $\mathcal{S} = \{S_1, \ldots, S_m\}$ of sets, where $S_i \subseteq U$, with $1 \leqslant i \leqslant m$
Output: A minimum cardinality collection $\mathcal{S}' \subseteq \mathcal{S}$ such that for each element $u_i \in U$, with $1 \leqslant i \leqslant n$, there exists a set of $\mathcal{S}'$ containing u_i.

Minimum Set Cover is known to be W[2]-hard when parameterized by the size of the solution [23].

Given an instance $(U, \mathcal{S})$ of Minimum Set Cover, we define a bipartite graph $G = (V, E)$, an instance of Min 2-Club Cover, where $V = V_1 \uplus V_2$, as follows:

$$V_1 = \{v_i : u_i \in U\} \cup \{z_1\} \qquad V_2 = \{w_i : S_i \in \mathcal{S}\} \cup \{z_2\}$$

$$E = \{\{v_i, w_j\} : u_i \in S_j\} \cup \{\{z_1, w_j\} : 1 \leqslant j \leqslant m\}\} \cup \{z_1, z_2\}$$

The graph G is bipartite, as there is no edge connecting two vertices of V_1 or two vertices of V_2. Next, we prove the main results on which the reduction is based.

Lemma 8. *Let $(U, \mathcal{S})$ be an instance of* Minimum Set Cover *and let $G = (V, E)$ be the corresponding instance of* Min 2-Club Cover. *Given a solution of* Minimum Set Cover *of size z, then we can compute in polynomial time a solution $\mathcal{C}$ of* Min 2-Club Cover *of size $z + 1$.*

Lemma 9. *Let $(U, \mathcal{S})$ be an instance of* Minimum Set Cover *and let $G = (V, E)$ be the corresponding instance of* Min 2-Club Cover. *Given a solution of* Min 2-Club Cover *of size h, with $h \geqslant 2$, we can compute in polynomial time a set cover of $(U, \mathcal{S})$ having size at most $h - 1$.*

From Lemmas 8 and 9, and from the W[2]-hardness of Minimum Set Cover [23] by h (the size of solution), we can conclude that Min 2-Club Cover is W[2]-hard on bipartite graphs.

Theorem 10. Min 2-Club Cover *is W[2]-hard on bipartite graphs when parameterized by the number of 2-clubs in the cover.*

As a consequence of Lemmas 8 and 9, we can prove also a bound on the approximation of Min 2-Club Cover on bipartite graphs.

Corollary 11. Min 2-Club Cover *is not approximable within factor* $\Omega(\log(|V|))$ *on bipartite graphs.*

6 An FPT Algorithm for Min 2-Club Cover on Graphs of Bounded Treewidth

In this section we show that Min 2-Club Cover is fixed parameter tractable when parameterized by the treewidth δ of the input graph G. First, we present the definitions of tree decomposition and of nice tree decomposition of a graph [17]. Given a tree decomposition of a graph G having width k, a nice tree decomposition of width k can be computed in linear time.

Given a graph $G = (V, E)$, a tree decomposition of G is a tree $T = (B, E_B)$ (we denote $|B| = l$), where each vertex $B_i \in B$, $1 \leqslant i \leqslant l$, is a bag (that is $B_i \subseteq V$), with $|B_i| \leqslant \delta + 1$, such that:

- $\bigcup_{i=1}^{l} B_i = V$
- For every $\{u, v\} \in E$, there is a bag $B_j \in B$, with $1 \leqslant j \leqslant l$, such that $u, v \in B_j$
- The bags of T containing a vertex $u \in V$ induce a subtree of T.

In a nice tree decomposition, each $B_i \in B$ can be an *introduce vertex* (B_i has a single child B_j, with $B_i = B_j \cup \{u\}$, where $u \in V$), a *forget vertex* (B_i has a single child B_j, with $B_i = B_j \setminus \{u\}$, where $u \in V$), or a *join vertex* (B_i has exactly two children B_l, B_r with $B_i = B_l = B_r$). Moreover, a nice tree decomposition is rooted, and each leaf-bag is associated with a single vertex of V. From now on, we will assume that we are given a nice tree decomposition.

We denote by T_i, with $1 \leqslant i \leqslant l$, the subtree of T rooted at B_i, and we denote by $V(T_i)$ the vertices contained in at least one bag of T_i. Moreover, $G_i = G[V(T_i)]$. Given a 2-club X of G such that $X \cap V(T_i) \neq \emptyset$, with $1 \leqslant i \leqslant l$, $X \cap T(V_i)$ is a *partial 2-club*. Notice that all the vertices of a partial 2-club have distance at most two in G and that two vertices $u, v \in X \cap (V(T_i) \setminus B_i)$, with $1 \leqslant i \leqslant l$, have distance at most 2 in G_i, since by the thrid property of tree decomposition $N(u) \cup N(v) \subseteq V(T_i)$.

Consider a solution S of Min 2-Club Cover on G and a 2-club X of S. Let $X_i = X \cap V(T_i)$, $1 \leqslant i \leqslant l$, be a partial 2-club of G, define $X[B_i] = X \cap B_i$. We define two tables associated with $X[B_i]$ and X_i, with $1 \leqslant i \leqslant l$, that will be useful in the rest of the section:

- A table $D(X[B_i])$ that, for two vertices $u, v \in X[B_i]$, stores their distance in $G[X_i]$. More precisely, $D(X[B_i])[u, v]$, with $u, v \in X[B_i]$, is defined as follows:

$$D(X[B_i])[u, v] = \begin{cases} 0 & \text{if } u = v \\ 1 & \text{if } d_{G[X_i]}(u, v) = 1 \\ 2 & \text{if } d_{G[X_i]}(u, v) = 2 \\ +\infty & \text{else} \end{cases}$$

– A table $H(X[B_i])$ that stores the distance (not greater than 2) in $G[X_i]$ of the vertices in $X_i \setminus B_i$ from each vertex $z \in X[B_i]$. Notice that this distance must be at most two, since $N(X_i \setminus B_i) \subseteq V(T_i)$. Table $H(X[B_i])$ contains rows of length $|X[B_i]|$ having values in $\{1, 2\}$. Given a vertex $u \in (X_i \setminus B_i)$, there exists a row $r \in \{1, 2\}^{|X[B_i]|}$ that belongs to $H(X[B_i])$ such that $d_{G[X_i]}(u, v) = r[v]$, for each $v \in X[B_i]$.

An *empty table* is a table that does not contain any row or column.

Before giving the algorithm, we show that we can restrict ourselves to a special kind of solutions. The proof of the next lemma follows from the properties of nice tree decomposition. Note that when we say $D(X[B_i]) = D(Y[B_i])$, we mean $D(X[B_i])[u, v] = D(Y[B_i])[u, v]$ for every $u, v \in B_i$ (similarly for H).

Lemma 12. *Given a subtree T_i of a nice tree decomposition of G, consider a set S' of distinct partial 2-clubs of T_i such that:*

a. *All the partial 2-clubs in S' contain the same subset B_i' of B_i (i.e. $X \cap B_i = Y \cap B_i = B_i'$ for every $X, Y \in S'$)*
b. *For every $X, Y \in S'$, $D(X[B_i]) = D(Y[B_i])$*
c. *For every $X, Y \in S'$, $H(X[B_i]) = H(Y[B_i])$.*

Then the following hold:

1. *Given a 2-club Z_X of G such that $Z_X \cap V(T_i) = X$, with $X \in S'$, $Z_X \setminus V(T_i) \neq \emptyset$ and $X \setminus B_i' \neq \emptyset$, then $Y \cup (Z_X \setminus X)$, with $Y \in S'$, is a 2-club of G.*
2. *An optimal solution of Min 2-Club Cover on instance G contains at most $2^{\delta+1}$ 2-clubs Z_X of G such that $Z_X \cap V(T_i) = X$, with $X \in S'$, $Z_X \setminus V(T_i) \neq \emptyset$ and $X \setminus B_i' \neq \emptyset$.*

Next, given a nice tree decomposition of G, we describe a dynamic programming recurrence to compute a solution of Min 2-Club Cover on G. We denote by $\langle B_i \rangle^t$, $1 \leq i \leq l$, a collection of t subsets of B_i, and we denote by $\langle B_i \rangle^t[j]$ the j-th subset, with $1 \leq j \leq l$, of $\langle B_i \rangle^t$. Similarly, we denote by $\langle D_i \rangle^t$ and $\langle H_i \rangle^t$, $1 \leq i \leq l$, two collections of t tables for B_i. We denote by $\langle U \rangle^t[j]$, with $U \in \{H_i, D_i\}$ the j-th table, with $1 \leq j \leq t$, of $\langle U \rangle^t$. Notice that $\langle U \rangle^t[j][a, b]$ denotes the entry associated with vertices a, b of table $\langle U \rangle^t[j]$. Each table $\langle D_i \rangle^t[j]$ consists of $|\langle B_i \rangle^t[j]|$ rows and columns, and is over values $\{0, 1, 2, +\infty\}$. Each table $\langle H_i \rangle^t[j]$ consists of $|\langle B_i \rangle^t[j]|$ columns. Moreover, we can define an upper bound on the number of rows of $\langle H_i \rangle^t[j]$. Notice that $\langle H_i \rangle^t[j]$ contains distinct rows and has values in $\{1, 2\}$, thus there can be at most $2^{|\langle B_i \rangle^t[j]|}$ rows in $\langle H_i \rangle^t[j]$. Informally, the tables $\langle H_i \rangle^t[j]$ and $\langle D_i \rangle^t[j]$ are used to guess the tables $D(X[B_i])$ and $H(X[B_i])$, for a partial 2-club X.

Define a function $C[\langle B_i \rangle^t, \langle D_i \rangle^t, \langle H_i \rangle^t, h]$, where $\langle B_i \rangle^t$ is a collection of t not necessarily disjoint subsets of B_i, $\langle D_i \rangle^t$ is a collection of t tables, where table $\langle D_i \rangle^t[j]$, with $1 \leq j \leq t$, is associated with subset $\langle B_i \rangle^t[j]$, $\langle H_i \rangle^t$ is a collection of t tables where table $\langle H_i \rangle^t[j]$, with $1 \leq j \leq t$, is associated with subset $\langle B_i \rangle^t[j]$.

Put $C[\langle B_i \rangle^t, \langle D_i \rangle^t, \langle H_i \rangle^t, h] = 1$ (else $C[\langle B_i \rangle^t, \langle D_i \rangle^t, \langle H_i \rangle^t, h] = 0$) if and only if there exists a collection S of h partial 2-clubs that covers $V(T_i)$ such that the following property holds:

Property 1. *For each partial 2-club S of $\mathcal{S}$, either $S \subseteq V(T_i) \setminus B_i$ or $S \cap B_i = \langle B_i \rangle^t[p]$, $1 \leqslant p \leqslant t$, and the following hold:*

1. $\bigcup_{p=1}^{t} \langle B_i \rangle^t[p] = B_i$
2. *For each S of $\mathcal{S}$ such that $S \cap B_i = \langle B_i \rangle^t[p]$ for some $1 \leq p \leq t$*
   (a) $D(S[B_i]) = \langle D_i \rangle^t[p]$
   (b) $H(S[B_i]) = \langle H_i \rangle^t[p]$

Consider a partial 2-club S of $\mathcal{S}$. Notice that by the definition of partial 2-clubs all the vertices in $S \setminus B_i$ have distance at most 2 in G_i. Moreover, if S does not contain vertices of B_i, it is indeed a 2-club of G_i.

Next, we describe the recurrence to compute $C[\langle B_i \rangle^t, \langle D_i \rangle^t, \langle H_i \rangle^t, h]$. Notice that, given a subset of B_i and two tables of $\langle D_i \rangle^t$, $\langle H_i \rangle^t$, there can exist multiple entries in C, that is $\langle B_i \rangle^t[x] = \langle B_i \rangle^t[y]$, $\langle D_i \rangle^t[x] = \langle D_i \rangle^t[y]$ and $\langle H_i \rangle^t[x] = \langle H_i \rangle^t[y]$, with $1 \leqslant x < y \leqslant t$. This is due to the fact that two partial 2-clubs may have been created in some bag of T_i and after the removal of some vertices they may have the same subset $\langle B_i \rangle^t[x]$ of B_i and the same tables $\langle D_i \rangle^t[x]$ and $\langle H_i \rangle^t[x]$. However, notice that, by Lemma 12, we consider at most $2^{\delta+1}$ partial 2-clubs having identical values of $\langle B_i \rangle^t[x]$, $\langle D_i \rangle^t[x]$ and $\langle H_i \rangle^t[x]$.

In the recurrence, we distinguish three cases depending on the fact that vertex B_i is an introduce vertex, a forget vertex or a join vertex. We present an informal description of the recurrence, before giving the details. First, we assume that each leaf bag contains a single vertex. In an introduce vertex B_i of the nice tree decomposition, with child B_j, where $B_i = B_j \cup \{u\}$, we consider the partial 2-clubs having vertices in B_i, and in particular we may add to the solution we are computing partial 2-clubs that contain u. The associated tables $\langle D_i \rangle^t[p]$ and $\langle H_i \rangle^t$ are computed. In a forget vertex of the nice tree decomposition, with child B_j, where $B_i = B_j \setminus \{u\}$, we update the partial 2-clubs that contains vertex u. Given a partial 2-club S that contains u, $S \cap B_i$ is obtained by removing u and if $S \cap B_j$ is $\{u\}$, the partial 2-club is removed, checking that is a 2-club. The tables $\langle D_i \rangle^t[p]$ and $\langle H_i \rangle^t[p]$ are updated, checking that u has distance at most two from the other vertices of S and adding a new row to table $\langle H_i \rangle^t[p]$ that stores the distance of u from the vertices of $S \cap B_i$. In a join vertex B_i with children B_l and B_r, it is considered the case that a partial 2-club S contains vertices of one or two subtrees of the nice tree decomposition. In the latter case, for two vertices u and v of S, table $\langle D_i \rangle^t[p]$ contains the minimum of the distances between u and v in the two subtrees. Moreover, it is checked that the vertices of S that belong to different subtrees have a common neighbour in S (this vertex by construction must be in $S \cap B_i$).

Introduce Vertex

Let B_i be an introduce vertex and let B_j be the only child of B_i, with $B_i = B_j \cup \{u\}$ (notice that u belongs to at least one set of $\langle B_i \rangle^t$). $C[\langle B_i \rangle^t, \langle D_i \rangle^t, \langle H_i \rangle^t, h]$ is the maximum over all possible combinations of $z \in \{0, 1, \ldots, t-1\}$, $\langle B_j \rangle^{t-z}$, $\langle D_j \rangle^{t-z}$, $\langle H_j \rangle^{t-z}$, of

$$C[\langle B_j \rangle^{t-z}, \langle D_j \rangle^{t-z}, \langle H_j \rangle^{t-z}, h-z]$$

where

- $\langle B_i \rangle^t[p]$, with $1 \leqslant p \leqslant t - z$, is either $\langle B_i \rangle^t[p] = \langle B_j \rangle^t[p]$ or $\langle B_i \rangle^t[p] = \langle B_j \rangle^t[p] \cup \{u\}$; $\langle B_i \rangle^t[p]$, with $t - z + 1 \leqslant p \leqslant t$ is either $\langle B_i \rangle^t[p] = \{u\}$ or $\langle B_i \rangle^t[p] = \langle B_j \rangle^{t-z}[y] \cup \{u\}$, with $1 \leqslant y \leqslant t - z$.
- If $\langle B_i \rangle^t[p] = \langle B_j \rangle^t[p]$, with $1 \leqslant p \leqslant t - z$, then $\langle D_i \rangle^t[p] = \langle D_j \rangle^t[p]$ and $\langle H_i \rangle^t[p] = \langle H_j \rangle^t[p]$
- If $\langle B_i \rangle^t[p] = \langle B_j \rangle^{t-z}[y] \cup \{u\}$, with $1 \leqslant p \leqslant t$ and $1 \leqslant y \leqslant t - z$, and $\langle H_i \rangle^t[p]$ and $\langle H_j \rangle^{t-z}[y]$ are empty tables, then $\langle D_i \rangle^t[p]$, with $t - z + 1 \leqslant p \leqslant t$, is a table containing the distances of the vertices of $\langle B_i \rangle^t[p]$ in $G[\langle B_i \rangle^t[p]]$
- If $\langle B_i \rangle^t[p] = \langle B_j \rangle^{t-z}[y] \cup \{u\}$, with $1 \leqslant p \leqslant t$ and $1 \leqslant y \leqslant t - z$, $\langle H_i \rangle^t[p]$ and $\langle H_j \rangle^{t-z}[y]$ are not empty tables, then
  - table $\langle D_i \rangle^t[p]$ is computed from $\langle D_j \rangle^{t-z}[y]$ by adding a new column and a new row (associated with u) and updating the values as follows:
    * For each $v \in \langle B_j \rangle^{t-z}[y]$, $\langle D_i \rangle^t[p][u, v] = d_{G[\langle B_i \rangle^t[p]]}(u, v)$, if $d_{G[\langle B_i \rangle^t[p]]}(u, v) \leqslant 2$, else $\langle D_i \rangle^t[p][u, v] = \infty$ (recall that u has no neighbor in $V(T_i) \setminus B_i$, so $\langle D_i \rangle^t[p]$ is set correctly for the new entries that include u).
    * Consider $v, w \in \langle B_j \rangle^t[p]$, with $\langle D_j \rangle^{t-z}[y][v, w] = \infty$; if $\{u, v\}$, $\{u, w\} \in E$, then $\langle D_i \rangle^t[p][v, w] = 2$, else $\langle D_i \rangle^t[p][v, w] = \langle D_j \rangle^{t-z}[y][v, w]$
  - $\langle H_i \rangle^t[p]$ is computed by adding a column associated with u to $\langle H_j \rangle^{t-z}[y]$, where $\langle H_i \rangle^t[p]$ is identical to $\langle H_j \rangle^{t-z}[y]$, except for the new column. The values of the new column are defined as follows:
    * For each row r of $\langle H_i \rangle^t[p]$, there must exist a vertex $v \in \langle B_i \rangle^t[p]$ such that $\{v, u\} \in E$ and $\langle H_j \rangle^{t-z}[y][r, v] = 1$; then $\langle H_i \rangle^t[p][r, u] = 2$. Notice if such a vertex v does not exist, there is a vertex $w \in X \cap V(T_j)$, where X is the partial 2-club such $\langle B_i \rangle^t[p] \subseteq X$, that has distance greater than two from u, since u and w do not have a common neighbour in $G[X]$.

Forget Vertex
Let B_i be a forget vertex and let B_j be the only child of B_i, with $B_i = B_j \setminus \{u\}$. $C[\langle B_i \rangle^t, \langle D_i \rangle^t, \langle H_i \rangle^t, h]$ is the maximum over all possible combinations of $z \in \{0, 1, \ldots, h - 1\}$, $\langle B_j \rangle^{t+z}$, $\langle D_j \rangle^{t+z}$, $\langle H_j \rangle^{t+z}$, of:

$$C[\langle B_j \rangle^{t+z}, \langle D_j \rangle^{t+z}, \langle H_j \rangle^{t+z}, h]$$

where

- $\langle B_i \rangle^t[p] = \langle B_j \rangle^{t+z}[p] \setminus \{u\}$, with $1 \leqslant p \leqslant t$
- $\langle B_j \rangle^{t+z}[p] = \{u\}$, with $t + 1 \leqslant p \leqslant t + z$

Tables $\langle D_i \rangle^t[p]$ and $\langle H_i \rangle^t[p]$, with $1 \leqslant p \leqslant t$, are computed as follows:

- If $\langle B_i \rangle^t[p] = \langle B_j \rangle^{t+z}[p]$, $1 \leqslant p \leqslant t$, then $\langle D_i \rangle^t[p]$ is identical to $\langle D_j \rangle^{t+z}[p]$ and $\langle H_i \rangle^t[p]$ is identical to $\langle H_j \rangle^{t+z}[p]$
- If $\langle B_i \rangle^t[p] = \langle B_j \rangle^{t+z}[p] \setminus \{u\}$, $1 \leqslant p \leqslant t$, then:
  - $\langle D_i \rangle^t[p]$ is computed by removing the row and the column of $\langle D_j \rangle^{t+z}[p]$ associated with u. Notice that we must have $\langle D_j \rangle^{t+z}[p][u,v] \neq \infty$, for each $v \in \langle B_i \rangle^t[p]$ (hence $v \in \langle B_j \rangle^{t+z}[p] \setminus \{u\}$), since each vertex adjacent to u in G belongs to $V(T_j)$; notice that no vertex of $V(T) \setminus V(T_j)$ is adjacent to both u and v.
  - $\langle H_i \rangle^t[p]$ is computed starting from $\langle H_j \rangle^{t+z}[p]$ as follows:
    * The column associated with u is removed from $\langle H_j \rangle^{t+z}[p]$.
    * A row r of length $|\langle B_i \rangle^t[p]|$ is added, where $r[v] = c$, with $1 \leqslant c \leqslant 2$, if $\langle D_j \rangle^t[p][u,v] = c$ (notice that we have checked that $\langle D_j \rangle^{t+z}[p][u,v] = c \leqslant 2$, for each v).

A Join Vertex

Let B_i be a join vertex and let B_l, B_r the left and right child, respectively, of B_i. Recall that $B_i = B_l = B_r$. $C[\langle B_i \rangle^t, \langle D_i \rangle^t, \langle H_i \rangle^t, h]$ is the maximum over all possible combinations of h_r, h_l, with $1 \leqslant h_l, h_r \leqslant h$, and s, q, with $1 \leqslant s \leqslant h_l$ and $1 \leqslant q \leqslant h_r$, $\langle B_l \rangle^s$, $\langle D_l \rangle^s$, $\langle H_l \rangle^s$, and $\langle B_r \rangle^q$, $\langle D_r \rangle^q$, $\langle H_r \rangle^q$ of:

$$C[\langle B_l \rangle^s, \langle D_l \rangle^s, \langle H_l \rangle^s, h_l] \wedge C[\langle B_r \rangle^q, \langle D_r \rangle^q, \langle H_r \rangle^q, h_r]$$

where

- $h = h_l + h_r - z$, for $0 \leqslant z \leqslant \min\{h_l, h_r\}$
- $\langle B_i \rangle^t[p] = \langle B_l \rangle^s[p] = \langle B_r \rangle^q[p]$, with $1 \leqslant p \leqslant z$ (if $z = 0$ this case does not hold; we assume without loss of generality that if $z \geqslant 1$ $\langle B_l \rangle^s[p]$ and $\langle B_r \rangle^q[p]$, with $1 \leqslant p \leqslant z$, are part of the same partial 2-club)
- $\langle B_i \rangle^t[p] = \langle B_l \rangle^s[p]$, with $z + 1 \leqslant p \leqslant h_l$
- $\langle B_i \rangle^t[h_l + p] = \langle B_r \rangle^q[z + p]$, with $1 \leqslant p \leqslant h_r - z$

Now, we describe how tables in $\langle D_i \rangle^t$ and $\langle H_i \rangle^t$ are constructed. Table $\langle D_i \rangle^t[p]$, with $1 \leqslant p \leqslant h_l + h_r - z$, is computed as follows:

- For p with $z + 1 \leqslant p \leqslant h_l$, $\langle D_i \rangle^t[p]$ is identical to $\langle D_l \rangle^s[p]$
- For p with $1 \leqslant p \leqslant h_r - z$, $\langle D_i \rangle^t[h_l + p]$ is identical to $\langle D_r \rangle^q[z + p]$
- For each $u, v \in \langle B_i \rangle^t[p]$, with $1 \leqslant p \leqslant z$, it holds

$$\langle D_i \rangle^t[p][u,v] = \min(\langle D_l \rangle^s[p][u,v], \langle D_r \rangle^q[p][u,v])$$

Table $\langle H_i \rangle^t[p])$, with $1 \leqslant p \leqslant h_l + h_r - z$ is computed as follows:

- table $\langle H_i \rangle^t[p]$, with $z + 1 \leqslant p \leqslant h_l$ is identical to $\langle H_l \rangle^s[p]$
- table $\langle H_i \rangle^t[h_l + p]$, with $1 \leqslant p \leqslant h_r - z$, is identical to $\langle H_r \rangle^q[z + p]$

– table $\langle H_i \rangle^t[p]$, with $1 \leqslant p \leqslant h_l - z$, is the union of rows of $\langle H_l \rangle^s[p]$ and rows $\langle H_r \rangle^q[p]$. Moreover, for each row a in $\langle H_l \rangle^s[p]$ and b in $\langle H_r \rangle^q[p]$, there must exist a column u associated with $u \in B_i$, such that $\langle H_l \rangle^s[p][a, u] = \langle H_r \rangle^q[p][b, u] = 1$ (if this does not hold, then there exist two vertices v and w of the partial 2-club X, with $\langle B_i \rangle^t[p] \subseteq X$, such that v belongs to $X \cap V(T_l)$, w belongs to $X \cap V(T_r)$ and w, v do not have a common neighbour in $G[X \cap V(T_i)]$).

In the base case, that is when B_i is a leaf of the tree decomposition and $B_i = \{u\}$, we put $C[\langle B_i \rangle^1, \langle D_i \rangle^1, \langle H_i \rangle^1, 1] = 1$, else $C[\langle B_i \rangle^t, \langle D_i \rangle^t, \langle H_i \rangle^t, 1] = 0$, because $B_i = \{u\}$ and . Each table in $\langle D_i \rangle^1$ contains exactly one row and one column associated with u (the only entry of this table has value 0), each table of $\langle H_i \rangle^t$ is an empty table.

Next, we prove the correctness of the recurrence.

Lemma 13. *Consider a nice tree decomposition (T, B) of a graph $G = (V, E)$ instance of* Min 2-Club Cover, *and let B_i be a vertex of T, with $1 \leqslant i \leqslant l$. Then*

$$C[\langle B_i \rangle^t, \langle D_i \rangle^t, \langle H_i \rangle^t, h] = 1$$

for some collection $\langle B_i \rangle^t$ of subsets of B_i, some collections of tables $\langle D_i \rangle^t$, $\langle H_i \rangle^t$, and two integers $t, h \geqslant 1$ if and only if there exists a covering S of the vertices in T_i consisting of h partial 2-clubs such that Property 1 holds for S.

Consider the root B_R of the nice tree decomposition. Then there exists a solution of Min 2-Club Cover over instance G consisting of h 2-clubs if and only if $C[\langle B_R \rangle^t, \langle D_R \rangle^t, \langle H_R \rangle^t, h] = 1$, where each table in $\langle D_R \rangle^t$ has values in $\{0, 1, 2\}$. We can conclude with the following result.

Theorem 14. *A solution of* Min 2-Club Cover *on a graph G having treewidth bounded by δ can be computed in time $O^*(2^{3(\delta+1)^2 + 9\delta + 9})$.*

References

1. Alba, R.D.: A graph-theoretic definition of a sociometric clique. J. Math. Sociol. **3**, 113–126 (1973)
2. Asahiro, Y., Doi, Y., Miyano, E., Samizo, K., Shimizu, H.: Optimal approximation algorithms for maximum distance-bounded subgraph problems. Algorithmica **80**(6), 1834–1856 (2018)
3. Balasundaram, B., Butenko, S., Trukhanov, S.: Novel approaches for analyzing biological networks. J. Comb. Optim. **10**(1), 23–39 (2005)
4. Bourjolly, J., Laporte, G., Pesant, G.: An exact algorithm for the maximum k-club problem in an undirected graph. Eur. J. Oper. Res. **138**(1), 21–28 (2002)
5. Cerioli, M.R., Faria, L., Ferreira, T.O., Martinhon, C.A.J., Protti, F., Reed, B.A.: Partition into cliques for cubic graphs: planar case, complexity and approximation. Discret. Appl. Math. **156**(12), 2270–2278 (2008)
6. Cerioli, M.R., Faria, L., Ferreira, T.O., Protti, F.: A note on maximum independent sets and minimum clique partitions in unit disk graphs and penny graphs: complexity and approximation. RAIRO-Theor. Inform. Appl. **45**(3), 331–346 (2011)

7. Chang, M., Hung, L., Lin, C., Su, P.: Finding large k-clubs in undirected graphs. Computing **95**(9), 739–758 (2013)
8. Dondi, R., Mauri, G., Sikora, F., Zoppis, I.: Covering a graph with clubs. J. Graph Algorithms Appl. **23**(2), 271–292 (2019)
9. Dondi, R., Mauri, G., Zoppis, I.: On the tractability of finding disjoint clubs in a network. Theor. Comput. Sci. (2019, to appear)
10. Dumitrescu, A., Pach, J.: Minimum clique partition in unit disk graphs. Graphs Comb. **27**(3), 399–411 (2011)
11. Fellows, M.R., Hermelin, D., Rosamond, F.A., Vialette, S.: On the parameterized complexity of multiple-interval graph problems. Theor. Comput. Sci. **410**(1), 53–61 (2009)
12. Garey, M.R., Johnson, D.S., Stockmeyer, L.J.: Some simplified NP-complete graph problems. Theor. Comput. Sci. **1**(3), 237–267 (1976)
13. Garey, M.R., Johnson, D.S.: Computers and Intractability: A Guide to the Theory of NP-Completeness. W. H. Freeman & Co., New York (1979)
14. Golovach, P.A., Heggernes, P., Kratsch, D., Rafiey, A.: Finding clubs in graph classes. Discrete Appl. Math. **174**, 57–65 (2014)
15. Hartung, S., Komusiewicz, C., Nichterlein, A.: Parameterized algorithmics and computational experiments for finding 2-clubs. J. Graph Algorithms Appl. **19**(1), 155–190 (2015)
16. Karp, R.M.: Reducibility among combinatorial problems. In: Miller, R.E., Thatcher, J.W. (eds.) Proceedings of a symposium on the Complexity of Computer Computations, held 20–22 March 1972, at the IBM Thomas J. Watson Research Center, Yorktown Heights, New York. The IBM Research Symposia Series, pp. 85–103. Plenum Press, New York (1972)
17. Kloks, T.: Treewidth, Computations and Approximations. LNCS, vol. 842. Springer, Heidelberg (1994). https://doi.org/10.1007/BFb0045375
18. Komusiewicz, C.: Multivariate algorithmics for finding cohesive subnetworks. Algorithms **9**(1), 21 (2016)
19. Komusiewicz, C., Sorge, M.: An algorithmic framework for fixed-cardinality optimization in sparse graphs applied to dense subgraph problems. Discrete Appl. Math. **193**, 145–161 (2015)
20. Laan, S., Marx, M., Mokken, R.J.: Close communities in social networks: boroughs and 2-clubs. Soc. Netw. Anal. Min. **6**(1), 20:1–20:16 (2016)
21. Mokken, R.: Cliques, clubs and clans. Qual. Quant.: Int. J. Methodol. **13**(2), 161–173 (1979)
22. Mokken, R.J., Heemskerk, E.M., Laan, S.: Close communication and 2-clubs in corporate networks: Europe 2010. Soc. Netw. Anal. Min. **6**(1), 40:1–40:19 (2016)
23. Paz, A., Moran, S.: Non deterministic polynomial optimization problems and their approximations. Theor. Comput. Sci. **15**, 251–277 (1981)
24. Pirwani, I.A., Salavatipour, M.R.: A weakly robust PTAS for minimum clique partition in unit disk graphs. Algorithmica **62**(3–4), 1050–1072 (2012)
25. Schäfer, A., Komusiewicz, C., Moser, H., Niedermeier, R.: Parameterized computational complexity of finding small-diameter subgraphs. Optim. Lett. **6**(5), 883–891 (2012)

On Cycle Transversals and Their Connected Variants in the Absence of a Small Linear Forest

Carl Feghali[1] , Matthew Johnson[2] , Giacomo Paesani[2(✉)] ,
and Daniël Paulusma[2]

[1] Department of Informatics, University of Bergen, Bergen, Norway
carl.feghali@uib.no
[2] Department of Computer Science, Durham University, Durham, UK
{matthew.johnson2,giacomo.paesani,daniel.paulusma}@durham.ac.uk

Abstract. A graph is H-free if it contains no induced subgraph isomorphic to H. We prove new complexity results for the two classical cycle transversal problems FEEDBACK VERTEX SET and ODD CYCLE TRANSVERSAL by showing that they can be solved in polynomial time for (sP_1+P_3)-free graphs for every integer $s \geq 1$. We show the same result for the variants CONNECTED FEEDBACK VERTEX SET and CONNECTED ODD CYCLE TRANSVERSAL. For the latter two problems we also prove that they are polynomial-time solvable for cographs; this was known already for FEEDBACK VERTEX SET and ODD CYCLE TRANSVERSAL.

1 Introduction

We consider three well-known graph transversals. To define the notion of a graph transversal, let $\mathcal{H}$ be a family of graphs, $G = (V, E)$ be a graph and $S \subseteq V$ be a subset of vertices of G. The graph $G - S$ is obtained from G by removing all vertices of S. We say that S is an $\mathcal{H}$-*transversal* of G if $G - S$ is $\mathcal{H}$-free, that is, $G - S$ contains no induced subgraph isomorphic to some graph of $\mathcal{H}$. In other words, S intersects every induced copy of every graph of $\mathcal{H}$ in G.

Due to their generality, graph transversals play a central role in Theoretical Computer Science. In this paper we focus on three classical transversal problems. Let C_r and P_r denote the cycle and path on r vertices, respectively, and let $G = (V, E)$ be a graph with a subset $S \subseteq V$. Then S is a *vertex cover, feedback vertex set*, or *odd cycle transversal* if S is an $\mathcal{H}$-transversal for, respectively, $\mathcal{H} = \{P_2\}$ (that is, $G - S$ is edgeless), $\mathcal{H} = \{C_3, C_4, \ldots\}$ (that is, $G - S$ is a forest), or $\mathcal{H} = \{C_3, C_5, \ldots\}$ (that is, $G - S$ is bipartite).

Usually the goal is to find a transversal of minimum size in some given graph. The corresponding decision problems for the three transversals given above are the classical VERTEX COVER, FEEDBACK VERTEX SET and ODD CYCLE TRANSVERSAL problems, which are to decide if a given graph has a

The research was supported by the Leverhulme Trust (RPG-2016-258).

L. A. Gąsieniec et al. (Eds.): FCT 2019, LNCS 11651, pp. 258–273, 2019.
https://doi.org/10.1007/978-3-030-25027-0_18

vertex cover, feedback vertex set or odd cycle transversal, respectively, of size at most k for some given positive integer k. Each of these three problems are well-studied and are well-known to be NP-complete.

We may add further constraints to a transversal. In particular, we may require a transversal of a graph G to be *connected*, that is, to induce a connected subgraph of G. The corresponding decision problems for the three above transversals are then called CONNECTED VERTEX COVER, CONNECTED FEEDBACK VERTEX SET and CONNECTED ODD CYCLE TRANSVERSAL, respectively. Garey and Johnson [13] proved that CONNECTED VERTEX COVER is NP-complete even for planar graphs of maximum degree 4 (see, for example, [11,29,33] for NP-completeness results for other graph classes). Grigoriev and Sitters [15] proved that CONNECTED FEEDBACK VERTEX SET is NP-complete for planar graphs with maximum degree 9. Chiarelli et al. [8] proved that CONNECTED ODD CYCLE TRANSVERSAL is NP-complete for graphs of arbitrarily large girth and for line graphs.

As all three decision problems and their connected variants are NP-complete, we may want to restrict the input to some special graph class in order to achieve tractability. Note that this approach is in line with the aforementioned results in the literature, where NP-completeness was proven for special graph classes, and also with, for instance, polynomial-time results for CONNECTED VERTEX COVER by Escoffier, Gourvès and Monnot [10] (for chordal graphs) and Ueno, Kajitani and Gotoh [32] (for graphs of maximum degree at most 3 and trees).

Just as in most of these papers, we consider *hereditary* graph classes, that is, graph classes closed under vertex deletion. Hereditary graph classes form a rich framework that captures many well-studied graph classes. It is not difficult to see that every hereditary graph class $\mathcal{G}$ can be characterized by a (possibly infinite) set $\mathcal{F}_\mathcal{G}$ of forbidden induced subgraphs. If $|\mathcal{F}_\mathcal{G}| = 1$, say $\mathcal{F} = \{H\}$, then $\mathcal{G}$ is said to be *monogenic*, and every graph $G \in \mathcal{G}$ is said to be *H-free*. Considering monogenic graph classes can be seen as a natural first step for increasing our knowledge on the complexity of an NP-complete problem in a *systematic* way.

The general strategy for obtaining complexity results for problems restricted to H-free graphs is to first try to prove that the restriction of each problem to H-free graphs is NP-complete whenever H contains a cycle or a claw. If this is the case, then we are left to consider the situation where H does not contain a cycle, implying that H is a forest, and does not contain a claw either, implying that H is a *linear forest*, that is, the disjoint union of one or more paths.

Indeed, when H contains a cycle or a claw, the problems CONNECTED VERTEX COVER [24], FEEDBACK VERTEX SET (respectively, via a folklore trick, see [3,22], and due to hardness for the subclass of line graphs of planar cubic bipartite graphs [24]), CONNECTED FEEDBACK VERTEX SET [8], ODD CYCLE TRANSVERSAL [8] and CONNECTED ODD CYCLE TRANSVERSAL [8] are all NP-complete for H-free graphs. Hence, for these five problems, we are then left to consider only the case where H is a linear forest. We note that the situation for VERTEX COVER is different. It follows from Poljak's construction [28] that VERTEX COVER is NP-complete for graphs of arbitrarily large girth, and thus for H-free graphs if H contains a cycle. However, VERTEX COVER is polynomial-time solvable for claw-free graphs [21,31].

In this paper we focus on proving new complexity results for FEEDBACK VERTEX SET, CONNECTED FEEDBACK VERTEX SET, ODD CYCLE TRANSVERSAL and CONNECTED ODD CYCLE TRANSVERSAL for H-free graphs. From the above we may assume that H is a linear forest. Below we first discuss the known polynomial cases. As we will use algorithms for VERTEX COVER and CONNECTED VERTEX COVER as subroutines for our new algorithms, we include these two problems in our discussion.

For each $s \geq 1$, VERTEX COVER (by combining the results of [1,30]) and CONNECTED VERTEX COVER [8] are polynomial-time solvable for sP_2-free graphs.[1] Moreover, VERTEX COVER is also polynomial-time solvable for $(sP_1 + P_6)$-free graphs, for every $s \geq 0$ [16], whereas CONNECTED VERTEX COVER is so for $(sP_1 + P_5)$-free graphs [19]. Their complexity for P_r-free graphs is unknown for $r \geq 7$ and $r \geq 6$, respectively.

The FEEDBACK VERTEX SET and ODD CYCLE TRANSVERSAL problems are polynomial-time solvable for permutation graphs [4], and thus for P_4-free graphs. Recently, Okrasa and Rzążewski [25] proved that ODD CYCLE TRANSVERSAL is NP-complete for P_{13}-free graphs. A small modification of their construction yields the same result for CONNECTED ODD CYCLE TRANSVERSAL. The complexity of FEEDBACK VERTEX SET and CONNECTED FEEDBACK VERTEX SET is unknown, when restricted to P_r-free graphs for $r \geq 5$. For every $s \geq 1$, both problems and their connected variants are polynomial-time solvable for sP_2-free graphs [8], using the price of connectivity for feedback vertex set [2,18].[2]

1.1 Our Results

We prove in Sect. 3 that CONNECTED FEEDBACK VERTEX SET and CONNECTED ODD CYCLE TRANSVERSAL are polynomial-time solvable for P_4-free graphs, just as FEEDBACK VERTEX SET and ODD CYCLE TRANSVERSAL are [4]. We then prove, in Sect. 4, that, for every $s \geq 1$, these four problems are all polynomial-time solvable for $(sP_1 + P_3)$-free graphs.

To prove our results, we rely on two proof ingredients. The first one is that we use known algorithms for VERTEX COVER and CONNECTED VERTEX COVER restricted to H-free graphs as subroutines in our new algorithms. The second one is that we consider the connected variant of the transversal problems in a more general form. For CONNECTED VERTEX COVER this variant is defined as follows:

---

[1] The graph $G + H$ is the disjoint union of graphs G and H and sG is the disjoint union of s copies of G; see Sect. 2.

[2] The price of connectivity concept was introduced by Cardinal and Levy [7] for vertex cover; see also, for example, [6,9].

CONNECTED VERTEX COVER EXTENSION
  Instance: a graph $G = (V, E)$, a subset $W \subseteq V$ and a positive integer k.
  Question: does G have a connected vertex cover S_W with $W \subseteq S_W$ and
        $|S_W| \leq k$?

Note that CONNECTED VERTEX COVER EXTENSION becomes the original problem if $W = \emptyset$. In the same way we define the problems CONNECTED FEEDBACK VERTEX SET EXTENSION and CONNECTED ODD CYCLE TRANSVERSAL EXTENSION. In fact we will prove all our results for connected feedback vertex sets and connected odd cycle transversals for the extension versions. This is partially out of necessity – the extension versions sometimes serve as auxiliary problems for some of our inductive arguments and may do so for future results as well – but it does also lead to slightly stronger results.

Remark 1. Note that one could also define extension versions for any original transversal problem. However, such extension versions will be polynomially equivalent. In particular we could solve any of them on input (G, W, k) by considering the original problem on input $(G - W, k - |W|)$ and adding W to the solution. However, due to the connectivity condition, we cannot use this approach for the connected variants and need to follow a more careful approach.

Remark 2. It is known that VERTEX COVER is polynomial-time solvable for $(P_1 + H)$-free graphs whenever it is so for H-free graphs. This follows from a well-known observation. see, e.g., [23]: one can solve the complementary problem of finding a maximum independent set in a $(P_1 + H)$-free graph by solving this problem on each H-free graph obtained by removing a vertex and all its neighbours. However, this trick does not work for CONNECTED VERTEX COVER. Moreover, it does not work for FEEDBACK VERTEX SET and ODD CYCLE TRANSVERSAL and their connected variants either.

2   Preliminaries

Let $G = (V, E)$ be a graph. For a set $S \subseteq V$, the graph $G[S]$ denotes the subgraph of G induced by S. We say that S is *connected* if $G[S]$ is connected. We write $G - S$ for the graph $G[V \setminus S]$. A subset $D \subseteq V$ is a *dominating* set of G if every vertex of $V \setminus D$ is adjacent to at least one vertex of D. An edge uv of a graph $G = (V, E)$ is *dominating* if $\{u, v\}$ is a dominating set. The *complement* of G is the graph $\overline{G} = (V, \{uv \mid uv \notin E \text{ and } u \neq v\})$. The *neighbourhood* of a vertex $u \in V$ is the set $N(u) = \{v \mid uv \in E\}$ and for $U \subseteq V$, we let $N(U) = \bigcup_{u \in U} N(u) \setminus U$. We denote the *degree* of a vertex $u \in V$ by $\deg(u) = |N(u)|$.

Let $G = (V, E)$ be a graph and let $S \subseteq V$. Then S is a *clique* if all vertices of S are pairwise adjacent and an *independent set* if all vertices of S are pairwise non-adjacent. A graph is *complete* if its vertex set is a clique. We let K_r denote the complete graph on k vertices. Let $T \subseteq V$ with $S \cap T = \emptyset$. Then S is *complete* to T if there is an edge between every vertex of S and every vertex of T, and S is *anti-complete* to T if there are no edges between S and T. In the first case we also say that S is *complete* to $G[T]$ and in the second case anticomplete to $G[T]$.

A graph is *bipartite* if its vertex set can be partitioned into at most two independent sets. A bipartite graph is *complete* if its vertex set can be partitioned into two independent sets X and Y such that there is an edge between every vertex of X and every vertex of Y. Note that every edge of a complete bipartite graph is dominating.

Let G_1 and G_2 be two vertex-disjoint graphs. The *union* operation creates the *disjoint union* $G_1 + G_2$ of G_1 and G_2, that is, the graph with vertex set $V(G_1) \cup V(G_2)$ and edge set $E(G_1) \cup E(G_2)$. We denote the disjoint union of r copies of G_1 by rG_1. The *join* operation adds an edge between every vertex of G_1 and every vertex of G_2. A graph G is a *cograph* if G can be generated from K_1 by a sequence of join and union operations. A graph is a cograph if and only if it is P_4-free (see, for example, [5]).

The following lemma is well-known.

Lemma 1. *Every connected P_4-free graph on at least two vertices has a spanning complete bipartite subgraph which can be found in polynomial time.*

Let $G = (V, E)$ be a graph. The *contraction* of an edge $uv \in E$ deletes the vertices u and v and replaces them by a new vertex made adjacent to precisely those vertices that were adjacent to u or v in G (without introducing self-loops or multiple edges). Recall that a linear forest is the disjoint union of one or more paths. The following lemma is a straightforward observation.

Lemma 2. *Let H be a linear forest and let G be a connected H-free graph. Then the graph obtained from G after contracting an edge is also connected and H-free.*

Recall that Grzesik et al. [16] proved that VERTEX COVER is polynomial-time solvable for P_6-free graphs. Using the folklore trick mentioned in Remark 2 (see also, for example, [19,23]) their result can be formulated as follows.

Theorem 1 [16]. *For every $s \geq 0$, VERTEX COVER can be solved in polynomial time for $(sP_1 + P_6)$-free graphs.*

We recall also that CONNECTED VERTEX COVER is polynomial-time solvable for $(sP_1 + P_5)$-free graphs [19]. We will need the extension version of this result. Its proof, which we omit, is based on a straightforward adaption of the proof for CONNECTED VERTEX COVER on $(sP_1 + P_5)$-free graphs [19].

Theorem 2 [19]. *For every $s \geq 0$, CONNECTED VERTEX COVER EXTENSION can be solved in polynomial time for $(sP_1 + P_5)$-free graphs.*

3   The Case $H = P_4$

Recall that Brandstädt and Kratsch [4] proved that FEEDBACK VERTEX SET and ODD CYCLE TRANSVERSAL can be solved in polynomial time for permutation graphs, which form a superclass of the class of P_4-free graphs. Hence, we obtain the following proposition.

Proposition 1 [4]. FEEDBACK VERTEX SET *and* ODD CYCLE TRANSVERSAL *can be solved in polynomial time for P_4-free graphs.*

In this section, we prove that the (extensions versions of the) connected variants of FEEDBACK VERTEX SET and ODD CYCLE TRANSVERSAL are polynomial-time solvable on P_4-free graphs as well. We use Proposition 1 for the proofs.

Theorem 3. CONNECTED FEEDBACK VERTEX SET EXTENSION *can be solved in polynomial time for P_4-free graphs.*

Proof. Let $G = (V, E)$ be a P_4-free graph and $W \subseteq V$. We may assume without loss of generality that G is connected. We search for a smallest connected feedback vertex set S_W of G that contains W. By Lemma 1, we find in polynomial time a spanning complete bipartite subgraph $G' = (X, Y, E')$, so every edge in G' is dominating. The set S_W that we are looking for can be distributed over X and Y in various ways. So we first compute, in Case 1, a smallest feedback vertex set of G that contains both vertices of X and Y. Then, in Case 2, we compute a smallest feedback vertex set of G that is a subset of X, and then a smallest one that is a subset of Y. Afterwards, we take the smallest sets over all sets computed as our final output; note that some sets may not exist. However, as $S = V$ is a feedback vertex set of G, at least one set will be computed.

Case 1. $S_W \cap X \neq \emptyset$ and $S_W \cap Y \neq \emptyset$. In this case, $G[S_W]$ will contain an edge uv of G' and hence S_W will be connected. Otherwise, in order to ensure connectivity and to satisfy the condition of the case, we "guess" an edge uv with $u \in X \cap S_W$ and $v \in Y \cap S_W$, respectively. As we need to consider all possibilities of choosing this edge, this extra step adds an $O(n^2)$-time factor to the running time. We are now left to find a smallest feedback vertex set S' in $G - (W \cup \{u, v\})$. This takes polynomial time due to Proposition 1. We remember $S' \cup W \cup \{u, v\}$.

Case 2. $S_W \subseteq X$ or $S_W \subseteq Y$. We first consider the possibility that $S_W \subseteq X$. Then we must have that $W \subseteq X$; otherwise this possibility will not happen. We start by examining the situation where $S_W = X$. This can only happen if $G[Y]$ is a forest, in which case we remember $|S_W| = X$.

We now examine the situation where $S_W \subsetneq X$. Then Y must be independent, as otherwise $G - S_W$ contains a triangle. So, if Y is not independent, then we discard this option. Assume that Y is an independent set. If $|Y| = 1$, then $G[X] - S_W$ is an independent set, as otherwise $G - S_W$ contains a triangle. Hence, we must compute a smallest connected vertex cover of $G[X]$ that contains W.

We can do this in polynomial time due to Theorem 2. We remember the output. If $|Y| \geq 2$, then $|X \setminus S_W| = 1$, as otherwise $G[Y \cup (X \setminus S_W)]$ contains a cycle. Hence, we check in polynomial time if there exists a vertex $x \in X \setminus W$, such that $X \setminus \{x\}$ is connected. If so we remember the size $|X| - 1$.

We now repeat the same procedure to examine the possibility that $S_W \subseteq Y$. In the end we then take the output of minimum size.

Finally, as mentioned, we compare the size of the set computed in Case 1 with the size of the one computed in Case 2, and we return the smallest set as a smallest connected feedback vertex set of G that contains W.   □

The second result of this section can be proven in exactly the same way as Theorem 3.

Theorem 4. CONNECTED ODD CYCLE TRANSVERSAL EXTENSION *can be solved in polynomial time for P_4-free graphs.*

Proof. We do the same as in the proof of Theorem 3. The differences are the following. In Case 1, we need to compute a smallest odd cycle transversal S' in $G - (W \cup \{u, v\})$ (which can be done using Proposition 1 as well). In Case 2 we again start by examining the situation where $S_W = X$. This can only happen if $G[Y]$ is bipartite, in which case we remember $|S_W| = X$. We then consider the situation where $S_W \subsetneq X$ in the same as in the proof of Theorem 3 except that we no longer distinguish between $|Y| = 1$ and $|Y| \geq 2$, that is, we follow the approach used in the proof of Theorem 3 for the case where $|Y| = 1$ for all values of $|Y|$.   □

4   The Case $\mathbf{H = sP_1 + P_3}$

We will prove that FEEDBACK VERTEX SET and ODD CYCLE TRANSVERSAL and their connected variants can be solved in polynomial time for $(sP_1 + P_3)$-free graphs. In order to do this we need a structural result first.

Lemma 3. *For every $s \geq 0$, let G be a bipartite $(sP_1 + P_3)$-free graph. If the smallest connected component of G contains at least c vertices where*

$$c = \begin{cases} 3 & \text{if } s \leq 1 \\ 2s - 1 & \text{if } s \geq 2 \end{cases}$$

then G has only one component.

Proof. Assume that G has two connected components C_1 and C_2 that each contain at least c vertices. As C_1 is bipartite and contains at least $2s - 1$ vertices, it contains a set of s independent vertices that induce sP_1. As $c \geq 3$, there is a vertex v in C_2 of degree at least 2, and, as C_2 is bipartite, the neighbours of v are independent so v and two of its neighbours induce a P_3. Thus G is not $(sP_1 + P_3)$-free. This contradiction completes the proof.   □

We now state our four results. For the connected variants we can show the extension versions. We only include the proof of Theorem 8, which is the most involved and shows all our techniques, and omit the other proofs.

Theorem 5. *For every $s \geq 0$,* FEEDBACK VERTEX SET *can be solved in polynomial time for $(sP_1 + P_3)$-free graphs.*

Theorem 6. *For every $s \geq 0$,* CONNECTED FEEDBACK VERTEX SET EXTENSION *can be solved in polynomial time for $(sP_1 + P_3)$-free graphs.*

Theorem 7. *For every $s \geq 0$,* ODD CYCLE TRANSVERSAL *can be solved in polynomial time for $(sP_1 + P_3)$-free graphs.*

Theorem 8. *For every $s \geq 0$,* CONNECTED ODD CYCLE TRANSVERSAL EXTENSION *can be solved in polynomial time for $(sP_1 + P_3)$-free graphs.*

Proof. Let $G = (V, E)$ be an $(sP_1 + P_3)$-free graph on n vertices and $W \subseteq V$. If G is bipartite and W is connected, then W is the unique minimum connected odd cycle transversal that contains W. If G is bipartite and W is not connected, then in polynomial time we find the smallest set $U \supset W$ such that $G[U]$ is connected by adding vertices of shortest paths connecting the different components of $G[W]$ (assuming that all vertices of W belong to the same component as G; otherwise we return a no-answer). If G is disconnected, then each of its connected components except for one must be bipartite; otherwise we return a no-answer. From now on, assume that G is non-bipartite and connected. This means that V is a connected odd cycle transversal of G. We can determine in polynomial time whether V is the minimum size connected odd cycle transversal that contains W by checking, for each vertex $u \in V$, whether or not $V \setminus \{u\}$ is a connected odd cycle transversal of G that contains W. Thus from we assume that V is not a minimum connected odd cycle transversal of G that contains W.

If $s = 0$, then we can use Theorem 4. So we assume that $s \geq 1$ and that we can solve CONNECTED FEEDBACK VERTEX SET EXTENSION in polynomial time for $((s-1)P_1 + P_3)$-free graphs. We show that we can find, in polynomial time, a smallest connected odd cycle transversal S_W of G that contains W. In fact, we shall solve the equivalent problem of finding, in polynomial time, a bipartite subgraph B_W of G such that $B_W \cap W = \emptyset$, $G - B_W$ is connected and, subject to these conditions, B_W is of maximum size. To do this, we consider two cases. Let $c = 3$ if $s = 1$ and $c = 2s - 1$ otherwise (the constant c comes from Lemma 3, which we will apply in Case 2). Our two cases derive from assuming, or not, that at least one connected component of B_W has size at most $c - 1$. In each case, we attempt to find, subject to our assumption, a bipartite subgraph B of G such that $B \cap W = \emptyset$ and $G - B$ is connected, and, if such a set B exists, we find the solution of maximum size. We will see that, taken together, we cover all possible cases so the largest B found has size $|B_W|$. In particular, we note that B_W is not empty by our assumption that $S_W \neq V$.

Case 1. At least one connected component L of B_W has size at most $c - 1$. In this case we take every possible choice for L under consideration, discarding all sets that do not induce a bipartite graph, or whose removal disconnects the graph, or that intersect W (as none of these sets can be a candidate set for L). As $|V(L)| \leq c - 1$, there are at most $O(n^{c-1})$ choices. For each choice of L we do as follows.

Let U be the set of neighbours of the vertices of L that belong to $G - L$. Note that U must belong to S_W if we guessed it correctly, and so we may contract any edge inside $G[U]$ to modify U into an independent set. This takes polynomial time and, by Lemma 2, the resulting graph, which we denote by G again, is still $(sP_1 + P_3)$-free. Moreover, $G - L$ is still connected.

As L contains at least one vertex and G is $(sP_1 + P_3)$-free, $G - (L \cup U)$ is $((s-1)P_1 + P_3)$-free. Let S be a connected odd cycle transversal that contains U. As U is an independent set, each of its vertices has at least one neighbour in $S \setminus U$. Thus there are sets in $S \setminus U$ that dominate U. Let R be a smallest such set.

We consider each possible choice. If $|U| = 1$, then $|R| \leq 1$. Suppose $|U| \geq 2$. As U is an independent set on at least two vertices, S_W must contain three vertices of $G - L$ that form an induced path, which we denote by P. As G is $(sP_1 + P_3)$-free and U is independent, $V(P)$ must dominate all but at most $s-1$ vertices of U. Let U' be the subset of vertices of U that are not dominated by $V(P)$. So $|U'| \leq s - 1$. Consider a set that contains $V(P)$ and, for each vertex u in U', a neighbour of u in S_W. This set dominates U so is at least the size of R. Thus $|R| \leq |P| + |U'| \leq 3 + s - 1 = s + 2$.

Hence there at most $O(n^{s+2})$ possible choices for R. We consider each possible choice, and for each we compute the size of a smallest odd cycle transversal S_R in $G - (L \cup U)$ that contains $R \cup (W \setminus U)$. (Recall that $W \cap V(L) = \emptyset$, so $W \setminus U$ belongs to $G - (L \cup U)$.) As $G - (L \cup U)$ is $((s-1)P_1 + P_3)$-free, we can find S_R in polynomial time using our algorithm for $((s-1)P_1 + P_3)$-free graphs. Then $S_R \cup U$ is a smallest connected odd cycle transversal of G that contains $U \cup R \cup W$.

Now, over all choices for R, we keep the smallest $S_R \cup U$, which we denote by S_L. Then S_L is a smallest connected odd cycle transversal of G that contains W such that $G - S$ has L as a connected component.

Finally, from all sets S_L, we keep the smallest set found, which we denote by S_1. Then S_1 is a smallest connected odd cycle transversal S of G that contains W such that $G - S$ has a connected component of size at most $c - 1$. We find S_1 in polynomial time, as the number of choices for R and L is polynomially bounded and each choice can be processed in polynomial time. Let $B_1 = G - S$.

Case 2. Every connected component of B_W has size at least c.
In this case we will compute the largest induced bipartite graph B of G such that $B \cap W = \emptyset$ and $S = V(G) \setminus V(B)$ is connected, and, moreover, every connected component of B has size at least c. Then by Lemma 3, the subgraph B we are looking for is connected.

Dealing with the case where one partition class is small.
First we suppose that B has a bipartition (X, Y) such that $|X| \leq s - 1$. To find the best solution in this case, we consider each of the $O(n^{s-1})$ sets X of at most

$s - 1$ vertices of G. For every such X, we check whether X is an independent set (in constant time) and whether X is disjoint from W. If both conditions are satisfied, we wish to find Y, the largest possible independent set that is in $G - X$ and disjoint from W such that $G - (X \cup Y)$ is connected. By Theorem 2 we can do this in polynomial time by computing a minimum connected vertex cover S_X of $G - X$ that contains W. Then we can let $G - (X \cup S_X)$ be Y. Note that $X \cup Y$ might not be connected, so we may have duplicated some polynomial-time work. We pick the best solution B, and set $B_2 = B$ (we will update B_2 in the remainder of Case 2 and afterwards compare its size with the size of B_1).

Dealing with the case where both bipartition classes are large.
Now we suppose that B is connected, contains at least $c \geq 3$ vertices and has a bipartition in which each partition class contains at least s vertices. In particular B contains an induced P_3. We consider each of the $O(n^{2s})$ pairs of disjoint sets X' and Y' each containing s of the vertices of G. We check whether X' and Y' are both independent sets and are disjoint from W and whether $G[X' \cup Y']$ has an induced P_3. If these conditions are not satisfied, we discard the case. Otherwise, our aim will be to try to construct from X' and Y', a bipartite graph $B = (X, Y)$ such that $X' \subseteq X$, $Y' \subseteq Y$, $(X \cup Y) \cap W = \emptyset$ and $G - B$ is connected and, subject to these conditions, B is of maximum size. We now define (see also Fig. 1) a partition of $V \setminus (X' \cup Y') = U \cup V_X \cup V_Y \cup Z$ where

$$U = (N(X') \cap N(Y')) \cup W,$$
$$V_X = N(X') \setminus (Y' \cup N(Y') \cup W),$$
$$V_Y = N(Y') \setminus (X' \cup N(X') \cup W),$$
$$Z = V \setminus (X' \cup Y' \cup N(X') \cup N(Y') \cup W).$$

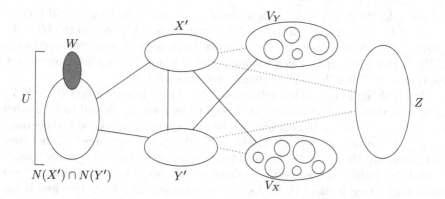

Fig. 1. The decomposition of G in Case 2, where a straight edge between two sets indicates that at least one edge must exist, a dotted edge indicates that no edges between the two sets exist, and the absence of an edge indicates that edges between the two sets could exist. The circles in V_X and V_Y represent disjoint unions of complete graphs.

No vertex $u \in U$ can be a member of B, as either u has at least one neighbour in X' and at least one neighbour in Y', or u belongs to W. We also know that $G[V_X]$ is P_3-free, as otherwise $Y' \cup V_X$ would induce an $sP_1 + P_3$. By the same argument, $G[V_Y]$ is also P_3-free. This means that both $G[V_X]$ and $G[V_Y]$ are the disjoint union of a set of complete graphs. Moreover, Z does not contain an independent set of size greater than $s - 1$ as otherwise, since $G[X' \cup Y']$ contains an induced P_3, we find that $X' \cup Y' \cup Z$ contains a subset of vertices that induce an $sP_1 + P_3$.

Step 1. Reduce Z to the Empty Set

We are going to reduce Z to the empty set via some branching. We consider all the possible ways the vertices of Z might be included in B. As Z does not contain an independent set of size greater than $s - 1$, every partition class of B contains at most $s - 1$ vertices of Z. Hence, we consider each of the $O(n^{2s-2})$ pairs of disjoint sets Z_X and Z_Y of size at most $s-1$ in Z. We check whether Z_X and Z_Y are independent sets. If they both are, we define X' to be $X' \cup Z_X$ and Y' to be $Y' \cup Z_Y$. We redefine U by adding to it the vertices of $Z \setminus (Z_X \cup Z_Y)$; note that U still contains W. Moreover, vertices of V_X with a neighbour in Z_Y cannot belong to Y' (recall that they cannot be in X' either). Similarly, the set of vertices of V_X with a neighbour in Z_X cannot be members of X' or Y' either. Thus we redefine U again by adding all these vertices to it, and V_X and V_Y by removing the vertices we placed in U.

We now have a partition $X' \cup Y' \cup U \cup V_X \cup V_Y$ where $G[X', Y']$ is bipartite, U contains vertices that have neighbours in both X' and Y' or vertices that belong to W, the vertices of V_X have neighbours in X' but not in Y', and the vertices of V_Y have neighbours in Y' but not in X'. Moreover, $G[V_X]$ and $G[V_Y]$ are still the disjoint union of a set of complete graphs.

Step 2. Reduce U to a Singleton Set

We are going to reduce U to a singleton set via some branching. Recall that no vertex of U will be placed in the final partition classes X and Y of the bipartite graph B we are searching for. We contract every edge between two vertices in $G[U]$. In the new graph, which we denote by G again, U is an independent set. By Lemma 2, G is still $(sP_1 + P_3)$-free and connected.

As U belongs to the connected complement of the bipartite graph $B = (X, Y)$ we are searching for, the vertices of U need to be made connected to each other via paths in $G - (X' \cup Y')$. Following the same arguments as in Case 1, there must exist a set R of size at most $s + 2$ in $G - (X' \cup Y')$ that dominates U (if not then we can discard the case). We guess R by considering all $O(n^{s+2})$ possibilities. If needed we consider all possibilities of making $R \cup U$ connected via adding shortest paths connecting vertices of R. As every connected $(sP_1 + P_3)$-free graph has diameter at most $2s + 2$, we need to guess a total of $(|R| - 1)2s \leq 2s^2 + 2s$ additional vertices, so must consider $O(n^{2s^2+2s})$ possibilities. In each branch we contract all edges in $G[R \cup U]$ into a single vertex, which we denote by u. By Lemma 2 the resulting graph, which we denote by G again, is $(sP_1 + P_3)$-free and connected.

We redefine V_X and V_Y by removing the vertices that were added to U. Then $G[V_X]$ and $G[V_Y]$ are still the disjoint unions of complete graphs, which we denote by, respectively, $K_X^1, \ldots, K_X^q$ (if $V_X \neq \emptyset$) and $K_Y^1, \ldots, K_Y^r$ (if $V_Y \neq \emptyset$).

Step 3. Adding Vertices from V_X to Y' and from V_Y to X'

To complete the construction of B, we need only describe how to add (in polynomial time) as many vertices in total from V_X to Y' and from V_Y to X' in such a way that the new sets X and Y remain independent sets and the graph induced by $V \setminus (X \cup Y)$ is connected. In particular $V \setminus (X \cup Y)$ contains the vertex u, to which all vertices of W have been contracted. Note that from each K_X^h and each K_Y^i we can add at most one vertex to $X' \cup Y'$, as otherwise we create a triangle in B. However, we must be careful, as by adding a vertex from $V_X \cup V_Y$ to $X' \cup Y'$, we may lose connectivity of the graph $G - (X \cup Y)$; recall that X and Y are the sets we are trying to construct. Recall that $S = V(G) \setminus (X \cup Y)$ is the corresponding connected odd cycle transversal that we are trying to create. We analyse the possible structure of S by distinguishing two cases.

Case (i). The graph $G[S]$ contains no edges between $V_X \cap S$ and $V_Y \cap S$. Recall that X and Y can contain at most one vertex from each K_X^h and K_Y^i. Hence, u must be adjacent to at least one vertex of every K_X^h of size at least 2 and at least one vertex of every K_Y^i of size at least 2. If not, then we can discard our current choice, as it will not lead to a set S that is connected. If u is complete to a set $V(K_X^h)$, for $h \in \{1, \ldots, q\}$, we pick an arbitrary vertex of $V(K_X^h)$, and else a non-neighbour of u, and add it to Y'. We do the same thing when considering $V(K_Y^i)$, for $i \in \{1, \ldots, r\}$, and add a vertex to X'. We also put the vertices of all singleton connected components of $G[V_X]$ and $G[V_Y]$ in Y' and X', respectively. If the resulting set $X' \cup Y'$ is larger than B_2, then we let $B_2 = X' \cup Y'$.

Case (ii). The graph $G[S]$ has an edge xy where $x \in V_X \cap S$ and $y \in V_Y \cap S$. We consider all $O(n^2)$ possibilities for choosing the edge xy. Let xy be such a choice. By definition, x has a neighbour $v_x \in X$ and y has a neighbour $v_y \in Y$. As no vertex of V_Y has a neighbour in X, the vertices y, x, v_x induce a P_3. As G is $(sP_1 + P_3)$-free, x must then be complete to all but at most $s - 1$ graphs K_Y^i. Similarly, y must be complete to all but at most $s - 1$ graphs K_X^h. A graph K_X^h or K_Y^i is *bad* if it is not complete to y or x, respectively, and *good* otherwise.

We first consider the bad complete graphs. Note that x, y could be in a bad complete graph. For each bad complete graph, we "guess" at most one vertex distinct from x and y that we will move to X' or Y' (so we update X' and Y'). This leads to $O(n^{2s-2})$ possible cases and we consider each of them as follows.

We first check if the remaining vertices from the bad complete graphs, the vertex u and the vertices x, y all belong to the same connected component of $G - (X' \cup Y')$. This must hold in order for all these vertices to end up in the connected graph $G - (X \cup Y)$ we are looking for (if the branch under consideration is correct, then all vertices of $G - (X \cup Y)$ belong to $G - (X' \cup Y')$). So, if this does not hold, then we discard the case. Otherwise we try to connect the remaining vertices of the bad components, u, x, and y by considering all possibilities for choosing a smallest connected set in $G - (X' \cup Y')$ that contains all of them.

Before doing this, we first contract any edges between vertices that belong to the union of the bad complete graphs and the set $\{u, x, y\}$. As xy is an edge, this leads to an independent set of size at most $2(s-1) + 1 + 1 = 2s$. By Lemma 2 the resulting graph is $(sP_1 + P_3)$-free again, so the connected component that we are looking for has diameter at most $2s + 2$. This means that we need to "guess" at most $(2s - 1)(2s + 1) = 4s^2 - 1$ vertices. Hence, the total number of possible choices is $O(n^{4s^2-1})$. We consider each choice. For each choice, it remains to pick for every good complete graph an arbitrary vertex (if it exists) that was not involved in the guessing and put it in X' or Y' in an appropriate way. We may pick these vertices arbitrarily, as we can only pick one vertex from each complete graph and all remaining vertices of the good complete graphs are adjacent to one of x, y, ensuring connectivity. If the resulting set $X' \cup Y'$ is larger than B_2, then we let $B_2 = X' \cup Y'$. This completes the description of Case 2.

Note that from all the bipartite graphs $B = (X, Y)$ we kept track of in Case 2, we stored a largest one B_2. We compare B_2 with B_1, picking the largest as B_W. Then $S_W = V(G) - B_W$ is a smallest connected odd cycle transversal of G that contains W. The correctness of our algorithm follows from its description. Moreover, as the number of branches is polynomial and each branch is processed in polynomial time, the running time of our algorithm is polynomial.   □

5   Conclusions

We proved polynomial results for FEEDBACK VERTEX SET and ODD CYCLE TRANSVERSAL and their connected variants for H-free graphs, where $H = P_4$ or $H = sP_1 + P_3$. Natural cases for future work are the cases where $H = sP_1 + P_4$ for $s \geq 1$ and $H = P_r$ for $r \geq 5$. Note that Lemma 3 does not hold for $(sP_1 + P_4)$-free graphs: the disjoint union of any number of arbitrarily large stars is even P_4-free. We also lose the crucial property of a connected $(sP_1 + P_3)$-free graph G that any independent set U is dominated by a set $R \subseteq V(G) \setminus U$ with $|R| \leq s+2$. Recall that VERTEX COVER and CONNECTED VERTEX COVER are polynomial-time solvable even for $(sP_1 + P_6)$-free graphs [16] and $(sP_1 + P_5)$-free graphs [19] for every $s \geq 0$. Recall that ODD CYCLE TRANSVERSAL and CONNECTED ODD CYCLE TRANSVERSAL are known to be NP-complete for P_{13}-free graphs [25]. However, no integer r is known, for which any of the other four problems is NP-complete for P_r-free graphs.

Independent Transversals. A similar complexity study is also undertaken for the independent variants of the problems FEEDBACK VERTEX SET and ODD CYCLE TRANSVERSAL.[3] In particular, INDEPENDENT FEEDBACK VERTEX SET and INDEPENDENT ODD CYCLE TRANSVERSAL are polynomial-time solvable for P_5-free graphs [3], but their complexity status is unknown for P_6-free graphs. No integer r is known either such that INDEPENDENT FEEDBACK VERTEX SET and INDEPENDENT ODD CYCLE TRANSVERSAL are NP-complete for P_r-free graphs.

---

[3] INDEPENDENT VERTEX COVER is 2-COLOURING, which is polynomially solvable.

Hence, to make any further progress, we must understand the structure of P_r-free graphs better. This topic has been well-studied in recent years, see also for example [14,17]. However, more research and new approaches are needed.

Two Other Generalizations. A well-known way of generalizing FEEDBACK VERTEX SET and ODD CYCLE TRANSVERSAL is to pre-specify a set T of *terminal* vertices in a graph $G = (V, E)$ and to ask for a corresponding transversal of size at most k that intersects all cycles or odd cycles, respectively, that contain at least one terminal vertex from T. These problems are called SUBSET FEEDBACK VERTEX SET and SUBSET ODD CYCLE TRANSVERSAL (choose $T = V$ to get the original problem back).

Both SUBSET FEEDBACK VERTEX SET and SUBSET ODD CYCLE TRANSVERSAL are NP-complete for H-free graphs if H contains a cycle or claw, due to the aforementioned NP-completeness for the original problems. Moreover, SUBSET FEEDBACK VERTEX SET is polynomial-time solvable for sP_1-free graphs [27] and for permutation graphs [26], and thus for P_4-free graphs, but NP-complete for split graphs [12], or equivalently, $(C_4, C_5, 2P_2)$-free graphs, and thus for P_5-free graphs.[4] It would be interesting to obtain full complexity dichotomies for SUBSET FEEDBACK VERTEX SET and SUBSET ODD CYCLE TRANSVERSAL for H-free graphs. For the former problem it remains to solve the cases where $H = P_r + sP_1$ for every pair (r, s) with $r = 2, s \geq 2$ or $3 \leq r \leq 4, s \geq 1$. The latter problem seems to be mainly studied from a parameterized point of view [20].

References

1. Balas, E., Yu, C.S.: On graphs with polynomially solvable maximum-weight clique problem. Networks **19**, 247–253 (1989)
2. Belmonte, R., van't Hof, P., Kaminski, M., Paulusma, D.: The price of connectivity for feedback vertex set. Discrete Appl. Math. **217**, 132–143 (2017)
3. Bonamy, M., Dabrowski, K.K., Feghali, C., Johnson, M., Paulusma, D.: Independent feedback vertex set for P_5-free graphs. Algorithmica **81**, 1342–1369 (2019)
4. Brandstädt, A., Kratsch, D.: On the restriction of some NP-complete graph problems to permutation graphs. In: Budach, L. (ed.) FCT 1985. LNCS, vol. 199, pp. 53–62. Springer, Heidelberg (1985). https://doi.org/10.1007/BFb0028791
5. Brandstädt, A., Le, V.B., Spinrad, J.: Graph Classes: A Survey. Society for Industrial and Applied Mathematics (SIAM). SIAM Monographs on Discrete Mathematics and Applications (1999)
6. Camby, E., Cardinal, J., Fiorini, S., Schaudt, O.: The price of connectivity for vertex cover. Discrete Math. Theor. Comput. Sci. **16**, 207–224 (2014)
7. Cardinal, J., Levy, E.: Connected vertex covers in dense graphs. Theor. Comput. Sci. **411**, 2581–2590 (2010)
8. Chiarelli, N., Hartinger, T.R., Johnson, M., Milanic, M., Paulusma, D.: Minimum connected transversals in graphs: new hardness results and tractable cases using the price of connectivity. Theor. Comput. Sci. **705**, 75–83 (2018)

---

[4] FEEDBACK VERTEX SET is polynomial-time solvable even for sP_2-free graphs [8].

9. Camby, E., Schaudt, O.: The price of connectivity for dominating set: upper bounds and complexity. Discrete Appl. Math. **17**, 53–59 (2014)
10. Escoffier, B., Gourvès, L., Monnot, J.: Complexity and approximation results for the connected vertex cover problem in graphs and hypergraphs. Theor. Comput. Sci. **8**, 36–49 (2010)
11. Fernau, H., Manlove, D.: Vertex and edge covers with clustering properties: complexity and algorithms. J. Discrete Algorithms **7**, 149–167 (2009)
12. Fomin, F.V., Heggernes, P., Kratsch, D., Papadopoulos, C., Villanger, Y.: Enumerating minimal subset feedback vertex sets. Algorithmica **69**, 216–231 (2014)
13. Garey, M.R., Johnson, D.S.: The rectilinear Steiner tree problem is NP-complete. SIAM J. Appl. Math. **32**, 826–834 (1977)
14. Golovach, P.A., Johnson, M., Paulusma, D., Song, J.: A survey on the computational complexity of colouring graphs with forbidden subgraphs. J. Graph Theory **84**, 331–363 (2017)
15. Grigoriev, A., Sitters, R.: Connected feedback vertex set in planar graphs. In: Paul, C., Habib, M. (eds.) WG 2009. LNCS, vol. 5911, pp. 143–153. Springer, Heidelberg (2010). https://doi.org/10.1007/978-3-642-11409-0_13
16. Grzesik, A., Klimošová, T., Pilipczuk, M., Pilipczuk, M.: Polynomial-time algorithm for maximum weight independent set on P_6-free graphs. In: Proceedings of the SODA 2019, pp. 1257–1271 (2019)
17. Groenland, C., Okrasa, K., Rzążewski, P., Scott, A., Seymour, P., Spirkl, S.: H-colouring P_t-free graphs in subexponential time, Discrete Applied Mathematics, to appear
18. Hartinger, T.R., Johnson, M., Milanic, M., Paulusma, D.: The price of connectivity for transversals. Eur. J. Comb. **58**, 203–224 (2016)
19. Johnson, M., Paesani, G., Paulusma, D.: Connected vertex cover for (sP_1+P_5)-free graphs. In: Brandstädt, A., Köhler, E., Meer, K. (eds.) Proceedings of WG 2018. LNCS, vol. 11159, pp. 279–291. Springer, Heidelberg (2018). https://doi.org/10.1007/978-3-030-00256-5_23
20. Lokshtanov, D., Misra, P., Ramanujan, M.S., Saurabh, S.: Hitting selected (odd) cycles. SIAM J. Discret. Math. **31**, 1581–1615 (2017)
21. Minty, G.J.: On maximal independent sets of vertices in claw-free graphs. J. Comb. Theory Ser. B **28**, 284–304 (1980)
22. Misra, N., Philip, G., Raman, V., Saurabh, S.: On parameterized independent feedback vertex set. Theor. Comput. Sci. **461**, 65–75 (2012)
23. Mosca, R.: Stable sets for $(P_6, K_{2,3})$-free graphs. Discuss. Math. Graph Theory **32**, 387–401 (2012)
24. Munaro, A.: Boundary classes for graph problems involving non-local properties. Theor. Comput. Sci. **692**, 46–71 (2017)
25. Okrasa, K., Rzążewski, P.: Subexponential algorithms for variants of homomorphism problem in string graphs. In: Proceedings of WG 2019. LNCS (to appear)
26. Papadopoulos, C., Tzimas, S.: Polynomial-time algorithms for the subset feedback vertex set problem on interval graphs and permutation graphs. In: Klasing, R., Zeitoun, M. (eds.) FCT 2017. LNCS, vol. 10472, pp. 381–394. Springer, Heidelberg (2017). https://doi.org/10.1007/978-3-662-55751-8_30
27. Papadopoulos, C., Tzimas, S.: Subset feedback vertex set on graphs of bounded independent set size. In: Proceedings of IPEC 2018. LIPIcs, vol. 115, pp. 20:1–20:14 (2018)
28. Poljak, S.: A note on stable sets and colorings of graphs. Comment. Math. Univ. Carol. **15**, 307–309 (1974)

29. Priyadarsini, P.K., Hemalatha, T.: Connected vertex cover in 2-connected planar graph with maximum degree 4 is NP-complete. Int. J. Math., Phys. Eng. Sci. **2**, 51–54 (2008)

30. Tsukiyama, S., Ide, M., Ariyoshi, H., Shirakawa, I.: A new algorithm for generating all the maximal independent sets. SIAM J. Comput. **6**, 505–517 (1977)

31. Sbihi, N.: Algorithme de recherche d'un stable de cardinalité maximum dans un graphe sans étoile. Discrete Math. **29**, 53–76 (1980)

32. Ueno, S., Kajitani, Y., Gotoh, S.: On the nonseparating independent set problem and feedback set problem for graphs with no vertex degree exceeding three. Discrete Math. **72**, 355–360 (1988)

33. Wanatabe, T., Kajita, S., Onaga, K.: Vertex covers and connected vertex covers in 3-connected graphs. In: Proceedings of the IEEE International Symposium on Circuits and Systems, pp. 1017–1020 (1991)

Maximum Rectilinear Convex Subsets

Hernán González-Aguilar[1], David Orden[2(✉)], Pablo Pérez-Lantero[3],
David Rappaport[4], Carlos Seara[5], Javier Tejel[6], and Jorge Urrutia[7]

[1] Facultad de Ciencias, Universidad Autónoma de San Luis Potosí,
San Luis Potosí, Mexico
hernan@fc.uaslp.mx

[2] Departamento de Física y Matemáticas, Universidad de Alcalá,
Alcala de Henares, Spain
david.orden@uah.es

[3] Departamento de Matemática y Ciencia de la Computación, USACH,
Santiago, Chile
pablo.perez.l@usach.cl

[4] School of Computing, Queen's University,
Kingston, Canada
daver@cs.queensu.ca

[5] Departament de Matemàtiques, Universitat Politècnica de Catalunya,
Barcelona, Spain
carlos.seara@upc.edu

[6] Departamento de Métodos Estadísticos, IUMA, Universidad de Zaragoza,
Zaragoza, Spain
jtejel@unizar.es

[7] Instituto de Matemáticas, Universidad Nacional Autónoma de México,
Mexico City, Mexico
urrutia@matem.unam.mx

Abstract. Let P be a set of n points in the plane. We consider a variation of the classical Erdős-Szekeres problem, presenting efficient algorithms with $O(n^3)$ running time and $O(n^2)$ space complexity that compute: (1) A subset S of P such that the boundary of the rectilinear convex hull of S has the maximum number of points from P, (2) a subset S of

D. Orden—Research supported by project MTM2017-83750-P of the Spanish Ministry of Science (AEI/FEDER, UE).

P. Pérez-Lantero—Research supported by project CONICYT FONDECYT/Regular 1160543 (Chile).

D. Rappaport—Research supported by NSERC of Canada Discovery Grant RGPIN/06662-2015.

C. Seara—Research supported by projects MTM2015-63791-R MINECO/FEDER and Gen. Cat. DGR 2017SGR1640.

J. Tejel—Research supported by projects MTM2015-63791-R MINECO/FEDER and Gobierno de Aragón E41-17R.

J. Urrutia—Research supported by PAPIIT grant IN102117 from UNAM.

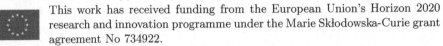 This work has received funding from the European Union's Horizon 2020 research and innovation programme under the Marie Skłodowska-Curie grant agreement No 734922.

L. A. Gąsieniec et al. (Eds.): FCT 2019, LNCS 11651, pp. 274–291, 2019.
https://doi.org/10.1007/978-3-030-25027-0_19

P such that the boundary of the rectilinear convex hull of S has the maximum number of points from P and its interior contains no element of P, (3) a subset S of P such that the rectilinear convex hull of S has maximum area and its interior contains no element of P, and (4) when each point of P is assigned a weight, positive or negative, a subset S of P that maximizes the total weight of the points in the rectilinear convex hull of S.

Keywords: Erdős-Szekeres problems · Convex subsets · Optimization · Orthoconvexity · Rectilinear convex hull

1   Introduction

Let P be a point set in general position in the plane. A subset S of P with k elements is called a *convex k-gon* if the elements of S are the vertices of a convex polygon, and it is called a *convex k-hole* of P if the interior of the convex hull of S contains no element of P. The study of convex k-gons and convex k-holes of point sets started in a seminal paper by Erdős and Szekeres [9] in 1935. Since then, numerous papers about both the combinatorial and the algorithmic aspects of convex k-gons and convex k-holes have been published. The reader can consult the two survey papers about so-called Erdős-Szekeres type problems [7,11].

There are recent papers studying the existence and number of convex k-gons and convex k-holes for finite point sets in the plane [1–3]. Papers dealing with the algorithmic complexity of finding largest convex k-gons and convex k-holes are, respectively, Chvátal and Kincsek [8] and Avis and Rappaport [5], which solve these problems in $O(n^3)$ time.

Erdős-Szekeres type problems have also been studied for colored point sets. Let P be a point set such that each of its elements is assigned a color, say red or blue. Bautista-Santiago et al. [6] studied the problem of finding a monochromatic subset S of P of maximum cardinality such that all of the elements of P contained in the convex hull of S have the same color. As a generalization, they also studied the problem in which each element of P has assigned a (positive or negative) weight. In this case, the goal is to find a subset S of P that maximizes the total weight of the points of P contained in the convex hull of S. Each of these problems was solved in $O(n^3)$ time and $O(n^2)$ space. Further, their algorithm can easily be adapted to find a subset S of P such that the convex hull of S is empty and of maximum area in $O(n^3)$ time and $O(n^2)$ space.

In this paper, we study Erdős-Szekeres type problems under a variation of convexity known as *rectilinear convexity*, or *orthoconvexity*: Let $P = \{p_1, \ldots, p_n\}$ be a set of n points in the plane in general position. A *quadrant* of the plane is the intersection of two open half-planes whose supporting lines are parallel to the x- and y-axes, respectively. We say that a quadrant Q is *P-free* if it does not contain any point of P. The *rectilinear convex hull* of P, denoted as $RCH(P)$, initially defined by Ottmann et al. [12], is defined as:

$$RCH(P) = \mathbb{R}^2 - \bigcup_{Q \text{ is } P\text{-free}} Q.$$

The rectilinear convex hull of a point set might be a simply connected set, yielding an intuitive and appealing structure (see Fig. 1a). However, in other cases the rectilinear convex hull can have several connected components (see Fig. 1b), some of which might be single points which we call *pinched* points. The *size* of $RCH(P)$ is the number of elements of P on the boundary of $RCH(P)$. The sizes of the rectilinear convex hulls in Figs. 1a and b are, respectively, thirteen and twelve.

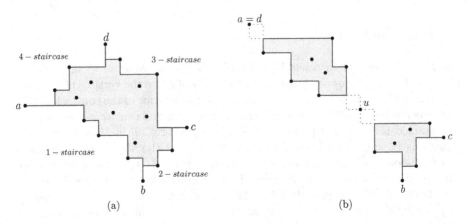

Fig. 1. (a) A point set with a connected rectilinear convex hull. (b) A point set whose rectilinear convex hull is disconnected, two of its components are pinched points.

Alegría-Galicia et al. [4] gave an optimal $\Theta(n \log n)$-time and $O(n)$-space algorithm to compute the orientation of the coordinate axes such that the rectilinear convex hull of a set P of n points in the plane has minimum area. The reader can refer to the literature for other results related to rectilinear convexity [4,10,14]. In this paper, we present efficient algorithms for the following geometric optimization problems:

MaxRCH: Given a set P of n points in the plane, find a subset $S \subseteq P$ such that the size of $RCH(S)$ is maximized.

MaxEmptyRCH: Given a set P of n points in the plane, find a subset $S \subseteq P$ such that the interior of $RCH(S)$ contains no point of P and the size of $RCH(S)$ is maximized.

MaxAreaRCH: Given a set P of n points in the plane, find a subset $S \subseteq P$ such that the interior of $RCH(S)$ contains no point of P and the area of $RCH(S)$ is maximized.

MaxWeightRCH: Given a set P of n points in the plane, such that each $p \in P$ is assigned a (positive or negative) weight $w(p)$, find a subset $S \subseteq P$ that maximizes $\sum_{p \in P \cap RCH(S)} w(p)$.

In Sect. 3, we give an $O(n^3)$-time $O(n^2)$-space algorithm to solve the MaxRCH problem. Then, in Sect. 4 we show how to adapt this algorithm to solve the other three problems, each in $O(n^3)$ time and $O(n^2)$ space. The complexities of our algorithms are the same as the complexities of the best-known algorithms to solve these problems with the usual definition of convexity.

2   Some Notation and Definitions

For the sake of simplicity, we assume that all point sets P considered in this paper are in *general position*, which means that no two points of P share the same x- or y-coordinate. Using a $O(n \log n)$-time preprocessing step, we can also assume when necessary that the points of a point set P are ordered by x-coordinate or y-coordinate. Given a point set P in the plane, we will use a, b, c, and d to denote the leftmost, bottommost, rightmost, and topmost points of P, respectively, unless otherwise stated. Note that a, b, c, and d are not necessarily different. In Fig. 1b, we have $a = d$.

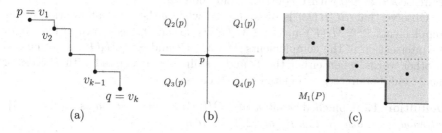

Fig. 2. (a) A 1-staircase. (b) The definition of the sets $Q_i(p)$. (c) A 7-point set P and its $M_1(P)$. The vertices of the boundary of $M_1(P)$ in P are the 1-extremal points of P. The thick polygonal line is the 1-staircase associated with P.

Given a point p of the plane, let p_x and p_y denote the x- and y-coordinates of p, respectively. For $p, q \in \mathbb{R}^2$, $p \neq q$, we write $p \prec q$ to denote that $p_x < q_x$ and $p_y < q_y$, and $p \prec' q$ to denote that $p_x < q_x$ and $p_y > q_y$. Let $p, q \in \mathbb{R}^2$, and consider a set $S = \{v_1, \ldots, v_k\}$ of k points such that $v_1 = p$, $v_k = q$, and $v_i \prec' v_{i+1}$ for $i = 1, 2, \ldots, k-1$. A 1-*staircase* joining p to q is an orthogonal polygonal chain, such that two consecutive elements of S are joined by an elbow consisting of a horizontal segment followed by a vertical segment. For an illustration, see Fig. 2a. A 3-*staircase* joining p to q is defined in a similar way, but using elbows whose first segment is vertical. Analogously, we define 2- and 4-staircases, except that we require $v_i \prec v_{i+1}$. The first segment is vertical in the 2-staircase and horizontal in the 4-staircase. Points of S are called the vertices of the staircase.

Any point p in the plane defines four open axis-aligned quadrants $Q_i(p)$, $i = 1, 2, 3, 4$, as follows (see Fig. 2b): $Q_1(p) = \{q \in \mathbb{R}^2 \mid p \prec q\}$, $Q_2(p) = \{q \in \mathbb{R}^2 \mid q \prec' p\}$, $Q_3(p) = \{q \in \mathbb{R}^2 \mid q \prec p\}$, and $Q_4(p) = \{q \in \mathbb{R}^2 \mid p \prec' q\}$. Given a point set P in the plane, for $i = 1, 2, 3, 4$, let

$$M_i(P) = \bigcup_{p \in P} \overline{Q_i(p)},$$

where $\overline{Q_i(p)}$ denotes the closure of $Q_i(p)$. The elements of P that belong to the boundary of $M_i(P)$, are called the (rectilinear) i-extremal points of P (see Fig. 2c). Note that the i-extremal points of P are the vertices of a i-staircase connecting all of them. This i-staircase, that we call the i-staircase associated with P, is the part of the boundary of $M_i(P)$ that connects all the i-extremal points of P (see Fig. 2c). For every $J \subseteq \{1, 2, 3, 4\}$, we say that $p \in P$ is J-extremal if p is j-extremal for every $j \in J$. The rectilinear convex hull of P is the set[1]

$$RCH(P) = \bigcap_{i=1,2,3,4} M_i(P),$$

see Fig. 1. The boundary of $RCH(P)$ is (a part of) the union of the 1-, 2-, 3- and 4-staircases associated with P. Observe that the endpoints of these four staircases are a, b, c and d, a is $\{1, 4\}$-extremal, b is $\{1, 2\}$-extremal, c is $\{2, 3\}$-extremal, and d is $\{3, 4\}$-extremal. In Fig. 1b, as $a = d$, then a is $\{1, 3, 4\}$-extremal and the 4-staircase associated with P consists of only point a.

Observe that $RCH(P)$ is disconnected when either the intersection of the complements $\mathbb{R}^2 \setminus M_1(P)$ and $\mathbb{R}^2 \setminus M_3(P)$ is not empty, as shown in Fig. 1b, or the intersection of the complements $\mathbb{R}^2 \setminus M_2(P)$ and $\mathbb{R}^2 \setminus M_4(P)$ is not empty. In other words, when either the 1- and 3-staircases associated with P cross or the 2- and 4-staircases associated with P cross.

Definition 1. *A pinched point u of $RCH(P)$ occurs when u is either $\{1, 3\}$-extremal, as shown in Fig. 1b, or $\{2, 4\}$-extremal.*

Definition 2. *The size of $RCH(P)$ is the number of points of P which are i-extremal for at least one $i \in \{1, 2, 3, 4\}$.*

From the definition of the staircases for P, the following observation is straightforward.

Observation 1. *Assume that the concatenation of the four i-staircases associated with P is traversed counter-clockwise. For two consecutive i-extremal points p and p', $Q_{i+2}(o)$ contains no element of P, where $i + 2$ is taken modulo 4 and $o = (p'_x, p_y)$ for $i = 1, 3$ or $o = (p_x, p'_y)$ for $i = 2, 4$.*

Given two points $u \neq v$ in the plane, let $B(u, v)$ denote the smallest open axis-aligned rectangle containing u and v, and let $P(u, v) = P \cap B(u, v)$. Note that u and v are two opposed corners of $B(u, v)$. If $u = v$, then we define $B(u, u)$ as point u.

Definition 3. *We say that $RCH(P)$ is vertically separable if rectangles $B(a, d)$ and $B(b, c)$ are separated by a vertical line. The two examples shown in Fig. 1 are vertically separable.*

---

[1] The notation $\mathcal{RH}(P)$ is also used for the rectilinear convex hull [4].

Given a point set S, and a horizontal line ℓ, let S' be the image of S under a reflection around ℓ. The following lemma is key for our algorithms:

Lemma 1. *For all $S \subseteq P$, $|S| \geq 2$, either $RCH(S)$ or $RCH(S')$ is vertically separable.*

Proof. Note that $d_x < b_x$ is necessary and sufficient for vertical separability of $B(a,d)$ and $B(b,c)$, that is, $RCH(S)$ is vertically separable. Suppose then that $b_x < d_x$, and let ℓ be a horizontal line. It is straightforward to see that, if we reflect the point set S around ℓ, then S becomes S' and we have that $RCH(S')$ is vertically separable.

We will assume in each of the problems MaxRCH, MaxEmptyRCH, MaxAreaRCH, and MaxWeightRCH that the optimal subset $S \subseteq P$ is such that $RCH(S)$ is vertically separable. To finish this section, we give one more definition. For every $p, q \in P$ such that $p \prec q$, let $R_{p\backslash q}$, $R_{q\backslash p}$, and $R_{p,q}$ be the subsets of P in the regions $Q_4(p) \backslash \overline{Q_4(q)}$, $Q_4(q) \backslash \overline{Q_4(p)}$, and $Q_4(p) \cap Q_4(q)$, respectively (see Fig. 3, left). Observe that if $r \in R_{p\backslash q}$ then $r \prec q$, if $r \in R_{q\backslash p}$ then $p \prec r$, and if $r \in R_{p,q}$ then $r \not\prec q$ and $p \not\prec r$. For every $p, q \in P$ such that $q \prec' p$, we define $R'_{p\backslash q}$, $R'_{q\backslash p}$ and $R'_{p,q}$ as the subsets of P in the regions $Q_4(q) \cap Q_3(p)$, $Q_4(q) \cap Q_1(p)$ and $Q_4(p)$, respectively.

3   Rectilinear Convex Hull of Maximum Size

In this section, we solve the MaxRCH problem. Given P, our goal is to combine four staircases in order to obtain a subset S of P whose rectilinear convex hull is of maximum size. All of this has to be done carefully, since the occurrence of pinched points may lead to overcounting.

Our algorithm to solve the MaxRCH problem proceeds in three steps: In the first step, we compute the 2-staircases of maximum size for every $p, q \in P$ such that $p \prec q$. In the second step, we compute what we call a *triple 1-staircase* and a *triple 3-staircase* of maximum sizes (yet to be defined). In the third and last step, we show how to combine a triple 1-staircase and a triple 3-staircase to solve the MaxRCH problem. In this step, we will make sure that the solution thus obtained is vertically separable. Our algorithm will run in $O(n^3)$ time and $O(n^2)$ space. We describe now in detail the steps of our algorithm.

The First Step: For every $p, q \in P$ such that $p \prec q$ or $p = q$, let $\mathcal{C}_{p,q}$ be a 2-staircase with endpoints p and q of maximum size, see Fig. 3, right. Let $C_{p,q}$ be the number of elements of P in $\mathcal{C}_{p,q}$. Note that $C_{p,q}$ equals the maximum number of 2-extremal points over all $S \cup \{p,q\}$ with $S \subseteq P(p,q)$. We can easily calculate $C_{p,q}$, for all $p, q \in P$ with $p \prec q$ or $p = q$, in $O(n^3)$ time and $O(n^2)$ space, using dynamic programming with the following recurrence:

$$C_{p,q} = \begin{cases} 1 & \text{if } p = q \\ \max\{1 + C_{r,q}\} \text{ over all } r \in P(p,q) \cup \{q\} & \text{if } p \neq q. \end{cases} \tag{1}$$

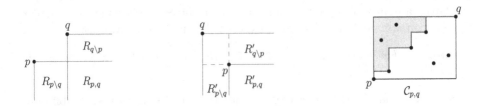

Fig. 3. Left: sets $R_{p\backslash q}$, $R_{q\backslash p}$, $R_{p,q}$, $R'_{p\backslash q}$, $R'_{q\backslash p}$ and $R'_{p,q}$. Right: example of $C_{p,q}$.

Using the elements $C_{p,q}$, it is a routine matter to determine $\mathcal{C}_{p,q}$, for any $p \prec q$.[2]

Definition 4. *Given a point set $S \subseteq P$, we define the* triple 1-staircase *(resp.,* triple 3-staircase*) associated with S as the concatenation of the 1-, 2- and 3-staircases (resp., the 3-, 4- and 1-staircases) associated with S.*

The Second Step: In this step, our goal is to obtain a triple 1-staircase and a triple 3- staircase of maximum cardinality, starting and ending at some pairs of points of P. Triple staircases allow us to conveniently manage pinched points and disconnections of the rectilinear convex hull. Notice that the boundary of $M_1(S) \cap M_2(S) \cap M_3(S)$ (except for its two infinite rays) always belongs to the triple 1-staircase associated with S.

Consider $p, q \in P$ such that $p \prec q$ or $p = q$. Let $Z(p,q) = Q_4(u)$, where $u = (p_x, q_y)$, and let $z(p,q) = Z(p,q) \cap P$. Let $\mathcal{T}_{p,q}$ be the triple 1-staircase of maximum cardinality among all subsets $S \cup \{p,q\}$ with $S \subseteq z(p,q)$. If $S' \subseteq z(p,q)$ is the set associated with $\mathcal{T}_{p,q}$, observe that $M_1(S' \cup \{p,q\}) \cap M_2(S' \cup \{p,q\}) \cap M_3(S' \cup \{p,q\})$ may contain points in $P(p,q)$, it may be disconnected, and it may have pinched points (see Fig. 4). Note that p and q are always the endpoints of $\mathcal{T}_{p,q}$. Let $X_{p,q}$ denote the set of extreme vertices of $\mathcal{T}_{p,q}$ (that is, the set of 1-, 2- and 3-extremal points of $S' \cup \{p,q\}$), and let $T_{p,q}$ be the cardinality of $X_{p,q}$.

We calculate all of the $T_{p,q}$'s by dynamic programming using Eqs. (2) and (3). We store all of the $T_{p,q}$'s in a table T. If $\alpha_{p,q} = 1$ when $p = q$, and $\alpha_{p,q} = 2$ when $p \neq q$, then:

$$T_{p,q} = \max \begin{cases} C_{p,q} & \textbf{(A)} \\ 1 + T_{r,q} \text{ over all } r \in R_{p\backslash q} & \textbf{(B)} \\ 1 + T_{p,r} \text{ over all } r \in R_{q\backslash p} & \textbf{(C)} \\ \alpha_{p,q} + T_{r,r} \text{ over all } r \in R_{p,q} & \textbf{(D)} \\ \alpha_{p,q} + U_{p,r} \text{ over all } r \in R_{p,q} & \textbf{(E)} \end{cases} \quad (2)$$

where for every pair $p, r \in P$ such that $p \prec' r$

$$U_{p,r} = \max\{T_{r,s}\} \text{ over all } s \in R'_{p\backslash r}. \quad (3)$$

---

[2] We note that using not so trivial methods, we can calculate all of the $C_{p,q}$'s in $O(n^2 \log n)$ time. However, this yields no improvement on the overall time complexity of our algorithms.

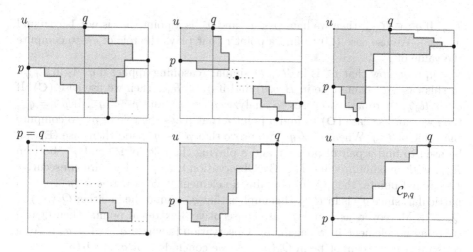

Fig. 4. Examples of triple 1-staircases $T_{p,q}$.

Values $U_{p,r}$ are stored in a table U. The next lemma shows the correctness of this recurrence.

Lemma 2. *The recurrence (2) correctly calculates $T_{p,q}$, the size of $X_{p,q}$, in $O(n^3)$ time and $O(n^2)$ space.*

Proof. Let $T_{p,q}$ be an optimal triple 1-staircase for a pair of points $p, q \in P$ such that $p \prec q$ and let $S' \subseteq z(p,q)$ be the point set associated with $T_{p,q}$. In the counter-clockwise traversal of the triple 1-staircase, let p^- and q^- be the elements of P that follow and precede p and q, respectively. Hence, $T_{p,q}$ can be obtained as an extension of T_{p,q^-}, $T_{p^-,q}$, or T_{p^-,q^-}.

If $p^-, q^- \in B(p,q)$, then necessarily $T_{p,q}$ is a 2-staircase (the 1 and 3-staircases of $T_{p,q}$ consist of only points p and q, respectively), so $T_{p,q} = C_{p,q}$ and case **(A)** is used to set $T_{p,q}$. Thus, we assume in the rest of the proof that at least one of p^- and q^- is not in $B(p,q)$. See Fig. 5 for the cases.

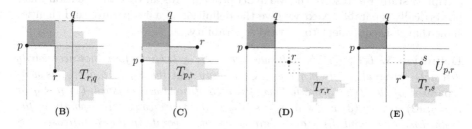

Fig. 5. Cases in the recursive computation of $T_{p,q}$.

If $p^- \in R_{p \backslash q}$, then we have $p^- \prec q$ and p^- is not 3-extremal in $T_{p^-,q}$. We use case **(B)** to find a point r that plays the role of p^- to compute the value of

$T_{p,q}$. If $q^- \in R_{q \backslash p}$, then we have $p \prec q^-$ and q^- is a point that is not 1-extremal in T_{p,q^-}. We use case **(C)** to find a point r that plays the role of q^- to compute the value of $T_{p,q}$.

Suppose now that p^- is in $R_{p,q}$ (a similar reasoning applies if q^- is in $R_{p,q}$). In this case, q^- cannot be in $B(p,q)$ and if $q^- \in R_{q \backslash p}$ then we use case **(C)**. If $q^- \in R_{p,q}$, there are two cases to analyze: $p^- = q^-$ and $p^- \neq q^-$. If $p^- = q^-$, then we can use case **(D)** to find a point r that plays the role of p^- to compute the value of $T_{p,q}$. When $p^- \neq q^-$, we prove that $p^- \prec q^-$, and then case **(E)** can be used to find a pair of points r and s playing the roles of p^- and q^-, both in $R_{p,q}$, with maximum value $T_{r,s}$. By Observation 1, as p and p^- are consecutive 1-extremal points, then $Q_3(o)$ contains no element in S', where $o = (p_x^-, p_y)$. In particular, since q^- is in $R_{p,q}$, this implies that q^- cannot be in either $Q_2(p^-)$ or $Q_3(p^-)$. Moreover, as q and q^- are consecutive 3-extremal points, then $Q_1(o')$ contains no element in S' again by Observation 1, where $o' = (q_x, q_y^-)$. As a consequence, q^- cannot be in $Q_4(p^-)$, so we conclude that $q^- \in Q_1(p^-)$.

To compute tables T and U, we scan the elements of P from right to left. Each time an element $p \in P$ is encountered, we scan all of the $q \in P$ such that $p_x < q_x$, again from right to left. When $p \prec q$ we compute $T_{p,q}$, and when $p \prec' q$ we compute $U_{p,q}$. Each entry of T and U is determined in $O(n)$ time. Thus, U and T can be computed in overall $O(n^3)$ time and $O(n^2)$ space. Cases **(A)** to **(D)** are in $O(n)$. We charge the work done in case **(E)** to constructing table U, which can be done in $O(n)$ time per entry. Thus the entire complexity is in $O(n^3)$ time and $O(n^2)$ space.

In a totally analogous way, we can calculate triple 3-staircases of maximum size. For $p \prec q$ or $p = q$, let $T'_{p,q}$ be the size of the triple 3-staircase $T'_{p,q}$ of maximum cardinality among all subsets $S' \cup \{p, q\}$, where S' is now a subset of points of P in $Q_2(v)$ with $v = (q_x, p_y)$. After rotating the coordinates by π, observe that the triple 3-staircase $T'_{p,q}$ is the triple 1-staircase $T_{q,p}$. Thus, by symmetry with the $T_{p,q}$'s, all the $T'_{p,q}$'s can also be calculated in $O(n^3)$ time and $O(n^2)$ space.

The Third Step: In this step, we show how to combine a triple 1-staircase and a triple 3-staircase to solve the MaxRCH problem. Recall that the solution must be vertically separable. Next we give the definition of a 4-separator and then we show that it is equivalent to vertical separability.

Definition 5. *Let $S \subseteq P$ be any subset with $|S| \geq 2$. Given four (not necessarily distinct) extremal points $p, q, r, s \in S$, we say that the tuple (p, q, r, s) is a 4-separator of $RCH(S)$ if the following five conditions are satisfied: (1) $p \prec q$ or $p = q$; (2) $q \prec' r$; (3) $r \prec s$ or $r = s$; (4) p and r are consecutive points in the 1-staircase of S; and (5) s and q are consecutive points in the 3-staircase of S (see Fig. 6).*

The next lemma shows the equivalence between vertical separability and the existence of 4-separators.

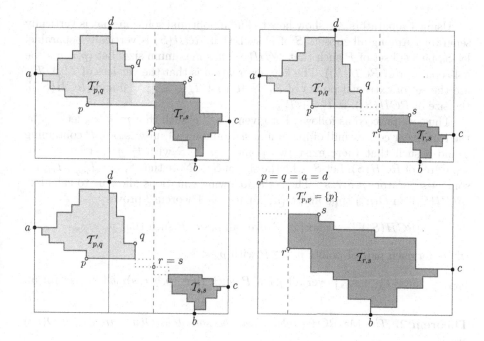

Fig. 6. Examples of 4-separators (p, q, r, s).

Lemma 3. *RCH(S) is vertically separable if and only if RCH(S) has a 4-separator.*

Proof. Let us first assume that $RCH(S)$ is vertically separable, that is, $B(a, d)$ and $B(b, c)$ are separated by a vertical line. Recall that a, b, c, and d are the leftmost, bottommost, rightmost, and topmost points of S, respectively. Then, we can argue the following: If $a \prec d$, then S has at least one 1-extremal point to each side of the vertical line through d. Otherwise, if $a = d$, then S has at least one 1-extremal point to the right side. Thus, covering both cases, let p and r be the two consecutive 1-extremal points of S such that $p_x \leq d_x < r_x$. Now, given that $d_x < r_x$, we have that: If r is also 3-extremal, thus a pinched point, then S has at least one 3-extremal point to the left side of the vertical line through r. Otherwise, if r is not 3-extremal, then S has at least one 3-extremal point to each side of this line. Thus, we can define s and q as the two consecutive 3-extremal points of S such that $q_x < r_x \leq s_x$. It is straightforward to see now that (p, q, r, s) is a 4-separator of $RCH(S)$. Note that, when $p = a = d$, then necessarily $q = p = a = d$.

Assume now that (p, q, r, s) is a 4-separator of $RCH(S)$. We then have that: $d \prec' q$ or $d = q$, and $r \prec' b$ or $r = b$. These conditions, together with $q \prec' r$, directly imply that $B(a, d)$ and $B(b, c)$ are separated by a vertical line, thus $RCH(S)$ is vertically separable.

Using 4-separators, we show how to find an optimal solution that is vertically separable. Among all subsets S of P such that $RCH(S)$ is vertically separable, let S_0 be a subset of P such that $RCH(S_0)$ has maximum size. Let (p, q, r, s) be a 4-separator of $RCH(S_0)$. The key observation is that the vertices of $T'_{p,q} \cup T_{r,s}$ are the set of extremal points of S_0. Note that $T'_{p,q} \cap T_{r,s} = \emptyset$ and $|RCH(S_0)|$, the size of $RCH(S_0)$, is $T'_{p,q} + T_{r,s}$.

Thus, we proceed as follows: For given $p, s \in P$ such that $p_x < s_x$, let $\mathcal{S}_{p,s}$ be the rectilinear convex hull of maximum size, among all subsets $S \subseteq P$ containing p and s such that there exist two points $q, r \in S$ with (p, q, r, s) being a 4-separator of $RCH(S)$. Let $S_{p,s}$ be the size of $\mathcal{S}_{p,s}$. Note that $S_{p,s} = T'_{p,q} + T_{r,s}$ for some 4-separator (p, q, r, s). Then, the following equations allow us to calculate $|RCH(S_0)|$ in $O(n^3)$ time and $O(n^2)$ space, as Theorem 2 proves:

$$|RCH(S_0)| = \max\{S_{p,s}\} \text{ over all } p, s \in P \text{ such that } p_x < s_x \tag{4}$$

where for each pair of points $p, s \in P$ with $p_x < s_x$

$$S_{p,s} = \max\{T'_{p,q} + T_{r,s}\} \text{ over all } q, r \in P \text{ such that } (p, q, r, s) \text{ } is\text{ } a \text{ } 4 \text{ -separator 5 .} \tag{5}$$

Theorem 2. *The* MaxRCH *problem can be solved in* $O(n^3)$ *time and* $O(n^2)$ *space.*

Proof. According to Eqs. (4) and (5), we only need to show how to compute $S_{p,s}$ in linear time, for given p and s. Let Q_p be the set of all points $q \in P$ such that $q \prec' s$, and $p \prec q$ or $p = q$. Let Q_s be the set of all points $r \in P$ such that $p \prec' r$, and $r \prec s$ or $r = s$. Note that $Q_p \cap Q_s = \emptyset$ and that p and s belong to Q_p and Q_s, respectively, only when $p_y > s_y$. Let $L_{p,s}$ be the list of the elements of $Q_p \cup Q_s$ sorted by x-coordinate. Observe that if (p, q, r, s) is a 4-separator of $RCH(\mathcal{S}_{p,s})$, then $r \in Q_s$, $q \in Q_p$, and q is the point $q^* \in Q_p$ from the beginning of $L_{p,s}$ to r such that $T'_{p,q^*} + T_{r,s} = \max\{T'_{p,q'} + T_{r,s}\}$ over all $q' \in Q_p$ from the beginning of $L_{p,s}$ to r.

We calculate $S_{p,s}$ by processing the elements of $L_{p,s}$ in order. For an element t in $L_{p,s}$, let q_t^* be the point in Q_p maximizing $T'_{p,q'}$ over all $q' \in Q_p$ from the beginning of $L_{p,s}$ to t (including t). When processing a point t, observe that if $t \in Q_p$, then q_t^* is either t or q_{t-1}^*. Otherwise, if $t \in Q_s$, then $q_t^* = q_{t-1}^*$. Moreover, if $t \in Q_s$, then we set $q = q_t^*$ and $r = t$, and consider (p, q, r, s) as a *feasible* 4-separator of $\mathcal{S}_{p,s}$. After processing the last element of $L_{p,s}$, among all the (linear number) feasible separators, we return the solution $\mathcal{S}_{p,s}$ induced by the feasible separator that maximizes $T'_{p,q} + T_{r,s}$. Thus, $S_{p,s}$ can be calculated in $O(n)$ time, once tables T and T' have been constructed.

4   Maximum Size/Area Empty Rectilinear Convex Hulls and Maximum Weight Rectilinear Convex Hull

In this section, we show how to adapt the algorithm of Sect. 3 to solve the MaxEmptyRCH, the MaxAreaRCH and the MaxWeightRCH problems. The first observation is that, as the optimal solution for any of these problems is the rectilinear convex hull of a subset S of P, then Lemmas 1 and 3 hold. This implies that we can assume that $B(a,d)$ and $B(b,c)$ are separated by a vertical line, so a 4-separator exists for the optimal solution in any of the problems. As a consequence, the algorithms to solve these three problems follow the same scheme as the algorithm described in the previous section, and we only need to show how to adapt in each problem the calculation of the 2-staircases, the triple 1- and 3-staircases and the rectilinear convex hulls $\mathcal{S}_{p,s}$ to fulfill the requirements on emptiness, area or weight.

We start by solving the MaxEmptyRCH problem, we continue with the MaxAreaRCH problem and we finish with the MaxWeightRCH problem.

4.1   Maximum Size Empty Rectilinear Convex Hull

To solve the MaxEmptyRCH problem in $O(n^3)$ time and $O(n^2)$ space, we modify the steps of our previous algorithm. These modifications ensure that the "interiors" of the triple 1- and 3-staircases and the rectilinear convex hulls are empty. Recall that in this problem we are looking for a subset $S \subseteq P$ such that $RCH(S)$ has maximum size and there is no element of P in the interior of $RCH(S)$.

The First Step: For a pair of points $p, q \in P$ such that $p \prec q$ or $p = q$, we say that the 2-staircase associated with a subset S of $P(p,q)$ is *empty* if no point of P is in the interior of $B(p,q) \cap M_2(S \cup \{p,q\})$, see Fig. 7, left.

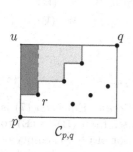
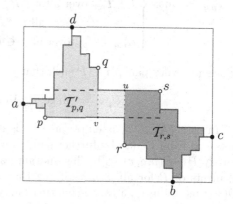

Fig. 7. Left: example of $\mathcal{C}_{p,q}$ when the 2-staircases must be empty. Right: third step of the algorithm when the interior must be empty of elements of P.

Let $\mathcal{C}_{p,q}$ be the empty 2-staircase of maximum cardinality over all subsets $S \cup \{p,q\}$ with $S \subseteq P(p,q)$, and let $C_{p,q}$ be the size of $\mathcal{C}_{p,q}$. Observe that if u is the point (p_x, q_y) and $r \in P$ is the vertex of the 2-staircase that follows p, then $P(r,u) = \emptyset$. Thus, values $C_{p,q}$ can be computed using the following recurrence:

$$C_{p,q} = \begin{cases} 1 & \text{if } p = q \\ \max\{1 + C_{r,q}\} \text{ over all } r \in P(p,q) \cup \{q\} \text{ such that } P(r,u) = \emptyset & \text{if } p \neq q. \end{cases} \tag{6}$$

As q and u are on the same horizontal line, $P(q,u)$ is not defined. In this case, we assume that $P(q,u)$ is the empty set. Using standard range counting techniques [13], we can preprocess the grid G generated by the vertical and horizontal lines through the elements of P in $O(n^2)$ time and space, so that for every pair of vertices u, v of G we can query the number of points in $P(u,v)$ in $O(1)$ time. Thus, we can decide whether $P(u,v) = \emptyset$ in $O(1)$ time. Therefore, values $C_{p,q}$ can be calculated in $O(n^3)$ time and $O(n^2)$ space.

The Second Step: For every $p, q \in P$ such that $p \prec q$ or $p = q$, we say that the triple 1-staircase $\mathcal{T}$ corresponding to a subset S of $z(p,q)$ is *empty* if the (disconnected) region $\mathcal{O}_{\mathcal{T}} = \overline{Z(p,q)} \cap M_1(S \cup \{p,q\}) \cap M_2(S \cup \{p,q\}) \cap M_3(S \cup \{p,q\})$ associated with $\mathcal{T}$ contains no element of P. Please, refer to Fig. 4, where the shaded areas correspond to $\mathcal{O}_{\mathcal{T}}$. Let $\mathcal{T}_{p,q}$ be the empty triple 1-staircase of maximum size among all subsets $S \cup \{p,q\}$, with $S \subseteq z(p,q)$. Let $E(p,q,r)$ denote the interior of $B(p,q) \cap Q_2(r)$ if $p \neq q$, and the empty set if $p = q$. In Fig. 5 examples of open rectangles $E(p,q,r)$ are shown as shaded rectangles . We show how to compute $T_{p,q}$, the cardinality of the set of extreme vertices of $\mathcal{T}_{p,q}$, using the following equations that are similar to Eqs. (2) and (3):

$$T_{p,q} = \max \begin{cases} C_{p,q} & \textbf{(A)} \\ 1 + T_{r,q} \text{ over all } r \in R_{p\backslash q} : P \cap E(p,q,r) = \emptyset & \textbf{(B)} \\ 1 + T_{p,r} \text{ over all } r \in R_{q\backslash p} : P \cap E(p,q,r) = \emptyset & \textbf{(C)} \\ \alpha_{p,q} + T_{r,r} \text{ over all } r \in R_{p,q} : P \cap E(p,q,r) = \emptyset & \textbf{(D)} \\ \alpha_{p,q} + U_{p,r} \text{ over all } r \in R_{p,q} : P \cap E(p,q,r) = \emptyset & \textbf{(E)} \end{cases} \tag{7}$$

where for every pair $p, r \in P$ such that $p \prec' r$

$$U_{p,r} = \max\{T_{r,s}\} \text{ over all } s \in R'_{p\backslash r}. \tag{8}$$

In case **(A)**, $\mathcal{O}_{p,q}$ is empty as $C_{p,q}$ is an empty 2-staircase. Equation (7) is obtained from Eq.(2) by further constraining r in the cases from **(B)** to **(E)** to satisfy $P \cap E(p,q,r) = \emptyset$. This guarantees that the interior of $\mathcal{O}_{p,q}$ is empty of elements of P for all p, q. Verifying that $P \cap E(p,q,r) = \emptyset$ can be decided in $O(1)$ time by using a range counting query. The proof of correctness of Eqs. (7) and (8) follows the same steps as in Lemma 2. Hence, computing the new table T can be done in $O(n^3)$ time and $O(n^2)$ space. By symmetry, values in $T'_{p,q}$, the sizes of the empty triple 3-staircases $\mathcal{T}'_{p,q}$, can also be calculated in $O(n^3)$ time and $O(n^2)$ space.

The Third Step: For given $p, s \in P$ such that $p_x < s_x$, let $\mathcal{S}_{p,s}$ be the *empty* rectilinear convex hull of maximum size, among all subsets $S \subseteq P$ containing p and s such that $RCH(S)$ is empty and there exist two points $q, r \in S$ with (p,q,r,s) being a 4-separator of $RCH(S)$. Let $S_{p,s}$ denote the size of $\mathcal{S}_{p,s}$. To compute $S_{p,s}$ we have to distinguish whether $p \prec' s$ or $p \prec s$. If $p \prec' s$ (see

Fig. 6, top-right, or any in the bottom), then $S_{p,s} = \max\{T'_{p,q} + T_{r,s}\}$ over all 4-separators (p,q,r,s). Otherwise, if $p \prec s$ (see Fig. 7, right), then we must ensure that each 4-separator (p,q,r,s) satisfies $P(u,v) = \emptyset$, where $u = (r_x, s_y)$ and $v = (q_x, p_y)$.

If S_0 is a subset of P such that $RCH(S_0)$ is empty, vertically separable and of maximum size, the new equations to compute $|RCH(S_0)|$ are:

$$|RCH(S_0)| = \max\{S_{p,s}\} \text{ over all } p, s \in P \text{ with } p_x < s_x \tag{9}$$

where for each pair of points $p, s \in P$ such that $p_x < s_x$

$$S_{p,s} = \begin{cases} \max\{T'_{p,q} + T_{r,s}\} : (p,q,r,s) \text{ is a 4 -separator} & p \prec' s \\ \max\{T'_{p,q} + T_{r,s}\} : (p,q,r,s) \text{ is a 4 -separator}, P(u,v) = \emptyset & p \prec s. \end{cases} \tag{10}$$

Theorem 3. *The* MaxEmptyRCH *problem can be solved in* $O(n^3)$ *time and* $O(n^2)$ *space.*

Proof. Again, we only need to show that given points p and s $S_{p,s}$ can be computed in linear time. When $p \prec' s$, we argue as in the proof of Theorem 2. However, when $p \prec s$, we need to only consider 4-separators such that $P(u,v) = \emptyset$. Let $Q_{p,s} = \{t \in P : p \prec t \prec s\}$, which satisfies $Q_{p,s} \cap Q_p = \emptyset$ and $Q_{p,s} \cap Q_s = \emptyset$. Recall that $Q_p \cap Q_s = \emptyset$ and that, when $p \prec s$, $Q_p = \{q \in P : p \prec q, q \prec' s\}$ and $Q_s = \{r \in P : r \prec s, p \prec' r\}$. Let $L_{p,s}$ be the list of the points of $Q_p \cup Q_s \cup Q_{p,s}$ sorted by x-coordinate. Assuming that we have already sorted P by x-coordinate $L_{p,s}$ is obtained in $O(n)$ time.

As before, we calculate $S_{p,s}$ by processing the elements of $L_{p,s}$ in order. For an element t in $L_{p,s}$, let q_t^* be the point in Q_p maximizing $T'_{p,q'}$ over all $q' \in Q_p$ from the beginning of $L_{p,s}$ to t (including t) subject to there being no elements of $Q_{p,s}$ in $L_{p,s}$ from q' to t. When processing a point $t \in Q_{p,s}$, we set $q_t^* = \textbf{nil}$ denoting that q_t^* is undefined. Observe that, when processing a point $t \in Q_p$, if $q_{t-1}^* = \textbf{nil}$ then $q_t^* = t$, and if $q_{t-1}^* \neq \textbf{nil}$ then q_t^* is either t or q_{t-1}^*. When processing a point $t \in Q_s$, then $q_t^* = q_{t-1}^*$, and if $q_{t-1}^* \neq \textbf{nil}$, then we set $q = q_t^*$ and $r = t$, and consider (p,q,r,s) as a feasible 4-separator of $S_{p,s}$. Note that for this 4-separator we have $P(u,v) = \emptyset$. After processing all elements in $L_{p,s}$, $S_{p,s}$ is determined by a feasible 4-separator that maximizes $T'_{p,q} + T_{r,s}$.

4.2 Maximum Area Empty Rectilinear Convex Hull

In the MaxAreaRCH problem, we determine an empty rectilinear convex hull of maximum area. To solve this problem, we proceed as in the previous subsection. The only difference is that we sum areas in all of our recurrences, instead of counting points. Given a bounded set $Z \subset \mathbb{R}^2$, we denote the area of Z as Area(Z).

Now, $\mathcal{C}_{p,q}$, $\mathcal{T}_{p,q}$, $\mathcal{T}'_{p,q}$ and $\mathcal{S}_{p,q}$ are as described in Sect. 4.1, with the difference that they maximize area instead of maximizing size. The areas are defined as follows. If $S = \{p, v_2, \ldots, v_{k-1}, q\}$ is the set of vertices of an empty 2-staircase, we define the area of this staircase as $\mathsf{Area}(B(p,q) \cap M_2(S))$. For an empty triple 1-staircase or an empty triple 3-staircase $\mathcal{T}$, its area is the area of its associated region $\mathcal{O}_{\mathcal{T}}$. The area of a rectilinear convex hull is the area of its interior.

The first step: For a pair of points $p, q \in P$ such that $p \prec q$ or $q = p$, we compute $\mathcal{C}_{p,q}$, the area of $\mathcal{C}_{p,q}$, using the following recurrence, which is a variant of Eq. (6) maximizing area:

$$
C_{p,q} = \begin{cases} 0 & \text{if } p = q \\ \max_{r \in P(p,q) \cup \{q\}} \{ \mathsf{Area}(B(r,u)) + C_{r,q} \} & : \ P(r,u) = \emptyset \ \text{if } p \neq q. \end{cases} \quad (11)
$$

where $u = (p_x, q_y)$. As $B(q,u)$ is not defined, we set $\mathsf{Area}(B(q,u)) = 0$.

The second step: For every $p, q \in P$ such that $p \prec q$ or $p = q$, let $T_{p,q}$ be the area of $\mathcal{T}_{p,q}$. All $T_{p,q}$'s can be calculated in $O(n^3)$ time and $O(n^2)$ space using the following equations, which are variants of Eqs. (7) and (8) maximizing area (recall that if $p = q$, then $E(p,q,r) = \emptyset$, so its area is 0):

$$
T_{p,q} = \max \begin{cases} C_{p,q} & \textbf{(A)} \\ \mathsf{Area}(E(p,q,r)) + T_{r,q} \text{ over all } r \in R_{p \setminus q} : P \cap E(p,q,r) = \emptyset & \textbf{(B)} \\ \mathsf{Area}(E(p,q,r)) + T_{p,r} \text{ over all } r \in R_{q \setminus p} : P \cap E(p,q,r) = \emptyset & \textbf{(C)} \\ \mathsf{Area}(E(p,q,r)) + T_{r,r} \text{ over all } r \in R_{p \setminus q} : P \cap E(p,q,r) = \emptyset & \textbf{(D)} \\ \mathsf{Area}(E(p,q,r)) + U_{p,r} \text{ over all } r \in R_{p,q} : P \cap E(p,q,r) = \emptyset & \textbf{(E)} \end{cases} \quad (12)
$$

where for every pair $p, r \in P$ such that $p \prec' r$

$$
U_{p,r} = \max\{T_{r,s}\} \text{ over all } s \in R'_{p \setminus r}. \quad (13)
$$

The areas $T'_{p,q}$ for the empty triple 3-staircases $\mathcal{T}'_{p,q}$ can be calculated in a similar way.

The Third Step: Let $S_{p,s}$ be the area of $\mathcal{S}_{p,s}$. Recall that, for given $p, s \in P$ such that $p_x < s_x$, $\mathcal{S}_{p,s}$ is the empty rectilinear convex hull of maximum area, among all subsets $S \subseteq P$ containing p and s such that $RCH(S)$ is empty and there exist two points $q, r \in S$ with (p, q, r, s) a 4-separator of $RCH(S)$. Observe that if $p \prec' s$, then $S_{p,s} = T'_{p,q} + T_{r,s}$ for some 4-separator (p, q, r, s). Otherwise, if $p \prec s$ then $S_{p,s} = \mathsf{Area}(B(u,v)) + T'_{p,q} + T_{r,s}$ for some 4-separator (p, q, r, s), subject to $P(u,v) = \emptyset$, where $u = u(r) = (r_x, s_y)$ and $v = v(q) = (q_x, p_y)$ (see Fig. 7, right).

Given that $\mathsf{Area}(B(u,v))$ depends on both r and q, using the inclusion/exclusion principle, we can then calculate $S_{p,s}$ as $S_{p,s} = T'_{p,q} + \mathsf{Area}(B(v,s)) + T_{r,s} + \mathsf{Area}(B(p,u)) - \mathsf{Area}(B(p,s))$. Since p and s are fixed, note that $U(p,q,s) = T'_{p,q} + \mathsf{Area}(B(v(q),s))$ depends only on q and $V(p,r,s) = T_{r,s} + \mathsf{Area}(B(p,u(r))) - \mathsf{Area}(B(p,s))$ depends only on r. Each of these two values can be computed in $O(1)$ time, once T and T' have been computed in the second step. If S_0 is a subset of P such that $RCH(S_0)$ is empty, vertically separable and of maximum area, the new equations to compute $RCH(S_0)$ are:

$$
|RCH(S_0)| = \max\{S_{p,s}\} \text{ over all } p, s \in P \text{ with } p_x < s_x, \quad (14)
$$

where for each pair of points $p, s \in P$ such that $p_x < s_x$

$$
S_{p,s} = \begin{cases} \max_{\{4\text{-separators } (p,q,r,s)\}}\{T'_{p,q} + T_{r,s}\} & p \prec' s \\ \max_{\{4\text{-separators } (p,q,r,s)\}}\{U(p,q,s) + V(p,r,s))\} : P(u(r),v(q)) = \emptyset & p \prec s. \end{cases}
$$
(15)

Theorem 4. *The* MaxAreaRCH *problem can be solved in* $O(n^3)$ *time and* $O(n^2)$ *space.*

Proof. The proof follows the proof of Theorem 3. The only difference is that, when processing an element t in $L_{p,s}$, q_t^* is the point in Q_p maximizing $T'_{p,q'} + \mathsf{Area}(B(v(q'), s))$, instead of maximizing $T'_{p,q'}$. After processing all elements in $L_{p,s}$, $S_{p,s}$ is determined by a feasible 4-separator that maximizes $T'_{p,q} + \mathsf{Area}(B(v, s)) + T_{r,s} + \mathsf{Area}(B(p, u)) - \mathsf{Area}(B(p, s))$.

4.3 Maximum Weight Rectilinear Convex Hull

In the MaxWeightRCH problem, each input point p of P is comes with a (positive or negative) weight $w(p)$. We determine a subset $S \subseteq P$ such that $RCH(S)$ has maximum weight, that is, such that $\sum_{p \in P \cap RCH(S)} w(p)$ is maximized. The algorithm to solve this problem combines the ideas of the previous algorithms and follows the same steps, however, now we add weights. We define $\mathsf{Weight}(Z) = \sum_{p \in P \cap Z} w(p)$ as the weight of a region $Z \subset \mathbb{R}^2$. Using the same range counting techniques [13] as in Sect. 3, we can preprocess the grid G generated by the vertical and horizontal lines through the elements of P in $O(n^2)$ time and space, so that for every pair of vertices u, v of G we can query $\mathsf{Weight}(B(u,v)) = \sum_{p \in P(u,v)} w(p)$ in $O(1)$ time.

Now, $C_{p,q}$, $T_{p,q}$, $T'_{p,q}$ and $S_{p,q}$ are as described in Sect. 3, except that weight is maximized. The weights are defined as follows. If $S = \{p, v_2, \ldots, v_{k-1}, q\}$ is the set of vertices of a 2-staircase, its weight is defined as $w(p) + w(q) + \mathsf{Weight}(B(p,q) \cap M_2(S))$. Note that the weights of all points in S are included in this formulae. For a triple 1-staircase or a triple 3-staircase T, its weight is the addition of the weights of the points of P that appear on the boundary or in the interior of $\mathcal{O}_T$, the region associated with T. Finally, the weight of a rectilinear convex hull is the addition of the points of P on its boundary or interior.

The First Step: If $C_{p,q}$ is the weight of $\mathcal{C}_{p,q}$, for a pair of points $p, q \in P$ such that $p \prec q$ or $q = p$, all $C_{p,q}$'s can be computed in $O(n^3)$ time and $O(n^2)$ space using the following recurrence:

$$
C_{p,q} = \begin{cases} w(p) & \text{if } p = q \\ w(p) + \max_{r \in P(p,q) \cup \{q\}}\{\mathsf{Weight}(B(r,u)) + C_{r,q}\} & \text{if } p \neq q. \end{cases}
$$
(16)

where $u = (p_x, q_y)$. We set $\mathsf{Weight}(B(q, u)) = 0$ as $B(q, u)$ is not defined.

The Second Step: If $T_{p,q}$ is the weight of $\mathcal{T}_{p,q}$, for every $p, q \in P$ such that $p \prec q$ or $p = q$, then all $T_{p,q}$'s (and, by symmetry, all $T'_{p,q}$'s) can be calculated in

$O(n^3)$ time and $O(n^2)$ space using the following equations, where $\alpha_{p,q} = w(p)$ if $p = q$, and $\alpha_{p,q} = w(p) + w(q)$ if $p \neq q$:

$$T_{p,q} = \max \begin{cases} C_{p,q} & \text{(A)} \\ w(p) + \text{Weight}(E(p,q,r)) + T_{r,q} \text{ over all } r \in R_{p \setminus q} & \text{(B)} \\ w(q) + \text{Weight}(E(p,q,r)) + T_{p,r} \text{ over all } r \in R_{q \setminus p} & \text{(C)} \\ \alpha_{p,q} + \text{Weight}(E(p,q,r)) + T_{r,r} \text{ over all } r \in R_{p,q} & \text{(D)} \\ \alpha_{p,q} + \text{Weight}(E(p,q,r)) + U_{p,r} \text{ over all } r \in R_{p,q} & \text{(E)} \end{cases} \quad (17)$$

where for every pair $p, r \in P$ such that $p \prec' r$

$$U_{p,r} = \max\{T_{r,s}\} \text{ over all } s \in R'_{p \setminus r}. \quad (18)$$

The Third Step: For given $p, s \in P$ such that $p_x < s_x$, let $\mathcal{S}_{p,s}$ be the weight of $\mathcal{S}_{p,s}$. Using similar reasoning as in the previous subsection, one can show that, if S_0 is a subset of P such that $RCH(S_0)$ is vertically separable of maximum weight, the following equations calculate $|RCH(S_0)|$:

$$|RCH(S_0)| = \max\{S_{p,s}\} \text{ over all } p, s \in P \text{ such that } p_x < s_x \quad (19)$$

where for each pair of points $p, s \in P$ such that $p_x < s_x$

$$S_{p,s} = \begin{cases} \max_{\{4 \text{-separators } (p,q,r,s)\}}\{T'_{p,q} + T_{r,s}\} & p \prec' s \\ \max_{\{4 \text{-separators } (p,q,r,s)\}}\{U(p,q,s) + V(p,r,s))\} & p \prec s. \end{cases} \quad (20)$$

Now $U(p,q,s) = T'_{p,q} + \text{Weight}(B(v(q),s))$ and $V(p,r,s) = T_{r,s} + \text{Weight}(B(p,u(r))) - \text{Weight}(B(p,s))$, with $u = u(r) = (r_x, s_y)$ and $v = v(q) = (q_x, p_y)$. The proof of the next theorem is a straightforward adaptation of the previous arguments.

Theorem 5. *The* MaxWeightRCH *problem can be solved in $O(n^3)$ time and $O(n^2)$ space.*

References

1. Aichholzer, O., et al.: On k-gons and k-holes in point sets. Comput. Geom. **48**(7), 528–537 (2015)
2. Aichholzer, O., et al.: 4-holes in point sets. Comput. Geom. **47**(6), 644–650 (2014)
3. Aichholzer, O., Fabila-Monroy, R., Hackl, T., Huemer, C., Pilz, A., Vogtenhuber, B.: Lower bounds for the number of small convex k-holes. Comput. Geom. **47**(5), 605–613 (2014)
4. Alegría-Galicia, C., Orden, D., Seara, C., Urrutia, J.: Efficient computation of minimum-area rectilinear convex hull under rotation and generalizations. CoRR, abs/1710.10888 (2017)
5. Avis, D., Rappaport, D.: Computing the largest empty convex subset of a set of points. In: Proceedings of the 1st Annual Symposium on Computational Geometry, pp. 161–167 (1985)

6. Bautista-Santiago, C., Díaz-Báñez, J.M., Lara, D., Pérez-Lantero, P., Urrutia, J., Ventura, I.: Computing optimal islands. Oper. Res. Lett. **39**(4), 246–251 (2011)
7. Brass, P., Moser, W., Pach, J.: Convex polygons and the Erdős-Szekeres problem. In: Chapter 8.2 in the Book Research Problems in Discrete Geometry. Springer (2005)
8. Chvátal, V., Klincsek, G.: Finding largest convex subsets. Congr. Numer. **29**, 453–460 (1980)
9. Erdős, P., Szekeres, G.: A combinatorial problem in geometry. Compos. Math. **2**, 463–470 (1935)
10. Fink, E., Wood, D.: Restricted-Orientation Convexity. Monographs in Theoretical Computer Science (An EATCS Series). Springer, Heidelberg (2004). https://doi.org/10.1007/978-3-642-18849-7
11. Morris, W., Soltan, V.: The Erdős-Szekeres problem on points in convex position-a survey. Bull. Am. Math. Soc. **37**(4), 437–458 (2000)
12. Ottmann, T., Soisalon-Soininen, E., Wood, D.: On the definition and computation of rectilinear convex hulls. Inf. Sci. **33**(3), 157–171 (1984)
13. Preparata, F.P., Shamos, M.I.: Computational Geometry: An Introduction. Springer, Heidelberg (2012). https://doi.org/10.1007/978-1-4612-1098-6
14. Rawlins, G.J.E., Wood, D.: Ortho-convexity and its generalizations. Mach. Intell. Pattern Recognit. **6**, 137–152 (1988)

Computing Digraph Width Measures on Directed Co-graphs

(Extended Abstract)

Frank Gurski, Dominique Komander[✉], and Carolin Rehs

Institute of Computer Science, Algorithmics for Hard Problems Group,
Heinrich-Heine-University Düsseldorf, 40225 Düsseldorf, Germany
dominique.komander@hhu.de

Abstract. In this paper we consider the digraph width measures directed feedback vertex set number, cycle rank, DAG-depth, DAG-width and Kelly-width. While the minimization problem for these width measures is generally NP-hard, we prove that it is computable in linear time for all these parameters, except for Kelly-width, when restricted to directed co-graphs. As an important combinatorial tool, we show how these measures can be computed for the disjoint union, series composition, and order composition of two directed graphs, which further leads to some similarities and a good comparison between the width measures. This generalizes and expands our former results for computing directed path-width and directed tree-width of directed co-graphs.

Keywords: DFVS-number · Cycle rank · DAG-depth · DAG-width · Kelly-width · Directed co-graphs

1 Introduction

Undirected width parameters are well-known and frequently used in computations. Many NP-hard graph problems admit polynomial-time solutions when restricted to graphs of bounded width, like for example bounded tree-width or bounded path-width. Computing both parameters is hard even for bipartite graphs and complements of bipartite graphs [2], while for co-graphs it has been shown [7] that the path-width equals the tree-width and how to compute this value in linear time.

During the last years, width parameters for directed graphs have received a lot of attention [18]. Among these are directed tree-width and directed path-width. In our paper [21] we proved that for directed co-graphs both parameters are equal and computable in linear time. But directed tree-width and directed path-width are not the only attempts to generalize undirected tree-width and path-width for directed graphs. Furthermore, there are the parameters directed

---

The paper is eligible for the best student paper award.

C. Rehs—The work of the third author was supported by the German Research Association (DFG) grant GU 970/7-1.

© Springer Nature Switzerland AG 2019
L. A. Gąsieniec et al. (Eds.): FCT 2019, LNCS 11651, pp. 292–305, 2019.
https://doi.org/10.1007/978-3-030-25027-0_20

feedback vertex set number, cycle rank, DAG-depth, DAG-width and Kelly-width, which have also been considered in [17]. In this paper, we extend our results from [21] and give linear time solutions to compute these width parameters for the disjoint union, series composition and, except for Kelly-width, as well for the order composition of two directed graphs. This leads to a constructive linear-time-algorithm to get the width and the according decompositions of directed co-graphs. For most of the parameters, we could even expand this algorithm to extended directed co-graphs, which are an extension of the directed co-graphs defined in [12] by an additional operation considered in [24].

Our algorithms lead to some tightened bounds for directed path-width, directed tree-width, directed feedback vertex set number, cycle rank, DAG-depth, DAG-width and Kelly-width of extended directed co-graphs and for some of the parameters, they even lead to equalities.

2   Preliminaries

We use the notations of Bang-Jensen and Gutin [3] for graphs and digraphs. When talking about digraphs, we always mean directed graphs with neither multi-edges nor loops. A digraph is a tournament if for all vertices $u \neq v$, there is exactly one of the edges (u, v) and (v, u). It is completely bidirectional if both of these edges are in the edge set.

Orientations. An *orientation* of an undirected graph G is a digraph, where all edges $\{u, v\}$ of G are replaced by either (u, v) or (v, u). For a *biorientation*, every edge $\{u, v\}$ is replaced by either (u, v) or (v, u) or both. For a *complete biorientation*, every edge $\{u, v\}$ is replaced by (u, v) and (v, u). The complete biorientation of an undirected graph G is denoted by $\overleftrightarrow{G}$.

Special Directed Graphs. We recall some special directed graphs. Let

$$\overleftrightarrow{K_n} = (\{v_1, \ldots, v_n\}, \{(v_i, v_j) \mid 1 \leq i \neq j \leq n\})$$

be a bidirectional complete digraph on n vertices. For $n \geq 2$ we denote by

$$\overrightarrow{P_n} = (\{v_1, \ldots, v_n\}, \{(v_1, v_2), \ldots, (v_{n-1}, v_n)\})$$

a directed path on n vertices and for $n \geq 2$ we denote by

$$\overrightarrow{C_n} = (\{v_1, \ldots, v_n\}, \{(v_1, v_2), \ldots, (v_{n-1}, v_n), (v_n, v_1)\})$$

a directed cycle on n vertices. A *directed acyclic digraph (DAG for short)* is a digraph without any $\overrightarrow{C_n}$, $n \geq 2$ as subdigraph. By $\overrightarrow{T_n}$ we denote the transitive tournament on n vertices.

2.1 Recursively Defined Digraphs

Co-graphs have been introduced in the 1970s by a number of authors under different notations. We recall the definition of directed co-graphs from [12]. The following operations have already been considered by Bechet in [4].

- The *disjoint union* of $G_1, \ldots, G_k$, denoted by $G_1 \oplus \ldots \oplus G_k$, is the digraph with vertex set $V_1 \cup \ldots \cup V_k$ and arc set $E_1 \cup \ldots \cup E_k$.
- The *series composition* of $G_1, \ldots, G_k$, denoted by $G_1 \otimes \ldots \otimes G_k$, is defined by their disjoint union plus all possible arcs between vertices of G_i and G_j for all $1 \le i, j \le k$, $i \ne j$.
- The *order composition* of $G_1, \ldots, G_k$, denoted by $G_1 \oslash \ldots \oslash G_k$, is defined by their disjoint union plus all possible arcs from vertices of G_i to vertices of G_j for all $1 \le i < j \le k$.

The class of *directed co-graphs* can be defined recursively. The one-vertex-digraph is a directed co-graph and every disjoint union, series composition and order composition of directed co-graphs is a directed co-graph.

The following transformation has been considered by Johnson et al. in [24] and generalizes the operations disjoint union and order composition.

- The *directed union* of $G_1, \ldots, G_k$, denoted by $G_1 \ominus \ldots \ominus G_k$, is a subdigraph of the order composition $G_1 \oslash \ldots \oslash G_k$ and contains the disjoint union $G_1 \oplus \ldots \oplus G_k$ as a subdigraph.

Including this operation to the definition of directed co-graphs, we obtain the class of *extended directed co-graphs*.

For every (extended) directed co-graph, we can define a tree structure, denoted as *di-co-tree*. The leaves of the di-co-tree represent the vertices of the digraph and the inner nodes of the di-co-tree correspond to the operations applied on the subexpressions defined by the subtrees. For every directed co-graph one can construct a di-co-tree in linear time, see [12].

3 Digraph Width Measures

In Table 1 we summarize some examples for the value of digraph width measures of special digraphs. Further examples can be found in [17, Table 1].

Table 1. The value of digraph width measures of special digraphs.

G	d-tw(G)	d-pw(G)	dfn(G)	cr(G)	ddp(G)	dagw(G)	kw(G)
$\overrightarrow{P_n}$	0	0	0	0	$\lfloor \log(n) \rfloor + 1$	1	0
$\overrightarrow{C_n}$	1	1	1	1	$\lfloor \log(n-1) \rfloor + 2$	2	1
$\overrightarrow{T_n}$	0	0	0	0	n	1	0
$\overleftrightarrow{P_n}$	1	1	$\lfloor \frac{n}{2} \rfloor$	$\lfloor \log(n) \rfloor$	$\lfloor \log(n) \rfloor + 1$	2	1
$\overleftrightarrow{K_n}$	$n-1$	$n-1$	$n-1$	$n-1$	n	n	$n-1$

3.1   Directed Tree-Width

We will use the directed tree-width introduced by Johnson et al. [24].[1]

An out-tree is a tree with a distinguished root such that all arcs are directed away from the root. For two vertices u, v of an out-tree T, the notation $u \leq v$ means that there is a directed path on ≥ 0 arcs from u to v and $u < v$ means that there is a directed path on ≥ 1 arcs from u to v.

Let $G = (V, E)$ be some digraph and $Z \subseteq V$. A vertex set $S \subseteq V - Z$ is Z-*normal* if there is no directed path in $G - Z$ with first and last vertices in S that uses a vertex of $G - (Z \cup S)$.

Definition 1 (Directed tree-width, [24]). *A (arboreal) tree-decomposition of a digraph $G = (V_G, E_G)$ is a triple $(T, \mathcal{X}, \mathcal{W})$. Here $T = (V_T, E_T)$ is an out-tree, $\mathcal{X} = \{X_e \mid e \in E_T\}$ and $\mathcal{W} = \{W_r \mid r \in V_T\}$ are sets of subsets of V_G, such that the following two conditions hold true.*

(dtw-1) $\mathcal{W} = \{W_r \mid r \in V_T\}$ *is a partition of V_G into non-empty subsets.*[2]
(dtw-2) *For every $(u, v) \in E_T$ the set $\bigcup \{W_r \mid r \in V_T, v \leq r\}$ is $X_{(u,v)}$-normal.*

The width *of a (arboreal) tree-decomposition $(T, \mathcal{X}, \mathcal{W})$ is*

$$\max_{r \in V_T} \left| W_r \cup \bigcup_{e \sim r} X_e \right| - 1.$$

Here, $e \sim r$ means that r is one of the two vertices of arc e. The directed tree-width *of G, $d\text{-}tw(G)$ for short, is the smallest integer k such that there is a (arboreal) tree-decomposition $(T, \mathcal{X}, \mathcal{W})$ for G of width k.*

Determining whether the directed tree-width of some given digraph is at most some given value w is NP-complete. On the other hand, determining whether the directed tree-width of some given digraph is at most some given value w is polynomial for directed co-graphs [21].

The results of [24] lead to an XP-algorithm[3] for directed tree-width w.r.t. the standard parameter which implies that for each constant w, it is decidable in polynomial time whether a given digraph has directed tree-width at most w.

Lemma 1 ([20,21]). *Let $G = (V_G, E_G)$ and $H = (V_H, E_H)$ be two vertex-disjoint digraphs, then the following properties hold.*

---

[1] There are also further directed tree-width definitions such as allowing empty sets W_r in [23], using sets W_r of size one only for the leaves of T in [29] and using strong components within (dtw-2) in [13, Chap. 6]. Further in works of Courcelle et al. [9–11] the directed tree-width of a digraph G is defined by the tree-width of the underlying undirected graph. One reason for this could be the algorithmic advantages of the undirected tree-width.

[2] A remarkable difference to the undirected tree-width from [30] is that the sets W_r have to be disjoint and non-empty.

[3] XP is the class of all parameterized problems that can be solved in a certain time, see [14] for a definition.

1. $d\text{-}tw(G \oplus H) = \max\{d\text{-}tw(G), d\text{-}tw(H)\}$
2. $d\text{-}tw(G \oslash H) = \max\{d\text{-}tw(G), d\text{-}tw(H)\}$
3. $d\text{-}tw(G \ominus H) = \max\{d\text{-}tw(G), d\text{-}tw(H)\}$
4. $d\text{-}tw(G \otimes H) = \min\{d\text{-}tw(G) + |V_H|, d\text{-}tw(H) + |V_G|\}$

3.2   Directed Path-Width

The notation of directed path-width was introduced by Reed, Seymour, and Thomas around 1995 and relates to directed tree-width introduced by Johnson, Robertson, Seymour, and Thomas in [24].

Definition 2 (Directed path-width). *A* directed path-decomposition *of some digraph* $G = (V, E)$ *is a sequence* $(X_1, \ldots, X_r)$ *of subsets of* V, *called* bags, *such that the following three conditions hold true.*

(dpw-1) $X_1 \cup \ldots \cup X_r = V$.
(dpw-2) *For each* $(u, v) \in E$ *there is a pair* $i \leq j$ *such that* $u \in X_i$ *and* $v \in X_j$.
(dpw-3) *If* $u \in X_i$ *and* $u \in X_j$ *for some* $u \in V$ *and two indices* i, j *with* $i \leq j$, *then* $u \in X_\ell$ *for all indices* ℓ *with* $i \leq \ell \leq j$.

The width *of a directed path-decomposition* $\mathcal{X} = (X_1, \ldots, X_r)$ *is*

$$\max_{1 \leq i \leq r} |X_i| - 1.$$

The directed path-width *of* G, $d\text{-}pw(G)$ *for short, is the smallest integer* w *such that there is a directed path-decomposition of* G *of width* w.

Determining whether the directed path-width of some given digraph with maximum semi-degree $\Delta^0(G) = \max\{\Delta^-(D), \Delta^+(D)\} \leq 3$ is at most some given value w is NP-complete by a reduction from undirected path-width for planar graphs with maximum vertex degree 3 [26].

Lemma 2 ([20,21]). *Let* $G = (V_G, E_G)$ *and* $H = (V_H, E_H)$ *be two vertex-disjoint digraphs, then the following properties hold.*

1. $d\text{-}pw(G \oplus H) = \max\{d\text{-}pw(G), d\text{-}pw(H)\}$
1. $d\text{-}pw(G \oslash H) = \max\{d\text{-}pw(G), d\text{-}pw(H)\}$
1. $d\text{-}pw(G \ominus H) = \max\{d\text{-}pw(G), d\text{-}pw(H)\}$
1. $d\text{-}pw(G \otimes H) = \min\{d\text{-}pw(G) + |V_H|, d\text{-}pw(H) + |V_G|\}$

3.3   Directed Feedback Vertex Set (DFVS) Number

Definition 3 (DFVS-number). *The* directed feedback vertex set number *of a digraph* $G = (V, E)$, *denoted by* $dfn(G)$, *is the minimum cardinality of a set* $S \subset V$ *such that* $G - S$ *is a DAG.*

Theorem 1 ($\bigstar^4$). *Let* $G = (V_G, E_G)$ *and* $H = (V_H, E_H)$ *be two vertex-disjoint digraphs, then the following properties hold.*

---

[4] The proofs of the results marked with a $\bigstar$ are omitted due to space restrictions.

1. $dfn(G \oplus H) = dfn(G) + dfn(H)$
2. $dfn(G \oslash H) = dfn(G) + dfn(H)$
3. $dfn(G \ominus H) = dfn(G) + dfn(H)$
4. $dfn(G \otimes H) = \min\{dfn(G) + |V_H|, dfn(H) + |V_G|\}$

3.4 Cycle Rank

Cycle rank was introduced in [15] and also appeared in [8] and [25].

Definition 4 (Cycle rank). *The* cycle rank *of a digraph* $G = (V, E)$*, denoted by* $cr(G)$*, is defined as follows.*

- *If* G *is acyclic,* $cr(G) = 0$.
- *If* G *is strongly connected, then* $cr(G) = 1 + \min_{v \in V} cr(G - \{v\})$.
- *Otherwise the cycle rank of* G *is the maximum cycle rank of any strongly connected component of* G.

Results on the cycle rank can be found in [19]. In this papers Gruber proved the hardness of computing cycle rank, even for sparse digraphs of maximum outdegree at most 2.

Proposition 1 ([19]). *For every digraph* G*, we have* $d\text{-}pw(G) \leq cr(G)$.

The cycle rank can be much larger than the directed path-width, which can be shown by a complete biorientation of a path graph $\overrightarrow{P_n}$ which has directed path-width 1 but arbitrary large cycle rank $\lfloor \log(n) \rfloor$, see [25].

Proposition 2 ([17]). *For every digraph* G*, we have* $cr(G) \leq dfn(G)$.

The DFVS-number can be much larger than the cycle rank, which can be shown by the disjoint union of $\frac{n}{3}$ directed cycles $\overrightarrow{C_3}$ which has cycle rank 1 but arbitrary large DFVS-number $\frac{n}{3}$.

Theorem 2 ($\bigstar$). *Let* $G = (V_G, E_G)$ *and* $H = (V_H, E_H)$ *be two vertex-disjoint digraphs, then the following properties hold.*

1. $cr(G \oplus H) = \max\{cr(G), cr(H)\}$
2. $cr(G \oslash H) = \max\{cr(G), cr(H)\}$
3. $cr(G \ominus H) = \max\{cr(G), cr(H)\}$
4. $cr(G \otimes H) = \min\{cr(G) + |V_H|, cr(H) + |V_G|\}$

3.5 DAG-depth

The DAG-depth of a digraph was introduced in [16] motivated by tree-depth for undirected graphs, given in [27].

For a digraph $G = (V, E)$ and $v \in V$, let G_v denote the subdigraph of G induced by the vertices which are reachable from v. The maximal elements in the partially ordered set $\{G_v \mid v \in V\}$ w.r.t. the graph inclusion order are the reachable fragments of G and will be denoted by $R(G)$.[5]

---

[5] In the undirected case, reachable fragments coincide with connected components.

Definition 5 (DAG-depth). *Let $G = (V, E)$ be a digraph. The DAG-depth of G, denoted by $ddp(G)$, is defined as follows.*

- *If $|V| = 1$, then $ddp(G) = 1$.*
- *If G has a single reachable fragment, then $ddp(G) = 1 + \min\{ddp(G - v) \mid v \in V\}$.*
- *Otherwise, $ddp(G)$ equals the maximum over the DAG-depth of the reachable fragments of G.*

Proposition 3 ([17]). *For every complete bioriented directed G, we have $ddp(G) = cr(G) + 1$.*

Theorem 3. *Let $G = (V_G, E_G)$ and $H = (V_H, E_H)$ be two vertex-disjoint digraphs, then the following properties hold.*

1. $ddp(G \oplus H) = \max\{ddp(G), ddp(H)\}$
2. $ddp(G \oslash H) = ddp(G) + ddp(H)$
3. $ddp(G \ominus H) \leq ddp(G) + ddp(H)$
4. $ddp(G \otimes H) = \min\{ddp(G) + |V_H|, ddp(H) + |V_G|\}$

Proof. 1. Since there is no edge in $G \oplus H$ between a vertex from V_G and a vertex from V_H, every reachable fragment is a subset of V_G or a subset of V_H.

2. First, we observe that the set of reachable fragments for $G \oslash H$ can be obtained by $R(G \oslash H) = \{f \cup V_H \mid f \in R(G)\}$.

$ddp(G \oslash H) \leq ddp(G) + ddp(H)$

First, we remove the vertices of G from $G \oslash H$ in the same order as from G when verifying the depth of $ddp(G)$ using Definition 5. Afterwards, we remove the vertices of H from $G \oslash H$ in the same order as from H when verifying the depth of $ddp(H)$ using Definition 5. The observation above allows to use this ordering.

$ddp(G \oslash H) \geq ddp(G) + ddp(H)$

First suppose that it is optimal to begin removing vertices from V_G of $G \oslash H$. Then it is no drawback to remove all vertices from V_G of $G \oslash H$ first and all vertices from V_H afterwards, since every vertex of V_H is reachable from every vertex of V_G. Since none of the vertices of V_G is reachable from a vertex of V_H the vertices of V_H do not effect the number of fragments, reachable from V_G. Next, suppose that it is optimal to begin removing vertices from V_H of $G \oslash H$. Then it is no drawback to remove all vertices from V_H of $G \oslash H$ first and all vertices from V_G afterwards, since none of the vertices of V_G is reachable from a vertex of V_H and thus the vertices of V_G do not effect the number of fragments, reachable from V_H.

3. $ddp(G \ominus H) \leq ddp(G) + ddp(H)$ holds, since the equality of 2. does not hold true in this case, since for a small number of edges $ddp(G \ominus H)$ is much smaller than $ddp(G) + ddp(H)$. Note that a lower bound is $ddp(G \ominus H) \geq \max\{ddp(G), ddp(H)\}$, since $G \ominus H$ is equal to the disjoint union if no edges emerge.

4. $\text{ddp}(G \otimes H) \leq \min\{\text{ddp}(G) + |V_H|, \text{ddp}(H) + |V_G|\}$

Since $G \otimes H$ has only one reachable fragment as long as it contains vertices from V_G and vertices from V_H, we can apply the second case of Definition 5 to verify an upper bound of $\text{ddp}(G) + |V_H|$ by removing the vertices of H one by one from $G \otimes H$ and to verify an upper bound of $\text{ddp}(H) + |V_G|$ by removing the vertices of G one by one from $G \otimes H$.

$\text{ddp}(G \otimes H) \geq \min\{\text{ddp}(G) + |V_H|, \text{ddp}(H) + |V_G|\}$

Since in $G \otimes H$ every vertex of V_G has an edge to and from every vertex of V_H, $G \otimes H$ has only one reachable fragment as long as it contains vertices from V_G and V_H. Thus, we have to apply the second case of Definition 5 as long we have vertices from V_G and vertices from V_H. This either leads to a subdigraph induced by $V_G - V_G'$ for some $V_G' \subset V_G$ or to a subdigraph induced by $V_H - V_H'$ for some $V_H' \subset V_H$. Thus, we have

$$\text{ddp}(G \otimes H) \geq \min\{|V_H| + |V_G'| + \text{ddp}(G - V_G'),$$
$$|V_G| + |V_H'| + \text{ddp}(H - V_H')\}$$
$$\geq \min\{|V_H| + \text{ddp}(G), |V_G| + \text{ddp}(H)\}.$$

This completes the proof.   □

Note that $\text{ddp}(G \ominus H)$ cannot be computed from $\text{ddp}(G)$ and $\text{ddp}(H)$ by a simple formula, since the disjoint union and the order operation behave differently.

3.6   DAG-width

The DAG-width is a graph parameter which describes how close a digraph is to a directed acyclic graph (DAG). It has been defined in [5,6,28].

Let $G = (V_G, E_G)$ be a acyclic digraph. The partial order $\preccurlyeq_G$ on G is the reflexive, transitive closure of E_G. A *source* or *root* of a set $X \subseteq V_G$ is a $\preccurlyeq_G$-minimal element of X, that is, $r \in X$ is a root of X, if there is no $y \in X$, such that $y \preccurlyeq_G r$ and $y \neq x$. Analogously, a *sink* or *leaf* of a set $X \subseteq V_G$ is a $\preccurlyeq_G$-maximal element.

Let $V' \subseteq V_G$, then a set $W \subseteq V_G$ *guards* V' if for all $(u,v) \in E_G$ it holds that if $u \in V'$ then $v \in V' \cup W$.

Definition 6 (DAG-width). *A DAG-decomposition of some digraph* $G = (V_G, E_G)$ *is a pair* $(D, \mathcal{X})$ *where* $D = (V_D, E_D)$ *is a DAG and* $\mathcal{X} = \{X_u \mid X_u \subseteq V_G, u \in V_D\}$ *is a family of subsets of* V_G *such that:*

(dagw-1) $\bigcup_{u \in V_D} X_u = V_G$.

(dagw-2) *For all vertices* $u, v, w \in V_D$ *with* $u \succcurlyeq_D v \succcurlyeq_D w$, *it holds that* $X_u \cap X_w \subseteq X_v$.

(dagw-3) *For all edges* $(u,v) \in E_D$ *it holds that* $X_u \cap X_v$ *guards* $X_{\succcurlyeq_v} \setminus X_u$, *where* $X_{\succcurlyeq_v} = \bigcup_{v \succcurlyeq_D w} X_w$. *For any source* u, $X_{\succcurlyeq_u}$ *is guarded by* $\emptyset$.

The width of a DAG-decomposition $(D, \mathcal{X})$ is the number

$$\max_{u \in V_D} |X_u|.$$

The DAG-width of a digraph G, dagw(G) for short, is the smallest width of all possible DAG-decompositions for G.

We use the restriction to nice DAG-decompositions from [6, Theorem 24].

Proposition 4 ([6]). *For every graph G, we have dagw$(\overleftrightarrow{G}) = tw(G) + 1$.*

Proposition 4 implies that the NP-hardness of tree-width carries over to DAG-width.

There are even digraphs on n vertices whose optimal DAG-decompositions have super-polynomial many bags w.r.t n [1]. Furthermore, it has been shown that deciding whether the DAG-width of a given digraph is at most a given value is PSPACE-complete [1].

Proposition 5 ([17]). *For every digraph G, we have dagw$(G) \leq d\text{-}pw(G) + 1$.*

Proposition 6 ([6]). *For every digraph G, we have d-tw$(G) \leq 3 \cdot dagw(G) + 1$.*

Lemma 3 (★). *Let $G = (V, E)$ be a digraph of DAG-width at most k, such that $V_1 \cup V_2 = V$, $V_1 \cap V_2 = \emptyset$, and $\{(u, v), (v, u) \mid u \in V_1, v \in V_2\} \subseteq E$. Then there is a DAG-decomposition $(D, \mathcal{X})$, $D = (V_D, E_D)$, of width at most k for G such that for every $v \in V_D$ holds $V_1 \subseteq X_v$ or for every $v \in V_D$ holds $V_2 \subseteq X_v$.*

Obviously, this lemma also holds for a nice DAG-decomposition.

Theorem 4. *Let $G = (V_G, E_G)$ and $H = (V_H, E_H)$ be two vertex-disjoint digraphs, then the following properties hold.*

1. *dagw$(G \oplus H) = \max\{dagw(G), dagw(H)\}$*
2. *dagw$(G \oslash H) = \max\{dagw(G), dagw(H)\}$*
3. *dagw$(G \ominus H) = \max\{dagw(G), dagw(H)\}$*
4. *dagw$(G \otimes H) = \min\{dagw(G) + |V_H|, dagw(H) + |V_G|\}$*

Proof. Let G and H be two vertex-disjoint digraphs and let further $(D_G, \mathcal{X}_G)$ and $(D_H, \mathcal{X}_H)$ be their nice DAG-decompositions with minimum DAG-width. Let r_H be the root of D_H and let l_G be a leaf of D_G.

1. For $J = G \oplus H$, we first define a legit DAG-decomposition $(D_J, \mathcal{X}_J)$ for J and show that it is of minimum width afterwards. Let D_J be the disjoint union of D_G and D_H with an additional arc (l_G, r_H). Further, $\mathcal{X}_J = \mathcal{X}_G \cup \mathcal{X}_H$. $(D_J, \mathcal{X}_J)$ is a valid DAG-decomposition because it satisfies the conditions as follows. It holds that (dagw-1) is satisfied by $(D_G, \mathcal{X}_G)$ and $(D_H, \mathcal{X}_H)$ it is also satisfied by $(D_J, \mathcal{X}_J)$ because all vertices of J are included. As we do not add any vertices to the X-sets and G and H are vertex-disjoint, (dagw-2) is satisfied for $(D_J, \mathcal{X}_J)$. Further, (dagw-3) is satisfied for all arcs in D_G and

D_H. In D_J there is only one additional arc, (l_G, r_H). Since it holds that for r_H, $X_{\succeq r_H}$ is guarded by $\emptyset$ and we do not add any outgoing vertices to H and $X_{l_G} \cap X_{r_H} = \emptyset$, (dagw-3) is satisfied for $(D_J, \mathcal{X}_J)$. Thus, the DAG-width of the decomposition is limited by the larger width of G and H, such that $\mathrm{dagw}(G \oplus H) \le \max\{\mathrm{dagw}(G), \mathrm{dagw}(H)\}$.

The lower bound holds as G and H are both induced subdigraphs of J and a graph cannot have lower DAG-width than its induced subdigraphs. Hence $\mathrm{dagw}(J) \ge \max\{\mathrm{dagw}(G), \mathrm{dagw}(H)\}$ applies, which leads to $\mathrm{dagw}(J) = \max\{\mathrm{dagw}(G), \mathrm{dagw}(H)\}$.

2. Holds by the same arguments as given in (1.).
3. Holds by the same arguments as given in (1.).
4. For $J = G \otimes H$, set $D_J = D_G$ and $\mathcal{X}_J = \{X_u \cup V_H \mid X_u \in \mathcal{X}_G\}$. Then $(D_J, \mathcal{X}_J)$ is a DAG-decomposition for J: Obviously, (dagw-1) is satisfied. (dagw-2) and (dagw-3) are satisfied since they are satisfied for $\mathcal{X}_G$ and we add V_H to every vertex set in $\mathcal{X}_G$. Further, it holds that the width of $(D_J, \mathcal{X}_J)$ is $\mathrm{dagw}(G) + |V_H|$. In the same way, we get a DAG-decomposition of width $\mathrm{dagw}(H) + |V_G|$, so we have $\mathrm{dagw}(G \otimes H) \le \min\{\mathrm{dagw}(G) + |V_H|, \mathrm{dagw}(H) + |V_G|\}$.

For the lower bound, we use Lemma 3. Assume that $\mathrm{dagw}(G \otimes H) < \min\{\mathrm{dagw}(G) + |V_H|, \mathrm{dagw}(H) + |V_G|\}$. Let $(D_J, \mathcal{X}_J)$ be a minimal DAG-decomposition of J of size $k < \min\{\mathrm{dagw}(G) + |V_H|, \mathrm{dagw}(H) + |V_G|\}$. By Lemma 3 we have $V_H \subseteq X_v$ for all $X_v \in \mathcal{X}_J$ or $V_G \subseteq X_v$ for all $X_v \in \mathcal{X}_J$. Without loss of generality assume $V_H \subseteq X_v$ for all $X_v \in \mathcal{X}_J$ (because $V_G \subseteq X_v$ for all $X_v \in \mathcal{X}_J$, respectively). Then $(D'_G, \mathcal{X}'_G)$ with $D'_G = D_J$, $\mathcal{X}'_G = \{X_u \setminus V_H \mid X_u \in \mathcal{X}_J\}$ is a DAG-decomposition of size $k - |V_H|$ of G:
- (dagw-1) is satisfied since

$$\bigcup_{u \in V_{D'_G}} X_u = \bigcup_{u \in V_{D_J}} (X_u \setminus V_H) = \left(\bigcup_{u \in V_{D_J}} X_u \right) \setminus V_H$$
$$= V_J \setminus V_H$$
$$= (V_G \cup V_H) \setminus V_H$$
$$= V_G.$$

- (dagw-2) is satisfied since for all $u, v, w \in V_{D'_G}$ with $u \succeq_{D'_G} v \succeq_{D'_G} w$ and X_u^J, X_v^J and X_w^J the corresponding sets in $(D_J, \mathcal{X}_J)$ it holds that $X_u \cap X_w = (X_u^J \setminus V_H) \cap (X_w^J \setminus V_H) = (X_u^J \cap X_w^J) \setminus V_H \subseteq X_v^J \setminus V_H = X_v$ as $u \succeq_{D_J} v \succeq_{D_J} w$.
- (dagw-3) is satisfied since for all edges $(u, v) \in E_{D'_G}$, we have $(u, v) \in E_{D_J}$ and as $X_u \cap X_v = (X_u^J \cap X_v^J) \setminus V_H$ which guards $X_{\succeq_{D'_G} v} \setminus X_u = X_{\succeq_{D_J} v} \setminus X_u^J$. For the root, the condition is trivially satisfied.

But it holds that $k - |V_H| < \min\{\mathrm{dagw}(G) + |V_H|, \mathrm{dagw}(H) + |V_G|\} - |V_H| \le \mathrm{dagw}(G) + |V_H| - |V_H| = \mathrm{dagw}(G)$. This is a contradiction, as it is not possible to create a DAG-decomposition of size smaller than $\mathrm{dagw}(G)$.

It follows that $\mathrm{dagw}(G \otimes H) \ge \min\{\mathrm{dagw}(G) + |V_H|, \mathrm{dagw}(H) + |V_G|\}$ and thus that $\mathrm{dagw}(G \otimes H) = \min\{\mathrm{dagw}(G) + |V_H|, \mathrm{dagw}(H) + |V_G|\}$.

This completes the proof.   □

3.7 Kelly-Width

The Kelly-width is also led from directed acyclic graphs, which leads to the idea that it is very similar to the DAG-width. It has been defined in [22].

Definition 7. (Kelly-width). *A Kelly decomposition of a digraph* $G = (V_G, E_G)$ *is a triple* $(\mathcal{W}, \mathcal{X}, D)$ *where* D *is a directed acyclic graph,* $\mathcal{X} = \{X_u \mid X_u \subseteq V_G, u \in V_D\}$ *and* $\mathcal{W} = \{W_u \mid W_u \subseteq V_G, u \in V_D\}$ *are families of subsets of* V_G *such that:*

1. $\mathcal{W}$ *is a partition for* V_G.
2. *For all vertices* $v \in V_G$, X_v *guards* $W_{\succcurlyeq v}$.
3. *For all vertices* $v \in V_G$, *there is a linear order* $u_1, \ldots, u_s$ *on the children of* v *such that for every* u_i *it holds that* $X_{u_i} \subseteq W_i \cup X_i \cup \bigcup_{j<i} W_{\succcurlyeq u_j}$. *Similarly, there is a linear order* $r_1, r_2, \ldots$ *on the roots of* D *such that for each root* r_i *it holds that* $W_{r_i} \subseteq \bigcup_{j<i} W_{\succcurlyeq r_j}$.

The width *of a Kelly decomposition* $(\mathcal{W}, \mathcal{X}, D)$ *is the number*

$$\max_{u \in V_D} |X_u| + |W_u|.$$

The Kelly-width *of a digraph* G, *denoted with* $kw(G)$, *is the smallest width of all possible Kelly decompositions for* G.

Definition 8. (Directed elimination ordering). *Let* $G = (V, E)$ *be a digraph. A directed elimination ordering* $\lhd$ *on* G *is a linear ordering on* V. *For* $\lhd = (v_0, v_1, \ldots, v_{n-1})$ *we define*

- $G_0^{\lhd} = G$
- $G_{i+1}^{\lhd} = (V_{i+1}^{\lhd}, E_{i+1}^{\lhd})$ *with* $V_{i+1}^{\lhd} = V_i^{\lhd} \setminus \{v_i\}$ *and* $E_{i+1}^{\lhd} = \{(u, v) \mid (u, v) \in E_i^{\lhd}$ *and* $u, v \neq v_i$ *or* $(u, v_i), (v_i, v) \in E_i^{\lhd}, u \neq v\}$

$G_i^{\lhd}$ *is the directed elimination graph at step* i *according to* $\lhd$.
The width *of* $\lhd$ *is the maximum out-degree of* v_i *in* $G_i^{\lhd}$ *over all* i.

Lemma 4 ([22]). *Let* G *be a digraph. The following are equivalent:*

1. G *has Kelly-width at most* $k + 1$
2. G *has a directed elimination ordering of width* $\leq k$

Proposition 7 ([22]). *For every digraph* G, *we have* $d\text{-}tw(G) \leq 6 \cdot kw(G) - 2$.

Proposition 8 ([17]). *For every digraph* G, *we have* $kw(G) \leq d\text{-}pw(G) + 1$.

Theorem 5. *Let* $G = (V_G, E_G)$ *and* $H = (V_H, E_H)$ *be two vertex-disjoint digraphs, then the following properties hold.*

1. $kw(G \oplus H) = \max\{kw(G), kw(H)\}$
2. $kw(G \oslash H) = \max\{kw(G), kw(H)\}$
3. $kw(G \ominus H) = \max\{kw(G), kw(H)\}$

4. $kw(G \otimes H) \leq \min\{kw(G) + |V_H|, kw(H) + |V_G|\}$

Proof. We use the fact that by Lemma 4, a digraph has Kelly-width $k+1$ if and only if it has a directed elimination ordering of width k. Let $G = (V_G, E_G)$ and $H = (V_H, E_H)$ be two vertex-disjoint digraphs with $kw(G) = k_G$ and $kw(H) = k_H$. Then, there exists a directed elimination ordering $\lhd_G$ of G of width $k_G - 1$ and a directed elimination ordering $\lhd_H$ of H of width $k_H - 1$.

1. For $J = G \oplus H$, we obtain a linear ordering $\lhd_J$ of J by adding first all vertices from $\lhd_H$ and from $\lhd_G$ to $\lhd_J$ afterwards. As no edges from H to G are inserted to J, this is a directed elimination ordering of width $\max\{k_H - 1, k_G - 1\}$. As G and H are both induced subdigraphs of J, there cannot exist a directed elimination ordering of smaller width. By Lemma 4 it follows that $kw(J) = \max\{k_H, k_G\}$, such that $kw(G \oplus H) = \max\{kw(G), kw(H)\}$.
2. Holds by the same arguments as in (1.).
3. Holds by the same arguments as in (1.).
4. For $J = G \otimes H$, we obtain a linear ordering $\lhd_J$ of J by adding first all vertices from $\lhd_H$ and afterwards from $\lhd_G$ to $\lhd_J$ (first $\lhd_G$, then $\lhd_H$ respectively). As there are exactly V_G (V_H) more outgoing edges for every vertex in V_H (V_G), this is a directed elimination ordering of J of width $k_H - 1 + |V_G|$ ($k_G - 1 + |V_H|$, respectively).

This completes the proof. □

Remark 1 (★). The value $\min\{kw(G) + |V_H|, kw(H) + |V_G|\}$ is not a lower bound for $kw(G \otimes H)$, even not if G and H are directed co-graphs.

3.8 Comparison

Theorem 6. *For every extended directed co-graph G, we have*

$$kw(G) \leq d\text{-}pw(G) = d\text{-}tw(G) = cr(G) = dagw(G) - 1 \leq ddp(G) - 1 \leq dfn(G).$$

For DFVS-Number, DAG-depth and Kelly-width equality is not possible by the following examples. For the disjoint union of two $\overleftrightarrow{K_n}$, it holds that $d\text{-}pw(2\overleftrightarrow{K_n}) = n - 1 < 2n - 2 = dfn(2\overleftrightarrow{K_n})$. For transitive tournaments $\overrightarrow{T_n}$, it holds that $d\text{-}pw(\overrightarrow{T_n}) = 0 < n = ddp(\overrightarrow{T_n})$. Further, let K'_n be the $2n$ vertex graph which is obtained by a complete graph K_n on n vertices and adding a pendant vertex for every of the n vertices of K_n, then for the complete biorientation $\overleftrightarrow{K'_n}$ it holds that $kw(\overleftrightarrow{K'_n} \otimes \overleftrightarrow{K'_n}) = 2n - 1 < 3n - 1 = d\text{-}pw(\overleftrightarrow{K'_n} \otimes \overleftrightarrow{K'_n})$.

But by Theorem 5 Kelly-width is always smaller or equal to path-width and its equal parameters and by Theorem 3 DAG-depth is always greater or equal to path-width and its equal parameters.

Theorem 7. *For every extended directed co-graph $G = (V, E)$ which is given by a binary di-co-tree the directed path-width, directed tree-width, directed feedback vertex set number, cycle rank, and DAG-width can be computed in time $\mathcal{O}(|V|)$.*

4   Conclusion and Outlook

In this paper, we are able to give linear time algorithms for the directed feedback set number, cycle rank and DAG-width of extended directed co-graphs and a linear-time algorithm for the DAG-depth of directed co-graphs. Further, we provided a comparison of all considered parameters for extended directed co-graphs and obtained equality for directed path-width, directed tree-width, cycle rank and DAG width. Further, we showed for bounds for the class of directed co-graphs for the directed vertex set number, DAG-depth and Kelly-width. This widely extends our results for directed path-width and directed tree-width from [21].

A further issue could be to find a linear or polynomial time algorithm to compute Kelly-width on directed co-graphs. Furthermore, it would be interesting for which superclasses of directed co-graphs it is still possible to find polynomial algorithms to get the considered parameters and for which superclasses these problems become NP-hard.

References

1. Amiri, S.A., Kreutzer, S., Rabinovich, R.: DAG-width is PSPACE-complete. Theor. Comput. Sci. **655**, 78–89 (2016)
2. Arnborg, S., Corneil, D.G., Proskurowski, A.: Complexity of finding embeddings in a k-tree. SIAM J. Algebr. Discrete Methods **8**(2), 277–284 (1987)
3. Bang-Jensen, J., Gutin, G.: Digraphs. Theory, Algorithms and Applications. Springer, Heidelberg (2009). https://doi.org/10.1007/978-1-84800-998-1
4. Bechet, D., de Groote, P., Retoré, C.: A complete axiomatisation for the inclusion of series-parallel partial orders. In: Comon, H. (ed.) RTA 1997. LNCS, vol. 1232, pp. 230–240. Springer, Heidelberg (1997). https://doi.org/10.1007/3-540-62950-5_74
5. Berwanger, D., Dawar, A., Hunter, P., Kreutzer, S.: DAG-width and parity games. In: Durand, B., Thomas, W. (eds.) STACS 2006. LNCS, vol. 3884, pp. 524–536. Springer, Heidelberg (2006). https://doi.org/10.1007/11672142_43
6. Berwangera, D., Dawar, A., Hunter, P., Kreutzer, S., Obdrzálek, J.: The dag-width of directed graphs. J. Comb. Theory Ser. B **102**(4), 900–923 (2012)
7. Bodlaender, H.L., Möhring, R.H.: The pathwidth and treewidth of cographs. SIAM J. Disc. Math. **6**(2), 181–188 (1993)
8. Cohen, R.S.: Transition graphs and the star height problem. In: Proceedings of the 9th Annual Symposium on Switching and Automata Theory, pp. 383–394. IEEE Computer Society (1968)
9. Courcelle, B.: From tree-decompositions to clique-width terms. Discrete Appl. Math. **248**, 125–144 (2018)
10. Courcelle, B., Engelfriet, J.: Graph Structure and Monadic Second-Order Logic. A Language-Theoretic Approach. Encyclopedia of Mathematics and its Applications. Cambridge University Press, Cambridge (2012)
11. Courcelle, B., Olariu, S.: Upper bounds to the clique width of graphs. Discrete Appl. Math. **101**, 77–114 (2000)
12. Crespelle, C., Paul, C.: Fully dynamic recognition algorithm and certificate for directed cographs. Discrete Appl. Math. **154**(12), 1722–1741 (2006)

13. Dehmer, M., Emmert-Streib, F. (eds.): Quantitative Graph Theory: Mathematical Foundations and Applications. CRC Press Inc., New York (2014)
14. Downey, R.G., Fellows, M.R.: Fundamentals of Parameterized Complexity. Springer, Heidelberg (2013). https://doi.org/10.1007/978-1-4471-5559-1
15. Eggan, L.E.: Transition graphs and the star height of regular events. Michigan Math. J. **10**, 385–397 (1963)
16. Ganian, R., Hliněný, P., Kneis, J., Langer, A., Obdržálek, J., Rossmanith, P.: On digraph width measures in parameterized algorithmics. In: Chen, J., Fomin, F.V. (eds.) IWPEC 2009. LNCS, vol. 5917, pp. 185–197. Springer, Heidelberg (2009). https://doi.org/10.1007/978-3-642-11269-0_15
17. Ganian, R., Hlinený, P., Kneis, J., Langer, A., Obdrzálek, J., Rossmanith, P.: Digraph width measures in parameterized algorithmics. Discrete Appl. Math. **168**, 88–107 (2014)
18. Ganian, R., et al.: Are there any good digraph width measures? J. Comb. Theory Ser. B **116**, 250–286 (2016)
19. Gruber, H.: Digraph complexity measures and applications in formal language theory. Discret. Math. Theor. Comput. Sci. **14**(2), 189–204 (2012)
20. Gurski, F., Rehs, C.: Computing directed path-width and directed tree-width of recursively defined digraphs. ACM Computing Research Repository (CoRR), abs/1806.04457, 16 p. (2018)
21. Gurski, F., Rehs, C.: Directed path-width and directed tree-width of directed co-graphs. In: Wang, L., Zhu, D. (eds.) COCOON 2018. LNCS, vol. 10976, pp. 255–267. Springer, Cham (2018). https://doi.org/10.1007/978-3-319-94776-1_22
22. Hunter, P., Kreutzer, S.: Digraph measures: Kelly decompositions, games, and orderings. Theor. Comput. Sci. **399**(3), 206–219 (2008)
23. Johnson, T., Robertson, N., Seymour, P.D., Thomas, R.: Addentum to "Directed tree-width" (2001)
24. Johnson, T., Robertson, N., Seymour, P.D., Thomas, R.: Directed tree-width. J. Comb. Theory Ser. B **82**, 138–155 (2001)
25. McNaughton, R.: The loop complexity of regular events. Inf. Sci. **1**(3), 305–328 (1969)
26. Monien, B., Sudborough, I.H.: Min cut is NP-complete for edge weighted trees. Theor. Comput. Sci. **58**, 209–229 (1988)
27. Nešetřil, J., de Mendez, P.O.: Tree-depth, subgraph coloring and homomorphism bounds. Eur. J. Comb. **27**, 1022–1041 (2006)
28. Obdrzálek, J.: Dag-width: connectivity measure for directed graphs. In: Proceedings of the ACM-SIAM Symposium on Discrete Algorithms (SODA), pp. 814–821. ACM-SIAM (2006)
29. Reed, B.: Introducing directed tree width. Electron. Notes Discret. Math. **3**, 222–229 (1999)
30. Robertson, N., Seymour, P.D.: Graph minors II. Algorithmic aspects of tree width. J. Algorithms **7**, 309–322 (1986)

Fault-Tolerant Parallel Scheduling of Arbitrary Length Jobs on a Shared Channel

Marek Klonowski[1], Dariusz R. Kowalski[2,3], Jarosław Mirek[4(✉)], and Prudence W. H. Wong[4]

[1] Wrocław University of Science and Technology, Wrocław, Poland
Marek.Klonowski@pwr.edu.pl
[2] School of Computer and Cyber Sciences, Augusta University, Augusta, USA
[3] SWPS University of Social Sciences and Humanities, Warsaw, Poland
dariusz.kowalski@swps.edu.pl
[4] University of Liverpool, Ashton Building, Ashton Street, Liverpool L69 3BX, UK
{J.Mirek,pwong}@liverpool.ac.uk

Abstract. We study the problem of scheduling n jobs on m identical, fault-prone machines f of which are prone to crashes by an adversary, where communication takes place via a multiple access channel without collision detection. Performance is measured by the total number of available machine steps during the whole execution (work). Our goal is to study the impact of preemption (i.e., interrupting the execution of a job and resuming it later by the same or different machine) and failures on the work performance of job processing. We identify features that determine the difficulty of the problem, and in particular, show that the complexity is asymptotically smaller when preemption is allowed.

1 Introduction

We examine the problem of performing n jobs by m machines reliably on a multiple access shared channel (MAC). This problem, originated by Chlebus et al. [6], has already been studied for unit length jobs, whereas this paper extends it by considering jobs with arbitrary lengths and studying the impact of features, such as preemption and the severity of failures, on the work performance.

The notion of preemption may be understood as the possibility of not performing a particular job in one attempt. This means that a single job may be interrupted partway through performing it and then resumed later by the same machine or even by another machine. Intuitively, the model without preemption is more general, yet both have subtleties that distinguish them noticeably.

This work is supported by the Polish National Science Center (NCN) grant UMO-2017/25/B/ST6/02553, and by Networks Sciences & Techologies (NeST), School of EEECS, University of Liverpool.

L. A. Gąsieniec et al. (Eds.): FCT 2019, LNCS 11651, pp. 306–321, 2019.
https://doi.org/10.1007/978-3-030-25027-0_21

Communication takes place on a shared channel, also known as multiple access channel, without collision detection. A message is transmitted successfully only when one machine transmits, and when more than one message is transmitted then it is no different from background noise (i.e., no transmission on the channel). Therefore, we say that a job completion is confirmed only when the corresponding machine transmits such message successfully (before any failure it may suffer).

We consider the impact of an *adaptive f-bounded* adversary on the performance of algorithms for the problem. This kind of adversary may decide to crash up to f machines at any time of the execution. We use work as the effectiveness measure, i.e., the total number of machine steps available for computations. It is related (after dividing by m) to the average time performance of machines. Work may also correspond to energy consumption - an operational machine generates (consumes) a unit of work (energy) in every time step.

Previous and Related Results. Our work can be seen as an extension of the *Do-All* problem defined in a seminal work by Dwork, Halpern and Waarts [9]. This line of research was followed up by several other papers [3–5,8,10] which considered the message-passing model with point-to-point communication. In all these papers the authors assumed that performing a single job contributes a unit to work complexity. Paper [8] introduced a model, wherein the performance measure for *Do-All* solutions is extended to the *available machine steps* (i.e., including idle rounds). Authors in [8] developed an algorithm solving the problem with work $\mathcal{O}(n+(f+1)m)$ and message complexity $\mathcal{O}((f+1)m)$. We adopt this kind of work measurement in our paper.

The authors of [4] developed a deterministic algorithm with effort (i.e. sum of work and messages sent during the execution) $\mathcal{O}(n+m^a)$, for a specific constant $1 < a < 2$, against the unbounded adversary which may crash all but one machine. They presented the first algorithm for this type of adversary with both work and communication $o(n + m^2)$, where communication is understood as the total number of point-to-point messages sent during the execution. They also gave an algorithm achieving both work and communication $\mathcal{O}(n+m\log^2 m)$ against a strongly-adaptive linearly-bounded adversary.

In [12] there is an algorithm based on a gossiping protocol, solving the problem with work $\mathcal{O}(n+m^{1+\varepsilon})$, for any fixed constant ε. In [16] *Do-All* is studied for an asynchronous message-passing model, where executions are restricted in a way that the delay of each message is at most d. The authors proved $\Omega(n+md\log_d m)$ bound on the expected work. They also developed several algorithms - among them a deterministic one with work $\mathcal{O}((n + md)\log m)$. Further developments and comprehensive related literature can be found in the book by Georgiou and Shvartsman [13].

The line of research closest to ours is the *Do-All* problem of unit length jobs on a multiple access channel with no collision detection, which was first studied in [6], where the authors have shown a lower bound of $\Omega(n + m\sqrt{n})$ on work when there are no crashes, and $\Omega(n+m\sqrt{n}+m\min\{f,n\})$ in presence of crashes caused by an adaptive f-bounded adversary. They also proposed an algorithm, called Two-Lists, with optimal performance $\Theta(n + m\sqrt{n} + m\min\{f,n\})$. A recent paper [15] discusses different models of adversaries on a multiple access channel in the context of performing unit length jobs.

Extending the study of unit length jobs to arbitrary length jobs has taken place in a different context, c.f., [17], where one can find recent results and references to such extensions in the context of fault-tolerant centralized scheduling. There has been a long history of studying the preemptive vs non-preemptive model, e.g., [7,18,19], to which our work also contributes.

The problem we study finds application in other areas, e.g. in window scheduling. It is common in applications (like broadcast systems [1,14]) that requests need to be served periodically in the form of windows. Our problem could be applied to help selecting a window length based on controlling work (aka the number of available machine steps).

Our Results. In this paper, we consider fault-tolerant scheduling of n arbitrary length jobs on m identical machines reliably over a multiple access channel. Our contributions are threefold, on: deterministic preemptive setting and deterministic non-preemptive setting. In the setting with job preemption, we prove a lower bound on work $\Omega(L + m\sqrt{L} + m\min\{f,L\} + m\alpha)$, where L is the sum of lengths of all jobs and α is the maximum length of a job, which holds for both deterministic and randomized algorithms against an adaptive *f-bounded* adversary. We design a corresponding deterministic distributed algorithm, called ScaTri, achieving work $\mathcal{O}(L + m\sqrt{L} + m\min\{f,L\} + m\alpha)$, which matches the lower bound asymptotically with respect to an overhead.

In the model without job preemption, we show a slightly higher lower bound on work $\Omega\left(L + \frac{L}{n}m\sqrt{n} + m\min\{f,n\} + m\alpha\right)$, implying a separation between the two settings: with and without preemption. Similarly as in the previous setting, it holds for both deterministic and randomized algorithms against an adaptive *f-bounded* adversary, thus showing that randomization does not help against an adaptive adversary regardless of the (non-)preemption setting. We develop a corresponding deterministic distributed algorithm, called DefTri, achieving work $\mathcal{O}(L + \alpha m\sqrt{n} + \alpha m\min\{f,n\})$. Results are discussed and compared in detail in Sect. 5.

2   Technical Preliminaries

In this section we formally describe the model for the considered problem and also provide a high-level specification of a black-box procedure TaPeBB that we use in our algorithms. Additionally we present definitions of performing jobs and technicalities regarding preemption.

Machines. In our model we assume having m identical machines, with unique identifiers from the set $\{1, \ldots, m\}$. The distributed system of these machines is synchronized with a global clock, and time is divided into synchronous time slots, called *rounds*. All machines start simultaneously at a certain moment. Furthermore every machine may halt voluntarily. Note that a halted machine does not restart. Every operational machine generates a unit of work per round (even when it is idle), that has to be included while computing the complexity of an algorithm. In this paper by M we denote the number of operational, i.e., not crashed, machines. M may change during the execution.

Communication Model. The communication channel for machines is the, widely considered in literature, *multiple access channel* [2,11], also called a *shared channel*, where a broadcasted message reaches every operational machine. We do not allow simultaneous transmissions of several messages, what means that we consider a channel without collision detection. Hence when more than one message is transmitted at a certain round, then machines hear a signal indifferent from background noise. A job is said to be *confirmed* if a machine, completing the job, successfully communicates this fact via the channel.

Adversarial Model. Machines may fail by crashing, which happens because of an adversary activity. One of the factors describing the adversary is its power f, ie., the total number of failures that it may enforce. We assume that $0 \leq f \leq m - 1$, thus at least one machine remains operational until an algorithm terminates. Machines that were crashed neither restart nor contribute to work.

We consider two adversarial models. An *adaptive f-bounded adversary* can observe the execution and make decisions about up to f crashes arbitrarily, at any moment of the execution. A *non-adaptive f-bounded adversary*, in addition to choosing the faulty subset of f machines, has to declare prior the execution in which rounds crashes will occur, i.e., for every machine declared as faulty, there must be a corresponding round number in which the fault will take place. It is worth noticing that in the context of deterministic algorithms such an adversary is consistent with the adaptive adversary that may decide online which machines will be crashed at any time, as the algorithm may be simulated by the adversary in advance, before its execution, providing knowledge about the best decisions to be taken. Finally, an $(m - 1)$-bounded adversary is also called an *unbounded* adversary.

Complexity Measure. The complexity measure to be used in our analysis is, as mentioned before, *work*, also called the *total number of available machine steps*. It is the number of machine steps available for computations. This means that each operational machine that did not halt contributes a unit of work in each round, even if it stays idle.

Precisely, assume that an execution starts when all the machines begin simultaneously in some fixed round r_0. Let r_v be the round when machine v halts (or is crashed). Then its work contribution is equal to $r_v - r_0$. Consequently the algorithm complexity is the sum of such expressions over all machines i.e.: $\sum_{1 \leq v \leq m} (r_v - r_0)$.

Jobs and Reliability. We assume that the list of jobs is known to all machines and we expect that machines perform all n jobs as a result of executing an algorithm. We assume that *jobs have arbitrary lengths*, are *independent* (may be performed in any order) and *idempotent* (may be performed many times, even concurrently by different machines).

Furthermore a *reliable* algorithm satisfies the following conditions in any execution: *all jobs are eventually performed, if at least one machine remains non-faulty* and *each machine eventually halts, unless it has crashed.*

We assume that jobs have some minimal (atomic or unit) length. Consequently, we can also assume that each job's length is a multiple of the minimal length. As the model that we consider is synchronous, this minimal length may be justified by the round duration required for local computations for each machine. By ℓ_a we denote the length of job a. We also use L to denote the sum of lengths of all jobs, i.e., $L = \sum_i \ell_i$. Finally, by α we denote the length of the longest job.

Preemptive vs Non-preemptive Model. By the means of *preemption* we define the possibility of performing jobs in several pieces. Precisely, consider job a, of length ℓ_a (for simplicity we assume that ℓ_a is even) and machine v is scheduled to perform job a at some time of the algorithm execution. Assume that v performs $\ell_a/2$ units of job a and then reports such progress. When preemption is available the remaining $\ell_a/2$ units of job a may be performed by some machine w where $w \neq v$.

Length ℓ_a of job a means that job a requires ℓ_a rounds to be fully performed. Such view allows to think that job a is a chain of ℓ_a *tasks*. Hence we conclude that all jobs form a set of chains of unit length tasks.

We further denote by a_k task k of job a. However, when we refer to a single job, disregarding its tasks, we refer to it simply as job a.

We define two types of jobs regarding how intermediate progress is handled:

Oblivious Job—it is sufficient to have knowledge that previous tasks were done, in order to perform remaining tasks from the same job. In other words, any information from in between progress does not have to be announced.

Non-oblivious Job—any intermediate progress needs to be reported through the channel when interrupting the job to resume it later, and possibly pass the job to another machine.

In the preemptive oblivious model a job may be abandoned by machine v on some task k without confirming progress up to this point on the channel and then continued from the same task k by any machine w.

As an example of the preemptive oblivious model, consider a scenario that there is a job that a shared array of length x needs to be erased. If a machine stopped performing this job at some point, another one may reclaim that job without the necessity of repeating previous steps by simply reading to which point it has been erased.

For the preemptive non-oblivious model, consider that a machine executes Dijkstra's algorithm for finding the shortest path. If it becomes interrupted, then another machine cannot reclaim this job otherwise than by performing the job from the beginning, unless intermediate computations have been shared. In other words preemption is available with respect to maintaining information about tasks.

On the contrary to the preemptive model, we also consider the model without preemption i.e. where each job can be performed by a machine only in one piece - when a machine is crashed while performing such job, the whole progress is lost, even if it was reported on the channel before the crash took place.

The Task-Performing Black Box (TaPeBB). The algorithms we design in this paper employ a black-box procedure for arbitrary length jobs that is able to reliably perform a subset of input (in the form of jobs consisting of chains of consecutive tasks) or report that something went wrong. In what follows, we specify this procedure, called the *Task-Performing Black Box* (TAPEBB for short), and argue that it can be implemented and employed to our considerations.

We can use the procedure in both deterministic and randomized solutions. Precisely, all our algorithms use TAPEBB, despite the fact that they perform differently in the sense of work complexity.

Most important ideas of our results lie within how to preprocess the input rather than how to actually perform the jobs, thus employing such a black-box could improve the clarity of presentation.

General Properties of TaPeBB. A synchronous system is characterized by time being divided into synchronous slots, that we already called rounds. In what follows, each round is a possibility for machines to transmit.

The nature of arbitrary length jobs leads, however, to a concept that the time between consecutive broadcasts needs to be adjusted. Specifically, for sets containing long jobs in the non-preemptive configuration of the problem, it may be better to broadcast the fact of performing the job fully, rather than semi-broadcasting multiple times. Only the final transmission brings valuable information about progress in performing jobs and any intermediate transmissions congest the channel, indicating only that the broadcasting machine is still operational.

Therefore, we assume that TAPEBB has a feature of changing the duration between consecutive broadcasts. We will call the actual time step between consecutive broadcasts a *phase*. Denote the length of a phase by ϕ. Unless stated otherwise, we assume that $\phi = 1$, i.e., the duration of a phase and a round is consistent, thus machines may transmit in any round.

Input: TAPEBB(v, d, *JOBS, MACHINES*, ϕ)

- v represents the id of a machine executing TAPEBB.
- TAPEBB takes a list of machines MACHINES and a list of jobs JOBS as the input, yet from the task perspective, i.e., jobs are provided as a chain of tasks. All necessary information about tasks is available through list JOBS, including their id's, and how, as well as, in what order they build jobs. It may happen that a job is not done fully and only some initial segment of tasks forming that job is performed. Because list JOBS maintains information about tasks, such a situation can be successfully handled.
- Additionally, the procedure takes an integer value d. It specifies the number of machines that will be used in the procedure for performing provided job input.

The procedure works in such a way that each working machine is responsible for performing a number of jobs. For clarity we assume that TAPEBB always uses the initial d machines from the list of machines. Using a certain number of machines in the procedure allows us to set an upper bound on the amount of work accrued during a single execution.

– We call a single execution an *epoch* and the parameter d is called the length of an epoch (i.e., the number of phases that form an epoch).
– ϕ is the length of a phase i.e. the duration between consecutive broadcasts.

Output: what jobs/tasks have been done and which machines have crashed.
– Having explained what is the length of an epoch in TAPEBB we now describe the capability of performing tasks in a single epoch, which is understood as the maximal number of jobs that may be confirmed in an epoch, when it is executed fully without any adversarial distractions. Firstly, let us note, that TAPEBB allows to confirm j tasks in some round j. This comes from the fact that if a machine worked for j rounds and was able to perform one task per round, then it can confirm at most j tasks when it comes to broadcasting in round j. Therefore, the capability of performing tasks in an epoch is at most $\sum_{j=1}^{d} j$ which is the sum of an arithmetic series with common difference equal 1 over all rounds of an epoch.
– As a result of running a single epoch we have an output information about which tasks were actually done and whether there were any machines identified as crashed, when machines were communicating progress (a crash is consistent with a machine being silent when scheduled to broadcast).

A candidate algorithm to serve as TAPEBB is the TWO-LISTS algorithm from [6] and we refer readers to the details therein. Nevertheless, we assume that it may be substituted with an arbitrary algorithm fulfilling the requirements that we stated above.

3   Preemptive Model

In this section we consider the scheduling problem in the model with preemption, which is, intuitively, an easier model to tackle. We show a lower bound for oblivious jobs (Sect. 3.1) and then present the algorithm for non-oblivious jobs (Sect. 3.2).

3.1   Lower Bound

We first recall the minimal work complexity introduced in [6].

Lemma 1 ([6], Lemma 2). *A reliable, distributed algorithm, possibly randomized, performs work $\Omega(n + m\sqrt{n})$ in an execution in which no failures occur.*

Recall that L denotes the total length of jobs and α denotes the length of the longest job. As jobs are built with unit length tasks, we can look at this result from the task perspective. Precisely, as L can be considered to represent the number of tasks needed to be performed by the system, then the lower bound for our model translates in a straightforward way. Furthermore, in our model there is a certain bottleneck dictated by the longest job. Reason being is that there may be an execution with one long job, in comparison to others being short. Thus the longest job determines the magnitude of work in the complexity formula.

We conclude with the following theorem, setting the lower bound for the considered problem.

Theorem 1. *The adaptive f-bounded adversary, for $0 \leq f < m$, can force any reliable, possibly randomized and centralized algorithm and a set of oblivious jobs of arbitrary lengths with preemption, to perform work $\Omega(L + m\sqrt{L} + m\min\{f, L\} + m\alpha)$.*

3.2 Algorithm ScaTri

In this section we present our algorithm Scaling-Triangle MAC Scheduling (SCATRI for short). In SCATRI machines have access to all the jobs and the corresponding tasks that build those jobs. This means that they know their id's and lengths.

We assume that each machine maintains three lists: MACHINES, TASKS and JOBS. The first list is a list of operational machines and is updated according to the information broadcasted through the communication channel. If there is information that some machine was recognized as crashed, then this machine is removed from the list. In the context of the TAPEBB procedure, recall that this is realized as the output information. TAPEBB returns the list of crashed machines so that operational machines may update their lists.

List TASKS represents all the tasks that are initially computed from the list of jobs. Every task has its unique identifier, which allows to discover to which job it belongs and its position in that job. This allows to preserve consistency and coherency: task k cannot be performed before task $k - 1$. If some tasks are performed then this fact is also updated on the list.

List JOBS represents the set of jobs—it is a convenient way to know what are the consecutive parts of input for the TAPEBB procedure. Jobs are assigned to each machine at the beginning of an epoch (TAPEBB execution). We assume that machines have instant access to jobs lengths. Information on list JOBS may be updated directly from information maintained on list TASKS. To ease the discussion, by $|XYZ|$ we denote the length of list XYZ and by $M = |\text{MACHINES}|$ the actual number of operational machines.

The capability of performing tasks in a TAPEBB epoch is at most $\sum_{j=1}^{d} j$ (cf. Sect. 2) which is the sum of an arithmetic series with common difference equal 1 over all rounds of an epoch. If we take a geometric approach and draw lines or boxes of increasing lengths one next to the other, then drawing this

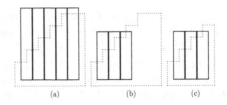

Fig. 1. A very general idea about how SCATRI works. The vertical stripes represent jobs and the dotted-line triangle is the capability of the algorithm to perform jobs in a single epoch (or parts of them i.e. some consecutive tasks). Consequently, if there are enough jobs to pack into an epoch, then it is executed (a). Otherwise (b), the length of the epoch is reduced (c). This helps preventing excessive idle work.

sequence would form a triangle. Hence, providing a subset of jobs as the input is consistent with packing them into such a triangle, cf., Fig. 1.

In TAPEBB executed with $\phi = 1$ broadcasts take place in each round. This means that in round 1 machine 1 is scheduled to broadcast, in round 2 machine 2 and, in general, in round j machine j is scheduled to broadcast. Consequently, j tasks can be confirmed as performed in round j, unless machine j is crashed and hence silence is heard in round j. Therefore, the capability of performing tasks by TAPEBB is referred to as the triangle.

We assume d describes the current length of an epoch. Initially it is set to m, but it may be reduced while the algorithm is running, cf. Fig. 1(b) and (c). In what follows, let us assume that the length of the epoch d is set to $m/2^i$ for some i. We need to fill in a triangle of size $\sum_{j=1}^{\frac{m}{2^i}} j$. Initially we need to provide $m/2^i$ jobs, that will form the base of the triangle. Jobs are provided as the input in such a way that the shortest ones are preferred for machines with lower id's. If there are several jobs with same lengths, then those with lower id's are preferred, cf. Fig. 2. After having the base filled, it is necessary to look whether there are any gaps in the triangle, see Fig. 2(b). If so, another layer of jobs is placed on top of the base layer, and the procedure is repeated, see Fig. 2(c) and (d). Otherwise, we are done and ready to execute TAPEBB.

This approach allows to "trim" longer jobs preferably—these will have more tasks completed after executing TAPEBB than shorter ones, because they are scheduled to be done by machines with higher id's, which are broadcasting in further rounds. As machines with higher id's are broadcasting in further rounds, then they can confirm more tasks with a single broadcast.

One can observe that performing a transmission is an opportunity to confirm on the channel that the tasks that were assigned to a machine are done. Additionally this confirms that a certain machine is still operational. In what follows these two types of aggregated pieces of information: which jobs were done, and which machines were crashed, are eventually provided as the output of TAPEBB.

When $d = m/2^i$ for some $i = 1, \ldots, \log m$ it may happen that there are not enough tasks (jobs) to fill in a maximal triangle. If this happens we will, in

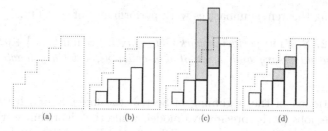

Fig. 2. An illustration of job input. (a) Initially there is a certain capability of TAPEBB (triangle size) for performing jobs. (b) The initial layer is filled with jobs (white blocks), yet there are still some gaps. (c) An additional layer (gray blocks) is placed to fill in the triangle entirely. (d) In fact those additional jobs will only be done partially.

some cases, reduce the job-schedule triangle (and simultaneously the length of the epoch) to $m/2^{i+1}$ and try to fill in a smaller triangle, cf., Fig. 1(b) and (c).

Finally, we use $\text{TAPEBB}(v, d, \text{JOBS}, \text{MACHINES}, \phi)$ to denote which machine executes the procedure, the size of the schedule and the length of the epoch, the list of jobs from which the input will be provided, the list of operational machines, and the phase duration. We emphasize that we described the process of assigning jobs to machines from the algorithm perspective, i.e., we illustrated that the system needs to provide input to TAPEBB. Nevertheless, for the sake of clarity we assume that TAPEBB collects the appropriate input by itself according to the rules described above, after having lists JOBS and MACHINES provided as the input. Figs. 1 and 2 illustrate the idea standing behind SCATRI.

Algorithm 1. SCATRI; code for machine v

---

1   - initialize MACHINES to a sorted list of all m names of machines;
2   - initialize JOBS to a sorted list of all n names of jobs;
3   - initialize TASKS to a sorted list of all tasks according to the information from JOBS;
4   - initialize variable d representing the length of an epoch;
5   - initialize variable $\phi := 1$ representing the length of a phase;
6   - initialize $i := 0$;
7   **repeat**
8       $d := m/2^i$;
9   **if** $|\mathit{TASKS}| \geq d(d+1)/2$ **then**
10   execute $\text{TAPEBB}(v, d, \text{JOBS}, \text{MACHINES}, \phi)$;
11   update JOBS, MACHINES, TASKS according to the output information from TAPEBB;
12   **end**
13   **else**
14          $i := i + 1$;
15   **end**
16 **until** $|\mathit{JOBS}| = 0$;

The following theorem summarizes work performance of SCATRI:

Theorem 2. SCATRI *performs work* $\mathcal{O}(L + m\sqrt{L} + m\min\{f, L\} + m\alpha)$ *against the adaptive adversary and a set of non-oblivious, arbitrary length jobs with preemption.*

As mentioned in the beginning of Sect. 3, the lower bound here is proved for oblivious jobs in the preemptive model, while the algorithm works reliably for non-oblivious jobs in the same model, because TAPEBB procedure provides any intermediate job performing progress as the output information and all the information is updated sequentially after each epoch. This implies that the distinction between oblivious and non-oblivious jobs in the preemptive model and against an adaptive adversary does not matter, thus we finish this section with the following corollary:

Corollary 1. SCATRI *is optimal in asymptotic work efficiency for jobs with arbitrary lengths with preemption for both oblivious and non-oblivious jobs.*

4   Non-preemptive Model

In this section we consider the complementary problem of performing jobs when preemption is not available. We begin with the lower bound and then proceed to the algorithm description.

4.1   Lower Bound

In the non-preemptive model each job can only be performed by a machine in one piece. This reflects an all-or-nothing policy, i.e., a machine cannot make any intermediate progress while performing a job. If it starts working on a job then either it must be done entirely or it is abandoned without any tasks performed.

In what follows any intermediate broadcasts while performing a job are not helpful for the system. The only meaningful transmissions are those which allow to confirm certain jobs being done.

Theorem 3. *A reliable, distributed algorithm, possibly randomized, performs work* $\Omega\left(L + \frac{L}{n}m\sqrt{n} + m\min\{f, n\} + m\alpha\right)$ *in an execution with at most f failures against an adaptive adversary.*

4.2   Algorithm DefTri

The key reason to consider the notion of a phase (i.e. time between consecutive broadcasts), introduced in the model section, is that it allows to assume broadcasts are done after jobs are done entirely. The only question is how to set the phase parameter appropriately.

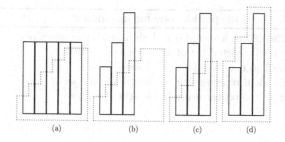

Fig. 3. Most important features of DEFTRI. Similarly to SCATRI we apply the method of reducing idle work by reducing the input size (epoch length) for the task performing procedure ((a), (b), (c)). Additionally we change the duration on phases (time between consecutive broadcasts) in order to be able to fill in jobs entirely - in the model without preemption jobs cannot be done partially (d).

We analyze our algorithm from the average length of the current set of jobs perspective. To justify this setting let us consider two scenarios. For the first one, when the phase parameter is set to 1, which is consistent with transmissions taking place each round we already mentioned above that most of the transmissions have no effect on progressing with performing jobs. However, setting the phase parameter to the length of the longest job α can generate excessive idle work.

In what follows, setting the phase parameter to the average length of the current set of jobs allows us to estimate the number of jobs that can be performed in a certain period. What is more it proves that this setting always allows to schedule a significant number of jobs to be done, while preventing from excessive idle work. Hence we begin with a simple fact showing that the number of jobs with length twice the average does not exceed half of the total number of jobs.

Fact 1. *Let n be the number of jobs, L be the sum of all the lengths of jobs and let $\frac{L}{n}$ represent the average length of a job. Then we have that $\left|\{a : \ell_a > 2\frac{L}{n}\}\right| \leq \frac{n}{2}$.*

Algorithm Deforming-Triangle MAC Scheduling (DEFTRI for short) is similar to SCATRI introduced in the previous section and on the top of its design there is the TAPEBB procedure for performing jobs. Roughly speaking, the algorithm, repetitively, tries to choose jobs that could be packed into a specific triangle (parameters of which are controlled by the algorithm) and feed them to TAPEBB. Furthermore, once again the main goal of the algorithm is to avoid redundant or idle work. Hence, the main feature is to examine whether there is an appropriate number of jobs to fill in an epoch of TAPEBB. However, as we are dealing with the non-preemptive model, jobs cannot be done in pieces.

Hence, apart from checking the number of jobs (and possibly reducing the size of an epoch) the algorithm also changes the phase parameter, i.e., the duration between consecutive broadcasts—this is somehow convenient in order to know how the input jobs should be fed into TAPEBB. It settles a "framework" (epoch) with "pumped rounds" that should be filled in appropriately.

Algorithm 2. DefTri; pseudocode for machine v

```
1  - initialize MACHINES to a sorted list of all m names of machines;
2  - initialize JOBS to a sorted list of all n names of jobs;
3  - initialize TASKS to a sorted list of all tasks;
4  - initialize variable d representing the length of an epoch;
5  - initialize variable φ representing the length of a phase;
6  - initialize i := 0;
7  repeat
8  │   d := m/2^i;
9  │   if |JOBS| ≥ d(d + 1)/2 then
10 │   │   φ := |TASKS|/|JOBS| // set φ to current average length of a job
11 │   │   execute TaPeBB(v, d, JOBS, MACHINES, φ);
12 │   │   update JOBS, MACHINES, TASKS according to TaPeBB output;
13 │   end
14 │   else
15 │   │   i := i + 1;
16 │   end
17 until |JOBS| = 0;
```

Using Fact 1 assures that if the size of an epoch is reduced accordingly to the number of jobs and the phase parameter is set accordingly to the actual average length of the current set of jobs (i.e., $|\text{TASKS}|/\text{JOBS}|$ in the code of Algorithm 2), then we are able to provide the input for TaPeBB appropriately and expect that at least $\sum_{j=1}^{\frac{m}{2^i}} j$ jobs will be performed for some size parameter i as a result of executing TaPeBB, yet without taking crashes into account.

Figure 3 illustrates the idea of DefTri.

Theorem 4. DefTri *performs work* $\mathcal{O}(L + \alpha m\sqrt{n} + \alpha m \min\{f, n\})$ *against the adaptive f-bounded adversary.*

5   Comparison of Results for the Two Models

In this section we compare the upper bound for preemptive scheduling from Theorem 2 with the lower bound in the model without preemption from Theorem 3. This comparison shows the range of dependencies between the model parameters for which both models are separated, i.e., the upper bound in the model with preemption is asymptotically smaller than the lower bound in the model without preemption. We settle the scope on the, intuitively greater, bound for the model without preemption and examine how the ranges of the parameters influence the magnitude of the formulas. Let us recall both formulas:

Preemptive, upper bound: $\mathcal{O}(L + m\sqrt{L} + m \min\{f, L\} + m\alpha)$;
Non-preemptive, lower bound: $\Omega(L + \frac{L}{n}m\sqrt{n} + \min\{f, n\} + m\alpha)$.

If L is the factor that dominates the bound in the non-preemptive model, then by simple comparison to other factors we have that L also dominates the bound in the preemptive model. In what follows both formulas are asymptotically equal when the total number of tasks is appropriately large.

When $\frac{L}{n}m\sqrt{n}$ dominates the non-preemptive formula, then by a simple observation we have that $1 \leq \sqrt{\frac{L}{n}}$, because the number of jobs is greater than the number of tasks, and applying a square root does not affect the inequality. Multiplying both sides of the inequality by $m\sqrt{L}$ gives $m\sqrt{L} \leq \frac{L}{n}m\sqrt{n}$. Thus, if $\frac{L}{n}m\sqrt{n}$ dominates the bound in the non-preemptive formula then the non-preemptive formula is asymptotically greater than the preemptive one.

On the other hand when L dominates the bound then they are asymptotically equal, as both formulas linearize for such magnitude. These results confirm that in the channel without collision detection the non-preemptive model is more demanding for most settings of parameters.

6   Conclusions

We addressed the problem of performing jobs of arbitrary lengths on a multiple-access channel without collision detection. Specifically, we analysed two scenarios. In the preemptive model, we showed a lower bound for the considered problem that is $\Omega(L + m\sqrt{L} + m\min\{f, L\} + m\alpha)$ and designed an algorithm that meets the proved bound, hence settled the problem. We also answered the question of how to deal with jobs without preemption, for which we showed a lower bound $\Omega(L + \frac{L}{n}m\sqrt{n} + m\min\{f, n\} + m\alpha)$ and developed a solution, basing on the one for preemptive jobs which work complexity is $\mathcal{O}(L + \alpha m\sqrt{n} + \alpha m\min\{f, n\})$.

Considering open directions for research considered in this paper, it is natural to address the question of channels with collision detection. Furthermore it is worth considering whether randomization could help improving the results, as it took place in similar papers considering the Do-All problem on a shared channel [15]. We conjecture that the use of randomization against non-adaptive adversaries leads to a solution with expected work $\mathcal{O}(L + m\sqrt{L} + m\alpha)$, and thus could prove that randomization helps against the non-adaptive adversary.

Finally, the primary goal of this work was to translate scheduling from classic models to the model of a shared channel, in which it was not considered in depth; therefore, a natural open direction is to extend the model further with other features considered in scheduling literature.

References

1. Bar-Noy, A., Naor, J., Schieber, B.: Pushing dependent data in clients-providers-servers systems. Wirel. Netw. **9**(5), 421–430 (2003)
2. Chlebus, B.S.: Randomized communication in radio networks. In: Pardalos, P.M., Rajasekaran, S., Reif, J.H., Rolim, J.D.P. (eds.) Handbook on Randomized Computing, vol. 1. Kluwer Academic Publisher (2001)
3. Chlebus, B.S., De Prisco, R., Shvartsman, A.A.: Performing tasks on synchronous restartable message-passing processors. Distrib. Comput. **14**(1), 49–64 (2001)
4. Chlebus, B.S., Gasieniec, L., Kowalski, D.R., Shvartsman, A.A.: Bounding work *and* communication in robust cooperative computation. In: Malkhi, D. (ed.) DISC 2002. LNCS, vol. 2508, pp. 295–310. Springer, Heidelberg (2002). https://doi.org/10.1007/3-540-36108-1_20
5. Chlebus, B.S., Kowalski, D.R.: Randomization helps to perform independent tasks reliably. Random Struct. Algorithms **24**(1), 11–41 (2004)
6. Chlebus, B.S., Kowalski, D.R., Lingas, A.: Performing work in broadcast networks. Distrib. Comput. **18**(6), 435–451 (2006)
7. Coffman, E.G., Garey, M.R.: Proof of the 4/3 conjecture for preemptive vs. non-preemptive two-processor scheduling. In: Proceedings of the Twenty-Third Annual ACM Symposium on Theory of Computing, STOC 1991, pp. 241–248. ACM, New York (1991)
8. De Prisco, R., Mayer, A., Yung, M.: Time-optimal message-efficient work performance in the presence of faults. In: Proceedings of the Thirteenth Annual ACM Symposium on Principles of Distributed Computing, PODC 1994, pp. 161–172. ACM, New York (1994)
9. Dwork, C., Halpern, J.Y., Waarts, O.: Performing work efficiently in the presence of faults. SIAM J. Comput. **27**(5), 1457–1491 (1998)
10. Galil, Z., Mayer, A., Yung, M.: Resolving message complexity of byzantine agreement and beyond. In: Proceedings of the 36th Annual Symposium on Foundations of Computer Science, FOCS 1995, p. 724. IEEE Computer Society, Washington, DC (1995)
11. Gallager, R.G.: A perspective on multiaccess channels. IEEE Trans. Inf. Theory **31**, 124–142 (1985)
12. Georgiou, C., Kowalski, D.R., Shvartsman, A.A.: Efficient gossip and robust distributed computation. Theor. Comput. Sci. **347**(1–2), 130–166 (2005)
13. Georgiou, C., Shvartsman, A.A.: Cooperative Task-Oriented Computing: Algorithms and Complexity. Morgan & Claypool Publishers, San Rafael (2011)
14. Gondhalekar, V., Jain, R., Werth, J.: Scheduling on airdisks: efficient access to personalized information services via periodic wireless data broadcast. Technical report TR-96-25, Department of Computer Science, University of Texas, Austin, TX (1996)
15. Klonowski, M., Kowalski, D.R., Mirek, J.: Ordered and delayed adversaries and how to work against them on a shared channel. Distrib. Comput. (2018)
16. Kowalski, D.R., Shvartsman, A.A.: Performing work with asynchronous processors: message-delay-sensitive bounds. In: Proceedings of the Twenty-Second Annual Symposium on Principles of Distributed Computing, PODC 2003, pp. 265–274. ACM, New York (2003)

17. Kowalski, D.R., Wong, P.W., Zavou, E.: Fault tolerant scheduling of tasks of two sizes under resource augmentation. J. Sched. **20**(6), 695–711 (2017)
18. Lucarelli, G., Srivastav, A., Trystram, D.: From preemptive to non-preemptive scheduling using rejections. In: Dinh, T.N., Thai, M.T. (eds.) COCOON 2016. LNCS, vol. 9797, pp. 510–519. Springer, Cham (2016). https://doi.org/10.1007/978-3-319-42634-1_41
19. McNaughton, R.: Scheduling with deadlines and loss functions. Manag. Sci. **6**(1), 1–12 (1959)

Rare Siblings Speed-Up Deterministic Detection and Counting of Small Pattern Graphs

Mirosław Kowaluk[1(✉)] and Andrzej Lingas[2]

[1] Institute of Informatics, University of Warsaw, Warsaw, Poland
kowaluk@mimuw.edu.pl
[2] Department of Computer Science, Lund University, Lund, Sweden
Andrzej.Lingas@cs.lth.se

Abstract. We consider a class of pattern graphs on $q \geq 4$ vertices that have $q - 2$ distinguished vertices with equal neighborhood in the remaining two vertices. Two pattern graphs in this class are siblings if they differ by some edges connecting the distinguished vertices.

In particular, we show that if induced copies of siblings to a pattern graph in such a class are rare in the host graph then one can detect the pattern graph relatively efficiently. For example, we infer that if there are N_d induced copies of a diamond (i.e., a graph on four vertices missing a single edge to be complete) in the host graph, then an induced copy of the complete graph on four vertices, K_4, as well as an induced copy of the cycle on four vertices, C_4, can be deterministically detected in $O(n^{2.75} + N_d)$ time. Note that the fastest known algorithm for K_4 and the fastest known deterministic algorithm for C_4 run in $O(n^{3.257})$ time. We also show that if there is a family of siblings whose induced copies in the host graph are rare then there are good chances to determine the numbers of occurrences of induced copies for all pattern graphs on q vertices relatively efficiently.

1 Introduction

The problems of detecting and counting subgraphs or induced subgraphs of a host graph that are isomorphic to a pattern graph are basic in graph algorithms. They are generally termed as *subgraph isomorphism* and *induced subgraph isomorphism* problems, respectively. Several well-known NP-hard problems can be regarded as their special cases.

More recent examples of applications of different variants of subgraph isomorphism include bio-molecular networks [1], social networks [17], automatic design of processor systems [19], and network security [18]. In the aforementioned applications, the pattern graphs are typically of fixed size which allows for polynomial-time solutions.

For a pattern graph on k vertices and a host graph on n vertices, the fastest known general algorithms for subgraph isomorphism and induced subgraph

© Springer Nature Switzerland AG 2019
L. A. Gąsieniec et al. (Eds.): FCT 2019, LNCS 11651, pp. 322–334, 2019.
https://doi.org/10.1007/978-3-030-25027-0_22

isomorphism run in time $O(n^{\omega(\lfloor k/3 \rfloor, \lceil (k-1)/3 \rceil, \lceil k/3 \rceil)})$ [5,10,15], where $\omega(p,q,r)$ denotes the exponent of fast matrix multiplication for rectangular matrices of size $n^p \times n^q$ and $n^q \times n^r$, respectively [13]. For convenience, we shall denote the latter upper time bound by $C(n.k)$, and $\omega(1,1,1)$ by just ω. It is known that $\omega \leq 2.373$ [14,21] and for example $\omega(1,2,1) \leq 3.257$ [13].

Kloks et al. [10] formulated equations in terms of the number of induced copies of 4-vertex pattern graphs in the input graph that allowed them to conclude that if the number is known for at least one of the pattern graphs then the remaining numbers can be computed in $O(n^\omega)$. Kowaluk et al. generalized the equations to include pattern graphs on more than four vertices in [11]. Using the equations, Vassilevska Williams et al. designed a randomized method for detecting an induced subgraph isomorphic to a pattern graph on four vertices different from K_4 and the four isolated vertices ($4K_1$) [20], subsuming the similar randomized approach from [6]. Their method runs in the same asymptotic time as that based on matrix multiplication for detecting triangles (i.e., K_3) from [9], i.e., in $O(n^\omega)$ time. The authors of [20] obtained a deterministic version of their algorithm also running in the triangle asymptotic time for the diamond, which is K_4 with one removed edge, denoted by $K_4 - e$. Recently, Bläser et al. in [2] and Dalirrooyfard et al. in [4] presented several improved upper time bounds for detection of induced copies of mostly small pattern graphs, in particular induced paths and cycles [2]. Both teams even exhibited infinite families of pattern graphs for which they could subsume the universal $C(n,k)$ bound. The main idea in [2] is to reduce the detection problem through graph polynomials to that of multilinear term detection. Dalirrooyfard et al. presented also interesting relative hardness results on induced subgraph detection in [4].

There are few known earlier examples of pattern graphs on four vertices, different from the diamond, for which isomorphic induced subgraphs can be deterministically detected in an n vertex host graph substantially faster than by the general method in $C(n,k)$ time. The examples include: (1) P_4, a path on four vertices, which can be detected in $O(n + m)$ time [3], where m is the number of edges in the host graph; (2) a paw, which is a triangle connected to the fourth vertex by an edge, denoted by $K_3 + e$., and can be detected in $O(n^\omega)$ time [16]; and (3) a claw (a star with three leaves), which can be also detected in $O(n^\omega)$ time [5]. Analogous upper bounds hold for the pattern graphs that are dual to any of the aforementioned pattern graphs. There are twelve pairwise non-isomorphic pattern graphs on four vertices. Only for the cycle on four vertices, C_4, and its complement, and K_4 and its complement, there are no known deterministic algorithms for the induced subgraph isomorphism that are asymptotically faster than the general method yielding the $O(n^{\omega(1,2,1)})$-time bound. For a relative lower bound for C_4 see [7].

Challenge 1. *Design a deterministic algorithm for detecting an induced copy of C_4 that is substantially asymptotically faster than the algorithm yielded by the general method running in $C(n,4) = O(n^{\omega(1,2,1)})$-time, i.e., $O(n^{3.257})$-time.*

The recent trade-off in [12] provides only a partial solution to this challenge. It shows that if the input graph does not contain a clique on $k+1$ vertices then C_4

can be deterministically detected in time $\tilde{O}(n^\omega k^\mu + n^2 k^2)$, where $\tilde{O}(f)$ stands for $O(f(\log f)^c)$ for some constant c, and $\mu \approx 0.46530$.

The problem of beating the $O(n^{\omega(1,2,1)})$-time bound for the detection of K_4 (or, an induced independent set on four vertices, equivalently) seems even more challenging, even if the usage of random bits is allowed. More generally, improving on the upper bound for the detection of K_l, for some fixed $l \geq 3$, yielded by the general method would be a breakthrough. It could yield faster algorithms for the detection of induced and non-necessarily induced copies of many other pattern graphs on at least l vertices.

Challenge 2. *For a fixed $l > 3$, design a deterministic or randomized algorithm for detecting a copy of K_l that is substantially asymptotically faster than the $C(n, l)$-time algorithm yielded by the general method.*

This is a very demanding challenge, it does not seem that it could be positively resolved by combining known approaches. The aforementioned trade-off from [12] implies that if the input graph does not contain an independent set on $k + 1$ vertices then K_4 can be deterministically detected in time $\tilde{O}(n^\omega k^\mu + n^2 k^2)$, where $\mu \approx 0.46530$.

In this paper, similarly as in [12], we provide partial results on induced subgraph isomorphism for small pattern graphs relevant to the aforementioned challenges. We consider quite different restrictions from those in [12]. They are concerned with the number of induced copies of pattern graphs very similar to the fixed pattern graph H, in the input graph. Surprisingly, if the aforementioned copies are rare, an induced copy of H (e.g., K_4 or C_4) can be detected (if any) relatively efficiently and even in some cases the numbers of induced copies for all pattern on the same number of vertices as H can be determined efficiently.

1.1 Our Contributions

We consider a class of pattern graphs on $q \geq 4$ vertices that have $q - 2$ distinguished vertices with equal neighborhood in the remaining two vertices. Two pattern graphs in this class are siblings if they differ solely by some edges connecting the distinguished vertices.

We show that if induced copies of siblings to a pattern graph in the class are rare in the host graph then one can detect the pattern graph relatively efficiently. In particular for $q = 4$, we show for a pair of siblings H_1, H_2 that an induced copy of H_2 in an n-vertex graph G can be deterministically detected in $O(n^{2.75} + N_1)$ time, where N_1 is the number of induced subgraphs of G isomorphic to H_1. We also show that if there is a family of siblings whose induced copies in the host graph G are rare then there are good chances to determine the numbers of occurrences of induced copies for all pattern graphs on q vertices relatively efficiently. In particular, we show for $q = 4$ that if N is the minimum of the sum of the numbers of induced copies of a pair of siblings taken over all pairs of siblings on four vertices, then we can compute for all pattern graphs H on four vertices the number of induced subgraphs isomorphic to H in $O(n^{2.725} + N)$ time.

2 Preliminaries

We shall consider only simple undirected graphs. Hence, in particular, the graph dual to a graph will be also simple.

A *subgraph* of the graph $G = (V, E)$ is a graph $H = (V_H, E_H)$ such that $V_H \subseteq V$ and $E_H \subseteq E$.

An *induced subgraph* of the graph $G = (V, E)$ is a graph $H = (V_H, E_H)$ such that $V_H \subseteq V$ and $E_H = E \cap (V_H \times V_H)$.

The *neighborhood* of a vertex v in a graph G is the set of all vertices in G adjacent to v.

Definition 1. *For $q \geq 4$, we distinguish the class $L_s(q)$ of pattern graphs H on q vertices $v_1, v_2, ..., v_q$ such that $v_1, ..., v_{q-2}$ have the same neighbors among the remaining two vertices v_{q-1}, v_q. A pair of graphs in $L_s(q)$ are called siblings if they differ only on the sets of their edges between $v_1, ..., v_{q-2}$. The maximal set of mutual siblings in $L_s(q)$ is called a sibling family in $L_s(q)$. If v_{q-1}, v_q. are adjacent (not adjacent) in H then H is in the subclass $L_s^+(q)$ ($L_s^-(q)$, respectively) of $L_s(q)$.*

Note that the subclasses $L_s^+(q)$ and $L_s^-(q)$ of $L_s(q)$ have a non-empty intersection.

In case of $L_s(4)$, a sibling family is just a pair of siblings. The following sibling pairs occur in $L_s(4)$: $(K_4, K_4 \backslash e)$, $(K_4 \backslash e, C_4)$, (K_3+e, K_3+K_1), $(3-star, P_3+K_1)$, $(2K_2, K_2 + 2K_1)$, $(K_2 + 2K_1, 4K_1)$ (Fig. 1).

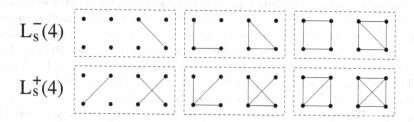

$L_s^-(4)$

$L_s^+(4)$

Fig. 1. The six pairs of siblings in $L_s(4)$ are marked by rectangles.

The *adjacency matrix* A of a graph $G = (V, E)$ is the $0-1$ $n \times n$ matrix such that $n = |V|$ and for $1 \leq i, j \leq n$, $A[i, j] = 1$ if and only if $\{i, j\} \in E$.

A *witness* for an entry $B[i, j]$ of the Boolean matrix product B of two Boolean matrices A_1 and A_2 is any index k such that $A_1[i, k]$ and $A_2[k, j]$ are equal to 1 [8].

Recall that for natural numbers p, q, r, $\omega(p, q, r)$ denotes the exponent of fast matrix multiplication for rectangular matrices of size $n^p \times n^q$ and $n^q \times n^r$, respectively. For convenience, ω stands for $\omega(1, 1, 1)$.

Fact 1. *The fast matrix multiplication algorithm runs in $O(n^\omega)$ time, where ω is not greater than 2.3728639 [14] (cf. [21]).*

In the following fact from [8], the upper bounds have been updated by incorporating the more recent upper bounds on fast rectangular matrix multiplication from [13].

Fact 2. *The deterministic k-witness algorithm from [8] takes as input an integer k and two $n \times n$ Boolean matrices, and returns a list of q witnesses for each positive entry of the Boolean matrix product of those matrices, where q is the minimum of k and the total number of witnesses for this entry. It runs in $n^{2.5719}k^{0.3176}$ time for $k < n^{0.394}$ and $O(n^{2.5}k^{0.5})$ time for $k \geq n^{0.394}$ [8].*

For convenience, $CW(n, k)$ will stand for the running time of the k-witness deterministic algorithm from Fact 2 for two input Boolean matrices of size $n \times n$.

Fact 3. *[10] If the number of induced subgraphs isomorphic to a pattern graph on four vertices in an n-vertex graph is known then the corresponding numbers for all other pattern graphs on four vertices can be computed in $O(n^\omega)$ time.*

The authors of [11] presented linear equations in terms of the numbers of induced copies of different pattern graphs on k vertices in an n-vertex graph G, generalizing those for 4-vertex pattern graphs from [10]. They also considered a system of equations formed by a subset of the aforementioned equations for pattern graphs satisfying additional conditions. For the definition of the system of equations termed $SEq(G, k, l)$ see Appendix. In the matrix formed by the left sides of the equations in $SEq(G, k, l)$, each row corresponds to a pattern graph H on k vertices, among which $k - l$ are independent. The columns in the matrix correspond to consecutive pattern graphs on k vertices. (We shall apply the equations for $k = q$ and $l = q - 2$.) Each row of the matrix contains non-zero coefficients corresponding to pattern graphs H' whose edge set is between those of H and K_k. It follows that the equations in $SEq(G, k, l)$ are independent. They can be formed in time $O(n^{\omega(\lceil(k-l)/2\rceil, 1, \lfloor(k-l)/2\rfloor)})$ [11]. See Appendix for details.

The following generalization of Fact 3 to include pattern graphs on $q \geq 4$ vertices is implicit in [11] and its proof requires adopting quite a lot of notation from there. Because it is not relevant for the main results in this paper we refer the reader for the proof including the aforementioned equations to Appendix. See also Theorem 4.1 in [11] and Theorem 7.5 in [2] for similar results with different kinds of restrictions.

Lemma 1. *Consider an n-vertex graph G. Let H be a pattern graph on $q \geq 4$ vertices such that the number of induced subgraphs isomorphic to H in G is known but it cannot be computed from a linear combination of the equations in $SEq(G, q, q - 2)$. Then the corresponding numbers for all other pattern graphs on q vertices can be found in $O(n^{\omega(\lceil(q-2)/2\rceil, 1, \lfloor(q-2)/2\rfloor)})$ time.*

Remark 1. For a pattern graph H on $q \geq 4$ vertices and an n-vertex graph G, the number of induced copies of H in G cannot be obtained from a linear combination of the equations in $SEq(G, q, q-2)$ iff the equations are independent after the removal of the unknown $x_{H,G}$ corresponding to H.

Proof. Suppose that the first condition does not imply the second one. Thus, there is a non-trivial combination of the equations after the removal of $x_{H,G}$ which yields zero. Applying the combination to the left sides of the original equations, by their independence, we can obtain the variable $x_{H,G}$ after rescaling, i.e., derive the number of induced copies of H from the original equations, a contradiction.

Conversely, suppose that the second condition does not imply the first Then, there is a combination of the left sides of the original equations which yields $x_{H,G}$. This combination applied to the left sides of the equations resulting from the removal of $x_{H,G}$ yields 0, a contradiction.   □

3   Main Results

The following lemma is folklore. It demonstrates that if induced copies of a pattern graph in an n vertex host graph are sufficiently frequent then an induced copy of the pattern graph can be detected in the host graph by sampling in sub-$C(n, q)$ time.

Lemma 2. *Let H be a pattern graph on q vertices. If an n-vertex graph G contains n^ψ induced subgraphs isomorphic to H then one can decide if G contains a copy of H in $O(C(n^{1-q^{-1}\psi} \ln^{q^{-1}} n, q) + n^2) \ln n)$ time with probability at least $1 - n^{-1}$, where $C(n, q)$ stands for the time taken by the standard algorithm to detect an induced subgraph in an n-vertex graph that is isomorphic to a given pattern graph on q vertices.*

Proof. Let $G = (V, E)$. Suppose first that the number of induced subgraphs of G isomorphic to H in G is known. Pick uniformly at random a subset S of $(\ln n)^{1/q} n^{1-q^{-1}\psi}$ vertices of G. For a given induced subgraph O of G isomorphic to H, the probability that all q vertices in O are in S is at least $\ln n n^{-\psi}$. Thus, the probability that none of the at least n^ψ induced subgraphs of G isomorphic to H is an induced subgraph of the subgraph $G[S]$ of G induced by S is at most $(1 - \ln n n^{-\psi})^{n^\psi}$ which is of order n^{-1}. Hence, it is sufficient to run a standard algorithm for the detection of an induced subgraph of $G[S]$ isomorphic to H on $G[S]$. If the number of induced subgraphs of G isomorphic to H in G is not known, we just apply a reversed exponential search starting from $\binom{n}{q}$.   □

Our main theorem is as follows.

Theorem 1. *Let G be an n-vertex graph, and let q be an integer not less than 4. Consider a pattern graph H in $L_s(q)$. Let $\bar{N}$ be the sum of the numbers of induced copies of siblings to H in G. There is a deterministic algorithm for detecting an induced subgraph of G isomorphic to H that runs in $O(CW(n, \lceil n^{\frac{1}{q-2}} \rceil) + \bar{N})$ time. There is also a deterministic algorithm for computing for all siblings in a sibling family the numbers of their induced copies in G that runs in $O(CW(n, \lceil n^{\frac{1}{q-2}} \rceil) + N^*)$ time, where N^* is the sum of the numbers of induced copies of the siblings in the family in G.*

Proof. Let $v_1, ..., v_q$ be vertices of H satisfying Definition 1. Suppose first that v_{q-1} is adjacent to v_q. Consider the following algorithm for the detection of an induced copy of H.

1. If v_1 is adjacent to v_{q-1} then set A_1 to the adjacency matrix of G otherwise set A_1 to the adjacency matrix of the graph dual to G. Similarly, if v_1 is adjacent to v_q then set A_2 to the adjacency matrix of G otherwise set A_2 to the adjacency matrix of the graph dual to G.
2. Compute the arithmetic product of A_1 with A_2 treated as an arithmetic $0-1$ matrices.
3. Set k to $\lceil n^{\frac{1}{q-2}} \rceil$ and run the algorithm for k-witness problem for the Boolean product B of A_1 with A_2.
4. For each edge e of G whose corresponding entry in B has at least $q - 2$ witnesses, iterate the following block.
   (a) If the entry has more than k witnesses then compute all its witnesseses in $O(n)$ time.
   (b) For each $(q-2)$-tuple of witnesses of the entry check if the vertices of the edge e and the $(q - 2)$-tuple jointly induce a subgraph of G isomorphic to H. If so then answer YES reporting that the subgraph induced by the edge and the $(q - 2)$-tuple of witnesses is isomorphic to H, and stop.
5. Answer NO

The correctness of the algorithm is obvious (see Fig. 2). If v_{q-1} and v_q are not adjacent in H then the for-loop in Step 4 is run for pairs of not adjacent vertices of G instead of edges of G.

The following time analysis holds in any of the aforementioned variants of the algorithm. Step 1 takes $O(n^2)$ time. Step 2 requires $O(n^\omega)$ time while Step 3 takes $CW(n, k)$ time. Observe that if Step 4(a) is performed then more than $\binom{k}{q-2}$ $(q-2)$-tuples are tested in the following Step 4(b). Step 4(a) is performed next time only if each of the aforementioned tests results in a discovery of an induced copy of a sibling to H. Therefore, Step 4(a) can be performed at most $\bar{N}/\binom{k}{q-2}+1$ times so it takes $O(n(\bar{N}/\binom{k}{q-2}+1))$ total time. In Step 4(b) at most $\bar{N}+1$ tuples can be tested in total. Hence, Step 4(b) requires at most $O(\bar{N}+n^2)$ total time. Note that since q is fixed, $\binom{k}{q-2} = \Theta(k^{q-2})$ holds. Consequently, $\binom{k}{q-2} = \Theta(n)$ holds by $k = \lceil n^{\frac{1}{q-2}} \rceil$. This completes the proof of the first part.

To prove the second part we assume similarly that $v_1, ..., v_q$ are vertices of any member in the sibling family satisfying Definition 1. We also assume first that v_{q-1} is adjacent to v_q. It is sufficient to remove Step 5 and to modify Step 4(b) in the algorithm in the proof of the first part to the following one:

For each $(q - 2)$-tuple of witnesses of the entry determine the pattern graph Q in the sibling family that is isomorphic to the subgraph of G induced by the vertices of the edge e and the $(q - 2)$-tuple, and increase the counter for Q (initially set to 0) by one.

Finally, when Step 4 is completed the numbers in the counters have to be divided by precomputed constants in order to obtain the true values of the number of induced copies of respective siblings in the family. These constants depend on the automorphisms of respective siblings, and are equal to the number of times each induced subgraph isomorphic to a sibling in the family respectively, is counted in the algorithm.

As previously, if v_{q-1} and v_q are not adjacent then the *for* loop in Step 4 is run for pairs of not adjacent vertices of G instead of edges of G.

The time analysis is analogous to that in the proof of the first part with the exception that $\bar{N}$ has to be replaced by N^* now.   □

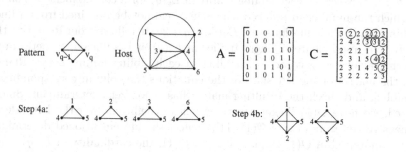

Fig. 2. An illustration of the crucial Step 4 in Algorithm 1.

Corollary 1. *Let N' be the number of induced subgraphs of an n-vertex graph G that are isomorphic to siblings of K_q in $L_s(q)$. We can decide if G contains a copy of K_q in $O(CW(n, \lceil n^{\frac{1}{q-2}} \rceil) + N')$ time.*

Lemma 3. *Let G be an n-vertex graph, let q be an integer not less than 4, and let $H_1, ..., H_r$ be a complete sibling family in $L_s(q)$. Suppose that for $i = 1, ..., r$ there are N_i induced subgraphs of G isomorphic to H_i in G. A Θ-approximation of $\sum_{i=1}^r N_i$ can be computed in $O(n^\omega)$ time.*

Proof. Let $G = (V, E)$. We may assume without loss of generality that the vertex set of $H_1, ..., H_r$ consists of vertices $v_1, ..., v_q$, where $v_1, ..., v_{q-2}$ have the same neighborhoods in $\{v_{q-1}, v_q\}$ in $H_1, ..., H_r$. Suppose first that v_{q-1} and v_q are adjacent in $H_1, ..., H_r$, i.e., $H_1, ..., H_r$ are in $L_s^+(q)$. Define the matrices A_1 and A_2 as in the proof of Theorem 1. Treat them as $0 - 1$ arithmetic matrices and compute their arithmetic product C. Then, we obtain the following equation:

$$\Theta(\sum_{i=1}^r N_i) = \sum_{\{i,j\} \in E \& C[i,j] > q-3} \binom{C[i,j]}{q-2}$$

since the right-hand side is equal to $\sum_{i=1}^r \alpha_i N_i$, where for $i = 1, ..., r$, α_i are the constants depending on the automorphisms of H_i equal to the number of times each induced subgraph isomorphic to H_i is counted on the right-hand side. If v_{q-1} and v_q are not adjacent in $H_1, ..., H_r$, i.e., $H_1, ..., H_r$ are in $L_s^-(q)$, then we obtain an analogous equation by just replacing $\{i, j\} \in E$ with $\{i, j\} \notin E$ in the summation. It remains to observe that the computation of C takes $O(n^\omega)$ time.   □

Corollary 2. *Let N_f be the minimum number of induced subgraphs in G isomorphic to a pattern graph in a sibling family over all families of mutual siblings in $L_s(q)$ that contain a pattern graph H such that the number of induced copies of H in G cannot be obtained from a linear combination of the equations in $SEq(G, q, q - 2)$ (see Appendix). If N_f is well defined then we can compute for all pattern graphs H on q vertices the number of induced subgraph isomorphic to H in G in time $O(n^{\omega(\lceil (q-2)/2 \rceil, 1, \lfloor (q-2)/2 \rfloor)} + CW(n, \lceil n^{\frac{1}{q-2}} \rceil) + N_f)$.*

Proof. First, we check each sibling family in $L_s(q)$ for a containment of a sibling for which the number of induced copies in G cannot be obtained from a linear combination of the equations in $SEq(G, q, q - 2)$. It follows from Remark 1 that to sieve out appropriate siblings families in $L_s(q)$, it is sufficient to perform the following algebraic test for each sibling in each sibling family $L_s(q)$: Remove from the matrix of the left sides of the equations the column corresponding to the sibling and check the resulting matrix has a non-zero determinant. Since q is fixed, we have $O(1)$ linear equations whose coefficients on the left side are nonnegative integers of size $O(1)$ [11]. Hence, the aforementioned determinant can be computed in $O(1)$ time. Since $q = O(1)$, the cardinality of $L_s(q)$ is also $O(1)$. Consequently, the total time taken by the checking is $O(1)$ provided the equations are given. The latter can be formed in $O(n^{\omega(\lceil (q-2)/2 \rceil, 1, \lfloor (q-2)/2 \rfloor)})$ time [11].

If there is at least one sibling family in $L_s(q)$ satisfying the aforementioned requirement then we proceed as follows. First, we compute the approximations of the sums of the numbers of induced copies for all such sibling families in $L_s(q)$ by using Lemma 3 in order to select such a sibling family with the lowest value of the approximation. Then, we apply the method of Theorem 1 just to this family. □

3.1   Siblings on Four Vertices

A complete sibling family in $L_s(4)$ is just a pair of siblings. Fact 2 yields the upper bound $O(n^{2.75})$ on $CW(n, \sqrt{n})$. Hence, we obtain the following corollaries from the results of the preceding section and the aforementioned fact. In particular, Theorem 1 yields the following corollary.

Corollary 3. *Let G be an n-vertex graph. Consider two sibling graphs H_1 and H_2 in $L_s(4)$. Suppose that for $i = 1, 2$ there are N_i induced subgraphs of G isomorphic to H_i. For $i = 1, 2$, there is a deterministic algorithm for detecting an induced subgraph of G isomorphic to H_i that runs in $O(n^{2.75} + N_{3-i})$ time. There is also a deterministic algorithm for computing N_1 and N_2 that runs in $O(n^{2.75} + N_1 + N_2)$ time.*

Lemma 2 and Corollaries 3 yield the following "comfort" corollary showing that for at least one in a pair of siblings one can subsume a known upper bound on detection or counting.

Corollary 4. *Let H_1, H_2 be a pair of siblings in $L_s(4)$, and let G be a graph on n vertices. Then at least one of the following statements holds:*

1. *One can determine the number of induced copies of H_1 in G as well as that of H_2 in G deterministically in $O(n^{2.75})$-time.*
2. *Induced copies of each of the siblings occur in G and one can find their representatives in $O(n^2 \ln n)$ randomized time with high probability by sampling.*
3. *Induced copies of at least one of the siblings occur in G and one can find their representatives in $O(n^{2.75})$ deterministic time and $O(n^2)$ randomized time with high probability by sampling.*

Proof. If there are $O(n^{2.75})$ induced copies of each of the siblings in G then the first statement holds by the second part of Corollary 3. If there are $\Omega(n^{2.75})$ induced copies of each of the siblings in G then the second statement holds by Lemma 2 and $O(C(n^{1-2.75/4} \ln^{4^{-1}} n, 4) + n^2) \ln n) \le O((n^{3.257(1-2.75/4)} \ln^{3.25/4} n + n^2) \ln n) \le O(n^2 \ln n)$. Finally, if one of the siblings H_i has $\Omega(n^{2.75})$ induced copies in G while the other has only $O(n^{2.75})$ induced copies in G then one can find deterministically an induced copy of H_i in G in $O(n^{2.75})$ time by the first part of Corollary 3 and one can find such a copy in $O(n^2)$ randomized time with high probability by sampling according to Lemma 2. □

Corollary 3 also yields the following corollary.

Corollary 5. *Let N_d be the number of induced subgraphs of an n-vertex graph G that are isomorphic to $K_4 \setminus e$. We can decide if G contains a copy of K_4 as well as if G contains an induced copy of C_4 in $O(n^{2.75} + N_d)$ time.*

To obtain the last corollary we specialize the proof of Corollary 2 using Fact 3 instead of Lemma 1.

Corollary 6. *Let N be the minimum number of induced subgraphs in G isomorphic to a pattern graph in a pair of siblings over all pairs of siblings in $L_s(4)$. We can compute for all pattern graphs H on four vertices the number of induced subgraph isomorphic to H in $O(n^{2.75} + N)$ time.*

4   Final Remark

By the "comfort" corollary (Corollary 4), the only case that we cannot get a speed-up for detecting a sibling $H_1 \in L_s(4)$ in the host graph G occurs if the induced copies of H_1 in G are relatively rare and on the contrary the induced copies of its sibling H_2 in G are frequent. But then, we can detect at least H_2 faster by sampling.

Acknowledgments. The authors are thankful to anonymous referees for valuable comments. The research has been supported in part by Swedish Research Council grant 621-2017-03750.

Appendix: Proof of Lemma 1

Notation

A set of single representatives of all isomorphism classes for graphs on k vertices is denoted by $\mathcal{H}_k$ while its subset consisting of graphs having an independent set on at least $k - l \geq 1$ vertices is denoted by $\mathcal{H}_k(l)$.

Let H be a graph on k vertices and let H_{sub} be an induced subgraph of H on l vertices such that the $k - l$ vertices in $H \setminus H_{sub}$ form an independent set. Consider the family of all supergraphs H' of H (including H) in $\mathcal{H}_k$ such that H' has the same vertex set as H, H_{sub} is also an induced subgraph of H', and the set of edges between H_{sub} and $H' \setminus H_{sub}$ is the same as that between H_{sub} and $H \setminus H_{sub}$. This family is denoted by $\mathcal{H}_k(H_{sub}, H)$, and its intersection with $\mathcal{H}_k$ is denoted by $S\mathcal{H}_k(H_{sub}, H)$.

For a graph $H \in \mathcal{H}_k$ and a host graph G on at least k vertices, the number of sets of k vertices in G that induce a subgraph of G isomorphic to H is denoted by $NI(H, G)$.

For $H \in \mathcal{H}_k(l)$, the set $Eq(H, l)$ consists of the following equations in one-to-one correspondence with induced subgraphs H_{sub} of H on l vertices

$$\sum_{H' \in S\mathcal{H}_k(H_{sub}, H)} B(H_{sub}, H') x_{H', G} = \sum_{H' \in S\mathcal{H}_k(H_{sub}, H)} B(H_{sub}, H') NI(H', G),$$

where $H \setminus H_{sub}$ is an independent set in H, and $B(H_{sub}, H')$ are easily computable coefficients. For our purposes, we need to define the coefficients only when $H \in \mathcal{H}_k(k-2)$, $H \setminus H_{sub}$ consists of two independent vertices and $H' \in S\mathcal{H}_k(H_{sub}, H)$. Then, the coefficient $B(H_{sub}, H')$ is just the number of automorphisms of H' divided by the number of automorphisms of H' that are identity on H_{sub} by Lemma 3.6 in [11]. By Lemma 3.5 in [11], for $H \in \mathcal{H}_k(l)$, the right-hand side of an equation in $Eq(H, l)$ can be evaluated in time $O(n^l(k - l) + T_l(n))$, where $T_l(n)$ stands for the time required to solve the so called l-neighborhood problem. By Theorem 6.1 in [11], $T_l(n) = O(n^{\omega(\lceil (k-l)/2 \rceil, 1, \lfloor (k-l)/2 \rfloor)})$.

By $SEq(G, k, l)$, we shall denote the system of equations obtained by picking, for each H in $\mathcal{H}_k(l)$, an arbitrary equation from $Eq(H, l)$. By Lemma 3.7 in [11], the resulting system of $|\mathcal{H}_k(l)|$ equations is linearly independent.

4.1   Proof

Consider Theorem 4.1 in [11] with fixed $k = q$, $l = q - 2$, and $O(n^{\omega(\lceil (q-2)/2 \rceil, 1, \lfloor (q-2)/2 \rfloor)})$ substituted for $T_{q-2}(n)$ according to Theorem 6.1 in [11]. Then, the theorem states that if for all $H \in \mathcal{H}_{q-2} \setminus \mathcal{H}_{q-2}(q - 2)$ the values $NI(H, G)$ are known then for all $H' \in \mathcal{H}_{q-2}$, the numbers $NI(H', G)$ and $N(H', G)$ can be determined in time $O(n^{\omega(\lceil (q-2)/2 \rceil, 1, \lfloor (q-2)/2 \rfloor)})$. Since $\mathcal{H}_q \setminus \mathcal{H}_q(q - 2) = \{K_q\}$, it follows directly from Theorem 4.1 in [11] that if the number of (induced) subgraphs isomorphic to K_q in the host graph is known then for all the pattern graphs on q vertices the corresponding numbers can be computed in the time specified in the lemma statement. The argumentation given in the proof of Theorem 4.1 in [11] works equally well when the number of induced copies of an arbitrary pattern graph H on q vertices is known.

Namely, following the proof of Theorem 4.1 in [11] with $k = q$ and $l = q-2$, we form $SEq(G, q, q-2)$. Since we assume that q is fixed, the coefficients $B(H_{sub}, H')$ on the left-sides of the equations in $SEq(G, q, q - 2)$ can be computed in $O(1)$ time. By Lemma 3.5 and Theorem 6.1 in [11], the right-sides of the equations can be computed in time $O(n^{\omega(\lceil (q-2)/2 \rceil, 1, \lfloor (q-2)/2 \rfloor)})$. Let us the graphs in $\mathcal{H}_k$ so that the number of edges is non-decreasing and the graphs in $\mathcal{H}_k(l)$ form a prefix of the sorted sequence. Let B be the $|\mathcal{H}_k(l)| \times |\mathcal{H}_k|$ matrix corresponding to the left-hand sides of the equations in $SEq(G, q, q - 2)$, with the rows of B corresponding to $H \in \mathcal{H}_k(q - 2)$ and the columns of B corresponding to $H' \in \mathcal{H}_q$ sorted in the aforementioned way. Consider the leftmost maximal square submatrix M of the matrix B. Since M has zeros below the diagonal starting from the leftmost top-left corner, we infer that the resulting $|H_q(q-2)|$ equations with $|H_q|$ unknowns are also linearly independent. Hence, when we substitute the known number of induced copies of H for the variable $x_{H,G}$ corresponding to H in the equations, we obtain a system S of $|H_q(q - 2)|$ equations with $|H_q| - 1$ unknowns. Since $H_q \setminus H_q(q - 2) = \{K_q\}$, the number of unknowns is equal to the number of equations. It follows from Remark 1 that the system S of equations resulting from the substitution is independent. Also, none of the resulting equations can disappear after the substitution. Hence, we can solve them in $O(|H_q(q - 2)|^3) = O(1)$ time. $\qquad\square$

References

1. Alon, N., Dao, P., Hajirasouliha, I., Hormozdiari, F., Sahinalp, S.C.: Biomolecular network motif counting and discovery by color coding. Bioinformatics (ISMB 2008) **24**(13), 241–249 (2008)
2. Bläser, M., Komarath, B., Sreenivasaiah, K.: Graph Pattern Polynomials. CoRR.abs/1809.08858 (2018)
3. Corneil, D.G., Perl, Y., Stewart, L.K.: A linear recognition algorithm for cographs. SIAM J. Comput. **14**(4), 926–934 (1985)
4. Dalirrooyfard, M., Duong Vuong, T., Virginia Vassilevska Williams, V.: Graph pattern detection: hardness for all induced patterns and faster non-induced cycles. In: Proceedings of STOC 2019 (2019, to appear)
5. Eisenbrand, F., Grandoni, F.: On the complexity of fixed parameter clique and dominating set. Theor. Comput. Sci. **326**, 57–67 (2004)
6. Floderus, P., Kowaluk, M., Lingas, A., Lundell, E.-M.: Detecting and counting small pattern graphs. SIAM J. Discrete Math. **29**(3), 1322–1339 (2015)
7. Floderus, P., Kowaluk, M., Lingas, A., Lundell, E.-M.: Induced subgraph isomorphism: are some patterns substantially easier than others? Theor. Comput. Sci. **605**, 119–128 (2015)
8. Gąsieniec, L., Kowaluk, M., Lingas, A.: Faster multi-witnesses for Boolean matrix product. Inf. Process. Lett. **109**, 242–247 (2009)
9. Itai, A., Rodeh, M.: Finding a minimum circuit in a graph. SIAM J. Comput. **7**, 413–423 (1978)

10. Kloks, T., Kratsch, D., Müller, H.: Finding and counting small induced subgraphs efficiently. Inf. Process. Lett. **74**(3–4), 115–121 (2000)
11. Kowaluk, M., Lingas, A., Lundell, E.-M.: Counting and detecting small subgraphs via equations and matrix multiplication. SIAM J. Discrete Math. **27**(2), 892–909 (2013)
12. Kowaluk, M., Lingas, A.: A fast deterministic detection of small pattern graphs in graphs without large cliques. In: Poon, S.-H., Rahman, M.S., Yen, H.-C. (eds.) WALCOM 2017. LNCS, vol. 10167, pp. 217–227. Springer, Cham (2017). https://doi.org/10.1007/978-3-319-53925-6_17
13. Le Gall, F.: Faster algorithms for rectangular matrix multiplication. In: Proceedings of 53rd Symposium on Foundations of Computer Science (FOCS), pp. 514–523 (2012)
14. Le Gall, F.: Powers of tensors and fast matrix multiplication. In: Proceedings of 39th International Symposium on Symbolic and Algebraic Computation, pp. 296–303 (2014)
15. Nešetřil, J., Poljak, S.: On the complexity of the subgraph problem. Comment. Math. Univ. Carol. **26**(2), 415–419 (1985)
16. Olariu, S.: Paw-free graphs. Inf. Process. Lett. **28**, 53–54 (1988)
17. Schank, T., Wagner, D.: Finding, counting and listing all triangles in large graphs, an experimental study. In: Proceedings of WEA, pp. 606–609 (2005)
18. Sekar, V., Xie, Y., Maltz, D.A., Reiter, M.K., Zhang, H.: Toward a framework for internet forensic analysis. In: Third Workshop on Hot Topics in Networking (HotNets-HI) (2004)
19. Wolinski, C., Kuchcinski, K., Raffin, E.: Automatic design of application-specific reconfigurable processor extensions with UPaK synthesis kernel. ACM Trans. Des. Autom. Electron. Syst. **15**(1), 1–36 (2009)
20. Vassilevska Williams, V., Wang, J.R., Williams, R., Yu H.: Finding four-node subgraphs in triangle time. In: Proceedings of SODA, pp. 1671–1680 (2015)
21. Vassilevska Williams, V.: Multiplying matrices faster than Coppersmith-Winograd. In: Proceedings of 44th Annual ACM Symposium on Theory of Computing (STOC), pp. 887–898 (2012)

Bivariate B-Splines from Convex Pseudo-circle Configurations

Dominique Schmitt[(✉)]

Institut IRIMAS, Université de Haute-Alsace, Mulhouse, France
Dominique.Schmitt@uha.fr

Abstract. An order-k univariate B-spline is a parametric curve defined over a set S of at least $k + 2$ real parameters, called knots. Such a B-spline can be obtained as a linear combination of basic B-splines, each of them being defined over a subset of $k + 2$ consecutive knots of S, called a configuration of S.

In the bivariate setting, knots are pairs of reals and basic B-splines are defined over configurations of $k + 3$ knots. Among these configurations, the Delaunay configurations introduced by Neamtu in 2001 gave rise to the first bivariate B-splines that retain the fundamental properties of univariate B-splines. An order-k Delaunay configuration is characterized by a circle that passes through three knots and contains k knots in its interior.

In order to construct a wider variety of bivariate B-splines satisfying the same fundamental properties, Liu and Snoeyink proposed, in 2007, an algorithm to generate configurations. Even if experimental results indicate that their algorithm generates indeed valid configurations, they only succeeded in proving it up to $k = 3$. Until now, no proof has been given for greater k.

In this paper we first show that, if we replace the circles in Neamtu's definition by maximal families of convex pseudo-circles, then we obtain configurations that satisfy the same fundamental properties as Delaunay configurations. We then prove that these configurations are precisely the ones generated by the algorithm of Liu and Snoeyink, establishing thereby the validity of their algorithm for all k.

Keywords: B-spline · Simplex spline · Convex pseudo-circles

1 Introduction

B-splines have been introduced by Schoenberg as extensions of the Bernstein polynomials that appear in Bézier curves. Both Bézier curves and B-splines have been extensively used in the context of curve modeling.

Given a set S of reals, called *knots* in spline theory, and an integer $k \geq 0$, an order-k (univariate) B-spline over S is a linear combination of order-k basic B-splines, each of them being defined over a subset of $k + 2$ consecutive knots of

© Springer Nature Switzerland AG 2019
L. A. Gąsieniec et al. (Eds.): FCT 2019, LNCS 11651, pp. 335–349, 2019.
https://doi.org/10.1007/978-3-030-25027-0_23

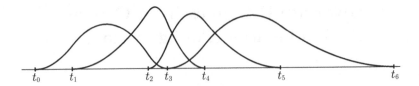

Fig. 1. Order-2 basic B-splines defined over subsets of 4 consecutive knots of the set $S = \{t_0, \ldots, t_6\}$.

S (see Fig. 1). The set of these basic B-splines spans an order-k B-spline space defined over S.

In order to define multivariate B-splines over a set S of knots in $\mathbb{R}^d$, notably if one wants to represent d-variate surfaces, one needs to extend both the definition of basic B-spline and the notion of "subset of consecutive knots of S", also called configurations of S. Different such generalizations have been proposed but Neamtu observed in 2001 that none preserves all fundamental properties of univariate B-splines [16]. One of these properties, called the *polynomial reproduction property*, states that the univariate B-spline space contains all degree-k polynomials. A spline space that does not satisfy this property cannot have optimal approximation properties [6,9]. Therefore, Neamtu proposed a new extension of univariate B-splines that satisfies this fundamental property [17]. First, basic B-splines are generalized using the simplex splines introduced by de Boor [5], where an order-k simplex spline is defined over a subset of $k + d + 1$ knots. Second, the configurations of $k + d + 1$ knots of S that are selected, are those for which there exists a sphere passing through $d + 1$ of the knots, the other k knots being inside the sphere, and the remaining knots of S being outside the sphere. Applications using these configurations can be found for example in [7,12].

In this paper, we propose a more general method to select configurations in the case $d = 2$, in that we replace circles by maximal families of convex pseudo-circles. Recall that a family of convex pseudo-circles is a set of convex Jordan curves that pairwise intersect at most twice. We show that any set of configurations defined by such a family, and which is maximal for inclusion, generates the basis of a bivariate spline space that satisfies the polynomial reproduction property.

In 2007, Liu and Snoeyink already pointed out that the generalization of Neamtu, while elegant, is restrictive in the types of splines that can be generated [14]. They proposed an algorithmic method to generate more general configurations in the case $d = 2$. They proved that their algorithm effectively constructs valid configurations up to $k = 3$. Even if experimental results indicate that the algorithm always works, this has never been proven in the general case (see in [11] a proof for some particular cases). Nevertheless, the configurations constructed by their algorithm appear to be efficient to represent surfaces with sharp features [13,20].

In this paper we prove that the configurations constructed by the algorithm of Liu and Snoeyink are precisely the ones defined here with maximal families of convex pseudo-circles. This proves that the algorithm always works.

2 B-Splines and Configurations

2.1 Notations

For a subset E of the plane $\mathbb{R}^2$, conv(E) denotes its convex hull, ∂E its boundary, $\overline{E}$ its closure, and relint(E) its relative interior, i.e., its interior in the subspace of $\mathbb{R}^2$ spanned by E.

If a and b are two distinct points in the plane, (ab) is the oriented straight line from a to b, $[a, b]$ is the closed segment connecting a and b, and $]a, b[$ is the open segment. For any oriented straight line ℓ, ℓ^+ and ℓ^- denote the open half planes respectively on the left and on the right of ℓ.

For a Jordan curve γ, disk(γ) denotes the bounded open component of $\mathbb{R}^2 \backslash \gamma$.

Unless otherwise stated, we consider S to be a finite set of $n > 2$ points in the plane, no three of them being collinear.

2.2 Basic B-Splines and Simplex Splines

Given a non-negative integer k and a sequence $t_0 < t_1 < t_2 \ldots < t_{k+1}$ of $k + 2$ reals, called knots, the order-k B-spline over $t_0, \ldots, t_{k+1}$ can be defined for any $x \in \mathbb{R}$ using the following variant of the Cox-de Boor recurrence formula. When $k = 0$, let

$$B(x|t_0, t_1) = \begin{cases} \frac{1}{t_1 - t_0} & \text{if } t_0 \le x < t_1 \\ 0 & \text{otherwise.} \end{cases}$$

When $k > 0$, let $\lambda_x = \frac{x - t_0}{t_{k+1} - t_0}$ and $1 - \lambda_x = \frac{t_{k+1} - x}{t_{k+1} - t_0}$ be the barycentric coordinates of x with respect to t_0 and t_{k+1}, and let

$$B(x|t_0, \ldots, t_{k+1}) = \lambda_x B(x|t_0, \ldots, t_k) + (1 - \lambda_x) B(x|t_1, \ldots, t_{k+1}). \tag{1}$$

This definition extends to higher dimension using the recurrence formula of Micchelli [15]. Let k be a non-negative integer and let T be a set of $k+3$ knots in the plane $\mathbb{R}^2$, not all collinear. For any point $x \in \mathbb{R}^2$, let $\{\lambda_{t,x} : t \in T\}$ be a set of reals such that $\sum_{t \in T} \lambda_{t,x} t = x$ and $\sum_{t \in T} \lambda_{t,x} = 1$, i.e., the $\lambda_{t,x}$ are barycentric coordinates of x with respect to the knots $t \in T$. Denoting by area$(\text{conv}(T))$ the area of the convex hull of T, the order-k bivariate simplex spline over T satisfies the following relations at every point $x \in \mathbb{R}^2$:

$$M(x|T) = \begin{cases} \frac{1}{\text{area}(\text{conv}(T))} & \text{if } x \in \text{conv}(T) \\ 0 & \text{otherwise} \end{cases} \qquad \text{when } |T| = 3$$

$$M(x|T) = \sum_{t \in T} \lambda_{t,x} M(x|T \setminus \{t\}) \qquad \text{when } |T| > 3.$$

Since this definition leaves some freedom in the choice of the reals $\lambda_{t,x}$ when $|T| > 3$, we can select any subset Q of three non-collinear knots in T and set $\lambda_{t,x}$ to zero for all knots t in $P = T \setminus Q$. The non-zero $\lambda_{t,x}$ are then the unique barycentric coordinates of x with respect to Q satisfying $\sum_{q \in Q} \lambda_{q,x} q = x$ and $\sum_{q \in Q} \lambda_{q,x} = 1$. Hence, for $|T| > 3$,

$$M(x|T) = M(x|P \cup Q) = \sum_{q \in Q} \lambda_{q,x} M(x|P \cup (Q \setminus \{q\})). \tag{2}$$

2.3 Valid Configurations

Given a set S of more than $k+3$ points in the plane, an order-k bivariate B-spline over S is a linear combination of simplex splines defined over subsets of $k + 3$ points of S. Thus, we need a method to select subsets T of $k + 3$ points of S, such that the simplex splines defined over these subsets span a bivariate spline space over S. From (2), this comes actually to select pairs (P, Q) of subsets of S such that $P \cap Q = \emptyset$, $|Q| = 3$, and $|P| = k$. Such a pair (P, Q) is called an *order-k configuration* of S or, for short, a k-configuration of S. Furthermore, if the circle γ circumscribed to Q is such that $\gamma \cap S = Q$ and disk$(\gamma) \cap S = P$, (P, Q) is also called a *Delaunay k-configuration* of S.

A family of k-configurations of S is said to be *valid*, if the spline space that it spans satisfies the polynomial reproduction property, i.e., the spline space contains all degree-k polynomials. In [17], Neamtu proved that the family of Delaunay k-configurations is valid. To avoid technical details, he proved this result in the case where the set S is infinite and locally finite, i.e., when conv$(S) = \mathbb{R}^2$ and the intersection of S with any ball of $\mathbb{R}^2$ is finite. Furthermore, S was supposed to be in general position to ensure that no three points in S are collinear and no four points are cocircular. In order to show that every polynomial can be written as a linear combination of simplex splines whose knot-sets are Delaunay configurations, Neamtu used the polar form of the polynomial. Given a degree k polynomial function f of a variable $x \in \mathbb{R}^2$, the polar form of f is the unique function F of k variables $x_1, \ldots, x_k \in \mathbb{R}^2$ that satisfies the following properties:

- F is symmetric, i.e., if σ is a permutation of $(x_1, \ldots, x_k)$, $F(x_1, \ldots, x_k) = F(\sigma(x_1, \ldots, x_k))$,
- F is multi-affine, i.e., $\forall i \in \{1, \ldots, k\}$, F is linear in x_i when $x_1, \ldots, x_{i-1}$, $x_{i+1}, \ldots, x_k$ are fixed,
- $F(x_1, \ldots, x_k) = f(x)$ when $x_1 = x_2 = \ldots = x_k = x$.

For example, the polar form of $f(x) = ax^2 + bx + c$ is $F(x_1, x_2) = ax_1 x_2 + \frac{bx_1}{2} + \frac{bx_2}{2} + c$. Theorem 1 states Neamtu's result.

Theorem 1. *Let Δ_k be the family of Delaunay k-configurations of a set S of points in $\mathbb{R}^2$ that is infinite, locally finite, and in general position. For every degree k polynomial f with polar form F,*

$$f(x) = \sum_{(P,Q) \in \Delta_k} F(P) \operatorname{area}(\operatorname{conv}(Q)) M(x|P \cup Q), \quad x \in \mathbb{R}^2.$$

In [14], Liu and Snoeyink observed that the proof of this theorem uses only two properties of Delaunay configurations.

The first one, is that the family Δ_0 of Delaunay 0-configurations induces a triangulation of S, in the sense that the triangles $\{\text{conv}(Q) : (\emptyset, Q) \in \Delta_0\}$ form a triangulation of S. In fact, it is the well known Delaunay triangulation of S.

The second property used, is the *edge matching property*. The *edges of a configuration* (P, Q) are all the pairs $(P, Q \setminus \{r\})$ and $(P \cup \{r\}, Q \setminus \{r\})$ with $r \in Q$. The edge matching property states that:

- Every edge $(P, \{s, t\})$ of a Delaunay i-configuration, $i \in \{0, \ldots, k\}$, is the common edge of precisely two Delaunay configurations. These configurations are of the form $(P \setminus \{r\}, \{r, s, t\})$ and $(P \setminus \{r'\}, \{r', s, t\})$, with r, r' two distinct points of $S \setminus \{s, t\}$. Each configuration is either of order $|P|$ or of order $|P| - 1$, depending on whether r, r' belong to P or not.
- The two configurations are of distinct orders if and only if r and r' are on the same side of (st).

Since the next section will be based on enumeration results stated in [8] for finite point sets, we first need to check how the above results apply when S is finite. First, the property that Delaunay 0-configurations induce a triangulation of S obviously holds when S is finite. Secondly, the edge matching property holds for Delaunay configuration edges that are not too close to the boundary of $\text{conv}(S)$. More precisely, one can check that the edges that do not satisfy the property are those of the form $(P, \{s, t\})$ with P and $S \setminus P$ on both sides of (st). When $|P| \in \{0, \ldots, k\}$, these are the edges $(P, \{s, t\})$ such that $|(st)^- \cap S| \leq k$ (or, symmetrically, $|(st)^+ \cap S| \leq k$, since $\{s, t\}$ is not ordered). When these edges are the only edges of a family of configurations where the general edge matching property is not satisfied, we say that the considered family of configurations verifies the *finite edge matching property*.

If a family of configurations satisfies the finite edge matching property, it is not hard to prove that the relation given by Theorem 1 still holds, provided that the domain of definition of f is restricted to the *depth-k region* of S. The depth-k region of S is the intersection of all half-planes $(st)^+$, $s, t \in S$, such that $|(st)^- \cap S| \leq k$. Properties of such regions can be found, for example, in [1]. Hence, we can restate the result of Liu and Snoeyink in the finite case.

Theorem 2. *Let S be a finite set of points in the plane, no three of them being collinear. Let $\{\Delta_0, \ldots, \Delta_k\}$ be a set of families of configurations of S of order $0, \ldots, k$ that satisfies the finite edge matching property and is such that Δ_0 induces a triangulation of S. For every degree k polynomial f with polar form F and for every point x in the depth-k region of S, we have*

$$f(x) = \sum_{(P,Q) \in \Delta_k} F(P) \, \text{area}(\text{conv}(Q)) \, M(x | P \cup Q).$$

We will say that a family of k-configurations of S is valid (in the finite setting) when the spline space that it spans verifies the polynomial reproduction property over the subset of the plane restricted to the depth-k region of S.

From Theorem 2, a family Δ_k of k-configurations of S is valid if there exist families of configurations $\Delta_0, \ldots, \Delta_{k-1}$ of order $0, \ldots, k-1$ such that Δ_0 induces a triangulation of S and the set $\{\Delta_0, \ldots, \Delta_k\}$ satisfies the finite edge matching property.

3   Convex Pseudo-circle Configurations

Let us come back to the recurrence relations (1) and (2), which respectively define univariate basic B-splines and bivariate simplex splines. In (1), the knots $t_0, \ldots, t_{k+1}$ are supposed to be consecutive in the whole sequence of knots, i.e., they form an interval that contains no other knot. Furthermore, t_0 and t_{k+1} are the endpoints of this interval. Thus, relation (2) becomes a natural generalization of (1) if we chose for $T = P \cup Q$ a subset of points of S whose convex hull contains no other point of S, and such that the points of Q are extreme points of T.

Let us introduce the following more general definitions.

Definition 1. *(i) A subset T of S is called a convex subset of S if $\mathrm{conv}(T) \cap S = T$.*

(ii) An ordered pair (P, Q) of subsets of S is called a convex pair of S, if P and Q are disjoint, $\mathrm{conv}(P \cup Q) \cap S = P \cup Q$, and Q is a subset of the extreme points of $P \cup Q$.

(iii) When $|Q| = 3$ and $|P| = k$, the convex pair (P, Q) is also called a convex k-configuration of S.

As defined in the previous section, Delaunay configurations are particular convex configurations (P, Q), in which the circle γ circumscribed to Q satisfies $\mathrm{disk}(\gamma) \cap S = P$. Furthermore, the family of Delaunay k-configurations is valid. Our aim now is to find other sub-families of convex configurations that are valid.

Notice first that, for any convex pair (P, Q), there exists a convex Jordan curve γ such that $\gamma \cap S = Q$ and $\mathrm{disk}(\gamma) \cap S = P$. The curve γ is called a *separating curve* of (P, Q) in S. Conversely, every pair (P, Q) of subsets of S for which there exists such a convex Jordan curve, is a convex pair of S.

Convex subsets are particular cases of convex pairs: T is a convex subset if and only if $(T, \emptyset)$ is a convex pair. Furthermore, a convex subset T is strictly separable from $S \setminus T$ by a convex Jordan curve.

Definition 2. *(i) Two convex subsets T and T' of S are said to be compatible if $\mathrm{conv}(T \setminus T') \cap \mathrm{conv}(T' \setminus T) = \emptyset$.*

(ii) Two distinct convex pairs (P, Q) and (P', Q') of S are said to be compatible if $\mathrm{relint}(\mathrm{conv}((P \cup Q) \setminus P')) \cap \mathrm{relint}(\mathrm{conv}((P' \cup Q') \setminus P)) = \emptyset$.

The only common points of $(P \cup Q) \setminus P'$ and of $(P' \cup Q') \setminus P$ being the points of $Q \cap Q'$, the relative interiors in (ii) can be removed when $Q \cap Q' = \emptyset$. This notably implies that $(T, \emptyset)$ is compatible with $(T', \emptyset)$ if and only if T is compatible with T'. More generally, we will say that a convex subset T is compatible with a convex pair (P, Q) if the convex pair $(T, \emptyset)$ is compatible with (P, Q).

Because circles intersect in at most two points, it is easy to see that Delaunay configurations are compatible. Conversely, Proposition 1 below shows that compatibility between convex configurations can be stated in terms of convex pseudo-circle separation. Recall that a family $\mathcal{C}$ of convex Jordan curves is said to form a family of convex pseudo-circles, if any two curves in $\mathcal{C}$ either intersect properly at exactly two points, meet at exactly one point, or do not meet at all. Proposition 1 extends to convex pairs a result given in [8] for convex subsets (see also Fig. 2).

Proposition 1. *The elements of a family $\mathcal{G}$ of convex pairs of S are pairwise compatible if and only if there exists a family $\mathcal{C}$ of convex pseudo-circles such that every element in $\mathcal{G}$ admits a separating curve in $\mathcal{C}$.*

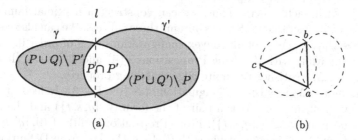

(a)   (b)

Fig. 2. (a) If two convex pairs (P, Q) and (P', Q') admit two separating curves γ and γ' that properly intersect in two points, then the subsets $(P \cup Q) \setminus P'$ and $(P' \cup Q') \setminus P$ are on both sides of (or on) the straight line l containing $\gamma \cap \gamma'$. The points of these subsets that are on l are the points of $Q \cap Q'$. The set $P \cap P'$ is in the open area $\mathrm{disk}(\gamma) \cap \mathrm{disk}(\gamma')$. (b) The relative interiors are necessary in Definition 2 (ii) to ensure that the pairs $(\emptyset, \{a, b, c\})$ and $(\emptyset, \{a, b\})$ are considered to be compatible.

The "only if part" of Proposition 1 is not trivial. It states that one only needs to consider the elements in $\mathcal{G}$ two by two, and to find two separating convex pseudo-circles for these two elements, to conclude that there exists one family $\mathcal{C}$ of convex pseudo-circles that are separating curves for all elements in $\mathcal{G}$.

Propositions 2 and 3 below come also from [8], and state inclusion relations between convex pairs.

Definition 3. *Let (P, Q) be a convex pair of S.*

(i) *Every subset T of S such that $P \subseteq T \subseteq P \cup Q$ is called a subset of (P, Q).*
(ii) *Every pair (P', Q') such that $P' \cap Q' = \emptyset$ and $P \subseteq P' \subseteq P' \cup Q' \subseteq P \cup Q$ is called a sub-pair of (P, Q).*

Clearly, T is a subset of (P, Q) if and only if $(T, \emptyset)$ is a sub-pair of (P, Q).

Proposition 2. (i) *If (P, Q) is a convex pair of S then every sub-pair (P', Q') of (P, Q) is also a convex pair of S.*

(ii) *Furthermore, if $|Q| \leq 3$ every sub-pair (P',Q') of (P,Q) distinct from (P,Q) is compatible with (P,Q) and with every convex pair that is compatible with (P,Q).*

(iii) *Two distinct convex pairs (P,Q) and (P',Q') are compatible if and only if the subsets of (P,Q) are compatible with the subsets of (P',Q').*

Proposition 3. *Given a family $\mathcal{F}$ of compatible convex subsets of S, for every convex pair $(P,\{s,t\})$ of S compatible with $\mathcal{F}$ such that $(st)^- \cap S \neq P$, there exists a point $r \in (P \cap (st)^+) \cup ((S \setminus P) \cap (st)^-)$ such that $(P \setminus \{r\}, \{r,s,t\})$ is a convex pair of S compatible with $\mathcal{F}$.*

This result is closely related to the finite edge matching property. Indeed, if Δ_k is a family of compatible convex k-configurations of S that is maximal for inclusion then, from Proposition 2, the subsets of the elements in Δ_k are compatible with each others. Thus, we can construct a maximal family $\mathcal{F}$ of compatible convex subsets of S that contains these subsets. We can also construct the families $\Delta_0, \ldots, \Delta_{k-1}$ of all convex configurations of S of order $0, \ldots, k-1$ that are compatible with $\mathcal{F}$. Using Propositions 2 (ii) and 3, we can then prove that $\{\Delta_0, \ldots, \Delta_k\}$ satisfies the finite edge matching property.

Furthermore, Δ_0 induces a triangulation of S (see Fig. 3). Indeed, from Definition 1, every element of Δ_0 is a pair of the form $(\emptyset, \{q,s,t\})$ and the triangle qst meets no point of $S \setminus \{q,s,t\}$. From Proposition 2 (iii), if $(\emptyset, \{q',s',t'\})$ is another pair in Δ_0, it is compatible with $(\emptyset, \{q,s,t\})$, i.e., from Definition 2, the triangles qst and $q's't'$ have disjoint interiors. Consider now an edge of such a triangle, for example the edge $[s,t]$ of rst. Without loss of generality, we can suppose that $q \in (st)^+$. From Proposition 2 (ii), $(\emptyset, \{s,t\})$ is a convex pair of S compatible with $\mathcal{F}$. From Proposition 3, if $(st)^- \cap S \neq \emptyset$, there exists $r \in (st)^-$ such that $(\emptyset, \{r,s,t\})$ is a convex 0-configuration of S compatible with $\mathcal{F}$. This means that, if $[s,t]$ is not an edge of $\text{conv}(S)$, it is a common edge of two triangles induced by Δ_0. By induction, the set of these triangles triangulates S.

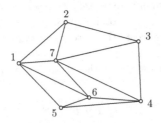

Fig. 3. The convex 3-subsets $\{1,7,2\}$, $\{1,6,7\}$, $\{1,5,6\}$, $\{2,7,3\}$, $\{3,7,4\}$, $\{4,7,6\}$, and $\{4,6,5\}$ of the point set $S = \{1, \ldots, 7\}$ are pairwise compatible. If they all belong to a same maximal family $\mathcal{F}$ of compatible convex subsets of S, then the family Δ_0 of convex 0-configurations of S compatible with $\mathcal{F}$ induces the represented triangulation, no matter which other convex subsets are in $\mathcal{F}$. Notice that by a result of [8], $\mathcal{F}$ contains precisely $\binom{7}{0} + \binom{7}{1} + \binom{7}{2} + \binom{7}{3} = 64$ elements.

The following theorem is an immediate consequence of the above results and of Theorem 2.

Theorem 3. *Every family Δ_k of compatible convex k-configurations of S that is maximal for inclusion is valid.*

4   Generation of Valid Configurations

Consider a family Δ_k of (arbitrary) k-configurations of S, and a $(k+1)$-subset T of S for which there exists a pair $(P, \{r, s, t\})$ in Δ_k with $P \cup \{r\} = T$. The subset T is called a *vertex* of $(P, \{r, s, t\})$, and is also called a vertex of Δ_k. The segment $[s, t]$ oriented such that $r \in (st)^+$ is called the *link* of T in the pair $(P, \{r, s, t\})$. The link of T in Δ_k is the set of links of T in all pairs of Δ_k from which T is a vertex.

Take, for example, a family Δ_0 of 0-configurations of S that induces a triangulation $\mathcal{T}$ of S. Every vertex t of Δ_0 is a vertex of $\mathcal{T}$. The link of t in Δ_0 is the polyline that links together, in counterclockwise order, the neighbors of t in $\mathcal{T}$. The polyline is open when t is a vertex of $\partial \mathrm{conv}(S)$, and is closed otherwise. In Fig. 3, the link of 7 is the oriented closed polyline $(1, 6, 4, 3, 2)$ and the link of 1 is the oriented open polyline $(5, 6, 7, 2)$. Figure 4 gives examples of links in a family of 1-configurations of the same point set.

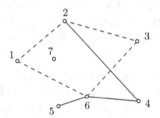

Fig. 4. One can check that the nine convex 1-configurations $(\{1\}, \{7, 5, 6\})$, $(\{2\}, \{1, 7, 3\})$, $(\{3\}, \{2, 7, 4\})$, $(\{4\}, \{7, 6, 3\})$, $(\{6\}, \{4, 7, 5\})$, $(\{6\}, \{1, 5, 7\})$, $(\{7\}, \{4, 3, 2\})$, $(\{7\}, \{4, 2, 1\})$, and $(\{7\}, \{4, 1, 6\})$ form a family Δ_1 of compatible convex 1-configurations of $S = \{1, \ldots, 7\}$. The dotted polygon is the link of $\{4, 7\}$ in Δ_1 and the full polyline oriented from 5 to 2 is the link of $\{1, 7\}$ in Δ_1.

Algorithm 1 below, due to Liu and Snoeyink [14], uses links to generate a family of $(k+1)$-configurations, starting from a family of k-configurations. The algorithm is described in the case of infinite and locally finite point sets. According to its authors, the algorithm is well-defined when every link $\mathcal{L}$, not reduced to a segment, forms a simple polygon. Furthermore, they proved that, if Δ_k is a *valid* family of k-configurations whose links satisfy this property, then the generated set Δ_{k+1} is also a *valid* family of $(k+1)$-configurations.

In order to apply Algorithm 1 to families of convex configurations of finite point sets, we need to characterize links in such families. For that, we first recall the definitions of a k-set polygon and of a centroid triangle.

Algorithm 1. To apply to a family Δ_k of k-configurations

---

initialize Δ_{k+1} to an empty set
for *every vertex T of Δ_k* **do**
    let $\mathcal{L}$ be the link of T in Δ_k
    if $\mathcal{L}$ *is not reduced to a segment (or to overlapping segments)* **then**
        compute a constrained triangulation of $\mathcal{L}$
        for *every triangle* $\mathrm{conv}(Q)$ *of this triangulation* **do**
            add the pair (T, Q) to Δ_{k+1}
return Δ_{k+1}

Definition 4. *(i) A subset T of k points of S is called a k-set of S if T can be strictly separated from $S \setminus T$ by a straight line.*
(ii) The k-set polygon of S, denoted by $\mathcal{Q}_k(S)$, is the convex hull of the centroids of all k-point subsets of S.

Notice that $\mathcal{Q}_1(S)$ is simply the convex hull of S and that the 1-sets of S are the vertices of this convex hull. The following characterization of the vertices and edges of $\mathcal{Q}_k(S)$ for all k is due to Andrzejak, Fukuda, and Welzl [2,3].

Proposition 4. *(i) If T is a k-set of S, then the centroid $g(T)$ of T is a vertex of $\mathcal{Q}_k(S)$. The centroid of any other k-point subset of S belongs to the interior of $\mathcal{Q}_k(S)$.*
(ii) Let T and T' be two k-sets of S. The centroid $g(T)$ is the predecessor of the centroid $g(T')$ in counterclockwise order on $\partial\mathcal{Q}_k(S)$ if and only if there exist $s, t \in S$ such that, setting $P = (st)^- \cap S$, one has $|P| = k - 1$, $T = P \cup \{s\}$, and $T' = P \cup \{t\}$.

Definition 5. *(i) For every convex configuration $(P, \{r, s, t\})$, the triangle whose vertices are the centroids $g(P \cup \{r\})$, $g(P \cup \{s\})$, $g(P \cup \{t\})$ is called the type-1 centroid triangle associated with $(P, \{r, s, t\})$. The triangle whose vertices are the centroids $g(P \cup \{r, s\})$, $g(P \cup \{s, t\})$, $g(P \cup \{r, t\})$ is called the type-2 centroid triangle associated with $(P, \{r, s, t\})$.*
(ii) For every convex pair $(P, \{s, t\})$ of S, the line segment $[g(P \cup \{s\}), g(P \cup \{t\})]$ is called the centroid segment associated with $(P, \{s, t\})$.

Centroid triangles have already been used in the literature with different kinds of configurations, as well in the plane [4,14,18], as in space [3,10,19]. In case of convex configurations, they satisfy the properties stated in Theorem 4, which were already proved in [8]. These properties generalize the fact that any maximal family of 0-configurations induces a triangulation of S.

Theorem 4. *Let $\mathcal{F}$ be a maximal family of compatible convex subsets of S and, for every $i \in \{0, \ldots, n - 3\}$, let Δ_i be the family of convex i-configurations of S compatible with $\mathcal{F}$. Set $\Delta_{-1} = \Delta_{n-2} = \emptyset$ and let $k \in \{0, \ldots, n - 2\}$.*

(i) The type-1 centroid triangles associated with the configurations in Δ_k and the type-2 centroid triangles associated with the configurations in Δ_{k-1} have pairwise disjoint interiors and form a triangulation $\mathcal{T}_{k+1}$ of the $(k + 1)$-set polygon of S (see Fig. 5).

(ii) *The centroids of the elements in $\mathcal{F}$ of size $k+1$ are pairwise distinct points and are the vertices of $\mathcal{T}_{k+1}$.*

(iii) *The open centroid segments associated with the convex pairs $(P, \{s,t\})$ of S compatible with $\mathcal{F}$ and such that $|P| = k$, are pairwise disjoint and are the edges of $\mathcal{T}_{k+1}$.*

The triangulation $\mathcal{T}_{k+1}$ defined by this theorem is called the order-$(k+1)$ centroid triangulation of S associated with $\mathcal{F}$.

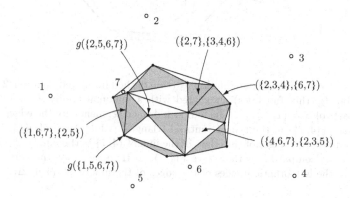

Fig. 5. A triangulation $\mathcal{T}_4$ of the 4-set polygon of the set $S = \{1, \ldots, 7\}$. The type-1 centroid triangles (in white) are associated with convex 3-configurations of S and the type-2 centroid triangles (in grey) are associated with convex 2-configurations of S. The type-2 centroid triangles form convex clusters called domains.

By definition, a $(k+1)$-point subset T of S is a vertex of a convex k-configuration $(P, \{r,s,t\})$ of S if, within a permutation of r, s, t, T is equal to $P \cup \{r\}$, i.e., $g(T)$ is a vertex of the type-1 centroid triangle associated with $(P, \{r,s,t\})$. Furthermore, the link $[s,t]$ of T in $(P, \{r,s,t\})$ is the image of the edge $[g(P \cup \{s\}), g(P \cup \{t\})]$ of this triangle by the homothety of center $g(P)$ and ratio $k+1$.

Let now Δ_k be a maximal family of compatible convex k-configurations of S, and let us characterize the links of the vertices of Δ_k. As in Sect. 3, we can construct a maximal family $\mathcal{F}$ of compatible convex subsets of S that contains all subsets of the configurations of Δ_k. Let $\mathcal{T}_{k+1}$ be the order-$(k+1)$ centroid triangulation of S associated with $\mathcal{F}$. From above, determining the link of a vertex T of Δ_k comes down to considering the set of type-1 centroid triangles that have $g(T)$ as a vertex in the triangulation $\mathcal{T}_{k+1}$, and to taking a homothetic image of the edge opposite to $g(T)$ in each triangle (see Fig. 6). To prove that the link of T forms a polyline, it suffices to prove that, for any two consecutive type-1 centroid triangles around $g(T)$, the two links determined by these triangles share a common endpoint. Now, when two type-1 centroid triangles are consecutive around $g(T)$, either they share a common edge, or they are separated from one

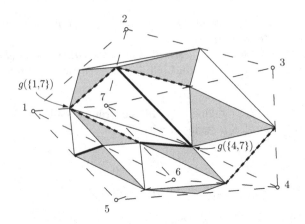

Fig. 6. An order-1 centroid triangulation $\mathcal{T}_1$ (thin dashed lines) and an order-2 centroid triangulation $\mathcal{T}_2$ (thin full lines) associated with a maximal family $\mathcal{F}$ of compatible convex subsets of $S = \{1, \ldots, 7\}$. The four sick dashed segments are the edges opposite to the vertex $g(\{4, 7\})$ in the type-1 centroid triangles (white) around $g(\{4, 7\})$ in $\mathcal{T}_2$. They are homothetic images of the edges of the link of $\{4, 7\}$ in the family Δ_1 of convex 1-configurations compatible with $\mathcal{F}$ (see Fig. 4). In the same way, the three sick full segments are the homothetic images of the edges of the link of $\{1, 7\}$ in Δ_1.

another by a sequence of type-2 centroid triangles with vertex $g(T)$. Let us give some properties of such clusters of type-2 centroid triangles.

Definition 6. *Given a $(k+2)$-subset T of S, consider the set composed of the type-2 centroid triangles in $\mathcal{T}_{k+1}$ and of the centroid segments in $\mathcal{T}_{k+1}$ that are all associated with convex pairs of S of the form (P, Q) with $P \cup Q = T$. If this set is nonempty, it is called the* domain *of T in $\mathcal{T}_{k+1}$.*

From the definition of centroid triangles and edges (Definition 5), it is clear that the edges of a type-2 centroid triangle of $\mathcal{T}_{k+1}$ belong to the same domain as the triangle. This implies that two type-2 centroid triangles of $\mathcal{T}_{k+1}$ that share a common edge belong also to the same domain. Hence, it is also the case for every cluster of type-2 centroid triangles around any vertex of $\mathcal{T}_{k+1}$. The following stronger property of domains has been given in [8].

Proposition 5. *Every domain in $\mathcal{T}_{k+1}$ is either reduced to a line segment or forms a triangulation of the vertices of a convex polygon.*

Using this property, we can now characterize the links in maximal families of convex configurations of a finite point set.

Theorem 5. *Let Δ_k be a maximal family of compatible convex k-configurations of S. Let T be a vertex of Δ_k and let $\mathcal{L}$ be the link of T.*

(i) *If T is not a $(k+1)$-set of S, either $\mathcal{L}$ is reduced to two overlapping line segments with opposite orientations, or $\mathcal{L}$ is a simple closed oriented polyline.*

(ii) If T is a $(k + 1)$-set of S, $\mathcal{L}$ is a simple open oriented polyline.

All that remains to be done to apply Algorithm 1 to convex configurations is to extend the notion of constrained triangulation to links that are not closed. Given a simple oriented polyline $\mathcal{L}$, closed or not, we call *constrained triangulation* of $\mathcal{L}$ any triangulation of the vertices of $\mathcal{L}$ that admits every edge of $\mathcal{L}$ as an edge, and that is restricted either to $disk(\mathcal{L})$ when $\mathcal{L}$ is closed, or to the subset of $conv(\mathcal{L})$ on the left side of $\mathcal{L}$ when $\mathcal{L}$ is open.

If $\mathcal{L}$ is open, any constrained triangulation of $\mathcal{L}$ is the union of constrained triangulations of simple closed polylines. The following lemma characterizes the triangles of such triangulations, in the case where $\mathcal{L}$ is a link in a family of convex configurations.

Lemma 1. *Let T be a nonempty subset of S. Let $\mathcal{L}$ be a simple closed polyline with at least three vertices, such that every vertex of $\mathcal{L}$ is in $S \setminus T$ and, for every edge $[u, v]$ of $\mathcal{L}$, $(T, \{u, v\})$ is a convex pair of S. Let r, s, t be three distinct vertices of $\mathcal{L}$ such that $conv(\{r, s, t\}) \subset \overline{disk(\mathcal{L})}$.*

(i) The pair $(T, \{r, s, t\})$ is a convex pair of S.
(ii) If (P, Q) is a convex pair of S such that $T \not\subseteq P$ and if (P, Q) is compatible with every convex pair $(T, \{u, v\})$ of S such that $[u, v]$ is an edge of $\mathcal{L}$, then (P, Q) is also compatible with $(T, \{r, s, t\})$.

Lemma 1 (i) shows that, when Algorithm 1 is applied to a family Δ_k of convex k-configurations, then the pairs (T, Q) constructed while triangulating the link of a vertex T of Δ_k, are convex configurations. Furthermore, from Definition 2, these configurations are pairwise compatible. Lemma 1 (ii) shows that these configurations are also compatible with the configurations of Δ_k. It also shows that, if T' is a vertex of Δ_k distinct from T, then the constructed pairs (T', Q) are compatible with (T, Q). Furthermore, it can be shown that the type-1 centroid triangles associated with Δ_k and the type-2 centroid triangles associated with Δ_{k+1} form a triangulation of the $(k + 2)$-set polygon of S. This implies that Δ_{k+1} is maximal and leads to the following theorem.

Theorem 6. *If Algorithm 1 is applied to a maximal family Δ_k of compatible convex k-configurations, then it generates a maximal family Δ_{k+1} of compatible convex $(k + 1)$-configurations.*

Liu and Snoeyink conjectured that, if their algorithm is first applied to any family Δ_0 of 0-configurations that induces a triangulation of S, and if the algorithm is then iteratively applied to the family of configurations generated by the algorithm itself, then all constructed configurations are valid. With the above results, we are now able to prove this conjecture.

Consider first any (classical) triangulation $\mathcal{T}$ of S. From Definition 1, for every triangle rst of $\mathcal{T}$, $(\emptyset, \{r, s, t\})$ is a convex 0-configuration of S. Let Δ_0 be the family of 0-configurations determined that way by $\mathcal{T}$. By Definition 5, rst is the type-1 centroid triangle associated with $(\emptyset, \{r, s, t\})$. If rst and $r's't'$ are two distinct triangles of $\mathcal{T}$, they have disjoint interiors and therefore, from Definition 2, the 0-configurations $(\emptyset, \{r, s, t\})$ and $(\emptyset, \{r', s', t'\})$ in Δ_0 are compatible. Since the type-1 centroid triangles associated with the configurations of Δ_0 form a triangulation of $\text{conv}(S)$, it follows from Theorem 4 (i), that Δ_0 is necessarily a maximal family of compatible convex 0-configurations.

Thus, from Theorem 6, if Algorithm 1 is first applied to Δ_0, and is then iteratively applied to the families Δ_k generated by the algorithm itself, Algorithm 1 constructs maximal families of compatible convex configurations. From Theorem 3, this proves that the configurations constructed by the algorithm are valid.

5 Conclusion

In this paper, we have shown that the algorithm of Liu and Snoeyink constructs maximal families of compatible convex k-configurations. Conversely, it can be proven that every maximal family of compatible convex k-configurations can be constructed by the algorithm. Thus, the algorithm can also be used to construct all centroid triangulations of a planar point set. These triangulations are interesting for themselves, and may serve as a useful tool in different contexts. They have already been used to count the number of subsets in a finite planar point set, that can be separated from the other points by a family of convex pseudo-circles [8]. It appears in this enumeration that, for a given point set S and a given k, all order-k centroid triangulations of S have the same number of type-1 centroid triangles. This implies that all maximal families of compatible convex k-configurations of S have the same number of elements. In particular, this number equals the number of Delaunay k-configurations. This shows that, if we replace Delaunay configurations by compatible convex configurations in the B-spline definition, we do not increase the size of the spline space basis.

If we want to extend the results of this paper in d-dimensional space, we can reuse the definition of convex configurations given by Definition 1 by just replacing $|Q| = 3$ by $|Q| = d + 1$. The definition of compatibility between convex configurations given by Definitions 2 can be used as such in every dimension. The definition of centroid triangles associated with convex configurations can also be easily extended. The real difficulty resides in the extension of the edge matching property and of the fact that the generalized centroid triangles cover the k-set polytope of S. For Delaunay configurations these results have already been proven respectively in [17] and in [19].

References

1. Agarwal, P.K., Sharir, M., Welzl, E.: Algorithms for center and Tverberg points. ACM Trans. Algorithms 5(1), 5:1–5:20 (2008)
2. Andrzejak, A., Fukuda, K.: Optimization over k-set polytopes and efficient k-set enumeration. In: Dehne, F., Sack, J.-R., Gupta, A., Tamassia, R. (eds.) WADS 1999. LNCS, vol. 1663, pp. 1–12. Springer, Heidelberg (1999). https://doi.org/10.1007/3-540-48447-7_1

3. Andrzejak, A., Welzl, E.: In between k-sets, j-facets, and i-faces: (i, j)-partitions. Discrete Comput. Geom. **29**, 105–131 (2003)
4. Aurenhammer, F., Schwarzkopf, O.: A simple on-line randomized incremental algorithm for computing higher order Voronoi diagrams. Internat. J. Comput. Geom. Appl. **2**, 363–381 (1992)
5. de Boor, C.: Splines as linear combinations of B-splines. a survey. In: Lorentz, G.G., Chui, C.K., Schumaker, L.L. (eds.) Approximation Theory II, pp. 1–47. Academic Press, New York (1976)
6. de Boor, C.: Topics in multivariate approximation theory. In: Turner, P.R. (ed.) Topics in Numerical Analysis. LNM, vol. 965, pp. 39–78. Springer, Heidelberg (1982). https://doi.org/10.1007/BFb0063200
7. Cao, J., Li, X., Wang, G., Qin, H.: Surface reconstruction using bivariate simplex splines on Delaunay configurations. Comput. Graph. **33**(3), 341–350 (2009)
8. Chevallier, N., Fruchard, A., Schmitt, D., Spehner, J.C.: Separation by convex pseudo-circles. In: proceedings of the 30th Annual Symposium on Computational Geometry, SOCG 2014, pp. 444–453. ACM, New York (2014)
9. Dahmen, W., Micchelli, C.A.: Recent progress in multivariate splines. In: Chui, C., Schumaker, L., Ward, J. (eds.) Approximation Theory IV, pp. 27–121. Academic Press, New York (1983)
10. Edelsbrunner, H., Nikitenko, A.: Poisson-Delaunay mosaics of order k. Discrete Comput. (2018). https://doi.org/10.1007/s00454-018-0049-2
11. El Oraiby, W., Schmitt, D., Spehner, J.C.: Centroid triangulations from k-sets. Internat. J. Comput. Geom. Appl. **21**(06), 635–659 (2011)
12. Hansen, M.S., Larsen, R., Glocker, B., Navab, N.: Adaptive parametrization of multivariate B-splines for image registration. In: 2008 IEEE Conference on Computer Vision and Pattern Recognition, pp. 1–8. IEEE (2008)
13. Liu, Y., Snoeyink, J.: Bivariate B-splines from centroid triangulations. In: Computational and Conformal Geometry (2007). http://www.math.sunysb.edu/~ccg2007/snoeyink.ppt
14. Liu, Y., Snoeyink, J.: Quadratic and cubic B-splines by generalizing higher-order Voronoi diagrams. In: Proceedings of ACM Symposium on Computational Geometry, pp. 150–157 (2007)
15. Micchelli, C.A.: A constructive approach to kergin interpolation in $\mathbb{R}^k$: multivariate B-splines and Lagrange interpolation. Rocky Mt. J. Math. **10**, 485–497 (1980)
16. Neamtu, M.: What is the natural generalization of univariate splines to higher dimensions? In: Mathematical Methods for Curves and Surfaces: Oslo 2000, pp. 355–392 (2001)
17. Neamtu, M.: Delaunay configurations and multivariate splines: a generalization of a result of B. N. Delaunay. Trans. Amer. Soc. **359**, 2993–3004 (2007)
18. Schmitt, D., Spehner, J.C.: On Delaunay and Voronoi diagrams of order k in the plane. In: Proceedings of 3rd Canadian Conference on Computational Geometry, pp. 29–32 (1991)
19. Schmitt, D., Spehner, J.C.: k-set polytopes and order-k Delaunay diagrams. In: Proceedings of the 3rd International Symposium on Voronoi Diagrams in Science and Engineering, pp. 173–185 (2006)
20. Zhang, Y., Cao, J., Chen, Z., Zeng, X.: Surface reconstruction using simplex splines on feature-sensitive configurations. Comput. Aided Geom. Design **50**, 14–28 (2017)

The Fault-Tolerant Metric Dimension
of Cographs

Duygu Vietz[(⊠)] and Egon Wanke

Heinrich-Heine-University Duesseldorf,
Universitaetsstr. 1, 40225 Duesseldorf, Germany
duygu.vietz@hhu.de

Abstract. A vertex set $U \subseteq V$ of an undirected graph $G = (V, E)$ is
a *resolving set* for G if for every two distinct vertices $u, v \in V$ there
is a vertex $w \in U$ such that the distance between u and w and the
distance between v and w are different. A resolving set U is *fault-tolerant*
if for every vertex $u \in U$ set $U \setminus \{u\}$ is still a resolving set. The *(fault-
tolerant) metric dimension* of G is the size of a smallest (fault-tolerant)
resolving set for G. The *weighted (fault-tolerant) metric dimension* for a
given cost function $c : V \longrightarrow \mathbb{R}_+$ is the minimum weight of all (fault-
tolerant) resolving sets. Deciding whether a given graph G has (fault-
tolerant) metric dimension at most k for some integer k is known to
be NP-complete. The weighted fault-tolerant metric dimension problem
has not been studied extensively so far. In this paper we show that the
weighted fault-tolerant metric dimension problem can be solved in linear
time on cographs.

Keywords: Graph algorithm · Complexity · Metric dimension ·
Fault-tolerant metric dimension · Resolving set · Cograph

1   Introduction

An undirected graph $G = (V, E)$ has *metric dimension at most k* if there is a
vertex set $U \subseteq V$ such that $|U| \leq k$ and $\forall u, v \in V$, $u \neq v$, there is a vertex $w \in U$
such that $d_G(w, u) \neq d_G(w, v)$, where $d_G(u, v)$ is the distance (the length of a
shortest path in an unweighted graph) between u and v. We call U a *resolving
set*. Graph G has *fault-tolerant metric dimension at most k* if for a resolving set
U with $|U| \leq k$ it holds that for every $u \in U$ set $U \setminus \{u\}$ is a resolving set for
G. The metric dimension of G is the smallest integer k such that G has metric
dimension at most k and the fault-tolerant metric dimension of G is the smallest
integer k such that G has fault-tolerant metric dimension at most k. The metric
dimension was independently introduced by Harary, Melter [13] and Slater [29].

If for three vertices $u, v \in V$, $w \in U$, we have $d_G(w, u) \neq d_G(w, v)$, then we
say that u and v are *resolved* by vertex w. The *metric dimension* of G is the size
of a minimum resolving set and the *fault-tolerant metric dimension* is the size
of a minimum fault-tolerant resolving set. In certain applications, the vertices of

© Springer Nature Switzerland AG 2019
L. A. Gąsieniec et al. (Eds.): FCT 2019, LNCS 11651, pp. 350–364, 2019.
https://doi.org/10.1007/978-3-030-25027-0_24

a (fault-tolerant) resolving set are also called *resolving vertices, landmark nodes* or *anchor nodes*. This is a common naming particularly in the theory of sensor networks.

Determining the metric dimension of a graph is a problem that has an impact on multiple research fields such as chemistry [3], robotics [22], combinatorial optimization [27] and sensor networks [18]. Deciding whether a given graph G has metric dimension at most k for a given integer k is known to be NP-complete for general graphs [12], planar graphs [6], even for those with maximum degree 6 and Gabriel unit disk graphs [18]. Epstein et al. showed the NP-completeness for split graphs, bipartite graphs, co-bipartite graphs and line graphs of bipartite graphs [7] and Foucaud et al. for permutation and interval graphs [10,11].

There are several algorithms for computing the metric dimension in polynomial time for special classes of graphs, as for example for trees [3,22], wheels [17], grid graphs [23], k-regular bipartite graphs [26], amalgamation of cycles [20], outerplanar graphs [6], cactus block graphs [19], chain graphs [9], graphs with a bounded number of resolving vertices in every EBC [31]. The approximability of the metric dimension has been studied for bounded degree, dense, and general graphs in [15]. Upper and lower bounds on the metric dimension are considered in [2,4] for further classes of graphs.

There are many variants of the metric dimension problem. The weighted version was introduced by Epstein et al. in [7], where they gave a polynomial-time algorithms on paths, trees and cographs. Oellermann et al. investigated the strong metric dimension in [24] and Estrada-Moreno et al. the k-metric dimension in [8], which is the same concept as the fault-tolerant metric dimension for the case $k = 2$.

The parameterized complexity was investigated by Hartung and Nichterlein. They showed that for the standard parameter the problem is $W[2]$-complete on general graphs, even for those with maximum degree at most three [14]. Foucaud et al. showed that for interval graphs the problem is FPT for the standard parameter [10,11]. Afterwards Belmonte et al. extended this result to the class of graphs with bounded treelength, which is a superclass of interval graphs and also includes chordal, permutation and AT-free graphs [1].

The fault-tolerant metric dimension has not been studied as intensive as the metric dimension so far. Hernando et al. introduced this concept in [16] and characterized the fault-tolerant resolving sets in a tree and showed that for general graphs the fault-tolerant metric dimension is bounded by a function of the metric dimension. Chaudhry et al. investigated in [5] the fault-tolerant metric and partition dimension of a graph and characterized the graphs with fault-tolerant metric dimension n. Raza et al. studied in [25] the fault-tolerant metric dimension of convex polytopes.

In this paper we show that the weighted fault-tolerant metric dimension problem can be solved in linear time on cographs and give an algorithm that computes a minimum weight fault-tolerant resolving set.

2   Definitions and Basic Terminology

We consider *graphs* $G = (V, E)$, where V is the set of *vertices* and E is the set of edges. We distinguish between *undirected graphs* with edge sets $E \subseteq \{\{u, v\} \mid u, v \in V, u \neq v\}$ and *directed graphs* with edge sets $E \subseteq V \times V$. Graph $G' = (V', E')$ is a *subgraph* of $G = (V, E)$, if $V' \subseteq V$ and $E' \subseteq E$. It is an *induced subgraph* of G, denoted by $G|_{V'}$, if $E' = E \cap \{\{u, v\} \mid u, v \in V'\}$ or $E' = E \cap (V' \times V')$, respectively. Vertex $u \in V$ is called a *neighbour* of vertex $v \in V$, if $\{u, v\} \in E$ in an undirected graph or $(u, v) \in E$ $((v, u) \in E)$ in a directed graph. With $N(u) = \{v \mid \{u, v\} \in E\}$ we denote the *open neighbourhood* of a vertex u in an undirected graph and with $N[u] = N(u) \cup \{u\}$ we denote the *closed neighbourhood* of a vertex u.

A sequence of $k+1$ vertices $(u_1, \ldots, u_{k+1})$, $k \geq 0$, $u_i \in V$ for $i = 1, \ldots, k+1$, is an *undirected path of length* k, if $\{u_i, u_{i+1}\} \in E$ for $i = 1, \ldots, k$. The vertices u_1 and u_{k+1} are the *end vertices* of undirected path p. The sequence $(u_1, \ldots, u_{k+1})$ is a *directed path of length* k, if $(u_i, u_{i+1}) \in E$ for $i = 1, \ldots, k$. Vertex u_1 is the start vertex and vertex u_{k+1} is the end vertex of the directed path p. A path p is a *simple path* if all vertices are mutually distinct.

An undirected graph G is *connected* if there is a path between every pair of vertices. An undirected graph G is *disconnected* if it is not connected. A *connected component* of an undirected graph G is a connected induced subgraph $G' = (V', E')$ of G such that there is no connected induced subgraph $G'' = (V'', E'')$ of G with $V' \subseteq V''$ and $|V'| < |V''|$. A vertex $u \in V$ is a *separation vertex* of an undirected graph G if $G|_{V \setminus \{u\}}$ (the subgraph of G induced by $V \setminus \{u\}$) has more connected components than G. Two paths $p_1 = (u_1, \ldots, u_k)$ and $p_2 = (v_1, \ldots, v_l)$ are *vertex-disjoint* if $\{u_2, \ldots, u_{k-1}\} \cap \{v_2 \ldots, v_{l-1}\} = \emptyset$. A graph $G = (V, E)$ with at least three vertices is *biconnected*, if for every vertex pair $u, v \in V$, $u \neq v$, there are at least two vertex-disjoint paths between u and v. A *biconnected component* $G' = (V', E')$ of G is an induced biconnected subgraph of G such that there is no biconnected induced subgraph $G'' = (V'', E'')$ of G with $V' \subseteq V''$ and $|V'| < |V''|$. The *distance* $d_G(u, v)$ between two vertices u, v in a connected undirected graph G is the smallest integer k such that there is a path of length k between u and v. The *distance* $d_G(u, v)$ between two vertices u, v such that there is no path between u and v in G is ∞. The *complement* of an undirected graph $G = (V, E)$ is the graph $\bar{G} = (V, \{\{u, v\} \mid u, v \in V, \{u, v\} \notin E\})$.

Definition 1 (Cograph). *An undirected Graph G is a cograph, if*

- $G = (\{u\}, \emptyset)$ *or*
- $G = (V_1 \cup V_2, E_1 \cup E_2)$ *for two cographs* $G_1 = (V_1, E_1)$ *and* $G_2 = (V_2, E_2)$ *or*
- $G = \bar{H}$ *for a cograph H.*

A cograph contains no induced P_4, therefore the diameter of a connected cograph G is at most 2. That is, the distance between two arbitrary vertices u, v in G is either 0 or 1 or 2. The concept of cographs was introduced independently by Jung [21], Seinsche [28] and Sumner [30].

Definition 2 (Resolving set, metric dimension). *Let $G = (V, E)$ be an undirected graph and let $c : V \longrightarrow \mathbb{R}_+$ be a function that assigns to every vertex a non-negative weight. A vertex set $R \subseteq V$ is a* resolving set *for G if for every vertex pair $u, v \in V$, $u \neq v$, there is a vertex $w \in R$ such that $d_G(u, w) \neq d_G(v, w)$. A resolving set $R \subseteq V$ has weight $k \in \mathbb{N}$, if $\sum_{v \in R} c(v) = k$. The set R is a* minimum resolving set *for G, if there is no resolving set $R' \subseteq V$ for G with $|R'| < |R|$. The set R is a* minimum weight resolving set *for G, if there is no resolving set $R' \subseteq V$ for G with $\sum_{v \in R'} c(v) < \sum_{v \in R} c(v)$. An undirected graph $G = (V, E)$ has* metric dimension $k \in \mathbb{N}$, *if k is the smallest positive integer such that there is a resolving set for G of size k. An undirected graph $G = (V, E)$ has* weighted metric dimension $k \in \mathbb{N}$ *if k is the smallest positive integer such that there is a resolving set for G of weight k.*

Definition 3 (Fault-tolerant resolving set, fault-tolerant metric dimension). *Let $G = (V, E)$ be an undirected graph and let $c : V \longrightarrow \mathbb{R}_+$ be a function that assigns to every vertex a non-negative weight. A vertex set $R \subseteq V$ is a* fault-tolerant resolving set *for G if for an arbitrary vertex $r \in R$ set $R \setminus \{r\}$ is a resolving set. A fault-tolerant resolving set $R \subseteq V$ has weight $k \in \mathbb{N}$, if $\sum_{v \in R} c(v) = k$. The set R is a* minimum fault-tolerant resolving set *for G, if there is no fault-tolerant resolving set $R' \subseteq V$ for G with $|R'| < |R|$. The set R is a* minimum weight fault-tolerant resolving set *for G, if there is no fault-tolerant resolving set $R' \subseteq V$ for G with $\sum_{v \in R'} c(v) < \sum_{v \in R} c(v)$. An undirected graph $G = (V, E)$ has* fault-tolerant metric dimension $k \in \mathbb{N}$, *if k is the smallest positive integer such that there is a fault-tolerant resolving set for G of size k. An undirected graph $G = (V, E)$ has* weighted fault-tolerant metric dimension $k \in \mathbb{N}$, *if k is the smallest positive integer such that there is a fault-tolerant resolving set for G of weight k.*

Equivalent to this definition one can say that a vertex set is a fault-tolerant resolving set if for every vertex pair there are two resolving vertices. Obviously every fault-tolerant resolving set is also a resolving set.

The concept of fault-tolerance can be extended easily on an arbitrary number of vertices, what is called the k-metric dimension in [8], $k \in \mathbb{N}$. The k-metric dimension is the size of a smallest k-resolving set. A k-resolving set resolves every pair of vertices at least k times. For $k = 1$ a k-resolving set is a resolving set and for $k = 2$ a k-resolving set is a fault-tolerant resolving set. One should note that for all $k > 2$ there are graphs that does not have a k-resolving set (for example graphs with twin vertices), whereas for $k \leq 2$ the entire vertex set is a k-resolving set.

Definition 4. *Let $G = (V, E)$ be an undirected graph and $u, v \in V$, $u \neq v$. For two vertices $u, v \in V$ we call $N(u) \triangle N(v) = (N(u) \cup N(v)) \setminus (N(u) \cap N(v))$ the* symmetric difference *of u and v. For a set $R \subseteq V$, we define the function*

$$h_R : V \times V \longrightarrow \mathbb{N}, \qquad h_R(u, v) = |(N(u) \triangle N(v) \cup \{u, v\}) \cap R|$$

$h_R(u, v)$ is the number of vertices in R that are u or v or a neighbour of u, but not of v or a neighbour of v, but not of u.

Definition 5 (neighbourhood-resolving). *Let* $G = (V, E)$ *be an undirected graph and* $u, v \in V$, $u \neq v$, *and* $R \subseteq V$. *Set* R *is called* neighbourhood-resolving *for* G, *if for every pair* $u, v \in V$, $u \neq v$, *we have* $h_R(u, v) \geq 1$.

A set R is neighbourhood-resolving for G, if for every two vertices $u, v \notin R$ there is a vertex $w \in R$ that is neighbour of exactly one of the vertices u and v. If $u \in R$ or $v \in R$ the value $h_R(u, v)$ is always at least 1. Obviously, every set that is neighbourhood-resolving for G is also a resolving set for G.

Definition 6 (2-neighbourhood-resolving). *Let* $G = (V, E)$ *be an undirected graph and* $u, v \in V$, $u \neq v$, *and* $R \subseteq V$. *Set* R *is called* 2-neighbourhood-resolving *for* G *if for every pair* $u, v \in V$, $u \neq v$, *we have* $h_R(u, v) \geq 2$.

A set R is 2-neighbourhood-resolving for G if

- for two vertices $u, v \in V \setminus R$ there are at least two vertices in R that are neighbour of exactly one of the vertices u and v and
- for two vertices u, v such that $u \in R$ and $v \notin R$ there is at least one vertex in R that is neighbour of exactly one of the vertices u and v.

For $u, v \in R$ the value $h_R(u, v)$ is always at least two. Obviously, every set that is 2-neighbourhood-resolving for G is also a fault-tolerant resolving set for G.

Lemma 1. *Let* $G = (V, E)$ *be a connected cograph and* $R \subseteq V$. *Vertex set* R *is a fault-tolerant resolving set for* G *if and only if* R *is 2-neighbourhood-resolving for* G.

Proof. "$\Rightarrow$": Assume that R is a fault-tolerant resolving set for G. We have to show that R is 2-neighbourhood-resolving for G, so let $u, v \in V$ and $r_1, r_2 \in R$ be the vertices that resolve u and v.

1. If $u, v \in R$, then obviously $h_R(u, v) \geq 2$.
2. If $u \in R$ and $v \notin R$, then either $d_G(u, r_1) \neq 0$ or $d_G(u, r_2) \neq 0$. Without loss of generality let $d_G(u, r_1) \neq 0$. Vertex $v \notin R$, so $d_G(v, r_1) \neq 0$. Since vertex r_1 resolves u, v and G is a connected cograph (and therefore the diameter is at most 2), r_1 has to be adjacent to exactly one of the vertices u, v. Thus, $r_1 \in u \triangle v \cap R$ and $u \in \{u, v\} \cap R$ and therefore $h_R(u, v) \geq 2$.
3. If $u, v \notin R$, then the distance between u and any vertex in R and the distance between v and any vertex in R is not 0. Since r_1 and r_2 resolve u and v both are adjacent to exactly one of the vertices u and v. Thus, $r_1, r_2 \in N(u) \triangle N(v)$ and therefore $h_R(u, v) \geq 2$.

"$\Leftarrow$": Assume that R is 2-neighbourhood-resolving for G. We have to show that R is a fault-tolerant resolving set for G. We do this by giving two resolving vertices for every vertex pair $u, v \in V$.

1. If $u, v \in R$, there are obviously two vertices in R, which resolve u and v.
2. If $u \in R$ and $v \in V \setminus R$, then u resolves u, v. Since $h_R(u, v) \geq 2$ and $|\{u, v\} \cap R| = 1$, we have $|N(u) \triangle N(v) \cap R| \geq 1$. Thus, there is a vertex $r \in R$, that is adjacent to exactly one of the vertices u, v, so r resolves u, v.
3. If $u, v \in V \setminus R$, then $|\{u, v\} \cap R| = 0$. Since $h_R(u, v) \geq 2$, it follows $|N(u) \triangle N(v) \cap R| \geq 2$. Thus, there are two vertices $r_1, r_2 \in R$, that are both adjacent to exactly one of the vertices u, v and so r_1, r_2 resolve u, v. □

Note that this equivalence does not apply to disconnected cographs, see Fig. 1.

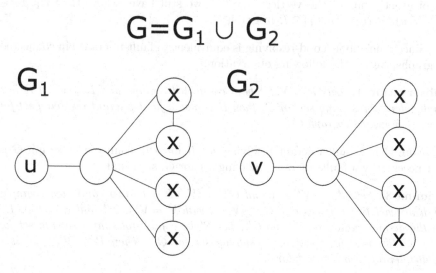

Fig. 1. The figure shows the disconnected cograph $G = G' \cup G''$, built by the union of the two connected cographs G' and G''. Let $R = R' \cup R''$ with $R' = \{r'_1, \ldots, r'_4\}$ and $R'' = \{r''_1, \ldots, r''_4\}$. R' is 2-neighbourhood-resolving and a fault-tolerant resolving set for G' and R'' is 2-neighbourhood-resolving and a fault-tolerant resolving set for G''. R is a fault-tolerant resolving set, but not 2NR for G, since $h_R(u', u'') = 0$. R is not a fault-tolerant resolving set for $\bar{G}$, since u' and u'' are neighbour of every resolving vertex in R in graph $\bar{G}$ and therefore cannot be resolved.

Thus, we state that 2-neighbourhood-resolving implies fault-tolerance in a cograph, fault-tolerance implies 2-neighbourhood-resolving in a connected cograph, but not in a disconnected cograph.

Lemma 2. *Let $G = (V, E)$ be a cograph and $R \subseteq V$. If R is 2-neighbourhood-resolving for G, then R is also 2-neighbourhood-resolving for $\bar{G}$.*

Proof. Let $R \subseteq V$ be 2-neighbourhood-resolving for G, i.e. for $u, v \in V$ we have $h_R(u, v) = |(N(u) \triangle N(v) \cup \{u, v\}) \cap R| \geq 2$. We distinguish between the following cases:

1. $u, v \in (N(u) \triangle N(v) \cup \{u, v\}) \cap R$: Obviously, $u, v \in (N(u) \triangle N(v) \cup \{u, v\}) \cap R$ in graph $\bar{G}$ and so $h_R(u, v) \geq 2$ in $\bar{G}$.

2. $u \in (N(u) \triangle N(v) \cup \{u, v\}) \cap R$ and $v \notin (N(u) \triangle N(v) \cup \{u, v\}) \cap R$: Since $h_R(u, v) \geq 2$ there has to be a vertex $w \in N(u) \triangle N(v) \cap R$, what implies that w is neighbour of either u or v. Without loss of generality let w be a neighbour of u. In graph $\bar{G}$ vertex w is not a neighbour of u, but a neighbour of v. So, we still have two vertices $u, w \in (N(u) \triangle N(v) \cup \{u, v\}) \cap R$ in graph $\bar{G}$.

3. $u, v \notin (N(u) \triangle N(v) \cup \{u, v\}) \cap R$: Since $h_R(u, v) \geq 2$ there has to be two vertices $w_1, w_2 \in N(u) \triangle N(v) \cap R$, what implies that both are neighbour of exactly one of the vertices u, v. Therefore, in graph $\bar{G}$ they are also neighbour of exactly one of the vertices u, v. So, we still have two vertices $w_1, w_2 \in (N(u) \triangle N(v) \cup \{u, v\}) \cap R$ in graph $\bar{G}$. $\qquad \square$

Since 2-neighbourhood-resolving is equivalent to fault-tolerance in connected cographs, we get the following observation:

Observation 1. *Let $G = (V, E)$ be a connected cograph and $R \subseteq V$. If R is a fault-tolerant resolving set for G, then R is also a fault-tolerant resolving set for the disconnected cograph $\bar{G}$.*

Note that a fault-tolerant resolving set R for a disconnected cograph G is not necessarily a fault-tolerant resolving set for $\bar{G}$, see Fig. 1.

Lemma 3. *Let $G' = (V', E')$ and $G'' = (V'', E'')$ be two connected cographs with at least two vertices and $G = (V, E)$ with $V = V' \cup V''$ and $E = E' \cup E''$ be the disjoint union of G' and G''. Let R' be a fault-tolerant resolving set for G' and R'' be a fault-tolerant resolving set for G''. Then $R = R' \cup R''$ is a fault-tolerant resolving set for G.*

Proof. We show that every pair $u, v \in V$ is resolved by two vertices in R. If $u, v \in V_1$ or $u, v \in V_2$ the pair is obviously resolved twice by vertices in $R_1 \subseteq R$ or $R_2 \subseteq R$. If $u \in V_1$ and $v \in V_2$ the pair is resolved by any two resolving vertices $r_1, r_2 \in R$, since either u or v will have distance ∞ to r_1 and r_2. $\qquad \square$

Note that R is not necessarily 2-neighbourhood-resolving for G (see Fig. 1).

Definition 7. *Let $G = (V, E)$ be a cograph and $R \subseteq V$ a fault-tolerant resolving set for G. A vertex $v \in V$ is called a k-vertex with respect to R, $k \in \mathbb{N}$, if $|N[v] \cap R| = k$.*

A vertex $v \in V$ is a k-vertex, if it has k vertices in its closed neighbourhood that are in R.

Lemma 4. *Let $G' = (V', E')$ and $G'' = (V'', E'')$ be two connected cographs and $G = (V, E)$ with $V = V' \cup V''$ and $E = E' \cup E''$ be the disjoint union of G' and G''. Let R' be 2-neighbourhood-resolving for G' and R'' be 2-neighbourhood-resolving for G''. Vertex set $R = R' \cup R''$ is 2-neighbourhood-resolving for G if and only if*

1. *there is at most one 0-vertex $v \in V$ with respect to R, i.e. there is no 0-vertex $v \in V'$ with respect to R' or there is no 0-vertex $v \in V''$ with respect to R'' and*
2. *there is no 0-vertex $v \in V'$ with respect to R', if there is a 1-vertex in V'' with respect to R'' and*
3. *there is no 1-vertex in V' with respect to R', if there is a 0-vertex in V'' with respect to R''.*

Proof. "$\Rightarrow$": Assume that R is 2-neighbourhood-resolving for G.

1. We show that there is at most one 0-vertex in V with respect to R. Assume there are two 0-vertices $u, v \in V$ with respect to R, i.e. $|N[u] \cap R| = 0$ and $|N[v] \cap R| = 0$. Then we have $h_R(u, v) = 0$, what contradicts the assumption that R is 2-neighbourhood-resolving.
2. We show that there is no 0-vertex in V' with respect to R' if there is a 1-vertex in V'' with respect to R''. Assume that there is a 0-vertex in $u \in V'$ with respect to R' and a 1-vertex in $v \in V''$ with respect to R''. Then we have $h_R(u, v) = 1$, what contradicts the assumption that R is 2-neighbourhood-resolving.
3. analogous to 2.

"$\Leftarrow$": Assume that the conditions 1., 2. and 3. hold. We show that R is 2-neighbourhood-resolving for G, i.e. for $u, v \in V$ we have $h_R(u, v) \geq 2$. For $u, v \in V'$ we have $h_{R'}(u, v) \geq 2$ and therefore also $h_R(u, v) \geq 2$. The same holds for $u, v \in V''$. Now let $u \in V'$ and $v \in V''$. $h_R(u, v) < 2$ if and only if $|N[u] \cap R| + |N[v] \cap R| < 2$, i.e. if

1. $|N[u] \cap R| = 0$ and $|N[v] \cap R| = 0$ or
2. $|N[u] \cap R| = 0$ and $|N[v] \cap R| = 1$ or
3. $|N[u] \cap R| = 1$ and $|N[v] \cap R| = 0$

Conditions 1. - 3. guarantee that none of these three cases appear. □

Theorem 2. *Let $G = (V, E)$ be a cograph. The weighted fault-tolerant metric dimension of G can be computed in linear time.*

Proof. We describe a linear time algorithm for computing the weighted fault-tolerant metric dimension of a connected cograph. For disconnected cographs we apply the algorithm for every connected component with at least two vertices. If there are isolated vertices, then each of them has to be in every weighted fault-tolerant resolving set, except for the case that there is exactly one isolated vertex. To get the weighted fault-tolerant metric dimension of the disconnected input graph, we build the sum of the weights of all isolated vertices if there are at least two, and the weighted fault-tolerant metric dimension for each connected component with at least two vertices.

To compute the weighted fault-tolerant metric dimension of a connected cograph $G = (V, E)$ it suffices to compute a set that is 2-neighbourhood-resolving for G and has minimal costs, since fault-tolerant resolving and 2-neighbourhood-resolving sets are equivalent in connected cographs (Lemma 1).

To compute a 2-neighbourhood-resolving set of minimum weight we use dynamic programming along the cotree $T = (V_T, E_T)$. The cotree T of G is a tree that describes the union and complementation of cographs. The inner nodes are either complementation-nodes or union-nodes. Every complementation-node has exactly one child and every union-node has exactly two children. The leaves of T are the vertices of G.

For every inner node of T we compute bottom up different types of minimum weight 2-neighbourhood-resolving sets for the corresponding subgraph of G. First we compute the 2-neighbourhood-resolving sets for the fathers of the leaves. For every other inner node $v \in V_T$ we compute the 2-neighbourhood-resolving sets from the 2-neighbourhood-resolving sets of all children of v. Finally, the minimum weight of all 2-neighbourhood-resolving sets at root r of T will be the minimum weight fault-tolerant metric dimension of G. From Lemma 2 we know that, if a set is 2-neighbourhood-resolving for a cograph G' then it is also 2-neighbourhood-resolving for $\bar{G}'$. The union of two fault-tolerant resolving sets is also a fault-tolerant resolving set (Lemma 3), but the union of two 2-neighbourhood-resolving sets is not necessarily a 2-neighbourhood-resolving set. We have to guarantee that the union of two 2-neighbourhood-resolving sets is also 2-neighbourhood-resolving, according to Lemma 4. For this, we have to keep track of the existence of 0- and 1-vertices in the 2-neighbourhood-resolving sets that we compute. Since a 0- or 1-vertex with respect to a set R becomes an $|R|$ or $(|R| - 1)$-vertex when complementing, we also have to keep track of $|R|$- and $(|R| - 1)$-vertices.

For a cograph $G = (V, E)$ we define 16 types of minimum weight 2-neighbourhood-resolving sets $R_{a,b,c,d}$, $a, b, c, d \in \{0, 1\}$.

For

- $a = 1$ we compute a minimum weight 2-neighbourhood-resolving set R for G such that there is a 0-vertex in G with respect to R and for $a = 0$ we compute a minimum weight 2-neighbourhood-resolving set for G such that there is no 0-vertex in G with respect to R.
- $b = 1$ we compute a minimum weight 2-neighbourhood-resolving set R for G such that there is a 1-vertex in G with respect to R and for $b = 0$ we compute a minimum weight 2-neighbourhood-resolving set for G such that there is no 1-vertex in G with respect to R.
- $c = 1$ we compute a minimum weight 2-neighbourhood-resolving set R for G such that there is a $(|R| - 1)$-vertex in G with respect to R and for $c = 0$ we compute a minimum weight 2-neighbourhood-resolving set for G such that there is no $(|R| - 1)$-vertex in G with respect to R.
- $d = 1$ we compute a minimum weight 2-neighbourhood-resolving set R for G such that there is a $|R|$-vertex in G with respect to R and for $d = 0$ we compute a minimum weight 2-neighbourhood-resolving set for G such that there is no $|R|$-vertex in G with respect to R.

Let $r_{a,b,c,d}$ be the weight of the corresponding minimum weight 2-neighbourhood-resolving sets $R_{a,b,c,d}$, i.e. the sum of the weights of all vertices in $R_{a,b,c,d}$. If there is no such 2-neighbourhood-resolving set for a certain a, b, c, d, we set $r_{a,b,c,d} = \infty$ and $R_{a,b,c,d} = undefined$.

Now we will analyze the 16 2-neighbourhood-resolving sets in more detail and describe, how they can be computed efficiently bottom up along the cotree. First one should note that $r_{1,1,c,d} = \infty$, $\forall c, d$, and $R_{1,1,c,d} = undefined$, since it is not possible to have a 0- and 1-vertex with respect to R in a 2-neighbourhood-resolving set (their symmetric difference would contain less than two resolving vertices), so it suffices to focus on the remaining 12 sets.

When complementing a graph G, the role of a 0-vertex and $|R|$-vertex with respect to R and the role of a 1-vertex and a $(|R| - 1)$-vertex with respect to R changes, that is $R_{a,b,c,d}$ for G is $R_{d,c,b,a}$ for $\bar{G}$. When unifying two cographs G_1 and G_2 we distinguish between the following three cases:

1. G_1 and G_2 both consist of a single vertex
2. G_1 consists of a single vertex and G_2 of at least two vertices
3. G_1 and G_2 both consist of at least 2 vertices

We will describe now how to compute $R_{a,b,c,d}$ for the three cases.

1. Let $G_1 = (\{v_1\}, \emptyset)$ and $G_2 = (\{v_2\}, \emptyset)$. Then there is exactly one valid 2-neighbourhood-resolving set for $G = G_1 \cup G_2$, namely $R = \{v_1, v_2\}$. In G we have no 0-vertex, two 1- and two $(|R| - 1)$-vertices and no $|R|$-vertex with respect to R. Therefore, $R_{0,1,1,0} = \{v_1, v_2\}$, $r_{0,1,1,0} = c(v_1) + c(v_2)$ and all other sets are infeasible, that is $r_{a,b,c,d} = \infty$ and $R_{a,b,c,d} = undefined$ for $a \neq 0 \lor b \neq 1 \lor c \neq 1 \lor d \neq 0$ (see Table 1).

Table 1. The table shows how $r_{a,b,c,d}$ is computed for $G = G_1 \cup G_2$.

	$G_1 = (\{v_1\}, \emptyset) \;\cup\; G_2 = (\{v_2\}, \emptyset)$
$r_{0,0,0,0}$	∞
$r_{0,0,0,1}$	∞
$r_{0,0,1,0}$	∞
$r_{0,0,1,1}$	∞
$r_{0,1,0,0}$	∞
$r_{0,1,0,1}$	∞
$r_{0,1,1,0}$	$c(v_1) + c(v_2)$
$r_{0,1,1,1}$	∞
$r_{1,0,0,0}$	∞
$r_{1,0,0,1}$	∞
$r_{1,0,1,0}$	∞
$r_{1,0,1,1}$	∞
$r_{1,1,0,0}$	∞
$r_{1,1,0,1}$	∞
$r_{1,1,1,0}$	∞
$r_{1,1,1,1}$	∞

2. Let $G_1 = (\{v_1\}, \emptyset)$ and $G_2 = (V_2, E_2)$ with $|V_2| \geq 2$. For some $a, b, c, d \in \{0, 1\}$ let $R''_{a,b,c,d}$ be the minimum weight 2-neighbourhood-resolving sets for G_2 and $r''_{a,b,c,d}$ be their weights. Let $G = G_1 \cup G_2$. $r_{0,0,c,d} = \infty$ and $R_{0,0,c,d} = undefined$, because vertex v_1 is either a 0-vertex (if it is not in the 2-neighbourhood-resolving set) or a 1-vertex (if it is in the 2-neighbourhood-resolving set) with respect to $R_{0,0,c,d}, \forall c, d$. $r_{0,1,c,1} = \infty$ and $R_{0,1,c,1} = undefined$, because it is crucial to put v_1 in the 2-neighbourhood-resolving set, if there should be no 0-vertex in G with respect to $R_{0,1,c,1}$, $\forall c$. If v_1 is in the 2-neighbourhood-resolving set, it is not possible to have a vertex that is neighbour of all resolving vertices, because v_1 has no neighbours. For $R_{0,1,0,0}$ and $R_{0,1,1,0}$ we have to put v_1 in the 2-neighbourhood-resolving set, so that there is no 0-vertex with respect to $R_{0,1,0,0}$ or $R_{0,1,1,0}$, what makes v_1 become a 1-vertex in G with respect to $R_{0,1,0,0}$ or $R_{0,1,1,0}$. We get $r_{0,1,0,0} = c(v_1) + \min\{r''_{0,0,0,0}, r''_{0,0,1,0}, r''_{0,1,0,0}, r''_{0,1,1,0}\}$ and thus $R_{0,1,0,0} = \{v_1\} \cup R_m$, whereas R_m is the set with the smallest weight out of $\{R''_{0,0,0,0}, R''_{0,0,1,0}, R''_{0,1,0,0}, R''_{0,1,1,0}\}$. For $R_{0,1,1,0}$ there has to be an $|R_{0,1,1,0}|$-vertex in G_2 with respect to $R_{0,1,1,0}$, so we get $r_{0,1,1,0} = c(v_1) + \min\{r''_{0,0,0,1}, r''_{0,0,1,1}, r''_{0,1,0,1}, r''_{0,1,1,1}\}$ and thus $R_{0,1,1,0} = \{v_1\} \cup R_m$, whereas R_m is the set with the smallest weight out of $\{R''_{0,0,0,1}, R''_{0,0,1,1}, R''_{0,1,0,1}, R''_{0,1,1,1}\}$. For $R_{1,0,c,d}$ it is not possible to put v_1 in the 2-neighbourhood-resolving set, because it would become a 1-vertex with respect to $R_{1,0,c,d}, \forall c, d$. Therefore, we get $r_{1,0,0,0} = r''_{0,0,0,0}$ and thus $R_{1,0,0,0} = R''_{0,0,0,0}$, $r_{1,0,0,1} = r''_{0,0,0,1}$ and thus $R_{1,0,0,1} = R''_{0,0,0,1}$, $r_{1,0,1,0} = r''_{0,0,1,0}$ and thus $R_{1,0,1,0} = R''_{0,0,1,0}$, $r_{1,0,1,1} = r''_{0,0,1,1}$ and thus $R_{1,0,1,1} = R''_{0,0,1,1}$ (see Table 2).

3. Let $G_1 = (V_1, E_1)$ and $G_2 = (V_2, E_2)$ with $|V_1| \geq 2$ and $|V_2| \geq 2$ and $G = G_1 \cup G_2$. For some $a, b, c, d \in \{0, 1\}$ let $R'_{a,b,c,d}$ be the minimum weight 2-neighbourhood-resolving sets for G_1 and $R''_{a,b,c,d}$ be the minimum weight 2-neighbourhood-resolving sets for G_2 and $r'_{a,b,c,d}$ and $r''_{a,b,c,d}$ be their weights. $r_{a,b,c,1} = \infty$ and $r_{a,b,1,d} = \infty$ and thus $R_{a,b,c,1} = undefined$ and $R_{a,b,1,d} = undefined$, $\forall a, b, c, d$, because G_1 and G_2 contain at least two resolving vertices in every 2-neighbourhood-resolving set. Therefore, it is not possible to have a vertex that is neighbour of all or of all except one of them. The three remaining sets are $R_{0,0,0,0}, R_{0,1,0,0}, R_{1,0,0,0}$. We get $r_{0,0,0,0} = \min\{r'_{0,0,c,d}|c, d \in \{0, 1\}\} + \min\{r''_{0,0,c,d}|c, d \in \{0, 1\}\}$ and thus $R_{0,0,0,0} = R'_m \cup R''_m$, whereas R'_m is the set with the smallest weight out of $\{R'_{0,0,c,d}|c, d \in \{0, 1\}\}$ and R''_m is the set with the smallest weight out of $\{R''_{0,0,c,d}|c, d \in \{0, 1\}\}$. We get $r_{0,1,0,0} = \min\{r'_{0,0,c,d} + r''_{0,1,c',d'}, r'_{0,1,c,d} + r''_{0,0,c',d'}|c, d, c', d' \in \{0, 1\}\}$ and thus $R_{0,1,0,0} = \min\{R'_{m_0} \cup R''_{m_1}, R'_{m_1} \cup R''_{m_0}, R'_{m_1} \cup R''_{m_1}\}$, whereas R'_{m_0} is the set with the smallest weight out of $\{R'_{0,0,c,d}|c, d \in \{0, 1\}\}$, R'_{m_1} is the set with the smallest weight out of $\{R'_{0,1,c,d}|c, d \in \{0, 1\}\}$, R''_{m_0} is the set with the smallest weight out of $\{R''_{0,0,c,d}|c, d \in \{0, 1\}\}$ and R''_{m_1} is the set with the smallest weight out of

Table 2. The table shows how $r_{a,b,c,d}$ is computed for $G = G_1 \cup G_2$.

	$G_1 = (\{v_1\}, \emptyset) \;\cup\; G_2 = (V_2, E_2)$
$r_{0,0,0,0}$	∞
$r_{0,0,0,1}$	∞
$r_{0,0,1,0}$	∞
$r_{0,0,1,1}$	∞
$r_{0,1,0,0}$	$c(v_1) + \min \begin{cases} r''_{0,0,0,0}, \\ r''_{0,0,1,0}, \\ r''_{0,1,0,0}, \\ r''_{0,1,1,0} \end{cases}$
$r_{0,1,0,1}$	∞
$r_{0,1,1,0}$	$c(v_1) + \min \begin{cases} r''_{0,0,0,1}, \\ r''_{0,0,1,1}, \\ r''_{0,1,0,1}, \\ r''_{0,1,1,1} \end{cases}$
$r_{0,1,1,1}$	∞
$r_{1,0,0,0}$	$r''_{0,0,0,0}$
$r_{1,0,0,1}$	$r''_{0,0,0,1}$
$r_{1,0,1,0}$	$r''_{0,0,1,0}$
$r_{1,0,1,1}$	$r''_{0,0,1,1}$
$r_{1,1,0,0}$	∞
$r_{1,1,0,1}$	∞
$r_{1,1,1,0}$	∞
$r_{1,1,1,1}$	∞

$\{R''_{0,1,c,d} | c, d \in \{0,1\}\}$. We get $r_{1,0,0,0} = \min\{r'_{1,0,c,d} + r''_{0,0,c',d'}, r'_{0,0,c,d} + r''_{1,0,c',d'} | c, d, c', d' \in \{0,1\}\}$ and thus $R_{1,0,0,0} = \min\{R'_{m_1} \cup R''_{m_0}, R'_{m_0} \cup R''_{m_1}\}$, whereas R'_{m_0} is the set with the smallest weight out of $\{R'_{0,0,c,d} | c, d \in \{0,1\}\}$, R'_{m_1} is the set with the smallest weight out of $\{R'_{1,0,c,d} | c, d \in \{0,1\}\}$, R''_{m_0} is the set with the smallest weight out of $\{R''_{0,0,c,d} | c, d \in \{0,1\}\}$ and R''_{m_1} is the set with the smallest weight out of $\{R''_{1,0,c,d} | c, d \in \{0,1\}\}$ (see Table 3).

For every node of the cotree T the computation of the 12 minimum weight 2-neighbourhood-resolving sets for the corresponding subgraph of G can be done in a constant number of steps. Since T has $\mathcal{O}(n)$ nodes, the overall runtime of our algorithm is linear to the size of the cotree. $\qquad\square$

Table 3. The table shows how $r_{a,b,c,d}$ is computed for $G = G_1 \cup G_2$.

	$G_1 = (\{v_1\}, \emptyset) \;\cup\; G_2 = (\{v_2\}, \emptyset)$
$r_{0,0,0,0}$	$\min\{r'_{0,0,c,d} \mid c,d \in \{0,1\}\} \;+\; \min\{r''_{0,0,c,d} \mid c,d \in \{0,1\}\}$
$r_{0,0,0,1}$	∞
$r_{0,0,1,0}$	∞
$r_{0,0,1,1}$	∞
$r_{0,1,0,0}$	$\min \begin{cases} r'_{0,0,c,d} + r''_{0,1,c',d'}, \\ r'_{0,1,c,d} + r''_{0,0,c',d'}, \\ r'_{0,1,c,d} + r''_{0,1,c',d'} \end{cases} \quad c,d,c',d' \in \{0,1\}$
$r_{0,1,0,1}$	∞
$r_{0,1,1,0}$	$c(v_1) + c(v_2)$
$r_{0,1,1,1}$	∞
$r_{1,0,0,0}$	$\min \begin{cases} r'_{1,0,c,d} + r''_{0,0,c',d'}, \\ r'_{0,0,c,d} + r''_{1,0,c',d'} \end{cases} \quad c,d,c',d' \in \{0,1\}$
$r_{1,0,0,1}$	∞
$r_{1,0,1,0}$	∞
$r_{1,0,1,1}$	∞
$r_{1,1,0,0}$	∞
$r_{1,1,0,1}$	∞
$r_{1,1,1,0}$	∞
$r_{1,1,1,1}$	∞

3 Conclusion

We showed that the weighted fault-tolerant metric dimension problem can be solved in linear time on cographs. Our algorithm computes the costs of a fault-tolerant resolving set with minimum weight as well as the set itself.

The complexity of computing the (weighted) fault-tolerant metric dimension is still unknown for many special graph classes like wheels and sun graphs. This is something that we will investigate in further work.

References

1. Belmonte, R., Fomin, F.V., Golovach, P.A., Ramanujan, M.: Metric dimension of bounded tree-length graphs. SIAM J. Discrete Math. **31**(2), 1217–1243 (2017)
2. Chappell, G., Gimbel, J., Hartman, C.: Bounds on the metric and partition dimensions of a graph. Ars Comb. **88**, 349–366 (2008)
3. Chartrand, G., Eroh, L., Johnson, M., Oellermann, O.: Resolvability in graphs and the metric dimension of a graph. Discrete Appl. Math. **105**(1–3), 99–113 (2000)
4. Chartrand, G., Poisson, C., Zhang, P.: Resolvability and the upper dimension of graphs. Comput. Math. Appl. **39**(12), 19–28 (2000)

5. Chaudhry, M.A., Javaid, I., Salman, M.: Fault-tolerant metric and partition dimension of graphs. Util. Math. **83**, 187–199 (2010)
6. Díaz, J., Pottonen, O., Serna, M., van Leeuwen, E.J.: On the complexity of metric dimension. In: Epstein, L., Ferragina, P. (eds.) ESA 2012. LNCS, vol. 7501, pp. 419–430. Springer, Heidelberg (2012). https://doi.org/10.1007/978-3-642-33090-2_37
7. Epstein, L., Levin, A., Woeginger, G.J.: The (weighted) metric dimension of graphs: hard and easy cases. Algorithmica **72**(4), 1130–1171 (2015)
8. Estrada-Moreno, A., Rodríguez-Velázquez, J.A., Yero, I.G.: The k-metric dimension of a graph. arXiv preprint arXiv:1312.6840 (2013)
9. Fernau, H., Heggernes, P., van't Hof, P., Meister, D., Saei, R.: Computing the metric dimension for chain graphs. Inf. Process. Lett. **115**, 671–676 (2015)
10. Foucaud, F., Mertzios, G.B., Naserasr, R., Parreau, A., Valicov, P.: Algorithms and complexity for metric dimension and location-domination on interval and permutation graphs. In: Mayr, E.W. (ed.) WG 2015. LNCS, vol. 9224, pp. 456–471. Springer, Heidelberg (2016). https://doi.org/10.1007/978-3-662-53174-7_32
11. Foucaud, F., Mertzios, G.B., Naserasr, R., Parreau, A., Valicov, P.: Identification, location-domination and metric dimension on interval and permutation graphs. i. bounds. Theor. Comput. Sci. **668**, 43–58 (2017)
12. Garey, M., Johnson, D.: Computers and Intractability: A Guide to the Theory of NP-Completeness. W. H. Freeman (1979)
13. Harary, F., Melter, R.: On the metric dimension of a graph. Ars Comb. **2**, 191–195 (1976)
14. Hartung, S., Nichterlein, A.: On the parameterized and approximation hardness of metric dimension. In: 2013 IEEE Conference on Computational Complexity (CCC), pp. 266–276. IEEE (2013)
15. Hauptmann, M., Schmied, R., Viehmann, C.: Approximation complexity of metric dimension problem. J. Discrete Algorithms **14**, 214–222 (2012)
16. Hernando, C., Mora, M., Slater, P.J., Wood, D.R.: Fault-tolerant metric dimension of graphs. Convexity Discrete Struct. **5**, 81–85 (2008)
17. Hernando, M., Mora, M., Pelayo, I., Seara, C., Cáceres, J., Puertas, M.: On the metric dimension of some families of graphs. Electron. Notes Discrete Math. **22**, 129–133 (2005)
18. Hoffmann, S., Wanke, E.: METRIC DIMENSION for Gabriel unit disk graphs Is NP-complete. In: Bar-Noy, A., Halldórsson, M.M. (eds.) ALGOSENSORS 2012. LNCS, vol. 7718, pp. 90–92. Springer, Heidelberg (2013). https://doi.org/10.1007/978-3-642-36092-3_10
19. Hoffmann, S., Elterman, A., Wanke, E.: A linear time algorithm for metric dimension of cactus block graphs. Theor. Comput. Sci. **630**, 43–62 (2016)
20. Iswadi, H., Baskoro, E., Salman, A., Simanjuntak, R.: The metric dimension of amalgamation of cycles. Far East J. Math. Sci. (FJMS) **41**(1), 19–31 (2010)
21. Jung, H.A.: On a class of posets and the corresponding comparability graphs. J. Comb. Theor. Series B **24**(2), 125–133 (1978)
22. Khuller, S., Raghavachari, B., Rosenfeld, A.: Landmarks in graphs. Discrete Appl. Math. **70**, 217–229 (1996)
23. Melter, R., Tomescu, I.: Metric bases in digital geometry. Comput. Vis. Graph. Image Process. **25**(1), 113–121 (1984)
24. Oellermann, O.R., Peters-Fransen, J.: The strong metric dimension of graphs and digraphs. Discrete Appl. Math. **155**(3), 356–364 (2007)
25. Raza, H., Hayat, S., Pan, X.F.: On the fault-tolerant metric dimension of convex polytopes. Appl. Math. Comput. **339**, 172–185 (2018)

26. Saputro, S., Baskoro, E., Salman, A., Suprijanto, D., Baca, A.: The metric dimension of regular bipartite graphs. arXiv/1101.3624 (2011). http://arxiv.org/abs/1101.3624
27. Sebö, A., Tannier, E.: On metric generators of graphs. Math. Oper. Res. **29**(2), 383–393 (2004)
28. Seinsche, D.: On a property of the class of n-colorable graphs. J. Comb. Theor. Ser. B **16**(2), 191–193 (1974)
29. Slater, P.: Leaves of trees. Congr. Numerantium **14**, 549–559 (1975)
30. Sumner, D.P.: Dacey graphs. J. Aus. Math. Soc. **18**(4), 492–502 (1974)
31. Vietz, D., Hoffmann, S., Wanke, E.: Computing the metric dimension by decomposing graphs into extended biconnected components. In: Das, G.K., Mandal, P.S., Mukhopadhyaya, K., Nakano, S. (eds.) WALCOM 2019. LNCS, vol. 11355, pp. 175–187. Springer, Cham (2019). https://doi.org/10.1007/978-3-030-10564-8_14

Retraction Note to: Complete Disjoint CoNP-Pairs but No Complete Total Polynomial Search Problems Relative to an Oracle

Titus Dose

Retraction Note to:
Chapter "Complete Disjoint CoNP-Pairs but No Complete Total Polynomial Search Problems Relative to an Oracle" in: L. A. Gąsieniec et al. (Eds.): *Fundamentals of Computation Theory*, LNCS 11651, https://doi.org/10.1007/978-3-030-25027-0_11

The author has retracted this chapter [1] because of a gap in the proof of the main theorem caused by an incorrect application of Claim 4. The author agrees to this retraction.

[1] Dose, T.: Complete disjoint CoNP-pairs but no complete total polynomial search problems relative to an Oracle. In: Gąsieniec, L., Jansson, J., Levcopoulos, C. (eds.) FCT 2019. LNCS, vol. 11651. Springer, Cham (2019). https://doi.org/10.1007/978-3-030-25027-0_11.

The retracted version of this chapter can be found at
https://doi.org/10.1007/978-3-030-25027-0_11

© Springer Nature Switzerland AG 2020
L. A. Gąsieniec et al. (Eds.): FCT 2019, LNCS 11651, p. C1, 2020.
https://doi.org/10.1007/978-3-030-25027-0_25

Author Index

Printed in the United States
By Bookmasters

Printed in the United States
By Bookmasters